FLUVIAL HYDRAULICS

FLUVIAL HYDRAULICS

S. Lawrence Dingman

OXFORD
UNIVERSITY PRESS
2009

OXFORD
UNIVERSITY PRESS

Oxford University Press, Inc., publishes works that further
Oxford University's objective of excellence
in research, scholarship, and education.

Oxford New York
Auckland Cape Town Dar es Salaam Hong Kong Karachi
Kuala Lumpur Madrid Melbourne Mexico City Nairobi
New Delhi Shanghai Taipei Toronto

With offices in
Argentina Austria Brazil Chile Czech Republic France Greece
Guatemala Hungary Italy Japan Poland Portugal Singapore
South Korea Switzerland Thailand Turkey Ukraine Vietnam

Published by Oxford University Press, Inc.
198 Madison Avenue, New York, New York 10016

www.oup.com

Library of Congress Cataloging-in-Publication Data

Dingman, S.L.
Fluvial hydraulics / S. Lawrence Dingman
p.cm
Includes bibliographical references and index
ISBN 978-0-19-517286-7
1. Streamflow. 2. Fluid mechanics I. Title
GB1207.D56 2008
551.48'3—dc22 2008046767

Quotation on p. ix from "A Man and His Dog" by Thomas Mann, in *Death in Venice and Seven Other Stories by Thomas Mann* (trans. H.T. Lowe-Porter), a Vintage Book © 1930, 1931, 1936 by Alfred A. Knopf, a division of Random House, Inc. Used by permission of Alfred A. Knopf.

9 8 7 6 5 4 3 2 1

Printed in the United States of America
on acid-free paper

Preface

The overall goal of this book is to develop a sound qualitative and quantitative understanding of the physics of natural river flows for practitioners and students with backgrounds in the earth sciences and natural resources who are primarily interested in understanding fluvial geomorphology. The treatment assumes an understanding of basic calculus and university-level physics.

Civil engineers typically learn about rivers in a course called Open-Channel Flow. There are many excellent books on open-channel flow for engineers [most notably the classic texts by Chow (1959) and Henderson (1961), and more recent works by French (1985) and Julien (2002)]. These courses and texts assume a foundation in fluid mechanics and differential equations, devote considerable attention to the aspects of flow involved in the design of structures, and generally provide only limited discussion of the geomorphic and other more "holistic" aspects of natural streams. By contrast, the usual curricula for earth, environmental, and natural resource sciences do not provide a thorough systematic introduction to the mechanics of river flows, despite its importance as a basis for understanding hydrologic processes, geomorphology, erosion, sediment transport and deposition, water supply and quality, habitat management, and flood hazards.

I believe that it is possible to build a sound understanding of fluvial hydraulics on the typical first-year foundation of calculus and calculus-based physics, and my hope is that this text will bridge the gap between these two approaches. It differs from typical engineering treatments of open-channel flow in its greater emphasis on natural streams and reduced treatments of hydraulic structures, and from most earth-science-oriented texts in its systematic development of the basic physics of river flows and its greater emphasis on quantitative analysis.

My first attempt to address this need was *Fluvial Hydrology*, published in 1984 by W.H. Freeman and Company. Although that book has been out of print for some time, comments from colleagues and students over the years made it clear that the need was real and that *Fluvial Hydrology* was useful in addressing it, and I continued to teach a course based on that text. Student and colleague interest, the publication of new databases, a number of theoretical and observational advances in the field, a growing interest in estimating discharge by remote sensing, the ready availability of powerful statistical-analysis tools, and my own growing discomfort with the Manning equation as the basic constitutive equation for open-channel flow, all led to a resurgence of my interest in river hydraulics (Dingman 1989, 2007a, 2007b; Dingman and Sharma 1997; Bjerklie et al. 2003, 2005b) and thoughts of revisiting the subject in a new textbook.

Although my goal remains the same, the present work is far more than a revision of *Fluvial Hydrology*. The guiding principles of this new approach are 1) a deeper foundation in basic fluid mechanics and 2) a broader treatment of the characteristics of natural rivers, including extensive use of data on natural river flows. The text itself has been drastically altered, and little of the original remains. However, I have tried to maintain, and enhance, the emphasis on the development of physical intuition—a sense of the relative magnitudes of properties, forces, and other quantities and relationships that are significant in a specific situation—and to emphasize patterns and connections.

The main features of this new approach include a more systematic review of the historical development of fluvial hydraulics (chapter 1); an extensive review of the morphology and hydrology of rivers (chapter 2); an expanded discussion of water properties, including turbulence (chapter 3); a more systematic development of fluid mechanics and the bases of equations used to describe river flows, including statistical and dimensional analysis (chapter 4); more complete treatment of velocity profiles and distributions, including alternatives to the Prandtl-von Kármán law (chapter 5); a more theoretically based treatment of flow resistance that provides new insights to that central topic (chapter 6); the use of published databases to quantitatively characterize actual magnitudes of forces and energies in natural river flows (chapters 7 and 8); more detailed treatment of rapidly varied flow transitions (chapter 10); a more detailed treatment of waves and an introduction to streamflow routing (chapter 11); and a more theoretically based and modern approach to sediment transport (chapter 12). Only the treatment of gradually varied flows (chapter 9) remains largely unchanged from *Fluvial Hydrology*. A basic understanding of dimensions, units, and numerical precision is still an essential, but often neglected, part of education in the physical sciences; the treatment of this, which began the former text, has been revised and moved to an appendix. The number of references cited has been greatly expanded as well as updated and now includes more than 250 items. A diligent attempt has been made to enhance understanding by regularizing the mathematical symbols and assuring that they are defined where used. I have used the "center dot" symbol for multiplication throughout so that multiletter symbols and functional notation can be read without ambiguity.

A course based on this text will be appropriate for upper level undergraduates and beginning graduate students in earth sciences and natural resources curriculums and will likely be taught by an instructor with an active interest in the field. Under these conditions, instructors will want to engage students in exploration of questions that arise and in discussion of papers from the literature, and to involve them in laboratory and/or field experiences. Therefore, I have not included exercises, but instead provide through the book's website (http://www.oup.com/fluvialhydraulics) an extensive database of flow measurements, a "Synthetic Channel" spreadsheet that can be used to explore the general nature of important hydraulic relations and the ways in which these relations change with channel characteristics, a simple spreadsheet for water-surface profile computations, links to other fluvial hydraulics and fluvial geomorphological websites that are available through the Internet, and a place for instructors and students to exchange ideas and questions.

I thank David Severn and Rachel Cogan of the Dimond Library at the University of New Hampshire (UNH) and Connie Mutel of the Iowa Institute of Hydraulic Research at the University of Iowa for assistance with references, permissions, and historical information. Data on world rivers were generously provided by Balazs Fekete of

UNH's Institute for the Study of Earth, Oceans, and Space. Cross-section survey data for New Zealand streams were provided by D.M. Hicks, New Zealand National Institute of Water and Atmospheric Research. I heartily thank Emily Faivre, John Stamm, David Bjerklie, Rob Ferguson, and Carl Bolster for reviews of various portions of the text at various stages in its development. Their comments were extremely helpful, but I of course am solely responsible for any errors and lack of clarity that remain.

This work would not have been possible without the encouragement and support of my parents in pursuing my undergraduate and graduate education; of the teachers who most inspired and educated me: John P. Miller at Harvard, Donald R.F. Harleman of the Massachusetts Institute of Technology, and Richard E. Stoiber at Dartmouth; and of Francis R. Hall and Gordon L. Byers, founders of UNH's Hydrology Program. I owe special thanks to my student Dave Bjerklie, now of the U.S. Geological Survey in Hartford, Connecticut, whose response to my initial research on the statistical analysis of resistance relations and subsequent discussions and research have been a major impetus for my continuing interest in fluvial hydraulics.

The love, support, and guidance of my wife, Jane Van Zandt Dingman, have sustained me in this work as in every aspect of my life.

Contents

I am very fond of brooks, as indeed of all water, from the ocean to the smallest weedy pool. If in the mountains in the summertime my ear but catch the sound of plashing and prattling from afar, I always go to seek out the source of the liquid sounds, a long way if I must; to make the acquaintance and to look in the face of that conversable child of the hills, where he hides. Beautiful are the torrents that come tumbling with mild thunderings down between evergreens and over stony terraces; that form rocky bathing-pools and then dissolve in white foam to fall perpendicularly to the next level. But I have pleasure in the brooks of the flatland too, whether they be so shallow as hardly to cover the slippery, silver-gleaming pebbles in their bed, or as deep as small rivers between overhanging, guardian willow trees, their current flowing swift and strong in the centre, still and gently at the edge. Who would not choose to follow the sound of running waters? Its attraction for the normal man is of a natural, sympathetic sort. For man is water's child, nine-tenths of our body consists of it, and at a certain stage the foetus possesses gills. For my part I freely admit that the sight of water in whatever form or shape is my most lively and immediate kind of natural enjoyment; yes, I would even say that only in contemplation of it do I achieve true self-forgetfulness and feel my own limited individuality merge into the universal. The sea, still-brooding or coming in on crashing billows, can put me in a state of such profound organic dreaminess, such remoteness from myself, that I am lost to time. Boredom is unknown, hours pass like minutes, in the unity of that companionship. But then, I can lean on the rail of a little bridge over a brook and contemplate its currents, its whirlpools, and its steady flow for as long as you like; with no sense or fear of that other flowing within and about me, that swift gliding away of time. Such love of water and understanding of it make me value the circumstance that the narrow strip of ground where I dwell is enclosed on both sides by water.

—Thomas Mann

FLUVIAL HYDRAULICS

1

Introduction to Fluvial Hydraulics

1.1 Rivers in the Global Context

Although rivers contain only 0.0002% of the water on earth (table 1.1), it is hard to overstate their importance to the functioning of the earth's natural physical, chemical, and biological systems or to the establishment and nutritional, economic, and spiritual sustenance of human societies.

1.1.1 Natural Cycles

The water flowing in rivers is the residual of two climatically determined processes, precipitation and evapotranspiration,[1] and the general **water-balance equation** for a region can be written as

$$Q = P - ET, \tag{1.1}$$

where Q is temporally averaged river flow (river discharge) from the region, P is spatially and temporally averaged precipitation, and ET is spatially and temporally averaged evapotranspiration.[2] The dimensions of the terms of equation 1.1 may be volume per unit time $[L^3\,T^{-1}]$ or volume per unit time per unit area $[L\,T^{-1}]$. (See appendix A for a review of dimensions and units.)

At the largest scale, the time-integrated global hydrological cycle can be depicted as in figure 1.1. The world's oceans receive about 458,000 km^3/year in precipitation and

Table 1.1 Volume of water in compartments of the global hydrologic cycle.

Compartment	Area covered (1,000 km^2)	Volume (km^3)	Percentage of total water	Percentage of freshwater
Oceans	361,300	1,338,000,000	96.5	—
Groundwater	134,800	23,400,000	1.7	—
Fresh		10,530,000	0.76	30.1
Soil water		16,500	0.001	0.05
Glaciers and permanent snow	16,227	24,064,000	1.74	68.7
Antarctica	13,980	21,600,000	1.56	61.7
Greenland	1,802	2,340,000	0.17	6.68
Arctic Islands	226	83,500	0.006	0.24
Mountains	224	40,600	0.003	0.12
Permafrost	21,000	300,000	0.022	0.86
Lakes	2,059	176,400	0.013	—
Fresh	1,236	91,000	0.007	0.26
Saline	822	85,400	0.006	—
Marshes	2,683	11,470	0.0008	0.03
Rivers	148,800	2,120	0.0002	0.006
Biomass	510,000	1,120	0.0001	0.003
Atmosphere	510,000	12,900	0.001	0.04
Total water	510,000	1,385,984,000	100	—
Total freshwater	148,800	35,029,000	2.53	100

The global cycle is diagrammed in figure 1.1. From Shiklomanov (1993), with permission of Oxford University Press.

lose 505,000 km^3/year in evaporation, while the continents receive 119,000 km^3/year in precipitation and lose 72,000 km^3/year via evapotranspiration.

> The water flowing in rivers—**river discharge**—is the link that balances the global cycle, returning about 47,000 km^3/year from the continents to the oceans.

Table 1.2 lists the world's largest rivers in terms of discharge. Note that the Amazon River contributes more than one-eighth of the total discharge to the world's oceans!

River discharge is also a major link in the global geological cycle, delivering some 13.5×10^9 T/year of particulate material and 3.9×10^9 T/year of dissolved material from the continents to the oceans (Walling and Webb 1987). Thus, "Rivers are both the means and the routes by which the products of continental weathering are carried to the oceans of the world" (Leopold 1994, p. 2). A portion of the dissolved material constitutes the major source of nutrients for the oceanic food web.

River discharge plays a critical role in regulating global climate. Its effects on sea-surface temperatures and salinities, particularly in the North Atlantic Ocean, drive the global thermohaline circulation that transports heat from low to high latitudes. The freshwater from river inflows also maintains the relatively low salinity of the Arctic Ocean, which makes possible the freezing of its surface; the reflection of the sun's energy by this sea ice is an important factor in the earth's energy balance.

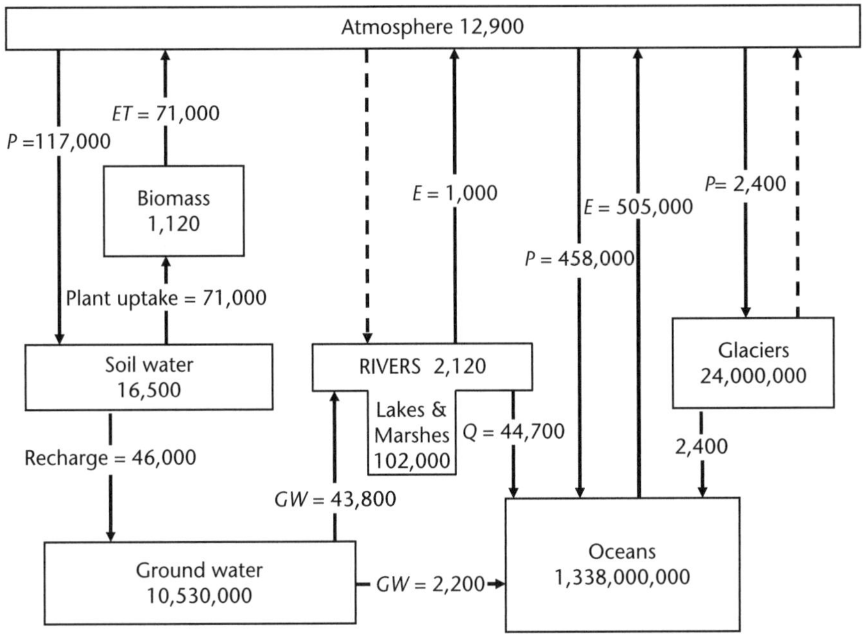

Figure 1.1 Schematic diagram of stocks (km^3) and annual fluxes (km^3/year) in the global hydrological cycle. *E*, evaporation; *ET*, evapotranspiration; *GW*, groundwater discharge; *P*, precipitation; *Q*, river discharge. Data on stocks, land and ocean precipitation, ocean evaporation, and river discharge are from Shiklomanov (1993) (see table 1.1); other fluxes are adjusted from Shiklomanov's values to give an approximate balance for each stock. Dashed arrows indicate negligible fluxes on the global scale

The drainage systems of rivers—river networks and their contributing watersheds—are the principal organizing features of the terrestrial landscape. These systems are nested hierarchies at scales ranging from a few square meters to 5.9×10^6 km^2 (the Amazon River drainage basin). The world's largest river systems in terms of drainage area are listed in table 1.3. At all scales, rivers are the links that collect the residual water (precipitation minus evapotranspiration and groundwater outflow) and its chemical and physical constituents and deliver them to the next level in the hierarchy or to the world ocean.

1.1.2 Human Significance

As indicated in figure 1.1, the immediate source of most of the water in rivers is groundwater. Conversely, virtually all groundwater is ultimately destined to become streamflow. River discharge is the rate at which nature makes water available for human use. Thus, at all scales, average river discharge is the metric of the water resource (Gleick 1993; Vörösmarty et al. 2000b).

Humans have been concerned with rivers as sources of water and food, as routes for commerce, and as potential hazards at least since the first civilizations developed along

Table 1.2 Average discharge from the world's 30 largest terminal drainage basins ranked by discharge.[a]

Rank	River	Discharge			
		km^3/year	% Total discharge to oceans	m^3/s	mm/year
1	Amazon	5,992	13.4	190,000	1,024
2	Congo	1,325	3.0	42,000	358
3	Chang Jiang	1,104	2.5	35,000	615
4	Orinoco	915	2.0	29,000	880
5	Ganges-Brahmaputra	631	1.4	20,000	387
6	Parana	615	1.4	19,500	231
7	Yenesei	561	1.3	17,800	217
8	Mississippi	558	1.2	17,700	174
9	Lena	514	1.1	16,300	213
10	Mekong	501	1.1	15,900	648
11	Irrawaddy	399	0.9	12,700	974
12	Ob	394	0.9	12,500	153
13	Zhujiang (Si Kiang)	363	0.8	11,500	831
14	Amur	347	0.8	11,000	119
15	Zambezi	333	0.7	10,600	167
16	St. Lawrence	328	0.7	10,400	259
17	Mackenzie	286	0.6	9,100	167
18	Volga	265	0.6	8,400	181
19	Shatt-el-Arab (Euphrates)	259	0.6	8,210	268
20	Salween	211	0.5	6,690	649
21	Indus	202	0.5	6,410	177
22	Danube	199	0.4	6,310	253
23	Columbia	191	0.4	6,060	264
24	Tocantins	168	0.4	5,330	218
25	Kolyma	128	0.3	4,060	192
26	Nile	96	0.2	3,040	25
27	Orange	91	0.2	2,900	97
28	Senegal	86	0.2	2,730	102
29	Syr-Daya	83	0.2	2,630	78
30	São Francisco	82	0.2	2,600	133

[a] "Terminal" means the drainage basin is not tributary to another stream.
Data are from web sites and various published sources.

the banks of rivers: the Indus in Pakistan, the Tigris and Euphrates in Mesopotamia, the Huang Ho in China, and the Nile in Egypt.

Water flowing in streams is used for a wide range of vital water resource management purposes, such as

- Human and industrial water supply
- Agricultural irrigation
- Transport and treatment of human and industrial wastes
- Hydroelectric power
- Navigation
- Food
- Ecological functions (wildlife habitat)

Table 1.3 Topographic data for the world's 30 largest terminal drainage basins ranked by drainage area.[a]

Rank	River	Area (10^6 km^2)	Length (km)	Elevation (m) Avg.	Max.	Min.[b]	Average slope ($\times 10^3$)
1	Amazon	5.854	4,327	430	6,600	0	1.66
2	Nile	3.826	5,909	690	4,660	0	1.45
3	Congo (Zaire)	3.699	4,339	740	4,420	0	1.11
4	Mississippi	3.203	4,185	680	4,330	0	1.66
5	Amur	2.903	5,061	750	5,040	0	1.80
6	Parana	2.661	2,748	560	6,310	0	1.59
7	Yenesei	2.582	4,803	670	3,500	0	1.94
8	Ob	2.570	3,977	270	4,280	0	1.28
9	Lena	2.418	4,387	560	2,830	0	1.83
10	Niger	2.240	3,401	410	2,980	0	0.94
11	Zambezi	1.989	2,541	1,050	2,970	0	1.60
12	Tamanrasett[c]	1.819	2,777	450	3,740	0	0.83
13	Chang Jiang (Yangtze)	1.794	4,734	1,660	7,210	0	3.27
14	Mackenzie	1.713	3,679	590	3,350	0	2.23
15	Ganges-Brahmaputra	1.638	2,221	1,620	8,848	0	6.00
16	Chari	1.572	1,733	510	3,400	260	1.10
17	Volga	1.463	2,785	1,710	1,600	0	0.52
18	St. Lawrence	1.267	3,175	310	1,570	0	1.22
19	Indus	1.143	2,382	1,830	8,240	0	5.50
20	Syr-Darya	1.070	1,615	650	5,480	0	2.84
21	Nelson	1.047	2,045	500	3,440	0	1.06
22	Orinoco	1.039	1,970	480	5,290	0	3.01
23	Murray	1.032	1,767	260	2,430	0	1.03
24	Great Artesian Basin	0.978	1,045	220	1,180	70	0.55
25	Shatt-el-Arab (Euphrates)	0.967	2,200	660	4,080	0	2.84
26	Orange	0.944	1,840	1,230	3,480	0	1.65
27	Huang He (Yellow)	0.894	4,168	2,860	6,130	0	2.93
28	Yukon	0.852	2,716	690	6,100	0	2.93
29	Senegal	0.847	1,680	250	10,700	0	0.43
30	Irharhar[c]	0.842	1,482	500	2,270	0	1.84

[a]Values were determined by analysis of satellite imagery at the 30-min scale (latitude and longitude) (average pixel is 47.4 km on a side). "Terminal" means the drainage basin is not tributary to another stream.
[b]A minimum elevation of 0 means the basin discharges to the ocean. A nonzero minimum elevation indicates that the basin discharges internally to the continent, usually to a lake. [c]River system mostly nondischarging under current climate.
Source: Data are from Vörösmarty et al. (2000).

- Recreation
- Aesthetic enjoyment[3]

Demand for water for all these purposes is growing with population, and roughly one-third of the world's peoples currently live under moderate to high water stress (Vörösmarty et al. 2000b). Water availability at a location on a river is assessed by analyses of the time distribution of river discharge at that location (discussed in section 2.5.6.2).

On the other hand, water flowing in rivers at times of flooding is one of the most destructive natural hazards globally. In the United States, flood damages total about $4 billion per year and are increasing rapidly because of the increasing concentration of people and infrastructure in flood-prone areas (van der Link et al. 2004). Assessment of this hazard and of the economic, environmental, and social benefits and costs of various strategies for reducing future flood damages at a riparian location is based on frequency analyses of extreme river discharges at that location (discussed in section 2.5.6.3).[4]

1.2 The Role of Fluvial Hydraulics

The term **fluvial** means "of, pertaining to, or inhabiting a river or stream." This book is about **fluvial hydraulics**—the internal physics of streams. In the civil engineering context, the subject is usually called **open-channel flow**; the term "fluvial" is used here to emphasize our focus on natural streams rather than design of structures.

An understanding of fluvial hydraulics underlies many important scientific fields:

- Because the terrestrial landscape is largely the result of fluvial processes, an understanding of fluvial hydraulics is an essential basis for the study of geomorphology.
- Fluvial hydraulics governs the movement of water through the stream network, so an understanding of fluvial hydraulics is essential to the study of hydrology.
- Stream organisms are adapted to particular ranges of flow conditions and bed material, so knowledge of fluvial hydraulics is the basis for understanding stream ecology.
- Knowledge of fluvial hydraulics is required for interpretation of ancient fluvial deposits to provide information about geological history.

Knowledge of fluvial hydraulics is also the basis for addressing important practical issues:

- Predicting the effects of climate change, land-use change (urbanization, deforestation, and afforestation), reservoir construction, water extraction, and sea-level rise on river behavior and dimensions.
- Forecasting the development and movement of flood waves through the channel system.
- Designing dams, levees, bridges, canals, bank protection, and navigation works.
- Assessing and restoring stream habitats.

One particularly important application of fluvial hydraulics principles is in the measurement of river discharge. Discharge measurement directly provides essential information about water-resource availability and flood hazards.

Because river discharge is concentrated in channels, it can in principle be measured with considerably more accuracy and precision than can precipitation, evapotranspiration, or other spatially distributed components of the hydrological cycle. Long-term average values of discharge typically have errors of ±5% (i.e., the true value is within 5% of the measured value 95% of the time). Errors in precipitation are generally at least twice that ($\geq$10%) and may be 30% or more depending on climate and the number and location of precipitation gages (Winter 1981; Rodda 1985; Groisman and Legates 1994). Areal evapotranspiration is virtually unmeasured, and in fact is usually estimated by solving equation 1.1 for *ET*. Thus, measurements of river discharge provide the most reliable information about regional water balances. And, because it is the space- and time-integrated residual of two climatically determined quantities (equation 1.1), river discharge is a sensitive indicator of climate change. Observations of long-term trends in precipitation and streamflow consistently show that changes in river discharge amplify changes in precipitation; for example, a 10% increase in precipitation may induce a 20% increase in discharge (Wigley and Jones 1985; Karl and Riebsame 1989; Sankarasubramanian et al. 2001). Discharge measurements are also invaluable for validating the hydrological models that are the only means of forecasting the effects of land use and climate change on water resources.

Fluvial hydraulics principles have long been incorporated in traditional measurement techniques that involve direct contact with the flow (discussed in section 2.5.3.1). New applications combining hydraulic principles, geomorphic principles, and empirical analysis are rapidly being developed to enable measurement of flows via remote-sensing techniques (Bjerklie et al. 2003, 2005a; Dingman and Bjerklie 2005; Bjerklie 2007) (see section 2.5.3.2).

1.3 A Brief History of Fluvial Hydraulics

In order to understand a science, it is important to have an understanding of how it developed. This section provides an overview of the evolution of the science of fluvial hydraulics, emphasizing the significant discrete contributions of individuals that combine to form the basis of our current understanding of the field. As with all science, each individual contribution is built on earlier observations and reasoning. The material in this section is based largely on Rouse and Ince (1963), and the quotes from earlier works are taken from that book. Their text gives a more complete sense of the ways in which individual advances are built upon earlier work than is possible in the present overview. You will find it fascinating reading, especially after you become familiar with the material in the present text.

As noted above, the first civilizations were established along major rivers, and it is clear that humans were involved in river engineering that must have been based on learning by trial and error since prehistoric times. The Chinese were building levees for flood protection and the people of Mesopotamia were constructing irrigation systems as early as 4000 B.C.E. In Egypt, irrigation was also practiced in lands adjacent to the Nile by 3200 B.C.E., and the earliest known dam was built at Sadd el Kafara (near Cairo) in the period 2950–2759 B.C.E..

However, science based on observation and reasoning and the written transmission of knowledge first emerged in Greece around 600 B.C.E. Thales of Miletus (640–546 B.C.E.) studied in Egypt. He believed that "water is the origin of all things," and both he and Hippocrates (460–380?) two centuries later articulated the philosophy that nature is best studied by observation. By far the most significant enduring hydraulic principles discovered by the ancient Greeks were Archimedes' (287–212 B.C.E.) laws of buoyancy:

> Any solid lighter than the fluid will, if placed in the fluid, be so far immersed that the weight of the solid will be equal to the weight of the fluid displaced.
>
> If a solid lighter than the fluid be forcibly immersed in it, the solid will be driven upwards by a force equal to the difference between its weight and the weight of the fluid displaced.
>
> A solid heavier than a fluid will, if placed in it, descend to the bottom of the fluid, and the solid will, when weighed in the fluid, be lighter than its true weight by the weight of the fluid displaced. (Rouse and Ince 1963, p. 17)

Hero of Alexandria (first century A.D.) wrote on several aspects of hydraulics, including siphons and pumps, and gave the earliest known expression of the law of continuity (discussed in section 4.3.2) for computing the flow rate (discharge) of a spring: "In order to know how much water the spring supplies it does not suffice to find the area of the cross section of the flow. … It is necessary also to find the speed of flow" (Rouse and Ince 1963, p. 22).

Although the writings of these and other Greek natural philosophers were preserved and transmitted to Europeans by Arabian scientists, there were no further scientific contributions to the field for some 1,500 years. The Romans built extensive and elaborate systems of aqueducts, reservoirs, and distribution pipes that are described in extensive surviving treatises by Vitruvius (first century B.C.E.) and Frontinus (40–103 A.D.). Although aware of the Greek writings on hydraulics, they did not add to them or even explicitly reflect them in their designs and computations. For example, although Frontinus understood that the rates of flow entering and leaving a pipe should be equal, he computed the flow rate based on area alone and did not seem to clearly understand, as Hero did, that velocity is also involved. Still, as Rouse and Ince (1963, p. 32) note, the Roman engineers must have sensed the effects of head, slope, and resistance on flow rates or their systems would not have functioned as well as they did.

There were no additions to scientific knowledge of hydraulics from the time of Hero until the Renaissance. However, during the Middle Ages, improvements in hydraulic machinery were made in the Islamic world, and a few scholars in Europe were considering the basic aspects of motion, acceleration, and resistance that laid the groundwork for subsequent advances in physics. During this period,

> the writings—and indeed the theories themselves—were numerous and complex, and … the background training of few scholars was sound enough to distinguish fallacy from truth. Progress was hence exceedingly slow and laborious, and not for centuries did the cumulative effect of many people in different lands clarify these elementary principles of mechanics on which the science of hydraulics was to be based. (Rouse and Ince 1963, p. 42)

In contrast to the dominant philosophies of the Middle Ages, the Italian Renaissance genius Leonardo da Vinci (1452–1519) wrote, "Remember when discoursing on the flow of water to adduce first experience and then reason." Da Vinci rediscovered the principle of continuity, stating that "a river in each part of its length in an equal time gives passage to an equal quantity of water, whatever the depth, the slope, the roughness, the tortuosity." He also correctly concluded from his observations of open-channel flows that "water has higher speed on the surface than on the bottom. This happens because water on the surface borders on air which is of little resistance, ... and water at the bottom is touching the earth which is of higher resistance.... From this follows that the part which is more distant from the bottom has less resistance than that below" and that "the water of straight rivers is the swifter the farther away it is from the walls, because of resistance" (discussed in sections 3.3, 5.3, and 5.4). From his observations of water waves, he correctly noted that "the speed of propagation of (surface) undulations always exceeds considerably that possessed by the water, because the water generally does not change position; just as the wheat in a field, though remaining fixed to the ground, assumes under the impulsion of the wind the form of waves traveling across the countryside" (Rouse and Ince 1963, p. 49) (discussed in sections 11.3–11.5).

Because da Vinci's observations were lost for several centuries, they did not contribute to the growth of science. For example, one of Galileo's pupils, Benedetto Castelli (1577?–1644?), again formulated the law of continuity more than a century after da Vinci, and it became known as Castelli's law. In 1697, another Italian, Domenico Guglielmini (1655–1710), published a major work on rivers, *Della Natura del Fiumi* (*On the Nature of Rivers*), which included among other things a description of uniform (i.e., nonaccelerating) flow very similar to that in the present text (see section 6.2.1, figure 6.2). In an extensive treatise on hydrostatics published posthumously in 1663, Blaise Pascal (1623–1662) showed that the pressure is transmitted equally in all directions in a fluid at rest (see section 4.2.2.2).

The major scientific advances of the seventeenth century were those of Sir Isaac Newton (1642–1727), who began the development of calculus, concisely formulated his three laws of motion based on previous ideas of Descartes and others, and clearly defined the concepts of mass, momentum, inertia, and force. He also formulated the basic relation of viscous shear (see equation 3.19), which characterizes Newtonian fluids. Newton's German contemporary, Gottfried Wilhelm von Leibniz (1646–1716), further developed the concepts of calculus and originated the concept of kinetic energy as proportional to the square of velocity (see section 4.5.2).

In the eighteenth century, the fields of theoretical, highly mathematical hydrodynamics and more practical hydraulics largely diverged. The foundations of hydrodynamics were formulated by four eighteenth-century mathematicians, Daniel Bernoulli (Swiss, 1700–1782), Alexis Claude Clairault (French, 1713–1765), Jean le Rond d'Alembert (French, 1717–1783), and especially Leonhard Euler (Swiss, 1708–1783). Bernoulli formulated the concept of conservation of energy in fluids (section 4.5), although the Bernoulli equation (equation 4.42) was actually derived by Euler. Euler was also the first to state the "microscopic" law of conservation of mass in derivative form (section 4.3.1, equation 4.16). The Frenchmen Joseph Louis Lagrange (1736–1813) and Pierre Simon Laplace (1749–1827) extended Euler's work in many areas of hydrodynamics. Although both Euler and Lagrange explored fluid motion by analyzing occurrences at a fixed point and by following a fluid "particle," the former approach has become known as Eulerian and the latter as

Lagrangian (section 4.1.4). One of Lagrange's contributions was the relation for the speed of propagation of a shallow-water gravity wave (equation 11.51); the Pole Franz Joseph von Gerstner (1756–1832) derived the corresponding expression for deep-water waves (equation 11.50).

Many of the advances in hydraulics in the eighteenth century were made possible by advances in measurement technology: Giovanni Poleni (Italian, 1683–1761) derived the basic equation for flow-measurement weirs (section 10.4.1) in 1717, and Henri de Pitot (French, 1695–1771) invented the Pitot tube in 1732, which uses energy concepts to measure velocity at a point. One of the most important and ultimately influential practical developments of this time was the work of Antoine Chézy (1718–1798), who reasoned that open-channel flow can usually be treated as uniform flow (section 6.2.1) in which "velocity ... is due to the slope of the channel and to gravity, of which the effect is restrained by the resistance of friction against the channel boundaries" (Rouse and Ince 1963, pp. 118–119). The equation that bears his name, derived in 1768 essentially as described in section 6.3 of this text, states that velocity (U) is proportional to the square root of the product of depth (Y) and slope (S), that is,

$$U = K \cdot Y^{1/2} \cdot S^{1/2}, \tag{1.2}$$

where K depends on the nature of the channel. The Chézy equation can be viewed as the basic equation for one-dimensional open-channel flow. Interestingly, Chézy's 1768 report was lost (although the manuscript survived), and his work was not published until 1897 by the American engineer Clemens Herschel (1842–1930) (Herschel 1897).

Although Chézy's work was generally unknown, others such as the German Johannn Albert Eytelwein (1764–1848) in 1801 proposed similar relations for open-channel flow. Interestingly, Gaspard de Prony (1755–1839) in 1803 proposed a formula for uniform open-channel flow identical to equation 7.42 of this text, which is identical to the Chézy relation for conditions usually encountered in rivers. In Italy, Giorgio Bidone (1781–1839) was the first to systematically study the hydraulic jump (section 10.1), in 1820, and Giuseppe Venturoli (1768–1846) made measurements confirming Eytelwein's formula and in 1823 was the first to derive an equation for water-surface profiles (section 9.4.1).

During this period, James Hutton's (English, 1726–1797) observations of streams and stream networks led him to conclude that the elements of the landscape are in a quasi-equilibrium state, implying relatively rapid mutual adjustment to changing conditions (section 2.6.2). This was a major philosophical advance in the understanding of the development of landscapes and the role of fluvial processes in that development.

Other hydraulic advances of the first half of the nineteenth century included a quantitative understanding of flow over broad-crested weirs (section 10.4.1.2), used in flow measurement, published in 1849 by Jean Baptiste Belanger (French, 1789–1874). Gaspard Gustave de Coriolis (French, 1792–1843) is best known for formulating the expression for the apparent force acting on moving bodies due to the earth's rotation (the Coriolis force, section 7.3.3.1), and also showed in 1836 the need for a correction factor (the Coriolis coefficient; see box 8.1) when using average velocity to calculate the kinetic energy of a flow. John Russell (English, 1808–1882) made observations of waves generated by barges in canals (1843), including the first descriptions of the solitary gravity wave (soliton; section 11.4.2). The first "modern" textbook on hydraulics (1845) was that of Julius Weisbach (German, 1806–1871), which included chapters on flow in canals and rivers and the measurement of water as well as the work on the resistance of fluids with which his name is associated—the Darcy-Weisbach friction factor (see box 6.2).

As described in sections 3.3.3 and 3.3.4 of this text, there are two states of fluid flow: laminar (or viscous) and turbulent. Despite the fact that flows in these two states have very different characteristics, explicit mention of this did not appear until 1839, in a paper by Gotthilf Hagen (German, 1797–1884). In a subsequent study (1854) Hagen clearly described the two states, anticipating by several decades the studies of Osborne Reynolds (see below), whose name is now associated with the phenomenon. Interest in scale models as an aid to the design of ships grew in this period, and it was in this context that Ferdinand Reech (French, 1805–1880) in 1852 first formulated the dimensionless ratio that relates velocities in models to those in the prototype. This ratio became known as the Froude number (sections 6.2.2.2 and 7.6.2) after William Froude (English, 1810–1879), who did extensive ship modeling experiments for the British government, though in fact he neither formulated nor even used the ratio.

Advances in the latter half of the nineteenth century, as with many earlier ones, were dominated by scientists and engineers associated with France's Corps des Ponts et Chaussées (Bridges and Highways Agency). Notable among these are Arsène Dupuit (1804–1866), Henri Darcy (1803–1858), Jacques Bresse (1822–1883), and Jean-Claude Barré de Saint-Venant (1797–1886). Dupuit's principal contributions to fluvial hydraulics were his 1848 analysis of water-surface profiles and their relation to uniform flow (section 9.2) and to variations in bed elevation and channel width (section 10.2), and his 1865 written discussion of the capacity of a stream to transport suspended sediment. Darcy, in addition to discovering Darcy's law of groundwater flow, studied flow in pipes and open channels and in 1857 demonstrated that resistance depended on the roughness of the boundary. Bresse in 1860 correctly analyzed the hydraulic jump using the momentum equation (section 10.1; equation 10.8). Saint-Venant in 1871 first formulated the general differential equations of unsteady flow, now called the Saint-Venant equations (section 11.1).

Dupuit's interest in sediment transport was followed by the work of Médéric Lachalas (1820–1904), which in 1871 discussed various types of sediment movement (figure 12.1), and the analysis of bed-load transport (1879) by Paul du Boys (1847–1924), which has been the basis for many approaches to the present day (section 12.5.1). Darcy's experimental work on flow resistance was carried on by his colleague Henri Bazin (1829–1917), whose measurements, published in 1865 and 1898, were analyzed by many later researchers hoping to discover a practical law of open-channel flow. Bazin's experiments also included measurements of the velocity distribution in cross sections (section 5.4) and of flow over weirs (section 10.4.1.1). Another Frenchman, Joseph Boussinesq (1842–1929), though not at the Corps des Ponts et Chaussées, made significant contributions in many aspects of hydraulics, including further insight in 1872 into the laminar-turbulent transition identified by

Hagen, the mathematical treatment of turbulence (section 3.3.4.3), and the formulation of the momentum equation (section 8.2.1, box 8.1).

There were also significant contemporary developments in England. These included Sir George Airy's (1801–1892) comprehensive treatment of waves and tides in 1845, including the derivation of the Airy wave equation (equation 11.46), and Sir George Stokes's (1819–1903) expansion in 1851 of Saint-Venant's equations to turbulent flow and his derivation of Stokes's law for the settling velocity of a spherical particle (equation 12.19). Combining experiment and analysis, Osborne Reynolds (1842–1912) made major advances in many areas, including the first demonstration of the phenomenon of cavitation (section 12.4.4.3), the seminal treatment in 1894 of turbulence as the sum of a mean motion plus fluctuations (section 3.3.4.2), and, most famously, the 1883 formulation of the Reynolds number quantifying the laminar-turbulent transition (section 3.4.2).

The names of Americans are conspicuously absent from the history of hydraulics until 1861, when two Army engineers, A. A. Humphreys and H. L. Abbot, published their *Report upon the Physics and Hydraulics of the Mississippi River*. In this they included a comprehensive review of previous European work on flow resistance and, finding that previous formulas did not consistently work on the lower Mississippi, attempted to develop their own. Their work prompted others to look for a universal resistance relation for open-channel flow. One significant contribution, in 1869, was that of two Swiss engineers, Emile Ganguillet (1818–1894) and Wilhelm Kutter (1818–1888), who accepted the basic form of the Chézy relation and proffered a complex formula for calculating the resistance as a function of boundary roughness, slope, and depth. Meanwhile, Phillipe Gauckler (1826–1905, also of the Corps des Ponts et Chaussées) in 1868 proposed two resistance formulas, one for rivers of low slope ($S < 0.0007$),

$$U = K \cdot Y^{4/3} \cdot S, \tag{1.3a}$$

and the other for rivers of high slope ($S > 0.0007$),

$$U = K \cdot Y^{2/3} \cdot S^{1/2}. \tag{1.3b}$$

Equation 1.3b was of particular significance because the Irish engineer Robert Manning (1816–1897) reviewed previous data on open-channel flow and stated in an 1889 report (although apparently without knowledge of Gauckler's work) that equation 1.3b fit the data better than others. However, Manning did not recommend that relation because it is not dimensionally correct (see appendix A), and proposed a modification that included a term for atmospheric pressure. Manning's proposed relation was never adopted, but ironically, equation 1.3b with K dependent on channel roughness has become the most widely used practical resistance relation and is called Manning's equation (section 6.8). As noted by Rouse and Ince (1963, p. 180), "What we now call the Manning formula was thus neither recommended nor even devised in full by Manning himself, whereas his actual recommendation received little further attention."

The first half of the twentieth century saw major advances in understanding real turbulent flows. In 1904, Ludwig Prandtl (German, 1875–1953) introduced the concept of the boundary layer (section 3.4.1), and in 1926 that of the mixing length (section 3.3.4.4) which tied Reynolds's statistical concepts of turbulence to physical phenomena. This laid the groundwork for a very significant breakthrough: the analytical derivation of the velocity distribution in turbulent boundary layers, which was developed by Prandtl and his student Theodore von Kármán (Hungarian who later emigrated to the United States, 1881–1963) and bears their names (section 5.3.1). This work, which grew out of studies of flow over airplane wings, was a major advance in understanding and modeling turbulent open-channel flows.

Meanwhile, the American Edgar Buckingham (1867–1940) introduced the concept of dimensional analysis (section 4.8.2) to English-speaking engineers in 1915; these concepts have guided countless fruitful investigations of flow phenomena. At the same time (1914) the American geologist Grove Karl Gilbert (1843–1918) carried out the first flume studies of the transport of gravel. Filip Hjulström (Swedish, 1902–1982) in 1935 and Albert Shields (German, 1908–1974) in 1936 provided analyses of data that form the basis for most subsequent studies of sediment entrainment (sections 12.4.1 and 12.4.2).

An influential text that appeared during this period was Hunter Rouse's (1906–1996) comprehensive and authoritative *Fluid Mechanics for Hydraulic Engineers* (Rouse 1938), which remains valuable to this day. In 1937, Rouse derived an expression for the vertical distribution of suspended sediment that is the basis for most analyses of this phenomenon (section 12.5.2.1), and in 1943 he concisely summarized experimental data on resistance–Reynolds number–roughness relations for the full range of flows in pipes in graphical form. A year later, Lewis F. Moody (American, 1880–1953) published a modified version of this graph (Moody 1944) that has been extended to open-channel flows and become known as the "Moody diagram" (see figure 6.8) (Ettema 2006).

The second half of the twentieth century saw significant advances in characterizing and understanding natural streams. Many of these advances were by Americans who applied the scientific and engineering insights described above and developed new approaches of analysis and measurement. One of these was the paper by Robert E. Horton (1875–1945) (Horton 1945), which was pivotal in turning the analysis of fluvial processes and landscapes from the qualitative approaches of geographers to a more quantitative scientific basis. A seminal conceptual contribution was the geologist J. Hoover Mackin's (1905–1968) clear articulation of Hutton's concept of dynamic equilibrium, the graded stream (Mackin 1948; see section 2.6.2). Building upon these developments, Luna Leopold (1915–2006) and several of his colleagues associated with the U.S. Geological Survey, most notably R. A. Bagnold (English, 1896–1990), W. B. Langbein (1907–1982), J. P. Miller (1923–1961), and M. G. Wolman (1924–), in the 1950s began an era of field research and innovative analysis that defined the field of fluvial processes and geomorphology for the rest of the century and beyond.

At the same time, V. T. Chow (American, 1919–1981) (Chow 1959) and Francis M. Henderson (Australian, 1921–) (Henderson 1966) distilled the advances described above to provide coherent and lucid engineering texts on open-channel hydraulics. These texts made the subject an essential part of civil engineering curricula and were a source of insights increasingly adopted and applied by earth scientists.

As the twenty-first century begins, two major problems of fluvial hydraulics remain far from completely solved: the a priori characterization of open-channel flow resistance/conductance (chapter 6) (the K in equation 1.2), and the prediction of sediment transport as a function of flow and channel characteristics (chapter 12). However, the coming years hold promise of major progress in understanding fluvial hydraulics and applying it to these and the critical problems described in section 1.2. This promise is largely the result of technological advances such as the ability to visualize and measure fluid and sediment motion, techniques for remote-sensing of streams, and advances in computer speed and storage that make possible the modeling of fluid flows. The measurements and insights of all the pioneering work described in the preceding paragraphs and in the remainder of this text will provide a sound basis for this progress.

1.4 Scope and Approach of This Book

The goal of the science of fluvial hydraulics is to understand the behavior of natural streams and to provide a basis for predicting their responses to natural and anthropogenic disturbances. The objective of this book is to develop a sound qualitative and quantitative basis for this understanding for practitioners and students with backgrounds in earth sciences and natural resources. This book differs from typical engineering treatments of open-channel flow in its greater emphasis on natural streams and reduced treatments of hydraulic structures. It differs from most earth-science-oriented texts in its greater emphasis on quantitative analysis based on the basic physics of river flows and its incorporation of analyses developed for engineering application.

The treatment here draws on your knowledge of basic mechanics (through first-year university-level physics) and mathematics (through differential and integral calculus) to develop a physical intuition—a sense of the relative magnitudes of properties, forces, and other quantities and relationships that are significant in a specific situation. Physical intuition consists not only of a store of factual knowledge, but also of a mental inventory of patterns that serve as guides to the parts of that knowledge that are relevant to the situation (Larkin et al. 1980). Thus, a special attempt is made in this book to emphasize patterns and connections.

The goal of chapter 2 is to provide a natural context for the analytical approach emphasized in subsequent chapters. It presents an overview of the characteristics of natural stream networks and channels and the ways in which geological, topographic, and climatic factors determine those characteristics. It also discusses the measurement and hydrological aspects of the flow within natural channels—its sources and temporal variability. The chapter concludes with an overview of the spatial and temporal variability of the variables that characterize stream channels, including the principle of dynamic equilibrium.

Water moves in response to forces acting on it, and its physical properties determine the qualitative and quantitative relations between those forces and the resulting motion. Chapter 3 begins with a description of the atomic and molecular structure of water that gives rise to its unique properties, and the nature of water substance in its three phases. The bulk of the chapter uses a series of thought experiments to elucidate the properties of liquid water that are crucial to understanding its behavior in open-channel flows: density, surface tension, and viscosity. Included here is an introduction to turbulence, flow states, and boundary layers, concepts that are central to understanding flows in natural streams.

Chapter 4 completes the presentation of the foundations of the study of open-channel flows by focusing on the physical and mathematical concepts that underlie the basic equations relating fluid properties and hydraulic variables. The objective here is to provide a deeper understanding of the origins, implications, and applicability of those equations. The chapter develops fundamental physical equations based on the concepts of mass, momentum, energy, force, and diffusion in fluids. The powerful analytical tool of dimensional analysis is described in some detail. Also discussed are approaches to developing equations *not* derived from fundamental physical laws: empirical and heuristic relations, which must often be employed due to the analytical and measurement difficulties presented by natural streamflows. Although most of this book is concerned with one-dimensional (cross-section-averaged "macroscopic") analysis, this chapter develops many of the equations initially at the more fundamental three-dimensional "microscopic" level.

The central problem of open-channel flow is the relation between cross-section-average velocity and flow resistance. The main objective of chapter 5 is to develop physically sound quantitative descriptions of the distribution of velocity in cross sections. The chapter focuses on the derivation of the Prandtl-von Kármán vertical velocity profile based on the characteristics of turbulence and boundary layers developed in chapter 3. Understanding the nature of this profile provides a sound basis for "scaling up" the concepts introduced at the "microscopic" level in chapter 4 and for determining (and measuring) the cross-section-averaged velocity.

Chapter 6 begins by reviewing the basic geometric features of river reaches and reach boundaries presented in chapter 2. It then adapts the definition of uniform flow as applied to a fluid element in chapter 4 to apply to a typical river reach and derives the Chézy equation, which is the basic equation for macroscopic uniform flows. This derivation allows formulation of a simple definition of resistance. The chapter then examines the factors that determine flow resistance, which involves applying the principles of dimensional analysis developed in chapter 4 and the velocity-profile relations derived in chapter 5. Chapter 6 concludes by exploring resistance in nonuniform flows and practical approaches to determining resistance in natural channels.

The goals of chapter 7 are to develop expressions to evaluate the magnitudes of the driving and resisting forces at the macroscopic scale, to examine the relative magnitudes of the various forces in natural streams, and to show how these forces change as a function of flow characteristics. Understanding the relative magnitudes of forces provides a helpful perspective for developing quantitative solutions to practical problems.

Chapter 8 integrates the momentum and energy principles for a fluid element (introduced in chapter 4) across a channel reach to apply to macroscopic one-dimensional steady flows, and compares the theoretical and practical differences between the energy and momentum principles. These principles are applied to solve practical problems in subsequent chapters.

Starting with the premise that natural streamflows can usually be well approximated as steady uniform flows (chapter 7), chapter 9 applies the energy relations of chapter 8 with resistance relations of chapter 6 to develop the equations of gradually varied flow. These equations allow prediction of the elevation of the water surface over extended distances (water-surface profiles), given information about discharge and channel characteristics. Gradually varied flow computations play an essential role in addressing several practical problems, including predicting areas subject to inundation by floods, locations of erosion and deposition, and the effects of engineering structures on water-surface elevations, velocity, and depth. Used in an inverse manner, they provide a tool for estimating the discharge of a past flood from high-water marks left by that flood.

Chapter 10 treats steady, rapidly varied flow, which is flow in which the spatial rates of change of velocity and depth are large enough to make the assumptions of gradually varied flow inapplicable. Such flow occurs at relatively abrupt changes in channel geometry; it is a common local phenomenon in natural streams and at engineered structures such as bridges, culverts, weirs, and flumes. Such flows are generally analyzed by considering various typical situations as isolated cases, applying the basic principles of conservation of mass and of momentum and/or energy as a starting point, and placing heavy reliance on dimensional analysis and empirical relations established in laboratory experiments. The chapter analyzes the three broad cases of rapidly varied flow that are of primary interest to surface-water hydrologists: the standing waves known as hydraulic jumps, abrupt transitions in channel elevation or width, and structures designed for the measurement of discharge (weirs and flumes).

The objective of chapter 11 is to provide a basic understanding of unsteady-flow phenomena, that is, flows in which temporal changes in discharge, depth, and velocity are significant. This understanding rests on application of the principles of conservation of mass and conservation of momentum to flows that change in one spatial dimension (the downstream direction) and in time. Temporal changes in velocity always involve concomitant changes in depth and so can be viewed as wave phenomena. Some of the most important applications of the principles of open-channel flow are in the prediction and modeling of the depth and speed of travel of waves such as flood waves produced by watershed-wide increases in streamflow due to rain or snowmelt, waves due to landslides or debris avalanches into lakes or streams, waves generated by the failure of natural or artificial dams, and waves produced by the operation of engineering structures.

Most natural streams are **alluvial**; that is, their channels are made of particulate sediment that is subject to entrainment, transport, and deposition by the water flowing in them. The goal of chapter 12 is to develop a basic understanding of these processes—a subject of immense scientific and practical import. The chapter begins by defining basic terminology and describes the techniques used to measure sediment in streams. It then explores empirical relations between sediment transport and streamflow and how these relations are used to understand some fundamental aspects of geomorphic processes. The basic physics of the forces that act on sediment particles in suspension and on the stream bed are formulated to provide an essential foundation for understanding entrainment and transport processes, and to gain some insight into factors that dictate the shape of alluvial-channel cross sections. The topic of bedrock erosion—a topic that is only beginning to be studied in detail—is also introduced. The chapter concludes by addressing the central issues of sediment transport: 1) the maximum size of sediment that can be entrained by a given flow (**stream competence**), and 2) the total amount of sediment that can be carried by a specific flow (**stream capacity**).

2

Natural Streams

Morphology, Materials, and Flows

2.0 Introduction and Overview

Stream is the general term for any body of water flowing with measurable velocity in a channel. Streams range in size from rills to brooks to rivers; there are no strict quantitative boundaries to the application of these terms. A given stream as identified by a name (e.g., Beaver Brook, Mekong River) is not usually a single entity with uniform channel and flow characteristics over its entire length. In general, the channel morphology, bed and bank materials, and flow characteristics change significantly with streamwise distance; changes may be gradual or, as major tributaries enter or the geological setting changes, abrupt. Thus, for purposes of describing and understanding natural streams, we focus on the stream reach:

> A **stream reach** is a stream segment with fairly uniform size and shape, water-surface slope, channel materials, and flow characteristics.

The length of a reach depends on the scale and purposes of a study, but usually ranges from several to a few tens of times the stream width. A reach should not include significant changes in water-surface slope and does not extend beyond the junctions of significant tributaries.

Each stream reach has a unique form and personality determined by the flows of water and sediment contributed by its drainage basin; its current and past geological, topographic, and climatic settings; and the ways it has been affected by humans. Thus, natural streams are complex, irregular, dynamic entities, and the characteristics of a given reach are part of spatial and temporal continuums. The **spatial continuum** extends upstream and downstream through the stream network and beyond to include the entire watershed; the **temporal continuum** may include the inheritance of forms and materials from the distant past (e.g., glaciations, tectonic movements, sea-level changes) as well as from relatively recent floods.

In subsequent chapters, this uniqueness and connection to spatial and temporal continuums will not always be apparent because we will simplify the channel geometry, materials, and flow conditions in order to apply the basic physical principles that are the essential starting point for understanding stream behavior. The purpose of this chapter is to present an overview of the characteristics of natural streams and some indication of the ways in which geological, topographic, and climatic factors determine those characteristics. This will provide a natural context for the analytical approach emphasized in subsequent chapters.

2.1 The Watershed and the Stream Network

2.1.1 The Watershed

A **watershed** (also called **drainage basin** or **catchment**) is topographically defined as the area that contributes all the water that passes through a given cross section of a stream (figure 2.1a). The surface trace of the boundary that delimits a watershed is called a **divide**. The horizontal projection of the area of a watershed is the **drainage area** of the stream at (or above) the cross section. The stream cross section that defines the watershed is at the lowest elevation in the watershed and constitutes the watershed outlet; its location is determined by the purpose of the analysis. For geomorphological analyses, the watershed outlet is usually where the stream enters a larger stream, a lake, or the ocean. Water-resources analyses usually require quantitative analyses of streamflow data, so for this purpose the watershed outlet is usually at a **gaging station** where streamflow is monitored (see section 2.5.3).

The watershed is of fundamental importance because the water passing through the stream cross section at the watershed outlet originates as precipitation on the watershed, and the characteristics of the watershed control the paths and rates of movement of water and the types and amounts of its particulate and dissolved constituents as they move through the stream network. Hence, watershed geology, topography, and land cover regulate the magnitude, timing, and sediment load of streamflow. As William Morris Davis stated, "One may fairly extend the 'river' all over its [watershed], and up to its very divides. Ordinarily treated, the river is like the veins of a leaf; broadly viewed, it is like the entire leaf" (Davis 1899, p. 495).

2.1.2 Stream Networks

The drainage of the earth's land surfaces is accomplished by **stream networks**—the veins of the leaf in Davis's metaphor—and it is important to keep in mind that stream reaches are embedded in those networks. Stream networks evolve in response

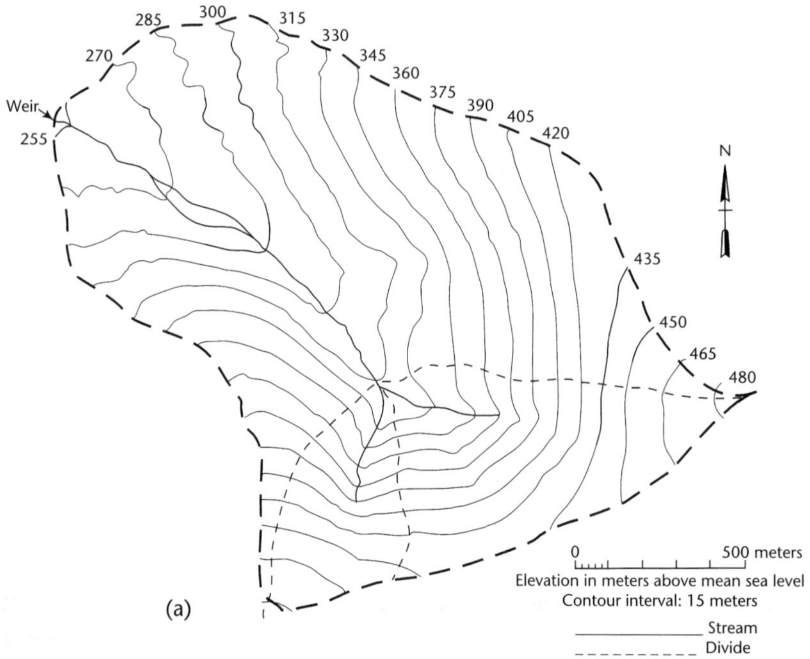

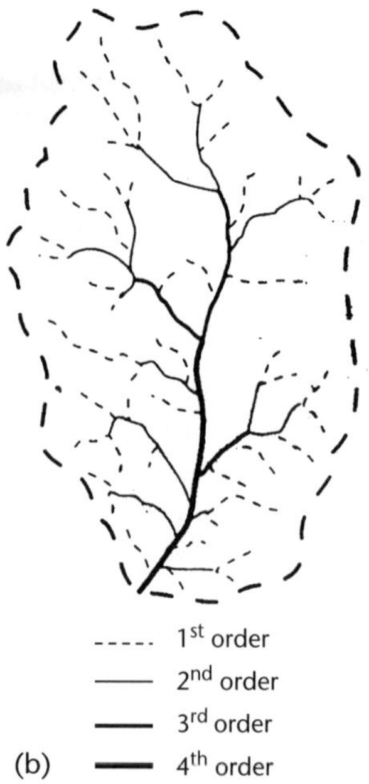

Figure 2.1 A watershed is topographically defined as the area that contributes all the water that passes through a given cross section of a stream. (a) The divide defining the watershed of Glenn Creek, Fox, Alaska, above a streamflow measurement site (weir) is shown as the long-dashed outline, and the divides of two tributaries as shorter-dashed lines. (b) The watershed of a fourth-order stream showing the Strahler system of stream-order designation.

to climate change, earth-surface processes, and tectonic processes, and network characteristics affect various dynamic aspects of stream response and geochemical processes. Knighton (1998) provided an excellent review of the evolution of stream networks, Dingman (2002) summarized their relation to hydrological processes, and Rodriguez-Iturbe and Rinaldo (1997) presented an exhaustive exploration of the subject.

2.1.2.1 Network Patterns

Network patterns, the types of spatial arrangement of river channels in the landscape, are determined by land slope and geological structure (Twidale 2004). Most drainage networks form a dendritic pattern like those of figures 2.1b and 2.2a: there is no preferred orientation of stream segments, and interstream angles at stream junctions are less than 90° and point downstream. The dendritic pattern occurs where there are no strong geological controls that create zones or directions of strongly varying susceptibility to chemical or physical erosion. Zones or directions more susceptible to erosion may display parallel, trellis, rectangular, or annular patterns (figure 2.2b–e). The distributary pattern (figure 2.2f) usually occurs where streams flow out of mountains onto flatter areas to form alluvial fans, or on deltas that form where streams enter lakes or the ocean. Regional geological structures may also cause patterns of any of these shapes to be arranged in radial or centripetal "metapatterns" (figure 2.2g,h). The presence of these patterns and metapatterns on maps, aerial photographs, or satellite images can provide useful clues for inferring the underlying geology (table 2.1).

2.1.2.2 Quantitative Description

Figure 2.1b shows the most common approach to quantitatively describing stream networks (Strahler 1952). Streams with no tributaries are designated **first-order streams**; the confluence of two first-order streams is the beginning of a **second-order stream**; the confluence of two second-order streams produces a **third-order stream**, and so forth. When a stream of a given order receives a tributary of lower order, its order does not change. The order of a drainage basin is the order of the stream at the basin outlet. The actual size of the streams designated a particular order depends on the scale of the map or image used,[1] the climate and geology of the region, and the conventions used in designating stream channels.

Within a given drainage basin, the numbers, average lengths, and average drainage areas of streams of successive orders usually show consistent relations of the form shown in figure 2.3. These relations are called the **laws of drainage-network composition** and are summarized in table 2.2. Networks that follow these laws—that is, that have bifurcation ratios, length ratios, and drainage-area ratios in the ranges shown—can be generated by random numbers, so it seems that the evolution of natural stream networks is essentially governed by the operation of chance (Leopold et al. 1964; Leopold 1994). Table 2.3 summarizes the numbers, average lengths, and average drainage areas of streams of various orders.

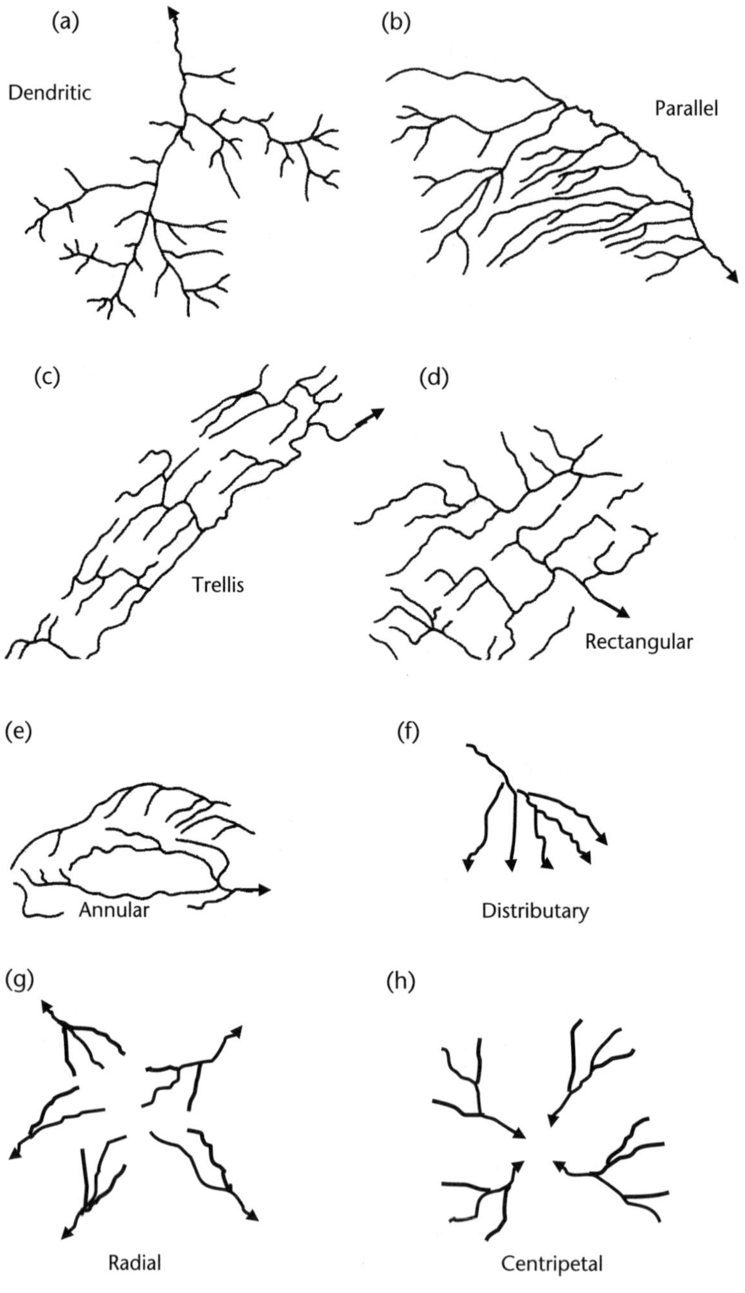

Figure 2.2 Drainage-network patterns (see table 2.1). Panels a–e are from Morisawa (1985).

Table 2.1 Stream-network patterns and metapatterns and their relation to geological controls.

Type	Description	Geological control	Figure
Dendritic	Treelike, no preferred channel orientation, acute interstream angles	None	2.2a
Parallel	Main channels regularly spaced and subparallel to parallel, very acute interstream angles	Closely spaced faults, monoclines, or isoclinal folds	2.2b
Trellis	Channels oriented in two mutually perpendicular directions, elongated in dominant drainage direction, nearly perpendicular interstream angles	Tilted or folded sedimentary rocks with alternating resistant/weak beds	2.2c
Rectangular	Channels oriented in two mutually perpendicular directions, lengths similar in both directions, nearly perpendicular interstream angles	Rectangular joint or fault system	2.2d
Annular	Main streams in approximately circular pattern, nearly perpendicular interstream angles	Eroded dome of sedimentary rocks with alternating resistant/weak beds	2.2e
Distributary	Single channel splits into two or more channels that do not rejoin	Thick alluvial deposits (alluvial fans, deltas)	2.2f
Radial (metapattern)	Stream networks radiate outward from central point	Volcanic cone or dome of intrusive igneous rock	2.2g
Centripetal (metapattern)	Stream networks flow inward to a central basin	Calderas, craters, tectonic basins	2.2h

After Summerfield (1991) and Twidale (2004).

A stream network can also be quantitatively described by designating the junctions of streams as **nodes** and the channel segments between nodes as **links**. Links connecting to only one node (i.e., first-order streams) are called **exterior links**; the others are **interior links**. The **magnitude** of a drainage-basin network is the total number of exterior links it contains; thus, the network of figure 2.1b is of magnitude 43. Typically, the number of links of a given order is about half the number for the next lowest order (Kirkby 1993).

The spatial intensity of the drainage network, or degree of dissection of the terrain by streams, is quantitatively characterized by the **drainage density**, D_D, which is the total length of streams draining that area, ΣX, divided by the area, A_D:

$$D_D \equiv \frac{\Sigma X}{A_D}. \tag{2.1}$$

Drainage density thus has dimensions $[\mathrm{L}^{-1}]$.

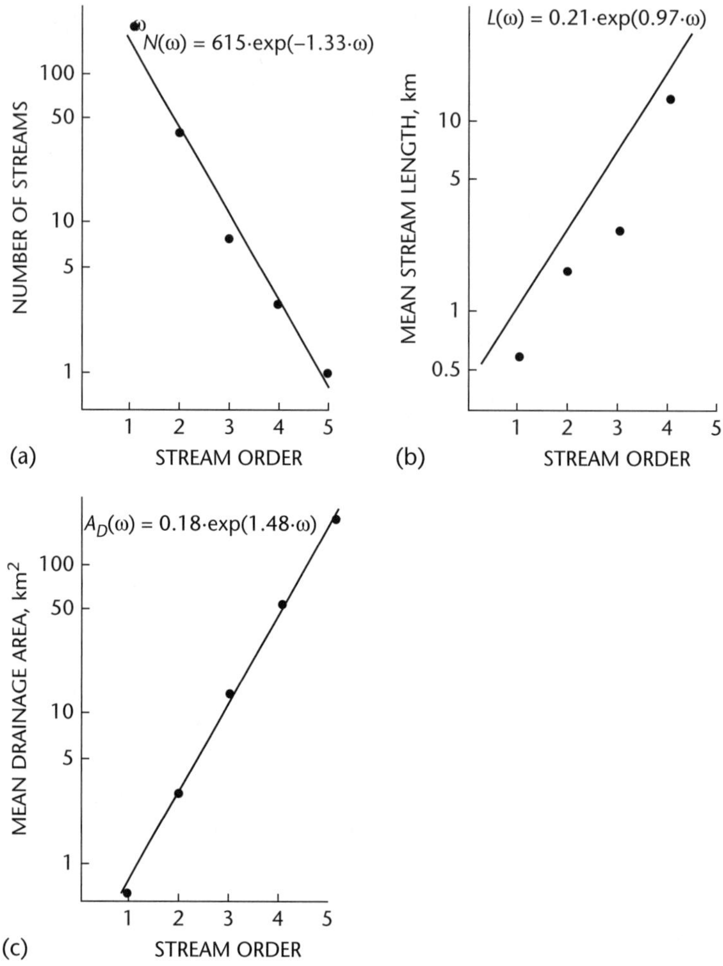

Figure 2.3 Plots of (a) numbers, $N(\omega)$, (b) average lengths, $L(\omega)$, and (c) average drainage areas, $A_D(\omega)$, versus order, ω, for a fifth-order drainage basin in England, illustrating the laws of drainage-network composition (table 2.2). After Knighton (1998).

Drainage density values range from less than $2\,\text{km}^{-1}$ to more than $100\,\text{km}^{-1}$. Drainage density has been found to be related to average precipitation, with low values in arid and humid areas and the largest values in semiarid regions (Knighton 1998). In a given climate, an area of similar geology tends to have a characteristic value; higher values of D_D are generally found on less permeable soils, where channel incision by overland flow is more common, and lower values on more permeable materials. However, it is important to understand that the value of D_D

Table 2.2 The laws of drainage-network composition.[a]

Law of	Definition	Mathematical form	Average value and usual range[b]
Stream numbers (Horton 1945)	$R_B = \frac{N(\omega)}{N(\omega+1)}$	$N(\omega) = \alpha_N \cdot \exp(-\beta_N \cdot \omega)$ $\alpha_N = N(1) \cdot R_B$ $\beta_N = \ln(R_B)$	$R_B = 3.70$ $3 < R_B < 5$
Stream lengths (Horton 1945)	$R_L = \frac{X(\omega+1)}{X(\omega)}$	$X(\omega) = \alpha_L \cdot \exp(\beta_L \cdot \omega)$ $\alpha_L = X(1)/R_L$ $\beta_L = \ln(R_L)$	$R_L = 2.55$ $1.5 < R_L < 3.5$
Drainage areas (Schumm 1956)	$R_A = \frac{A_D(\omega+1)}{A_D(\omega)}$	$A_D(\omega) = \alpha_A \cdot \exp(\beta_A \cdot \omega)$ $\alpha_A = A_D(1)/R_A$ $\beta_A = \ln(R_A)$	$R_A = 4.55$ $3 < R_A < 6$

[a] R_B, bifurcation ratio; R_L, length ratio; R_A, drainage-area ratio; $N(\omega)$, number of streams of order ω; $X(\omega)$, average length of streams of order ω; A_D, average drainage area of streams of order ω.
[b] Global average for orders 3–6 computed by Vörösmarty et al. (2000a, p. 23), considered to best "represent the geomorphic characteristics of natural basins."

Table 2.3 Orders, numbers, average lengths, and average areas of the world's streams.

Order[a]	Number	Average length (km)	Average area (km^2)
1	14,500,000	0.78	1.6
2	4,150,000	1.56	7.2
3	1,190,000	3.13	33
4	339,000	6.25	150
5	96,900	12.5	700
6	27,673	25.0	3,200
7	4,456	249	18,000
8	906	586	82,000
9	176	1,300	369,000
10	38	2,645	1,490,000
11	2	4,360	4,140,000

[a] Values for orders 6–11 taken from Vörösmarty et al. (2000c) assuming that first-order streams at the scale of their study correspond to "true" sixth-order streams (Wollheim 2005). Values for orders 1–5 are computed using the global average bifurcation, length, and area ratios computed by Vörösmarty et al. (2000c): $R_B = 3.70$; $R_L = 2.55$; $R_A = 4.55$.

for a given region will increase as the scale of the map on which measurements are made increases.

2.1.3 Watershed-Scale Longitudinal Profile

The **longitudinal profile** of a stream is a plot of the elevation of its channel bed versus streamwise distance. The profile can be represented as a relation between elevation (Z) and distance (X), or between slope, $S_0(\equiv -dZ/dX)$ and distance. Downstream distance can be used directly as the independent variable or may be replaced by drainage area, which increases with downstream distance, or by average or bankfull discharge, which usually increases with distance.

At the watershed scale, longitudinal profiles of streams from highest point to mouth are usually concave-upward, although some approach straight lines, and commonly there are some segments of the profile that are convex (figure 2.4).

The elevation at the mouth of a stream, usually where it enters a larger stream, a lake, or the ocean, is the stream's **base level**.

This level is an important control of the longitudinal profile because streams adjust over time by erosion or deposition to provide a smooth transition to base level.

The relation between **channel slope**, $S_0(X)$, and downstream distance, X, for a given stream can usually be represented by empirical relations of one of the following forms:

$$S_0(X) = S_0(0) \cdot \exp(-k_1 \cdot X), \tag{2.2a}$$

or

$$S_0(X) = k_2 \cdot X^{-m_2}, \tag{2.2b}$$

or by a relation between slope and drainage area, A_D,

$$S_0(X) = k_3 \cdot A_D{}^{-m_3}, \tag{2.2c}$$

where the coefficients and exponents vary from stream to stream depending on the underlying geology and the sediment size, sediment load, and water discharge provided by the drainage basin. Increasing values of k_1, $|m_2|$, or $|m_3|$ represent increasing concavity.

It is generally assumed that the smooth concave profiles modeled by equation 2.2a–c represent the "ideal" form that evolves over time in the absence of geological heterogeneities or disturbances. Deviations from this form that produce convexities in the profile are common and are due to 1) local areas of resistant rock formations, 2) introduction of coarser sediment or a large sediment deposit by a tributary or landslide, 3) tectonic uplift, or 4) a drop in base level. Pronounced steepenings due to these causes are called **knickpoints**.

Knighton (1998) reviewed many studies of longitudinal profiles and concluded,

Channel slope is largely determined by 1) the quantity of flow contributed by the drainage basin and 2) the size of the channel material.

In almost all river systems, bankfull (or average) discharge increases downstream as a result of increasing drainage area contributing flow; thus, channel slope can be estimated as

$$S_0(X) = k_4 \cdot Q(X)^{-m_4} \cdot d(X)^{m_5}, \tag{2.3a}$$

or

$$S_0(X) = k_5 \cdot A_D(X)^{-m_6} \cdot d(X)^{m_7}, \tag{2.3b}$$

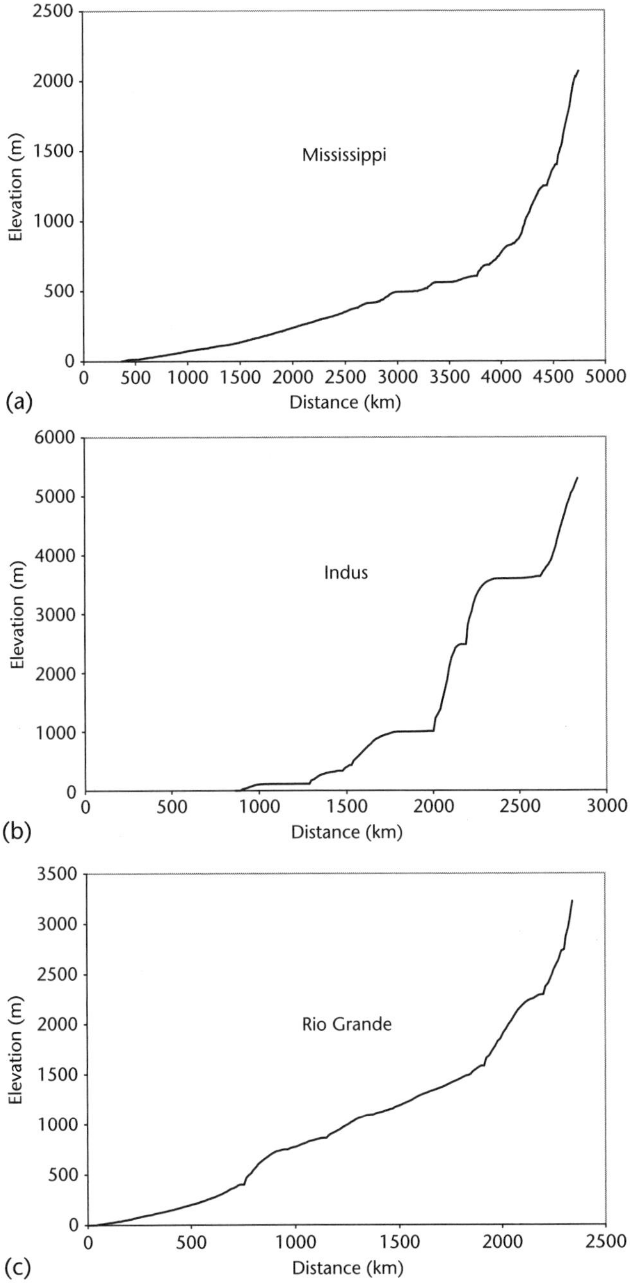

Figure 2.4 Examples of longitudinal profiles of large rivers. All examples are basically concave-upward, even those in which discharge does not increase downstream (lower Indus, Murray, Rio Grande), but some have convex reaches, especially pronounced for the Rio Grande and Indus. Data provided by B. Fekete, Water Systems Analysis Group, University of New Hampshire. *(continued)*

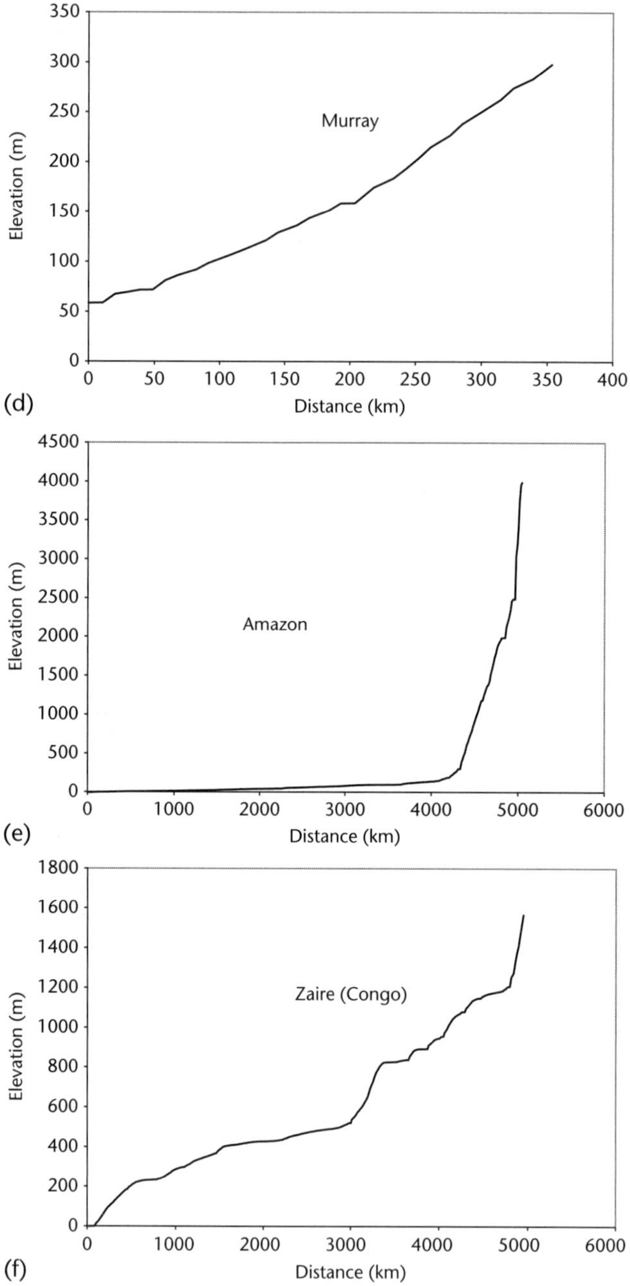

Figure 2.4 *Continued*

where Q is some measure of discharge (e.g., bankfull or average discharge), d is some measure of sediment size (e.g., median sediment diameter), X is downstream distance, and A_D is drainage area. The values of the empirical exponents m_4 through m_7 vary from region to region. As discussed in the following section, d tends to decrease downstream in most stream systems; thus, relations of the form of equation 2.3 predict that the more rapid the downstream increase in Q or A_D or the downstream decrease in d, the more concave the profile.

2.1.4 Downstream Decrease of Sediment Size

There is a general trend of downstream-decreasing bed-material sediment size in virtually all river systems (figure 2.5a), which is typically modeled as an exponential decay:

$$d(X) = d(0) \cdot \exp(-k_6 \cdot X), \tag{2.4}$$

where $d(0)$ is the grain size at $X = 0$ and k_6 is an empirical coefficient that varies from stream to stream (values for various streams are tabulated by Knighton 1998). In many river systems, the exponential decay is "reset" where tributaries contributing coarse material enter a main stream (figure 2.5b). Interestingly, the rate of size decrease is especially pronounced in gravel-bed streams, and an abrupt transition from gravel to sand is often observed.

Two physical processes produce the size decrease: grain breakdown by abrasion and selective transport of finer sizes. Experimental studies have shown that abrasion does not produce the downstream-fining rates observed in most rivers (see, e.g., Ferguson et al. 1996), so selective transport is almost always the dominant process producing downstream sediment-size decrease.

Hoey and Ferguson (1994) were able to simulate the rates of sediment-size decrease observed in a Scottish river using a physically based model. Their results supported the strong correlation between downstream rates of slope decrease and of particle size, as reflected in equation 2.3.

2.2 Channel Planform: Major Stream Types

2.2.1 Classification

Channel planform is the trace of a stream reach on a map.

The continuum of channel planforms in natural streams can be initially divided qualitatively into those with a single thread of flow and those with multiple threads. Channel planforms are further categorized quantitatively by their sinuosity:

The **sinuosity**, ζ, of a stream reach is defined as the ratio of its channel length, ΔX, to the length of its valley,[2] ΔX_v (figure 2.6).

$$\zeta \equiv \frac{\Delta X}{\Delta X_v}. \tag{2.5}$$

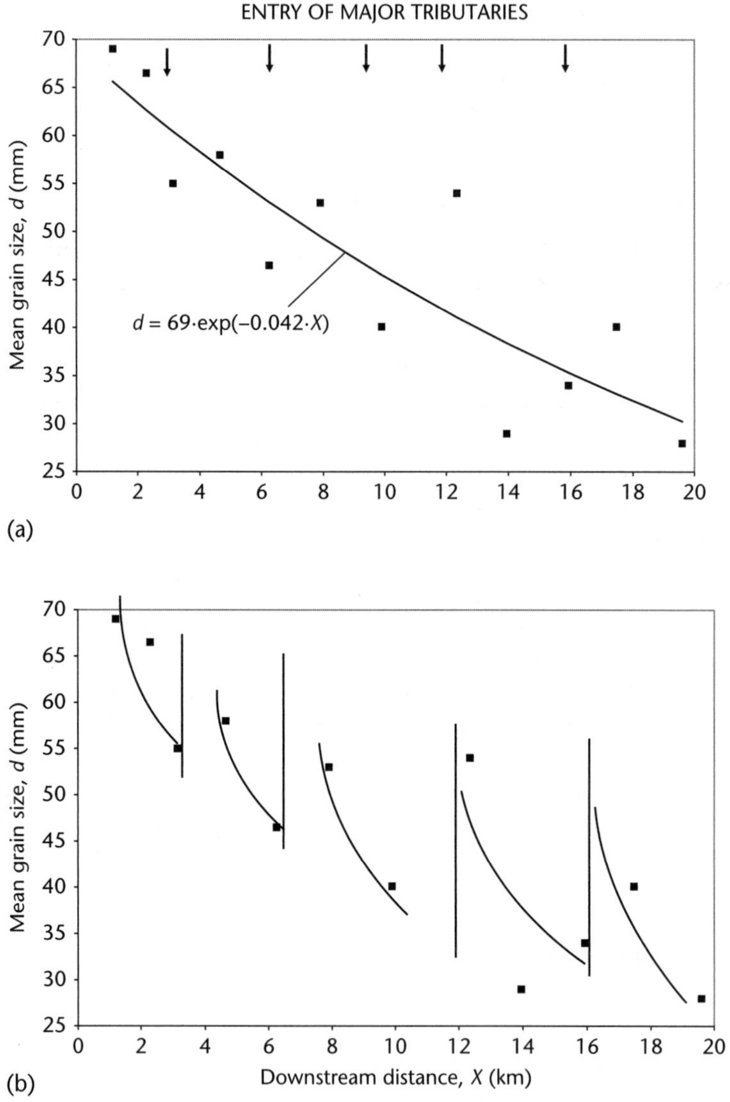

Figure 2.5 Downstream decrease in sediment size in the River Noe, England. Dots show measured values. (a) General trend modeled by exponential decay. (b) "Resetting" of exponential decay due to inputs of coarser material by tributaries. From *Fluvial Forms and Processes* (Knighton 1998), reproduced with permission of Edward Arnold Ltd.

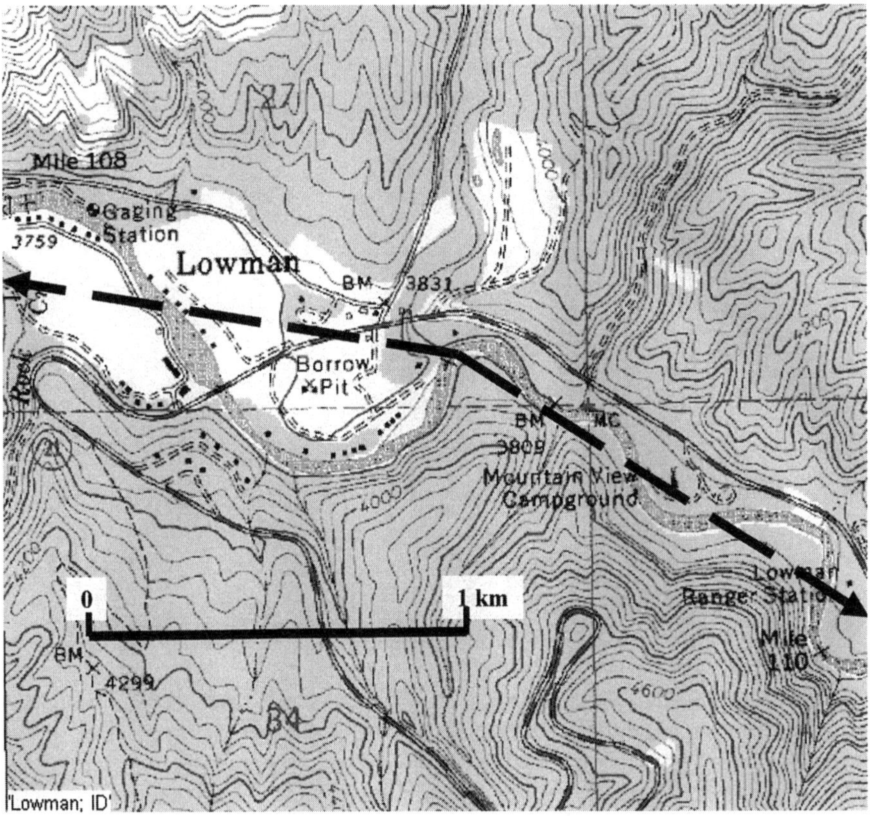

Figure 2.6 Sinuosity of a reach of the South Fork Payette River, Idaho. The dashed arrows represent the valley length, ΔX_v, which equals 2.61 km. The channel length, ΔX, is 3.53 km; thus, the reach sinuosity is 1.35. Contour interval is 40 ft. Solid and dashed parallel lines are roads.

Because $\Delta X \geq \Delta X_v$, it must be true that $\zeta \geq 1$. If the difference in elevation between the upstream and downstream ends of a reach is ΔZ, the channel slope, S_0, and valley slope, S_v, are given by

$$S_0 = \frac{\Delta Z}{\Delta X} \tag{2.6}$$

and

$$S_v = \frac{\Delta Z}{\Delta X_v}. \tag{2.7}$$

Therefore,

$$S_0 = \frac{\Delta Z}{\zeta \cdot \Delta X_v} = \frac{S_v}{\zeta} \leq S_v, \tag{2.8}$$

and we see that, for a given valley slope, channel slope depends on channel planform.

Figure 2.7 An intensely meandering stream in central Alaska. This stream has migrated extensively and left many abandoned channels. Photo by the author.

The most widely accepted qualitative categories of channel planforms, introduced by Leopold and Wolman (1957), are meandering, braided, and straight:

Meandering reaches contain single-thread flows characterized by high sinuosity ($\varsigma > 1.3$) with quasi-regular alternating bends (figure 2.7).

Braided reaches are characterized by flow within "permanent" banks in two or more converging and diverging threads around temporary unvegetated or sparsely vegetated islands made of the material being transported by the stream (figure 2.8). At near-bankfull flows, the islands are typically submerged and the flow becomes single thread.

Straight reaches contain single-thread flows that, while not strictly straight, do not exhibit the sinuosity or regularity of curvature of meandering channels.

In many cases the thread of deepest flow (called the **thalweg**) meanders within the banks of straight reaches. In nature, straight reaches on gentle slopes are rare, and their occurrence often indicates that the stream course has been artificially straightened.

A fourth basic category is often added to the three proposed by Leopold and Wolman (1957):

Anabranching (also called **anastomosing** or **wandering**) reaches contain multithread flows that converge and diverge around "permanent," usually

Figure 2.8 A braided glacial stream in interior Alaska. Photo by the author.

vegetated, islands. Individual threads may be single threads of varying sinuosity or braided.

These basic categories have been elaborated by Schumm (1981, 1985) to provide the classification shown in figure 2.9.

2.2.2 Relation to Environmental and Hydraulic Variables

Many empirical and theoretical studies have attempted to relate channel planform to channel slope, the size of material forming the bed and banks, and the timing and magnitude of flows of water and sediment provided by the drainage basin (Bridge 1993). The pioneering work of Leopold and Wolman (1957) showed that the presence of these patterns can be approximately predicted by where a given reach plots on a graph of channel slope versus bankfull discharge. They used empirical observations to define a discriminant line given by

$$S_0 = 0.012 \cdot {Q_{BF}}^{-0.44}, \tag{2.9}$$

where S_0 is channel slope and Q_{BF} is bankfull discharge in m^3/s. Braided reaches generally plot above the line given by equation 2.9, meandering reaches tend to plot below it, and straight reaches may plot on either side.

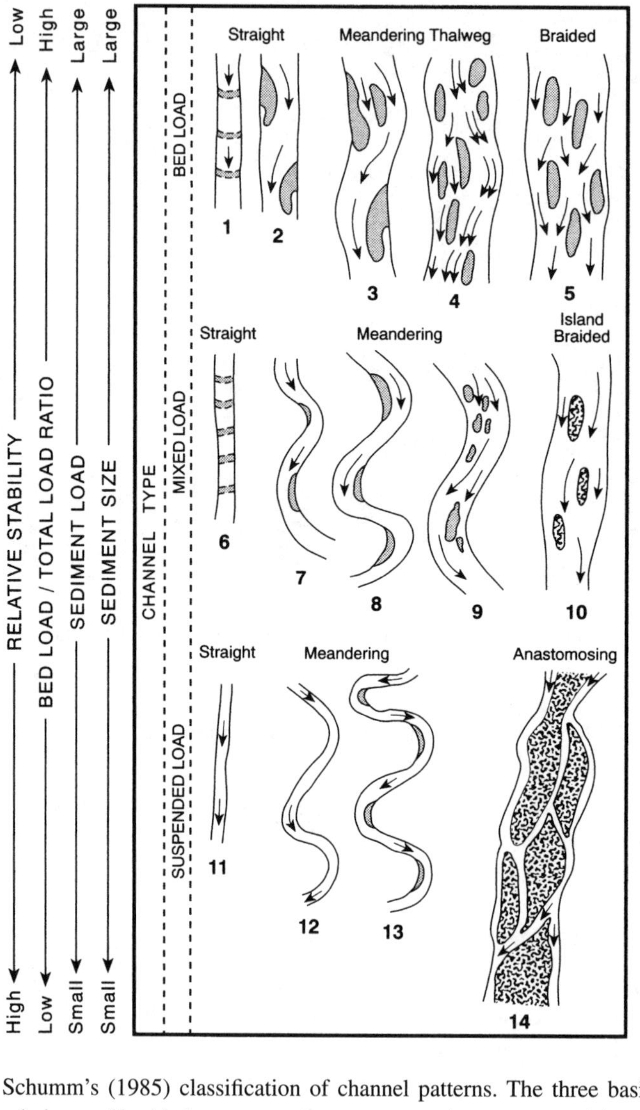

Figure 2.9 Schumm's (1985) classification of channel patterns. The three basic types are straight, meandering, and braided; anastomosing streams are shown as a special case of braided stream. The arrows on the left indicate typical associations of stream type with stability, the ratio of near-bed sediment transport ("bed load") to total sediment transport, total sediment transport, and sediment size. From *Fluvial Forms and Processes* (Knighton 1998), reproduced with permission of Edward Arnold Ltd.

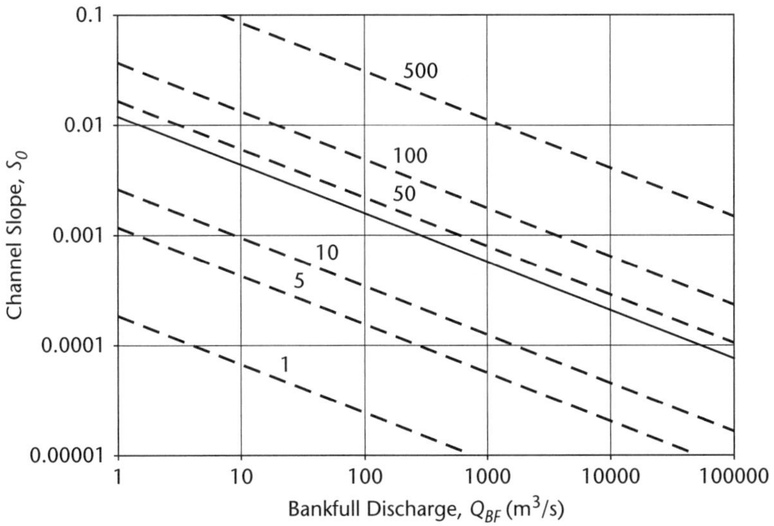

Figure 2.10 Braiding/meandering discriminant-function lines. Braided reaches plot above the lines; meandering reaches, below. Solid line is the discriminant function of Leopold and Wolman (1957) (equation 2.9); dashed lines are discriminant-function lines of Henderson (1961) (equation 2.10) labeled with values of d_{50} (mm).

The approach of Leopold and Wolman (1957) was refined by Henderson (1961), who found that the critical slope separating braided from meandering reaches was also a function of bed-material size and that the discriminant line could be expressed as

$$S_0 = 0.000185 \cdot d_{50}{}^{1.15} \cdot Q_{BF}{}^{-0.44}, \tag{2.10}$$

where d_{50} is the median diameter (mm) of bed material (measurement and characterization of bed material are discussed further in section 2.3.2). The discriminant functions given by equations 2.9 and 2.10 are plotted in figure 2.10; note that for Henderson's equation, both meandering and straight ($\zeta < 1.3$) channels plot below the lines given by equation 2.10, whereas braided channels plot above them. Henderson (1966) showed that an expression very similar to equation 2.10 could be theoretically derived from considerations of channel stability.

More recent studies have pursued similar theoretical approaches. For example, Parker (1976) derived a dimensionless stability parameter ε_P, which is calculated as

$$\varepsilon_P \equiv \frac{g^{1/2} \cdot S_0 \cdot Y_{BF}{}^{1/2} \cdot W_{BF}{}^2}{Q_{BF}}, \tag{2.11}$$

where g is gravitational acceleration and Y_{BF} and W_{BF} are bankfull depth and width, respectively. When $\varepsilon_P > 1$, a braided pattern develops in which the number of subchannels in the stream cross section is proportional to ε_P; when $\varepsilon_P \ll 1$, a meandering channel develops. Further theoretical justification of Parker's approach and support of discriminant functions of the form of equation 2.11 is given by Dade (2000).

However, the criterion of equation 2.11 has been criticized because it requires information about the channel dimensions (Y_{BF} and W_{BF}) and form (S_0, which depends in part on sinuosity as shown in equation 2.8) and so would be of little value for predicting channel planform. To avoid this problem, van den Berg (1995) developed a theory based on stream power (defined and discussed more fully in section 8.1.3) and proposed that a function relating valley slope, S_v, and bankfull discharge, Q_{BF}, to median bed-material size, d_{50}, can be used to discriminate between braided and single-thread reaches with $\zeta \geq 1.3$. He proposed two discriminant functions, one for sand bed streams ($d_{50} < 2$ mm),

$$S_v \cdot Q_{BF}{}^{0.5} = 0.0231 \cdot d_{50}{}^{0.42}, \tag{2.12a}$$

and one for gravel-bed streams ($d_{50} > 2$ mm),

$$S_v \cdot Q_{BF}{}^{0.5} = 0.0147 \cdot d_{50}{}^{0.42}, \tag{2.12b}$$

where Q_{BF} is in m^3/s and d_{50} is in mm. Reaches that plot above the line given by equation 2.12 are usually braided; those below are usually "meandering" (i.e., single thread with $\zeta \geq 1.3$) (figure 2.11). "Straight" reaches (i.e., single thread with $\zeta < 1.3$) plotted both above and below the discriminant lines, as also found by Leopold and Wolman (1957). Bledsoe and Watson (2001) refined van den Berg's approach by replacing the single discriminant equation 2.12 with a set of parallel lines that express the *probability* of being braided.

Van den Berg's discriminant functions (equation 2.12) appear to be a useful approach for predicting whether a given reach will be braided or meandering because 1) they give a correct prediction a high percentage of the time, 2) they have a theoretical justification, and 3) they involve variables that best reflect the

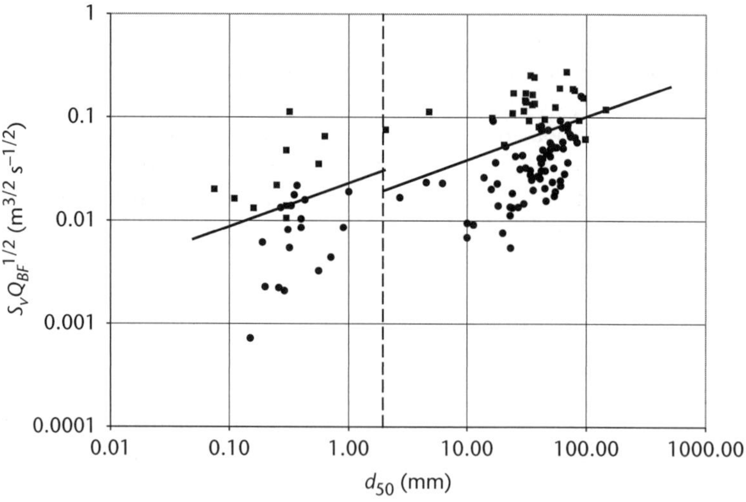

Figure 2.11 Braiding/meandering discriminant-function lines of van den Berg (1995) (equation 2.12). Squares, braided reaches; circles, meandering reaches.

topographic (S_v), hydrological (Q_{BF}), and geological (d_{50}) "givens" of a particular stream reach.

However, a number of recent studies have shown that the additional variable of bank vegetation can play a strong role in determining channel pattern (Huang and Nanson 1998; Tooth and Nanson 2004; Coulthard 2005; Tal and Paola 2007), and such effects are probably responsible for at least some of the misclassifications apparent in figure 2.11. To account for this effect, Millar (2000) formulated a discriminant relation for gravel-bed streams that explicitly includes the effect of bank vegetation:

$$S_0 = 3 \times 10^{-6} \cdot {d_{50}}^{0.51} \cdot {Q_{BF}}^{-0.25} \cdot \Phi^{1.75}, \tag{2.13}$$

where Φ is the maximum slope angle that the bank material can maintain in degrees. (This is the angle of repose, discussed further in section 2.3.3.) The value of Φ is about 40° for sparsely vegetated gravel banks, but may be as high as 80° for heavily vegetated banks because of the strength added by roots.

Note from equations 2.10, 2.12, and 2.13 that discharge, sediment size, and slope are major determinants of reach planform, and these are the same variables that largely determine the form of the longitudinal profile (equation 2.3).

2.2.3 Meandering Reaches

The quasi-regular alternating bends of stream meanders are described in terms of their wavelength, λ_m, their radius of curvature, r_m, and their amplitude, a_m (figure 2.12). Note that the radius of curvature of meander bends is not constant, so r_m is somewhat subjectively defined for the bend apex.

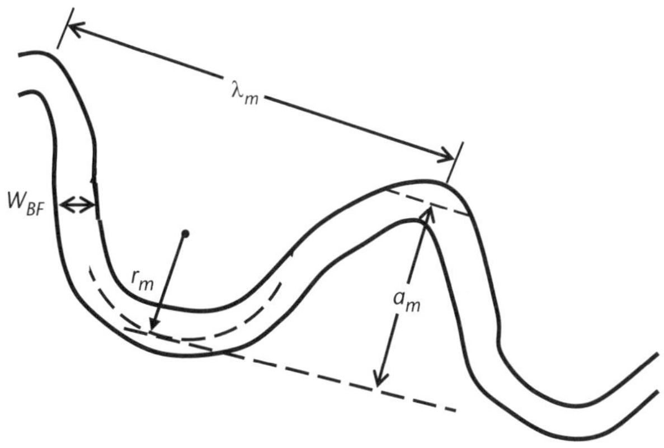

Figure 2.12 Planform of a meandering river showing definitions of meander wavelength, λ_m, radius of curvature, r_m, and amplitude, a_m. W_{BF} is bankfull channel width. Note that the radius of curvature of meander bends is not constant, so r_m is somewhat subjectively defined for the bend apex.

A large number of studies (see Leopold 1994; Knighton 1998), ranging from laboratory channels to the Gulf Stream, have shown that wavelength and radius of curvature are scaled to stream size as measured by bankfull width, W_{BF}:

$$\lambda_m \approx 11 \cdot W_{BF} \tag{2.14}$$

(the coefficient is almost always between 10 and 14), and

$$r_m \approx 2.3 \cdot W_{BF} \tag{2.15}$$

(the coefficient is usually between 2 and 3). The relation between amplitude and width is far less consistent, presumably because that dimension is controlled by bank erodibility, which is determined by local geology and, again, by bank vegetation. Because bankfull width is approximately proportional to the square root of bankfull discharge (see section 2.6.3.2), it is also generally true that $\lambda_m \propto Q_{BF}{}^{0.5}$ and $r_m \propto Q_{BF}{}^{0.5}$, with the coefficients dependent on the regional climate and geology (as well as the units of measurement).

Although it has been the subject of much investigation and speculation, there is no widely accepted complete physical theory of why meanders develop or why they display the observed scaling relationships to width. It does seem clear that the explanation is related to spatial regularities in helicoidal currents and horizontal eddies (for useful reviews, see Knighton 1998; Julien 2002). These currents and eddies are inherent aspects of turbulent open-channel flow and are present even in straight channels (as discussed further in section 6.2.2.3). Laboratory studies suggest that the flow resistance due to bends is minimized when the radius-of-curvature/bankfull-width ratio is 2 to 3 (Bagnold 1960), so this apparently accounts for the consistent empirical relations between those quantities (equation 2.15).

Within meandering reaches, planform features are directly linked to the longitudinal profile at the reach scale: Deeper zones with flatter beds, called **pools**, occur at the bends, whereas shallower, steeper **riffles** occur in the straight segments between the pools (figure 2.13). The transition from riffle to pool is a **run**, and from pool to riffle is a **glide**.

2.2.4 Braided Reaches

At flows below bankfull, braided reaches are characterized by two or more threads of flow that divide and rejoin within well defined, usually vegetated, banks. The islands that separate the threads are usually small relative to the overall channel width, temporary, and unvegetated. As indicated in equation 2.12a,b and figure 2.11, braiding tends to occur in reaches with relatively high bankfull discharge and steep valley slopes relative to the size of bed sediment. Braided reaches are further characterized by significant transport of bed material and by erodible channel banks.

The **degree of braiding** of a braided reach can be quantified as 1) the average number of active channels in the cross section or 2) the ratio of the sum of channel lengths to the length of the widest channel in the section (Knighton 1998). The relation between degree of braiding and flow and channel characteristics is not as clear as for meanders, in part because degree of braiding may vary considerably over short time periods. However, several studies have suggested that the number of braids increases with slope, and equation 2.11 is an attempt to quantify that relation.

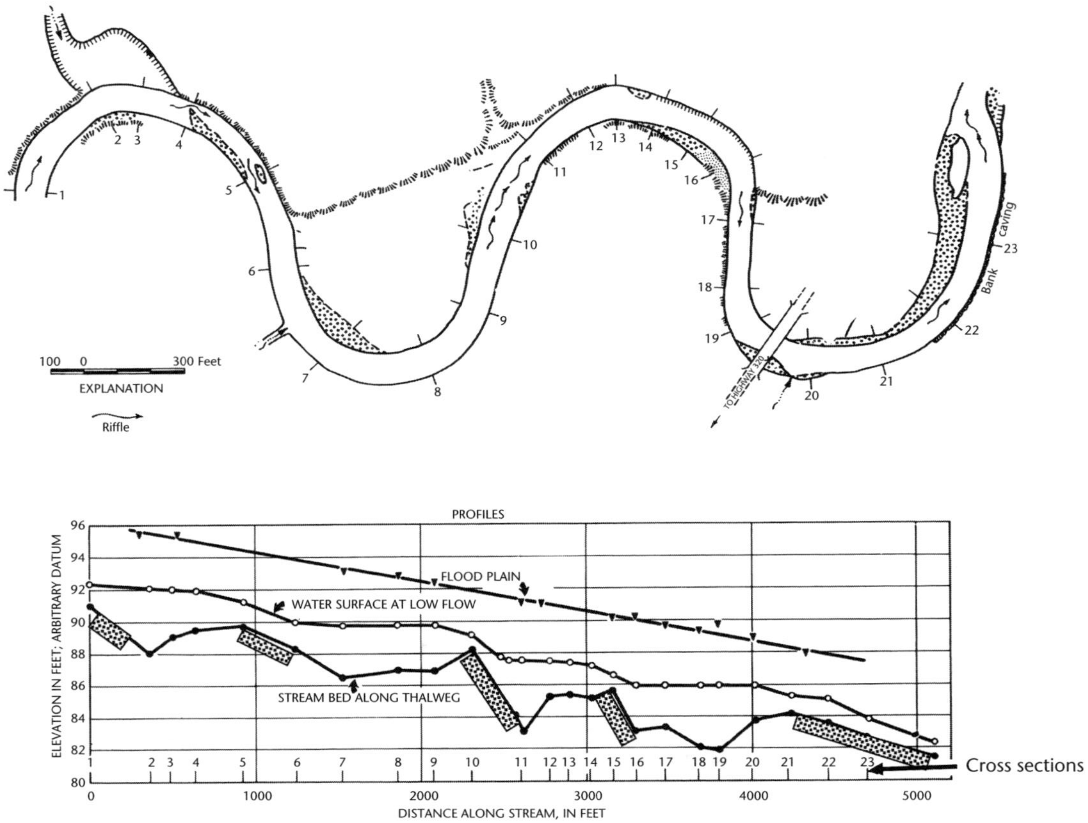

Figure 2.13 Local-scale plan and longitudinal profiles of channel bed (**thalweg** is deepest portion of bed), floodplain, and low-flow water surface of a meandering reach (Popo Agie River near Hudson, WY), showing typical spacing of pools and riffles (stippled areas on profile). Modified from Leopold and Wolman (1957).

2.2.5 "Straight" Reaches

Montgomery and Buffington (1997) developed a widely accepted classification of nonmeandering, single-thread mountain stream reaches that is based primarily on the form of the reach-scale longitudinal profile, which is related to the processes of sediment transport and storage. The characteristics of the stream subtypes they identified are summarized in table 2.4 and illustrated in figure 2.14. Note that the two categories found in valleys of low to moderate valley slope contain alternating pools and riffles or marginal bars with the same spacing as meandering reaches, that is, at five to seven widths (figure 2.13).

Wohl and Merritt (2005) conducted a study to identify the variables that are most influential in determining which of Montgomery and Buffington's channel subtypes occur. They found that slope was by far most important (as suggested in table 2.4), and that 69% of channels could be correctly classified based on slope, channel (bankfull) width, and bed-sediment size. Noting that bankfull width is highly correlated with bankfull discharge, we see that the same factors that determine whether a stream is meandering, braided, or "straight" also largely determine the subtype of "straight" reaches.

2.2.6 Anabranching Reaches

Anabranching reaches, like braided reaches, have flows in individual channels that diverge and converge around islands. They differ from braided reaches in that the individual channel threads are separated by stable, usually well-vegetated islands that are large relative to the channel width. Channel patterns in the individual channels may be meandering, braided, or straight depending on the local valley slope, sediment size, and discharge.

The anabranching river pattern is less common than the other three types, but is found in a wide range of climate settings. This pattern tends to occur where two conditions exist together: 1) flows are highly variable in time and floods are common, and 2) banks are resistant to erosion (Knighton 1998). Nanson and Knighton (1996) have proposed a further classification of anabranching streams based on slope, discharge, bed- and bank-sediment size, and other factors.

2.3 Channel Boundaries

2.3.1 Boundary Characteristics

The nature of the channel boundary, as well as its shape, affects the characteristics of flow. Figure 2.15 presents a classification of boundary characteristics and provides perspective for the discussions of stream hydraulics in subsequent chapters. Except for bedrock channels, natural stream channels consist of unconsolidated sediment particles that are not rigid and are subject to transport by the stream; these are called **alluvial channels**. In many cases, particularly in sand-bed streams, the particles that make up the channel bed are sculpted by the processes of sediment transport into wavelike bedforms with wavelengths and amplitudes ranging from a few centimeters

Table 2.4 Features and processes of mountain-stream reaches.

	Channel type						
Typical form/process	Alluvial dune ripple	Alluvial pool riffle	Alluvial plane bed	Alluvial step pool	Alluvial cascade	Colluvial	Bedrock
Slope[a]	Low	Low-moderate (0.003–0.02)	Moderate-steep (0.006–0.05)	Steep (0.03–0.2)	Steep (0.05–0.4)	Steep (0.15–0.5)	Moderate-steep (0.03–0.8)
Bed material[b]	Sand	Gravel	Gravel-cobble	Cobble-boulder	Cobble-boulder	Variable	Rock
Bedform pattern	Multilayered	Laterally oscillatory	Featureless	Vertically oscillatory	Random	Variable	Irregular
Dominant resistance elements[c]	Sinuosity, bedforms, sediment grains, banks	Bedforms, sediment grains, sinuosity, banks	Sediment grains, banks	Bedforms, sediment grains, banks	Sediment grains, banks	Sediment grains	Bed and bank irregularities
Confinement[d]	Unconfined	Unconfined	Variable	Confined	Confined	Confined	Confined
Pool spacing[e]	5 to 7	5 to 7	None	1 to 4	< 1	Unknown	Variable
Sediment sources	Fluvial,[f] bank failure	Fluvial,[f] bank failure	Fluvial,[f] bank failure, debris flows	Fluvial,[f] hillslope, debris flows	Fluvial,[f] hillslope, debris flows	Hillslope, debris flows	Fluvial,[f] hillslope, debris flows
Supply/transport limited	Transport limited	Variable	Supply limited	Supply limited	Supply limited	Transport limited	Supply limited
Sediment storage	Overbank, bedforms	Overbank, bedforms	Overbank	Bedforms	Upstream and downstream of flow obstructions	Bed	Pockets in bed

[a] Values in parentheses are ranges of slopes in a watershed in the Cascade Mountains of Washington State, USA.
[b] Grain-size diameters associated with these terms are given in section 2.3.2.1.
[c] Relation of channel features to resistance is discussed in detail in section 6.6.
[d] Refers to ability of channel to widen or migrate laterally into a floodplain.
[e] Number of channel widths.
[f] Transport from upstream.
Modified from Montgomery and Buffington (1997).

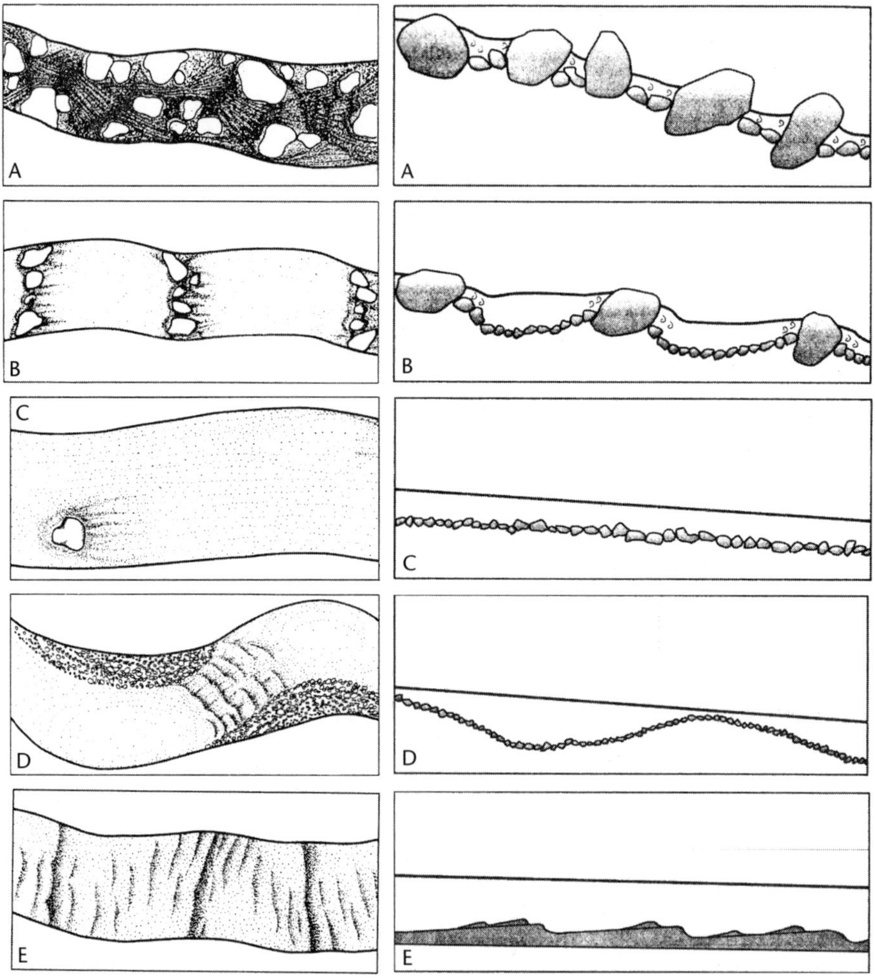

Figure 2.14 Planforms (left column) and local longitudinal profiles (right column) of "straight" single-thread mountain-stream types identified by Montgomery and Buffington (1997): (A) cascade, with nearly continuous highly turbulent flow around large sediment particles; (B) step pool, with alternating highly turbulent flow over steps and more tranquil flow in pools; (C) plane bed, with single boulder protruding through otherwise uniform flow; (D) pool riffle, showing exposed bars, highly turbulent flow over riffles, and tranquil flow in pools; (E) dune ripple, with ripples on stream-spanning dunes. From Montgomery and Buffington (1997), reproduced with permission.

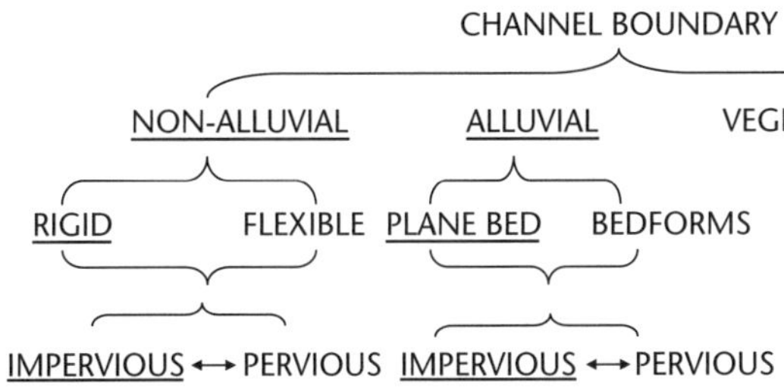

Figure 2.15 Classification of channel boundaries. "Alluvial" denotes boundaries that are subject to erosion, transport, and deposition. Most analytical relations are developed for channels characterized by underlined terms: rigid nonalluvial impervious or plane-bed alluvial impervious boundaries. However, many natural channels fall into other categories. After Yen (2002).

to a few meters (discussed further in chapters 6 and 12). Channel boundaries may also consist at least in part of vegetation (living and dead), ice, and artificial structures, and in many reaches the boundary is pervious and there may be significant hyporheic flow within the sediment that makes up the channel bed (see section 2.5.4).

All these factors complicate the application of theoretical analyses and laboratory experimental results to natural streams. We must keep in mind that most of the theoretical hydraulic relations and experimental results that we will encounter in subsequent chapters have been obtained for rigid, impervious, essentially plane boundaries, whereas many, if not most, natural channels fall into other categories.

The remainder of this section describes the characteristics of the sediment particles that most strongly affect the characteristics of natural channel boundaries.

2.3.2 Sediment Size and Shape

2.3.2.1 Particle Size

Boundaries of alluvial streams consist of a range of sizes of sediment particles, where "size" refers to some measure of the particle diameter. The shape of sediment particles is usually approximated as a triaxial ellipsoid, with the lengths of the three principal axes designated d_{max}, d_{int}, and d_{min} (figure 2.16). Three measures of particle size are commonly used:

1. **Sieve diameter:** The length of the intermediate axis of the particles, d_{int}; this is the dimension that determines the size of a sieve opening that the particle will pass through.
2. **Nominal diameter:** The diameter of a sphere that has the same volume as the particle, equal to $(d_{max} \cdot d_{int} \cdot d_{min})^{1/3}$.
3. **Fall diameter:** The diameter of a sphere with a density of 2,650 kg/m^3 having the same fall velocity (see section 12.3.2) in water at 24°C as the actual particle.

For particles of sand size and larger, the size distribution is directly measured in terms of sieve diameter. For sand-sized to small-gravel-sized particles, the sediment sample is passed through successively smaller sieves, and the weight of the particles caught

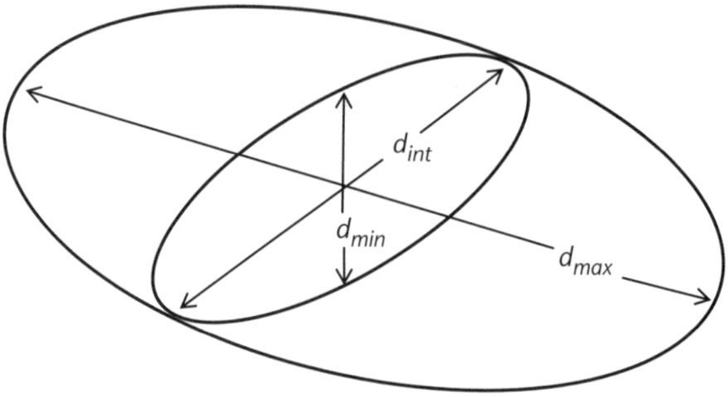

Figure 2.16 Sediment-particle shape idealized as a tri-axial ellipsoid with three mutually perpendicular principal axes designated d_{min}, d_{int}, and d_{max}. The three axes are not truly orthogonal in irregularly shaped natural particles. After Bridge (2003).

on each sieve is determined. For larger particles, the intermediate axis of individual sediment particles is directly measured by determining the longest dimension of the particle and then measuring the length of the longest axis perpendicular to that dimension. Several techniques for sampling and measuring the sizes of large particles, and for estimating the weights of such particles, are reviewed by Bunte and Abt (2001). The distribution of particles of silt size and smaller is usually measured by measuring the time distribution of the weight of material settling out of a suspension of sediment (fall velocity); thus, this technique actually measures the fall diameter.

Particles in various size ranges are categorized, for example, as "clay" at the smaller end of the scale all the way up to "boulders" (figure 2.17a). A complete picture of the size distribution of sediment present on a portion of a channel is given by the **sediment-size distribution**, a graph that relates the proportion (usually by weight) of sediment that is finer than a given diameter, d, to that diameter as shown in figure 2.17b.

For many purposes, the size of sediment in a given reach is often characterized by giving a single point on the sediment-size distribution, designated d_p; this is most commonly the median grain size, d_{50}, or the size that is larger than 84% of the sediment particles, d_{84}. The d_{84} value is usually assumed to characterize the effective height of channel-bed roughness elements that are major contributors to the frictional resistance that the channel exerts on the flowing water. This resistance is explored in detail in chapter 6.

In characterizing the sediment distribution in a reach, one must be aware that the layer of sediment at the surface is commonly significantly larger than the sediment below. This phenomenon, called **armoring**, is due to the selective transport of smaller particles and selective deposition of larger particles (Bunte and Abt 2001).

2.3.2.2 Particle Shape

Qualitative descriptions of basic particle shape are related to the axis ratios (figure 2.18a). One simple and commonly used quantitative descriptor of particle

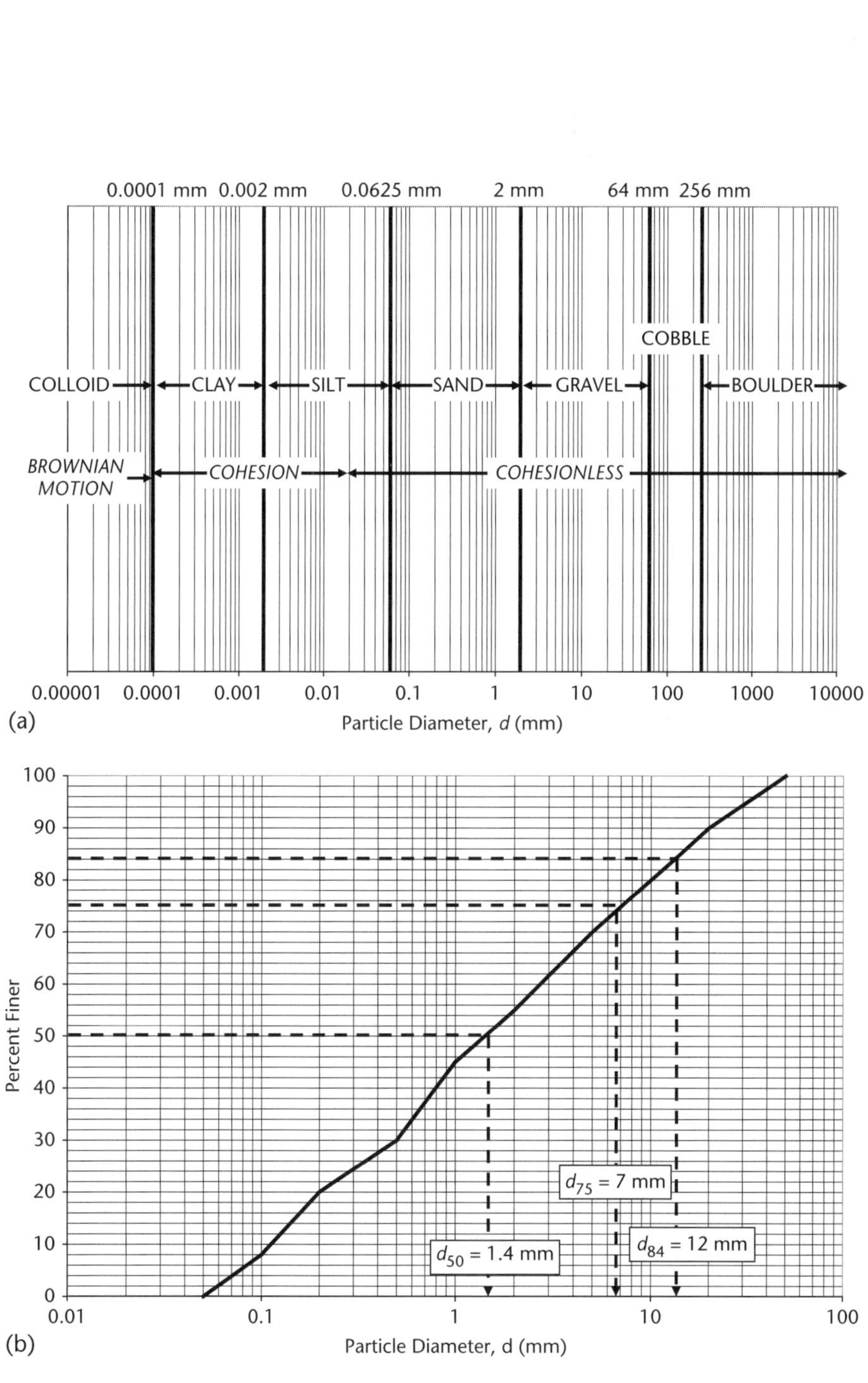

Figure 2.17 (a) Particle-size designations and physical behavior. (b) Typical sediment grain-size distribution. For this case, $d_{50} = 1.4$ mm, $d_{75} = 7$ mm, and $d_{84} = 12$ mm.

shape is the **Corey shape factor**, CSF:

$$CSF \equiv \frac{d_{\min}}{(d_{\max} \cdot d_{\text{int}})^{1/2}}. \tag{2.16}$$

The range of CSF is $0 < CSF \leq 1$, where $CSF = 1$ for a sphere or a cube, and the flatter the particle, the smaller the value of CSF (Dietrich 1982).

A second-order aspect of shape is the particle **roundness**, the degree to which the edges of a particle are rounded (figure 2.18b). Both aspects of shape are tedious to determine and, in the case of roundness, somewhat subjective.

2.3.2.3 Particle Weight

The weight of a particle in water is the gravitational force on the particle, which of course is an important determinant of sediment behavior. Particle weight, wt, is the product of the particle volume, V_p, and its submerged weight density,

$$wt = (\gamma_s - \gamma) \cdot V_p = (\rho_s - \rho) \cdot g \cdot V_p, \tag{2.17}$$

and, because the volume of a quasi-spherical particle is proportional to the cube of its radius,

$$wt = k_s \cdot (\gamma_s - \gamma) \cdot d^3 = k_s \cdot (\rho_s - \rho) \cdot g \cdot V_p, \tag{2.18}$$

where k_s is a shape-dependent proportionality constant ($k_s = \pi/6 = 0.524 \ldots$ for a sphere); g is gravitational acceleration; γ_s and ρ_s are the weight and mass densities, respectively, of the sediment particle, and γ and ρ are the weight and mass densities, respectively, of water.

The densities of water are approximately $\gamma = 9{,}800\ \text{N m}^{-3}$ and $\rho = 1{,}000\ \text{kg m}^{-3}$ and are weakly dependent on temperature (see section 3.3.1). Most sand- and silt-sized particles are made of the mineral quartz, and it is usually safe to assume that $\gamma_s = 2.65 \cdot \gamma$ and $\rho_s = 2.65 \cdot \rho$ in these size ranges. Large gravel and boulders are often rock fragments containing several minerals, and particles smaller than silt may consist of clay minerals; often one can assume these also have the density of quartz, but this may not be true in regions dominated by particular rock types.

2.3.3 Angle of Repose

The **angle of repose** is the maximum slope angle that the bank material can maintain. Angle of repose is an important determinant of channel cross-section shape (see section 12.6) and, as we saw in section 2.2.2, influences channel planform, as well.

Particles larger than about 0.015 mm are noncohesive, and the only forces determining their angle of repose are gravity and interparticle friction. Thus, for pure aggregations of sedimentary particles in that size range, the angle of repose is determined by particle size, shape, and roundness. Figure 2.19 shows angle of repose as a function of particle size and roundness for gravel and cobble particles that have high shape factor ($CSF > 0.8$). Typical values for sand are 30° to 32°, and for silt, about 30°.

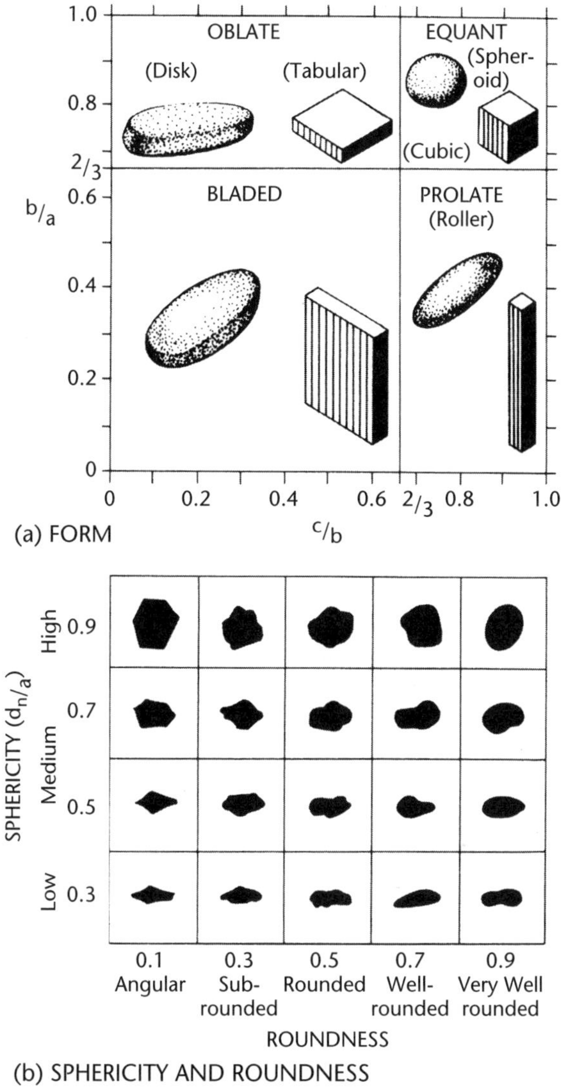

Figure 2.18 (a) Qualitative characterizations of particle shape based on principal-axis ratios. (b) Chart for converting qualitative assessments of particle sphericity and roundness to numerical values. From *Stratigraphy and Sedimentation* Zingg et al. (1963); reproduced with with permission.

Interparticle electrostatic forces become important for particles with diameters less than 0.015 mm (clays and fine silts); such materials are cohesive and can sustain angles of repose up to 90°. And, as noted in section 2.2.2, vegetation strongly affects strength of stream banks, and the angle of repose may be as high as 80° for heavily vegetated banks.

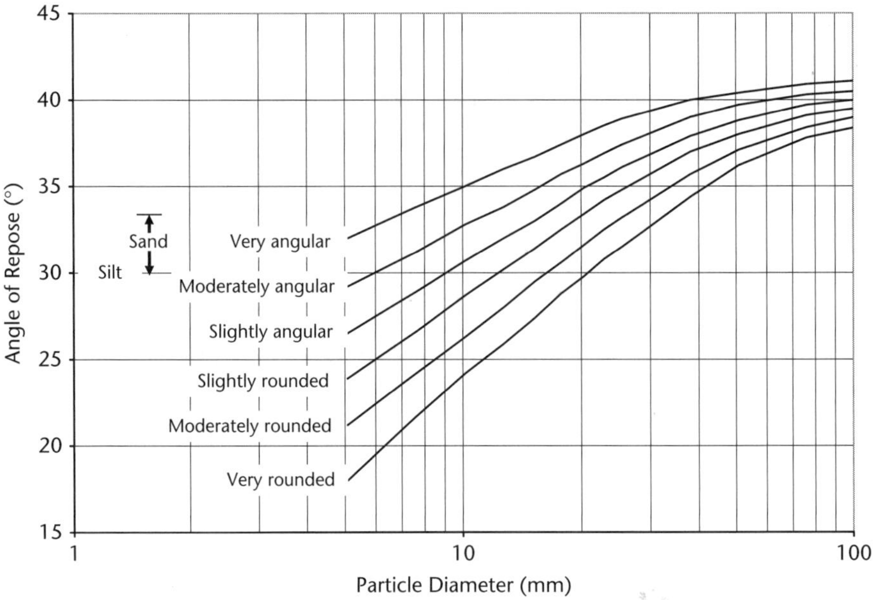

Figure 2.19 Angle of repose as a function of particle size and roundness for gravel and cobble particles, and typical values for sand and silt. Modified after Henderson (1961).

2.4 The Channel Cross-Section

2.4.1 General Characteristics and Definitions

Natural channel cross sections are, of course, generally concave-up, but usually irregular and more or less asymmetrical (figure 2.20a); cross sections in pronounced bends, especially meanders, have a characteristic highly asymmetrical form (figure 2.20b). The two ends of a channel cross section are defined by the **bankfull elevation**, or **bankfull stage**, which may be identified in many ways depending on local conditions (box 2.1). Channel cross-section geometry size and shape are described in terms of the bankfull parameters listed in table 2.5 and illustrated in figure 2.21.

Bankfull elevation is associated with the channel-forming discharge (also called bankfull discharge or dominant discharge). As discussed in section 2.5.6.3, this discharge is reached on average about once every one to two years in most places. Box 2.2 and figure 2.22 describe how channel size and shape parameters are determined from field measurements.

In general, the values of the size and shape parameters in a given cross-section change with the flow magnitude (discharge). The **hydraulic radius** (equations 2B2.6 and 2B2.12), defined as the cross-sectional area divided by the wetted perimeter, enters into important hydraulic formulas (discussed in section 6.3). The ratio of bankfull maximum depth to bankfull average depth, Ψ_{BF}/Y_{BF}, can be used to characterize channel shape (see section 2.4.2).

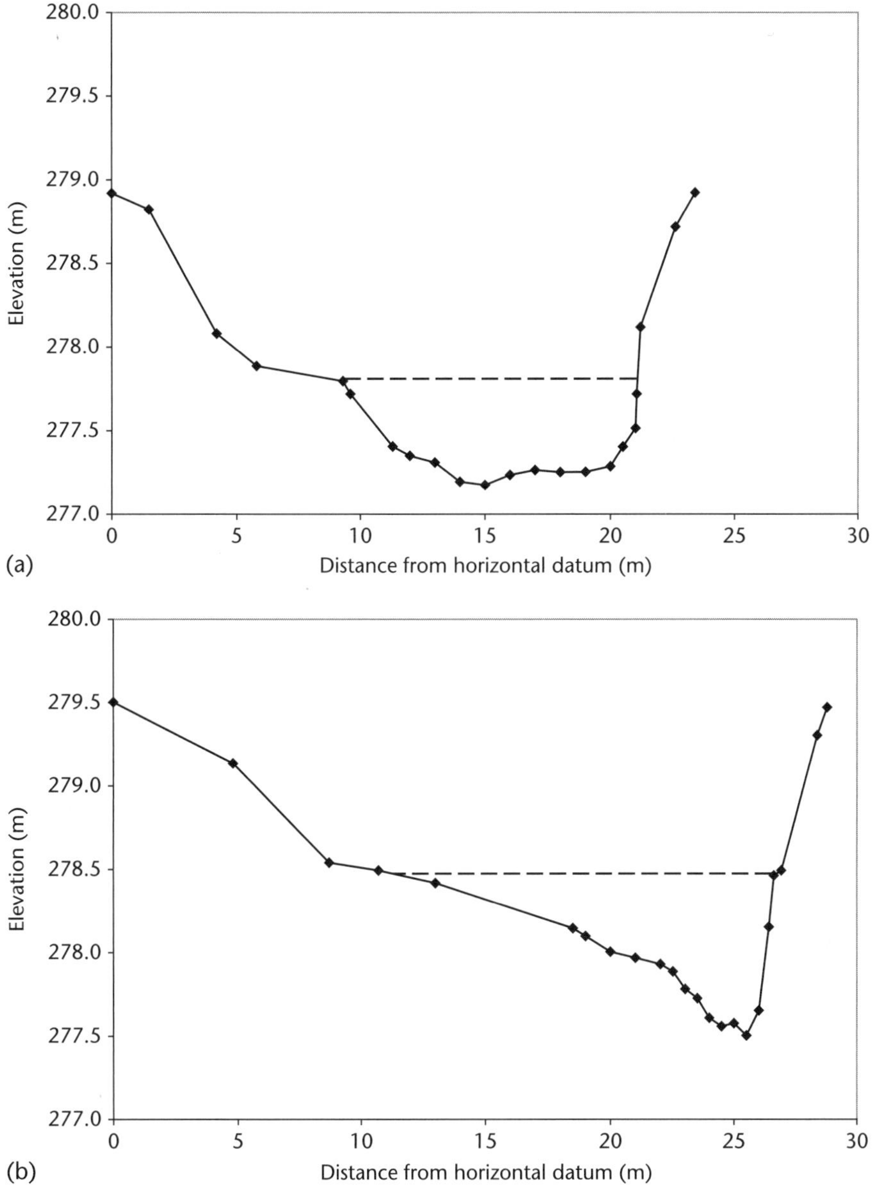

Figure 2.20 Surveyed cross sections of the Cardrona River at Albert Town, New Zealand, plotted at approximately 7-fold vertical exaggeration. (a) Quasi-symmetrical section in straight reach; (b) center of river bend to left showing asymmetry typical of pronounced river bends. Dashed lines show bankfull levels. Data provided by P.D. Mason, New Zealand National Institute of Water and Atmospheric Research (see Hicks and Mason, 1991, p. 125).

BOX 2.1 Field Determination of Bankfull Elevation

Ideally the bankfull elevation is apparent as a well-defined break in slope that separates the channel from the adjacent floodplain (see figure 2.25). However, it may not be easy to determine the bankfull elevation in the field. In many cases, particularly in smaller streams in mountainous areas, there may be no floodplain, or if present, a slope change is not always at the same elevation on both sides of the channel or may vary in elevation along the reach. Where a clear floodplain elevation is not present, Rosgen (1996) suggested the use of several alternative indicators of bankfull stage:

1. The elevation of the top of the highest active depositional features, such as gravel or sand bars along the banks or within the channel. (This elevation is usually considered to be the lowest possible elevation for bankfull stage.)
2. Change in the sediment size, because finer material is usually deposited by overbank flows.
3. The level of staining of rocks within or adjacent to the channel.
4. The level of exposed root hairs below an intact soil layer, indicating exposure to erosion by the stream.
5. The level below which lichens or certain riparian vegetation species (e.g., alders, willows) are absent.

Because of the inherent natural variability of the various bankfull indicators, the elevations of indicators should be determined along a reach, rather than at just a single cross section, and a "reach average" used for bankfull stage throughout the reach. In addition, Rosgen (1996) recommended that the following basic principles be applied in determining bankfull stage:

1. Attempt to identify which indicators in a region most closely correspond to the 1- to 2-year flood levels by calibrating bankfull-stage indicators to flow-frequency information at stream-gaging stations.
2. Use indicators that are appropriate for the stream type and location.
3. Use multiple indicators wherever possible.
4. Know the recent flood and drought history of the region to avoid being misled by recent flood deposits or encroachment of riparian vegetation during drought.

2.4.2 The Width/Depth Ratio and "Wide" Channels

The width/depth ratio, W/Y, is perhaps the most important shape parameter, because it is an inverse measure of the influence of the channel banks on the flow—the larger the value of W/Y, the smaller the frictional effects of the banks on the flow.

Table 2.5 Definitions of channel-geometry parameters (see figure 2.21).

Symbol	Definition
Size parameters	
A_{BF}	Bankfull cross-sectional area: the cross-sectional area at bankfull flow
A	Cross-sectional area at a particular in-channel flow; $A \le A_{BF}$
P_{wBF}	Bankfull wetted perimeter: the bankfull-to-bankfull distance measured along the channel bed
P_w	Wetted perimeter: the bank-to-bank distance measured along the channel bed at a particular in-channel flow; $P_w \le P_{wBF}$
R_{BF}	Bankfull hydraulic radius: $R_{BF} \equiv A_{BF}/P_{wBF}$
R	Hydraulic radius at a particular in-channel flow; $R \equiv A/P_w$
W_{BF}	Bankfull width: water-surface width at bankfull flow
W	Water-surface width at a particular in-channel flow; $W \le W_{BF}$
Ψ_{BF}	Bankfull maximum depth: maximum depth at bankfull flow
Ψ	Maximum depth at a particular in-channel flow; $\Psi \le \Psi_{BF}$
Y_{BF}	Bankfull average depth: average depth at bankfull flow; $Y_{BF} \equiv A_{BF}/W_{BF}$
Y	Average depth at a particular in-channel flow; $Y \equiv A/W$
Y_i	Depth at a particular location w_i in the cross section at a particular in-channel flow; $Y_i \le \Psi$
Shape parameters	
W_{BF}/Y_{BF}	Channel width/depth ratio
W/Y	Width/depth ratio at a particular flow
$A_{BF}/(W_{BF} \cdot \Psi_{BF}) = Y_{BF}/\Psi_{BF}$	Channel depth/maximum depth ratio
$(A_{BFR} - A_{BFL})/A_{BF}$[a]	Channel asymmetry index
$\max(A_{BFR}, A_{BFL})/\min(A_{BFR}, A_{BFL})$[a]	Channel asymmetry index

In natural channels, bankfull dimensions (identified by subscript "BF") are constant at a particular cross section; the other parameters vary with time as flow changes.

[a]A_{BFR} and A_{BFL} are the bankfull areas of the right and left halves of the cross section, respectively.

Figure 2.23 shows the ratios of wetted perimeter to width (P_w/W) and hydraulic radius to average depth (R/Y) as a function of W/Y for rectangular channels. Both ratios approach 1 as W/Y increases and are within 10% of 1 for W/Y values above 18. Thus, from a geometrical point of view, if W/Y is "large enough," we can simplify our analyses by assuming that 1) the wetted perimeter is equal to the water-surface width ($P_w = W$), and 2) the hydraulic radius is equal to the depth ($R = Y$).

From a dynamic point of view, data from flume studies (Cruff 1965) show that the P_w/W curve of figure 2.23 also represents the ratio of the actual channel friction to the friction that would exist without the banks. Thus, if W/Y is "large enough," we can simplify our analyses by neglecting the bank effects and considering only the frictional effects of the channel bed.

Figure 2.24a gives information on the bankfull width/depth ratios (W_{BF}/Y_{BF}) of natural channels. This is a cumulative-frequency diagram computed from a database of 499 measurements collected by Church and Rood (1983). It shows that more than 60% of the channels have $W_{BF}/Y_{BF} > 18$. Within a given channel, the width/depth ratio, W/Y, is a minimum at bankfull and is greater than W_{BF}/Y_{BF} for less-than-bankfull flows—this is illustrated in figure 2.24b for a parabolic channel with $W_{BF} = 25$ m

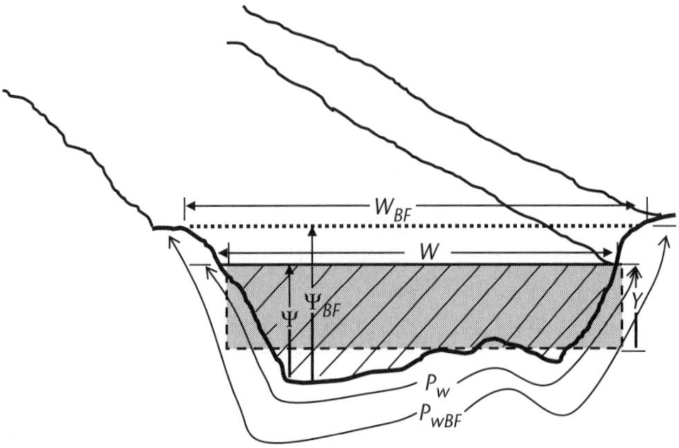

Figure 2.21 Diagram showing definitions of terms used to describe channel geometry. The subscript BF indicates bankfull values. The cross-hatched region denotes the cross-sectional area, A, and the shaded rectangle the average depth, $Y \equiv A/W$, of a subbankfull flow. Analogous quantities A_{BF} and $Y_{BF} \equiv A_{BF}/W_{BF}$ are defined for bankfull flow. Ψ indicates maximum depth. See box 2.2 and table 2.5.

BOX 2.2 Computation of Channel Cross-Section Geometry from Field Measurements

This box describes the basic approaches to measuring bankfull channel geometry and the geometry associated with subbankfull flows. Discharge-measurement techniques are described in detail in Herschy (1999a) and Dingman (2002). The reference by Harrelson et al. (1994) should be consulted as a basic guide to field techniques for stream measurements.

Channel (Bankfull) Geometry

Referring to figure 2.22, once the bankfull elevation z_{BF} is established (see box 2.1), a vertical datum ($z = 0$) is established at an elevation above z_{BF} across the channel by means of a tape, cable, or surveyor's level, and a horizontal datum ($w = 0$) is established on either the right or left bank ("left" and "right" are determined by an observer facing downstream). Then successive observations of distance from the horizontal datum, w_i, and vertical distance from the vertical datum downward to the channel bed, z_i, are made across the channel, usually by means of a surveyor's rod.

The first observation point (w_1, z_1) is established at the bankfull elevation on one bank, and the last (w_I, z_I) at the bankfull elevation on the other bank. Sufficient points are selected between the endpoints to characterize the cross-section shape.

1. At each point, compute the local bankfull depth, Y_{BFi}:

$$Y_{BFi} = (z_i - z_{BF}). \tag{2B2.1}$$

   Strictly speaking, depth is measured normal to the channel bottom rather than vertically, so equation 2B2.1 should be written as $Y_{BFi} = (z_i - z_{BF}) \cdot \cos(S)$, where S is the slope of the channel bottom and water surface (assumed parallel). However, slopes of natural channels virtually never exceed 0.1 [$= \tan(S)$], and because $\cos[\tan^{-1}(0.1)] = 0.995$, one can almost always assume $\cos(S) = 1$ without error. Then the bankfull quantities are computed by the formulas in steps 2–7.
2. Bankfull width, W_{BF}:

$$W_{BF} = w_I - w_1. \tag{2B2.2}$$

3. Bankfull cross-sectional area, A_{BF}:

$$A_{BF} = Y_{BFi} \cdot \left(\frac{w_2 - w_1}{2}\right) + \sum_{i=2}^{I-1} Y_{BFi} \cdot \left(\frac{w_{i+1} - w_{i-1}}{2}\right) + Y_{BFI} \cdot \left(\frac{w_I - w_{I-1}}{2}\right). \tag{2B2.3}$$

4. Bankfull average depth, Y_{BF}:

$$Y_{BF} = \frac{A_{BF}}{W_{BF}}. \tag{2B2.4}$$

5. Bankfull wetted perimeter, P_{wBF}:

$$P_{wBF} = \sum_{i=2}^{I} \frac{|Y_{BF_i} - Y_{BF_{i-1}}|}{\sin\left[\tan^{-1}\left(\frac{|Y_{BF_i} - Y_{BF_{i-1}}|}{w_i - w_{i-1}}\right)\right]}. \tag{2B2.5}$$

6. Bankfull hydraulic radius, R_{BF}:

$$R_{BF} = \frac{A_{BF}}{P_{wBF}}. \tag{2B2.6}$$

7. Bankfull maximum depth, Ψ_{BF}:

$$\Psi_{BF} = \max(Y_{BFi}). \tag{2B2.7}$$

Geometry at a Subbankfull Flow

As in figure 2.22, a horizontal datum ($w = 0$) is established on either right or left bank. Then successive observations of distance from the horizontal datum, w_i, and water depth, Y_i, are made across the channel. If the stream

(Continued)

BOX 2.2 *Continued*

can be waded, depth is usually measured by a graduated wading rod; if not, depth can be measured from a boat or bridge by weighted cable or sonar depth-sounding device. One can combine bankfull and flow-specific measurements by using the technique described in part 1 of this Box and measuring the water depth at each observation.

The first observation point (w_1, Y_1) is established at the intersection of the water surface and one bank, and the last (w_I, Y_I) at the intersection on the other bank. Measurements can begin on either bank; the endpoints are designated "left edge of water" (LEW) and "right edge of water" (REW) with respect to an observer facing downstream. Sufficient points are selected between the endpoints to characterize the cross-section shape.

1. At each point, measure the local water depth Y_i. Again, depth is defined as being normal to the channel bottom rather than vertical, so the height of the water measured on a vertically held device should strictly be multiplied by $\cos(S)$. However, as noted in part 1 in this box, one can virtually always assume $\cos(S) = 1$ without error. Then compute the following:
2. Width W, the distance between LEW and REW:

$$W = w_N - w_1. \tag{2B2.8}$$

3. Cross-sectional area, A:

$$A = Y_1 \cdot \left(\frac{w_2 - w_1}{2}\right) + \sum_{i=2}^{N-1} Y_i \cdot \left(\frac{w_{i+1} - w_{i-1}}{2}\right) + Y_N \cdot \left(\frac{w_N - w_{N-1}}{2}\right). \tag{2B2.9}$$

4. Average depth, Y:

$$Y = \frac{A}{W}. \tag{2B2.10}$$

5. Wetted perimeter, P_w:

$$P_w = \sum_{i=2}^{N} \frac{|Y_i - Y_{i-1}|}{\sin\left[\tan^{-1}\left(\frac{|Y_i - Y_{i-1}|}{w_i - w_{i-1}}\right)\right]}. \tag{2B2.11}$$

6. Hydraulic radius, R:

$$R = \frac{A}{P_w}. \tag{2B2.12}$$

7. Maximum depth, Ψ:

$$\Psi_{BF} = \max(Y_i). \tag{2B2.13}$$

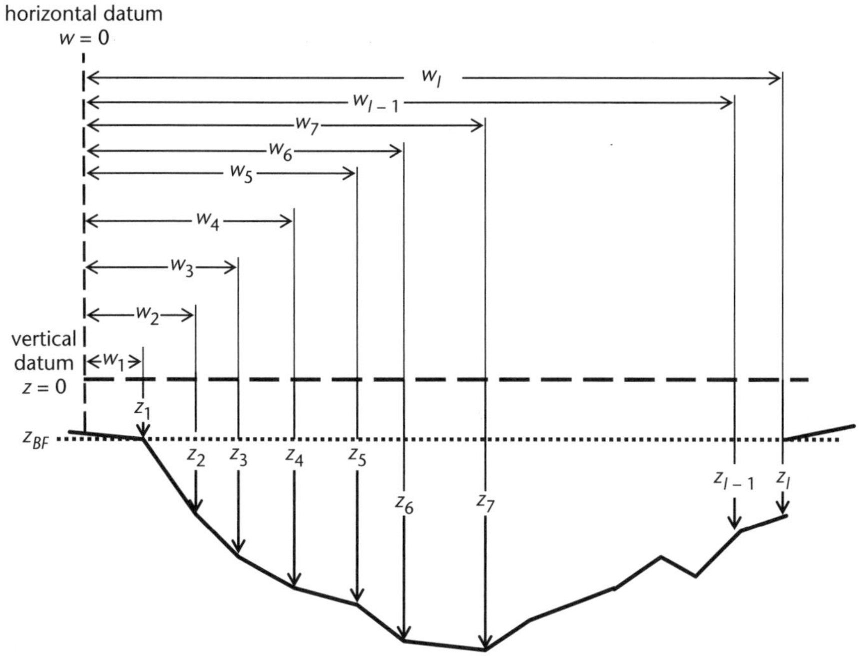

Figure 2.22 Diagram illustrating measurements used to characterize the bankfull channel cross section. See box 2.2.

and $Y_{BF} = 1$ m. Thus, the values plotted in figure 2.24a are minimum width/depth ratios for flows in natural channels, and we conclude that, for flows in natural channels, it is usually safe to assume that $P_w = W$ and $R = Y$. Cross sections or reaches for which $P_w = W$ and $R = Y$ are called **wide channels**.

2.4.3 Models of Cross-Section Shape

Reaches with constant cross-section shape and slope are **prismatic reaches**. Of course, natural channels are nonprismatic, but for analytical purposes it is useful to have prismatic models that approximate the shapes of natural river reaches. In practice, the three most common cross-section shapes encountered are the trapezoid, the rectangle, and the parabola. The trapezoid is the shape used for human-made canals and channels because it is relatively easy to construct and can approximate the shape of natural channels. The rectangle is obviously the simplest geometry, and is the shape of the laboratory flumes in which many of the experiments that are the basis for understanding open-channel flows are carried out. We will often use the rectangular model when deriving hydraulic relationships in later chapters. The parabola is also commonly used to approximate natural-channel cross sections (Chow 1959; Leopold et al. 1964), and we will sometimes use this model to develop analytical relations.

Many attempts have been made to derive expressions for the form of stream cross sections. In the remainder of this section we discuss two cross-section models, both

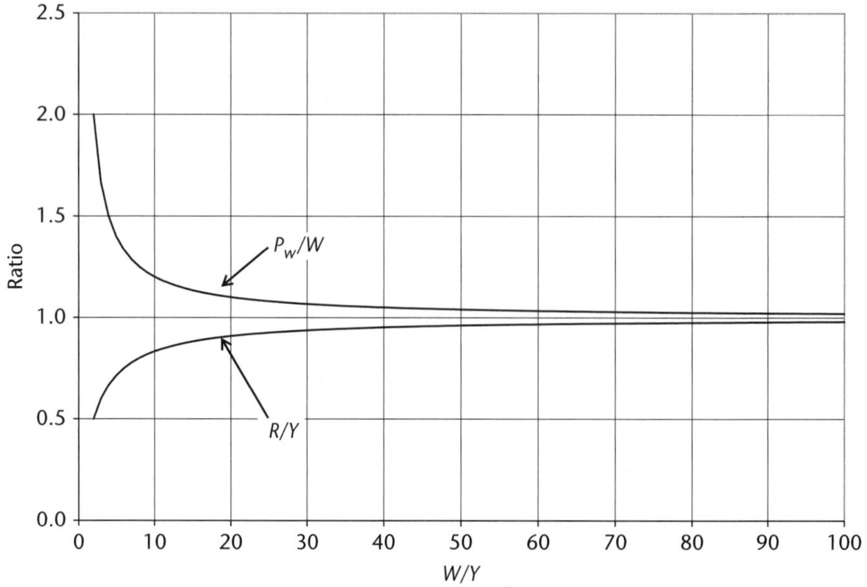

Figure 2.23 Ratios of wetted perimeter, P_w, to width, W, and hydraulic radius, R, to average depth, Y, for rectangular channels as functions of the width/depth ratio (W/Y). The P_w/W curve also represents the ratio of the frictional effects of the bottom and sides to the friction due to the bottom alone. Both curves are within $\pm 10\%$ of 1 for $W/Y > 18$. Similar curves can be drawn for other cross-section geometries.

of which assume a symmetrical section with the deepest point at the center: 1) a model derived from physical principles, called the "Lane stable channel," and 2) a flexible general model that includes the rectangle, the parabola, the Lane stable channel, and other forms. These are useful general models, but recall that they are not usually applicable to channel bends, where the cross section is typically strongly asymmetrical (figure 2.20b).

2.4.3.1 Lane's Stable Channel Cross-Section Model

The **Lane stable channel** model was derived by Lane (1955) assuming that the channel is made of noncohesive material that is just at the threshold of erosion when the flow is at bankfull elevation. This assumption dictates that the bank slope angle at the channel edge equals the angle of repose. (The complete development of the model, given in section 12.6, requires concepts that have not yet been introduced). Referring to figure 2.25a, Lane's relation giving the elevation of the channel bottom, z, as a function of distance from the center, w, is

$$z(w) = \Psi_{BF} \cdot \left[1 - \cos\left(\frac{\tan(\Phi)}{\Psi_{BF}} \cdot w\right)\right], 0 \leq w \leq W_{BF}/2, \qquad (2.19)$$

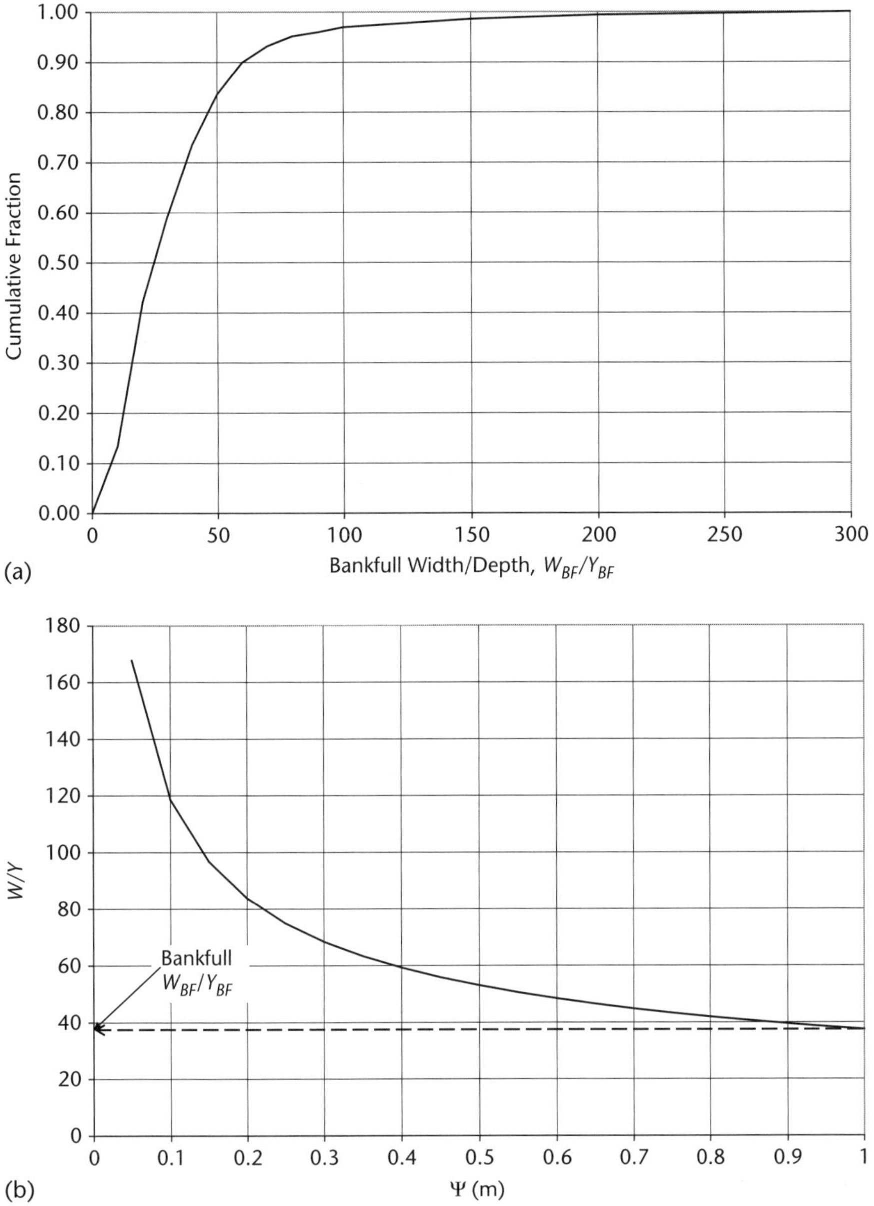

Figure 2.24 (a) Cumulative frequency of 499 measurements of bankfull width/depth ratios of natural channels by Church and Rood (1983). More than 60% have $W_{BF}/Y_{BF} > 18$. (b) Width/depth ratio as a function of maximum depth, Ψ, for a parabolic channel with $W_{BF} = 25$ m and $\Psi_{BF} = 1$ m showing that $W/Y \geq W_{BF}/Y_{BF}$.

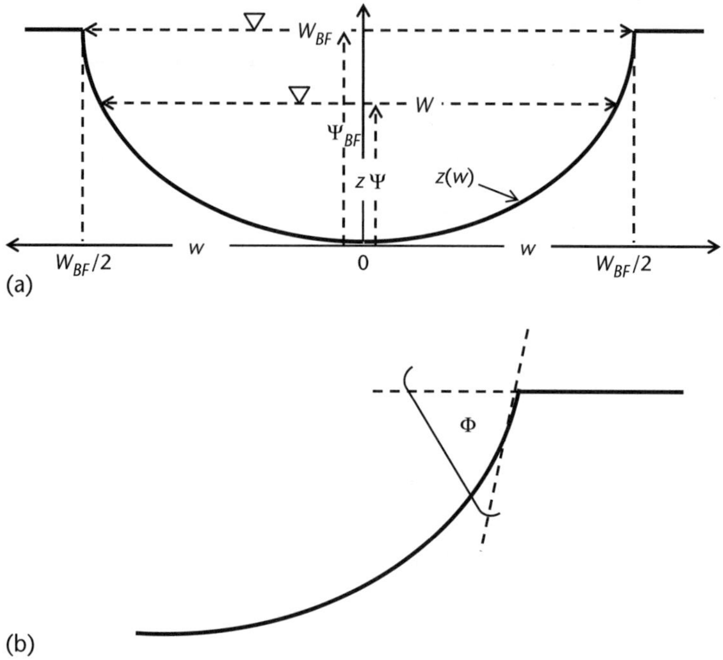

Figure 2.25 (a) Definitions of terms for equations 2.19 and 2.20. (b) In equation 2.19, the bank angle at the bankfull level equals the angle of repose of the bank material, Φ. In equation 2.20, the bank angle at the bankfull level $= \text{atan}[2 \cdot r \cdot (\Psi_{BF}/W_{BF})]$.

where Ψ_{BF} is the maximum (i.e., central) bankfull depth, W_{BF} is the bankfull width, and Φ is the angle of repose of the bank material (figure 2.25b). To use this model, Φ and either Ψ_{BF} or W_{BF} must be specified. This model implies the relations shown in table 2.6.

Using the range of Φ values from figure 2.19, equation 2.19 dictates that $5.7 \leq W_{BF}/Y_{BF} \leq 15.2$. However, we see from figure 2.24 that fewer than one-third of natural channels have bankfull width/depth ratios in this range, so the direct applicability of Lane's formula appears limited. We will examine the Lane model in more detail in section 12.6 and show that it can be made more flexible.

2.4.3.2 General Cross-Section Model

If we assume that channel cross sections are symmetrical and that bankfull maximum depth Ψ_{BF} and bankfull width W_{BF} are given, we can formulate a model for the shape of a channel cross section that includes the rectangle, the Lane model, and the parabola but is flexible enough to comprise a wider range of forms:

$$z(w) = \Psi_{BF} \cdot \left(\frac{2}{W_{BF}}\right)^r \cdot w^r, 0 \leq w \leq W_{BF}/2, \tag{2.20}$$

where r is an exponent that dictates the cross-section shape, and the other symbols are as in equation 2.19.

Table 2.6 Geometrical relations of the Lane stable channel model (equation 2.19) and general cross-section (equation 2.20) model.

Quantity	Lane stable channel (Ψ_{BF} and Φ specified)	General model (Ψ_{BF}, W_{BF}, and r specified)
Average depth, Y_{BF}	$\frac{2 \cdot \Psi_{BF}}{\pi} = 0.637 \cdot \Psi_{BF}$	$\left(\frac{r}{r+1}\right) \cdot \Psi_{BF}$
Cross-sectional area, A_{BF}	$\frac{2 \cdot \Psi_{BF}^2}{\tan(\Phi)} = \frac{\pi^2 \cdot Y_{BF}^2}{2 \cdot \tan(\Phi)} = \frac{4.93 \cdot Y_{BF}^2}{\tan(\Phi)}$	$\left(\frac{r}{r+1}\right) \cdot W_{BF} \cdot \Psi_{BF}$
Width, W_{BF}	$\frac{\pi \cdot \Psi_{BF}}{\tan(\Phi)} = \frac{\pi^2 \cdot Y_{BF}}{2 \cdot \tan(\Phi)} = \frac{4.93 \cdot Y_{BF}}{\tan(\Phi)}$	W_{BF}
Width/depth ratio, W_{BF}/Y_{BF}	$\frac{\pi^2}{2 \cdot \tan(\Phi)} = \frac{4.93}{\tan(\Phi)}$	$\left(\frac{r+1}{r}\right) \cdot \left(\frac{W_{BF}}{\Psi_{BF}}\right)$
Bank angle at channel edge, $\tan^{-1}\left(\left.\frac{dy}{dx}\right\rvert_{W_{BF}/2}\right)$	Φ	$\tan^{-1}\left(2 \cdot r \cdot \frac{\Psi_{BF}}{W_{BF}}\right)$

In equation 2.20, a triangle is represented by $r = 1$, the Lane channel by $r = 2/(\pi - 2) = 1.75$, a parabola by $r = 2$, and forms with increasingly flatter bottoms and steeper banks by increasing values of r. In the limit as $r \rightarrow \infty$, the channel is rectangular. Table 2.6 summarizes relationships implied by equation 2.20 and compares them with the Lane model. Although equation 2.20 is more general than the Lane model, it does not, in general, result in a bank angle equal to the angle of repose of the bank material. Table 2.7 summarizes formulas for computing geometrical attributes of cross sections modeled by equation 2.20.

The value of r that best approximates the form of a measured cross section can be determined from field measurements via the methods described in box 2.3. Using method 1, the value of r that best fits the natural channel of the Cardrona River (figure 2.20) is $r = 4.3$; figure 2.26 shows the actual and fitted cross sections.

2.5 Streamflow (Discharge)

2.5.1 Definition

Streamflow is quantified as **discharge**, Q, which is the volume rate of flow (volume per unit time) through a stream cross section (figure 2.27). Generally, discharge is an independent variable, imposed on a particular channel reach by meteorological events occurring over the watershed, modified by watershed topography, vegetation, and geology and upstream channel hydraulics.

Discharge is the product of the cross-sectional area of the flow, A, and the cross-sectional average velocity, U; A is the product of the water-surface width, W, and the cross-sectional average depth, Y. Thus,

$$Q = A \cdot U = W \cdot Y \cdot U. \quad (2.21a)$$

Table 2.7 Formulas for computing channel size and shape parameters as functions of bankfull channel width, W_{BF}, bankfull maximum depth, Ψ_{BF}, and maximum depth, Ψ, for the general cross-section model of equation 2.20.

Parameter	Flows < bankfull, $\Psi < \Psi_{BF}$	Bankfull flows, $\Psi = \Psi_{BF}$
Area	$A = \left(\frac{r}{r+1}\right) \cdot \left(\frac{W_{BF}}{\Psi_{BF}{}^{1/r}}\right) \cdot \Psi^{(r+1)/r}$	$A_{BF} = \left(\frac{r}{r+1}\right) \cdot W_{BF} \cdot \Psi_{BF}$
Average depth	$Y = \left(\frac{r}{r+1}\right) \cdot \Psi$	$Y_{BF} = \left(\frac{r}{r+1}\right) \cdot \Psi_{BF}$
Width	$W = W_{BF} \cdot \left(\frac{\Psi}{\Psi_{BF}}\right)^{1/r}$	W_{BF}
Wetted perimeter[a]	$P_w = 2 \cdot \int_0^{\Psi} \left(1 + \frac{4^r \cdot r^2 \cdot \Psi_{BF}{}^{2/r}}{W_{BF}{}^2} \cdot z^{2 \cdot (r-1)/r}\right)^{1/2} \cdot \mathrm{d}z$	$P_{wF} = 2 \cdot \int_0^{\Psi_{BF}} \left(1 + \frac{4^r \cdot r^2 \cdot \Psi_{BF}{}^{2/r}}{W_{BF}{}^2} \cdot z^{2 \cdot (r-1)/r}\right)^{1/2} \cdot \mathrm{d}z$
Width/depth ratio	$\frac{W}{Y} = \left(\frac{r+1}{r}\right) \cdot \left(\frac{W_{BF}}{\Psi_{BF}{}^{1/r}}\right) \cdot \Psi^{(1-r)/r}$	$\frac{W_{BF}}{Y_{BF}} = \left(\frac{r+1}{r}\right) \cdot W_{BF} \cdot \Psi_{BF}$

[a] In general, the integrals must be evaluated by numerical integration. For the parabola ($r = 2$) and $W/Y_m \geq 4$, P_w can be computed as $P_w = W + (8/3) \cdot (\Psi^2/W)$ (Chow 1959). For the rectangle ($r = \infty$), $P_w = W_{BF} + 2 \cdot \Psi$.

BOX 2.3 Estimating r from Field Measurements

For channel cross-sections that are approximately symmetrical, the cross-section shape can be mathematically described by measuring W_{BF} and Ψ_{BF} (box 2.2) and determining the appropriate value of r in equation 2.20. Here we describe three approaches that use the measurements described in box 2.2 to determine the "best-fit" value of r. In general, the three methods give different estimates; the first is the preferred approach because it finds the r value that minimizes the sum of the distances between the measured values and the estimated values across the channel.

1. Estimation via Minimization of Bankfull-Depth Differences

This trial-and-error method can be readily implemented in a spreadsheet. The bankfull width W_{BF}, maximum bankfull depth Ψ_{BF}, and bankfull depths Y_{BFi} at various distances X_i are determined as described in part 1 of box 2.2, and the location of the center of the channel, X_c, is determined as

$$X_c = X_1 + W_{BF}/2. \tag{2B3.1}$$

Compute the distance of each measurement point from the center, x_i, as

$$x_i = |X_i - X_c|. \tag{2B3.2}$$

The measured elevation of the channel bottom, z_i, at point x_i is

$$z_i = \max(Y_{BFi}) - Y_{BFi}. \tag{2B3.3}$$

If the cross-section is given by the model of equation 2.17, the elevation of the channel bottom $\hat{z}_i(r)$ at point x_i for a given value of r is given by

$$\hat{z}_i(r) = \Psi_{BF} \cdot \left(\frac{2}{W_{BF}}\right)^r \cdot x_i^r. \tag{2B3.4}$$

For a given r value, the sum of the squares of the differences between the measured and model values, $SS(r)$, can then be calculated as

$$SS(r) = \sum_{i=1}^{I} [\hat{z}_i(r) - z_i]^2. \tag{2B3.5}$$

The best-fit value of r that gives the smallest value of $SS(r)$ is then found by trial and error.

2. Estimation from Bankfull Depth and Average Depth

It can be shown from equation 2.17 that

$$A_{BF} = W_{BF} \cdot Y_{BF} = \left(\frac{r}{r+1}\right) \cdot W_{BF} \cdot \Psi_{BF}, \tag{2B3.6}$$

(Continued)

BOX 2.3 *Continued*

so

$$\hat{r} = \frac{\frac{A_{BF}}{W_{BF}\cdot\Psi_{BF}}}{1-\frac{A_{BF}}{W_{BF}\cdot\Psi_{BF}}} = \frac{\frac{Y_{BF}}{\Psi_{BF}}}{1-\frac{Y_{BF}}{\Psi_{BF}}}. \tag{2B3.7}$$

Thus, if W_{BF}, Ψ_{BF}, and A_{BF} are determined via cross-section surveys as described in box 2.2, the appropriate value of r can be estimated via equation 2B3.7.

3. Estimation via Regression of Bankfull Depth on Cross-Channel Distance

From equation 2B3.4,

$$\ln[\hat{z}(r)] = \ln(\Psi_{BF}) + \hat{r}\cdot\ln\left(\frac{2}{W_{BF}}\right) + \hat{r}\cdot\ln(x). \tag{2B3.8}$$

Thus, r can be estimated as the slope of the regression between $\ln[\hat{z}(r)]$ and $\ln(x)$. Note, however, that $\hat{r}$ should also equal

$$\hat{r} = \frac{B - \ln(\Psi_{BF})}{\ln\left(\frac{2}{W_{BF}}\right)}, \tag{2B3.9}$$

where B is the intercept of that regression. In general, the two values of $\hat{r}$ are not identical; the one given by the slope is preferable.

Equation 2.21a is used for computing reach discharge from measurements of width, depth, and velocity. However, for other situations it is probably preferable to write it as

$$W\cdot Y\cdot U = Q \text{ or } A\cdot U = Q \tag{2.21b}$$

to emphasize that Q is the independent variable, and the other factors adjust mutually in response to the discharge. The quantitative description of these mutual adjustments is called **hydraulic geometry**; this is discussed in section 2.6.3.

2.5.2 Relation to Channel Dimensions and Slope

As we will explore in more detail in chapter 6, a general expression relating the average velocity U of a flow in a wide channel to the local average depth Y and water-surface slope S_s can be derived from force-balance considerations:

$$U = K\cdot g^{1/2}\cdot Y^{1/2}\cdot S_s^{1/2}, \tag{2.22}$$

where K is the dimensionless reach **conductance**, which is a function of boundary roughness, channel curvature, and other factors; and g is gravitational acceleration.

Bjerklie et al. (2003) have shown that one can generally approximate the water-surface slope as the average channel slope, S_0. Thus, given a wide channel of specified size (bankfull width, W_{BF}, and bankfull maximum depth, Ψ_{BF}), shape (r),

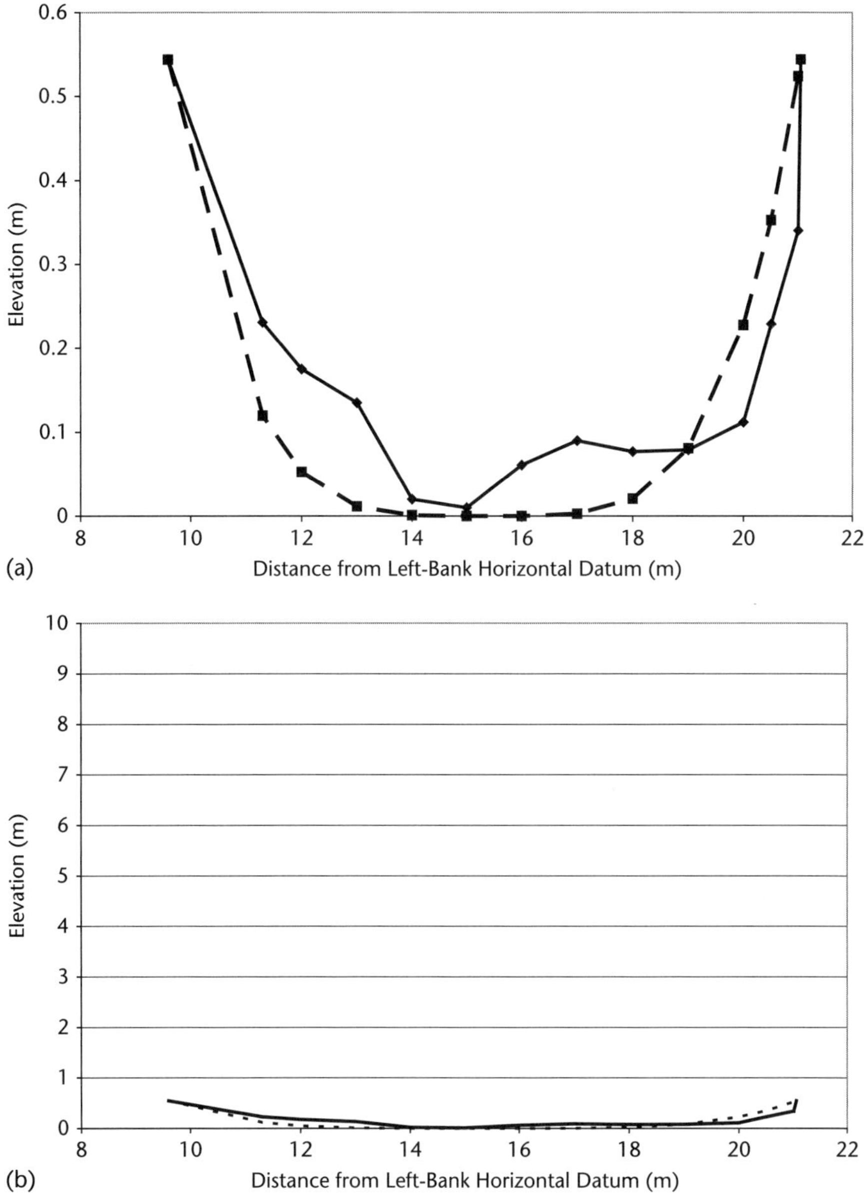

Figure 2.26 The Cardrona River cross section of figure 2.20a approximated by equation 2.20 with $r = 4.3$. (a) Section plotted at approximately 20-fold vertical exaggeration. (b) Section plotted with no vertical exaggeration. Solid line, actual cross section; dashed line, fitted cross section.

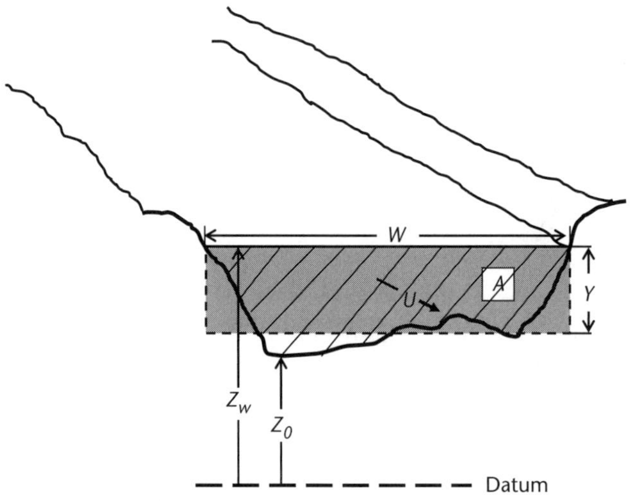

Figure 2.27 Definitions of terms defining discharge (equation 2.21) and stage (equation 2.25). Cross-hatched area is cross-sectional area of flow, A. Y is average depth, defined as $Y = A/W$; shaded area represents $A = W/Y$.

and slope (S_0), we can use equations 2.21 and 2.22 and relations for the general cross-section model (equation 2.20 and table 2.6) to derive an expression for discharge as a function of depth:

$$Q = K \cdot g^{1/2} \cdot \left(\frac{r+1}{r}\right)^{1/r} \cdot \left(\frac{W_{BF}}{\Psi^{1/r}}\right) \cdot Y^{3/2+1/r} \cdot S_0{}^{1/2}. \tag{2.23}$$

This relation indicates that discharge increases as the 3/2 power of depth for a rectangular channel ($r \to \infty$), as the square of depth for a parabolic channel ($r = 2$), up to 5/2 power for a triangular channel ($r = 1$).

2.5.3 Measurement

Methods for making instantaneous or quasi-instantaneous measurements of discharge include direct contact methods (volumetric measurement, velocity-area measurement, and dilution gaging) and indirect methods using stage (rating curve determined by natural control, weirs, and flumes). Remote-sensing methods can be classified as shown in table 2.8. The following subsections provide brief descriptions of each method.

2.5.3.1 Contact Methods

Contact methods involve instruments that touch the flowing water; these methods are described briefly below. "Direct" contact methods are those that measure discharge; "indirect" contact methods determine discharge by measuring the water-surface

Table 2.8 Determining stream discharge: Remote-sensing methods (Dingman and Bjerklie 2005).

Mode	Platform	Observable data types
Photography	Aircraft, satellites	Surface features including planform, sinuosity, etc.; bankfull and water-surface width; stereoscopy can provide slope
Visible and infrared digital imagery	Aircraft, satellites	Surface features including planform, sinuosity, etc.; bankfull and water-surface width
Synthetic aperture radar (SAR)	Aircraft, satellites, ground vehicles	Surface features including planform, sinuosity, etc.; bankfull and water-surface width; interferometry can provide slope; Doppler techniques can provide surface velocity
Radar altimetry	Aircraft, satellites	Water-surface elevation at discrete points, giving stage and possibly slope
Ground-penetrating radar	Ground vehicles, cableways, helicopters	Width and depth
Lidar	Aircraft, satellites	Surface velocity, stage, possibly slope
Topographic maps, digital-elevation models, geographic information systems	None	Static channel dimensions and morphology; ground slope

elevation and using empirical or theoretical relations between elevation and discharge. More detailed discussions of the various methodologies can be found in Herschy (1999a) and Dingman (2002).

Direct Measurement The **volumetric method** involves diverting the flow into a container of known volume and measuring the time required to fill it; clearly this is possible only for very small flows. The most commonly used direct-measurement method is the **velocity-area method**, which involves direct measurement of the average velocity U_i, depth Y_i, and width W_i of I subsections of the cross section and applying equation 2.21a to compute

$$Q = \sum_{i=1}^{I} U_i \cdot Y_i \cdot W_i. \tag{2.24}$$

The measurement locations may be accessed by wading, by boat, or from a stream-spanning structure. At least 20 subsections are usually required to get measurements of acceptable accuracy, spaced such that no more than 5% of the total discharge occurs in any one subsection. Because velocity varies with depth, measurements of velocity are made at prescribed depths and formulas based on hydraulic principles (see section 5.3.1.9) are invoked to compute U_i.

A recent modernization of the velocity-area method uses an acoustic Doppler current profiler (ADCP) to simultaneously measure and integrate the depth and velocity across a channel section, thereby obtaining all of the elements of equation 2.24 in one pass (Simpson and Oltman 1992; Morlock 1996). The ADCP unit is mounted on a boat or raft that traverses the cross section and measures depth via sonar and velocity via the Doppler shift of acoustic energy pulses. This system greatly reduces the time necessary to make a discharge measurement and allows measurements at stages when wading is precluded and at locations lacking stream-spanning structures.

In **dilution gaging**, a known concentration of a conservative tracer is introduced into the flow and the time distribution of its concentration is measured at a location far enough downstream to assure complete mixing. This technique is suitable for small, highly turbulent streams where complete mixing occurs over short distances (see White 1978; Dingman 2002).

Indirect Measurement At any cross section, the flow depth increases as discharge increases (equation 2.23). Thus, discharge can be measured indirectly by observing the water-surface elevation, or **stage**, Z_s, which is defined (figure 2.27) as

$$Z_s \equiv Z_w - Z_0, \tag{2.25}$$

where Z_w is the elevation of the water-surface, and Z_0 is the elevation of an arbitrary datum. The relation between stage and discharge is shown as a **rating curve** or **rating table**.

In a natural channel, the rating curve is established by repeated simultaneous measurements of discharge (usually via the velocity-area method), and the shape of the rating curve is determined by the configuration of the channel (equation 2.23). Because it is relatively easy to make continuous or frequent periodic measurements of Z_w by float or pressure transducer, the rating curve provides a means of obtaining a continuous record of discharge. However, to be useful, the rating curve must be established where dQ/dZ_w is large enough to provide the required accuracy. In most natural channels, the rating curve is subject to change over time due to erosion and/or deposition in the measurement reach, so periodic velocity-area measurements are required to maintain an accurate rating curve as well as to extend its range. Methods of stage measurement are described by Herschy (1999b).

In relatively small streams, discharge can be measured by constructing or installing artificial structures that provide a fixed rating curve. **Weirs** are structures that dam the flow and allow the water to spill over the weir crest, which is usually horizontal or V-shaped. At a point near the crest, the velocity U of the freely falling water is

$$U \propto (Z_w - Z_c)^{1/2}, \tag{2.26}$$

where Z_w is water-surface elevation, and Z_c is elevation of the weir crest. Because the constant of proportionality can be determined by measurement and the width of the flow is either constant or a known function of Z_w, equation 2.26 can be combined with equation 2.21b to give the discharge as a function of water-surface elevation, which is measured by float or pressure transducer. The hydraulics of weirs is discussed more fully in section 10.4.1.

Flumes are another type of flow-measurement structure; these constrict and thereby accelerate the flow to provide a known relation between discharge and stage. The exact form of the rating curve is determined by the flume geometry. Flume hydraulics is described more fully in section 10.4.2.

2.5.3.2 Remote-Sensing Methods

Using various combinations of active and passive imagery and visible-light, infrared, and radar sensors mounted on satellites or aircraft (table 2.8), it is possible to obtain direct quantitative information on channel planform and several hydraulic variables, including the area, width, elevation, and velocity of the water surface (Bjerklie et al. 2003). This information can be used in various combinations with hydraulic relations, statistical models, and topographic information (i.e., channel slope) to generate quantitative time- and location-specific estimates of discharge (Bjerklie et al. 2005a; Dingman and Bjerklie 2005), for some locations on relatively large rivers. Refinement of remote discharge-measurement techniques is an active area of research. However, because of accuracy limitations, it is likely that this capability will be useful only for locations that are remote or otherwise difficult to observe conventionally.

2.5.4 Sources

As noted in section 2.1.1, the ultimate source of all discharge in a stream reach is precipitation on the watershed that contributes flow to the reach. Typically, only a very small portion ($<5\%$) comes from precipitation falling directly on the channel network; the rest is water that has fallen on the nonchannel portions of the watershed and traveled to the stream network via subsurface or surface routes. In nonarid regions, most streamflow enters from the subsurface as groundwater outflow from "permanent" regional aquifers or from temporary aquifers that are present seasonally or as a result of heavy precipitation or snowmelt. These groundwater contributions are usually distributed more-or-less continuously along the stream network. Surface contributions occur as quantum inputs at tributary junctions and as overland flow; overland flow contributions are diffuse and occur only during or immediately following periods of significant rainfall or snowmelt.

A stream reach that receives groundwater flow is called a **gaining reach** because its discharge increases downstream (figure 2.28a). A **losing reach** is one in which discharge decreases downstream; such a reach may be connected to (figure 2.28b) or "perched" above (figure 2.28c) the general groundwater flow. A **flow-through reach** is one that simultaneously receives and loses groundwater (figure 2.28d).

Figure 2.29 shows an idealized relation between regional water-table contours and a stream reach. At any point, the regional groundwater flow vector, Q_G, is perpendicular to the contours but may be resolved into a down-valley, or **underflow component**, Q_{Gu}, and a riverward, or **baseflow component**, Q_{Gb}. Larkin and Sharp (1992) found that reaches can be classified as baseflow dominated ($Q_{Gb} > Q_{Gu}$), underflow dominated ($Q_{Gu} > Q_{Gb}$), or mixed flow ($Q_{Gb} \approx Q_{Gu}$) on the basis of river characteristics that can be readily determined from maps (table 2.9). Figure 2.30 shows examples of underflow- and baseflow-dominated rivers.

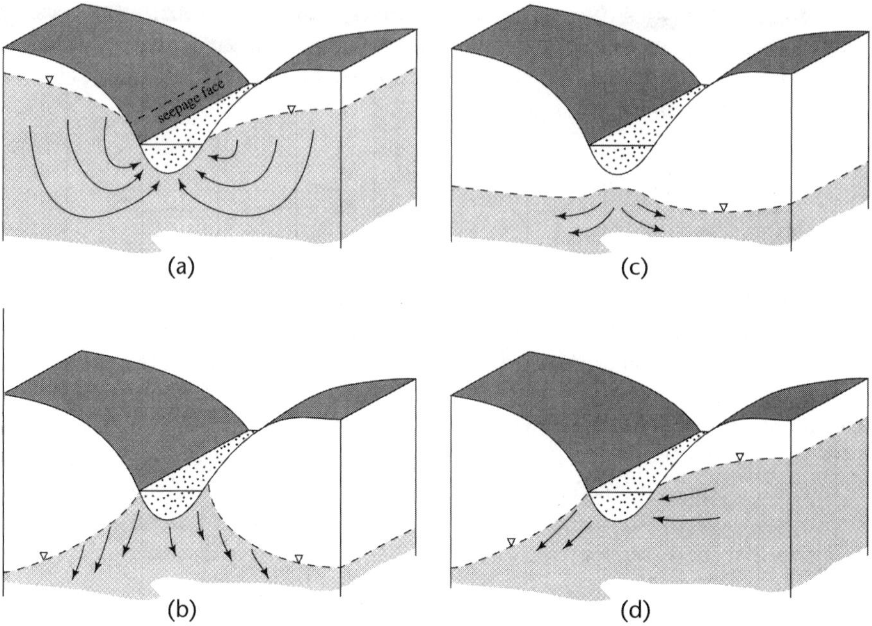

Figure 2.28 Groundwater–stream relations. A gaining reach (a) receives groundwater inputs from permanent, seasonal, or temporary aquifers. A losing reach lies above the local groundwater surface and may be connected (b) or unconnected (c) to it. In a flow-through reach (d), the groundwater enters on one bank and exits on the other.

At a more local scale, a stream bed typically is at least locally permeable and river water may exchange between the river and its bed and banks. The zone of down-river groundwater flow in the bed is called the **hyporheic zone**, and the importance of this zone to aquatic organisms, including spawning fish, is increasingly being recognized (e.g., Hakenkamp et al. 1993).

The lateral exchange of water between the channel and banks is commonly significant during high flows and is termed **bank storage** (figure 2.31). When flow generated by a rainfall or snowmelt event enters a gaining stream, a flood wave (the term is used even if no overbank flooding occurs) forms and travels downstream (described further in section 2.5.5). As the leading edge of the wave passes any cross section, the stream-water level rises above the water table in the adjacent bank, inducing flow from the stream into the bank (figure 2.31b). After the peak of the wave passes the section, the stream level declines and a streamward gradient is once again established (figure 2.31c).

2.5.5 Stream Response to Rainfall and Snowmelt Events

Figure 2.32a shows possible flow paths in a small upland watershed during a rainfall event. Rainfall rates are measured at one or more points on the watershed and spatially averaged; a graph of rainfall versus time is called a **hyetograph**. Watershed response

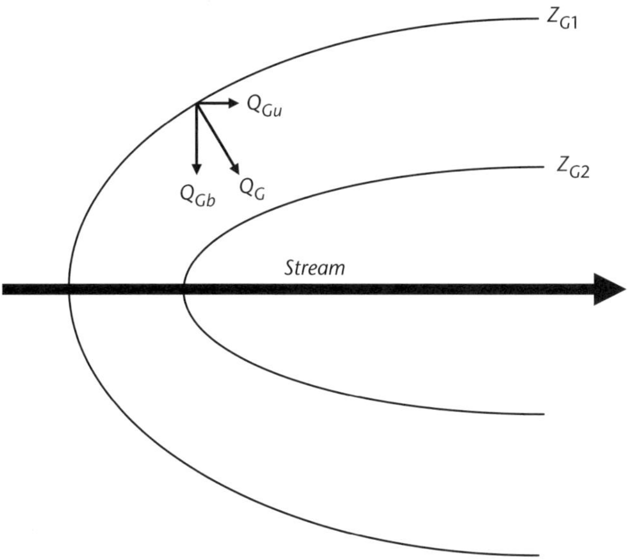

Figure 2.29 Idealized groundwater–stream relations. Curved lines represent contours of the groundwater table at elevations Z_{G1} and Z_{G2}; $Z_{G1} > Z_{G2}$. Q_G is the groundwater flow vector at an arbitrary point, which is resolved into an underflow component, Q_{Gu}, and a baseflow component, Q_{Gb}. Modified from Larkin and Sharp (1992).

Table 2.9 Relations between river–groundwater interaction and river type (see figures 2.29 and 2.30).

Dominant groundwater flow direction	Channel slope	Sinuosity	Width/depth ratio	Penetration[a]	Sediment load
Underflow	High (>0.0008)	Low (<1.3)	High (>60)	Low (<20%)	Mixed bedload
Baseflow	Low (<0.0008)	High (>1.3)	Low (<60)	High (>20%)	Suspended load
Mixed	≈ Valley slope; lateral valley slope flat				

[a]Degree of incision into valley fill.
From Larkin and Sharp (1992).

to the event (output) is characterized by measuring the stream discharge at a stream cross section whose location determines the extent of the watershed. A graph of discharge versus time is a **streamflow hydrograph**.

Figure 2.32b shows that the streamflow hydrograph is a spatially and temporally integrated response determined by 1) the spatially and temporally varying input rates, 2) the time required for each drop of water to travel from where it strikes the watershed surface to the stream network (determined by the length, slope, vegetative cover, soils, and geology of the watershed hillslopes), and 3) the time required for water to travel

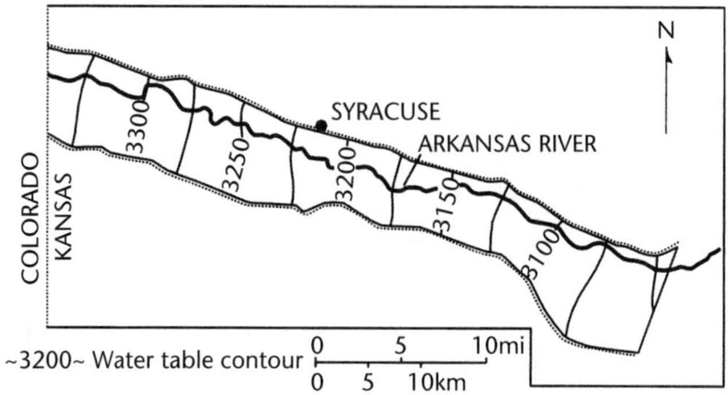

(a)

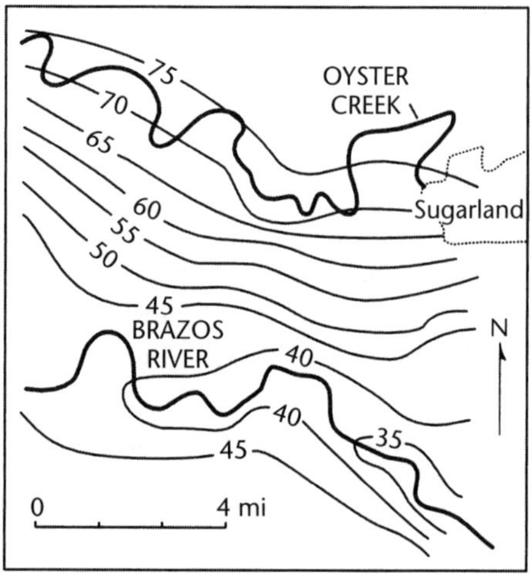

(b)

Figure 2.30 (a) An underflow-dominated stream: the Upper Arkansas River and its aquifer, in Kansas. (b) A baseflow-dominated stream: the Brazos River and its aquifer, in Texas. Contours are water-table elevations in feet above sea level. From Larkin and Sharp (1992); reproduced with permission of the Geological Society of America.

from its entrance into the channel to the point of measurement (determined by the length and nature of the channel network). In small watersheds (typically less than about 50 km^2 area), the travel time to the watershed outlet is determined mostly by the hillslope travel time; for larger watersheds, the travel time in the stream network becomes increasingly important.

Streamflow in response to a rainfall or snowmelt event takes the form of a flood wave that moves downstream through the stream network (figure 2.33). The observed hydrograph records the movement of the flood wave past the fixed point of measurement (figure 2.33, inset). Once the flood wave leaves the portion of the

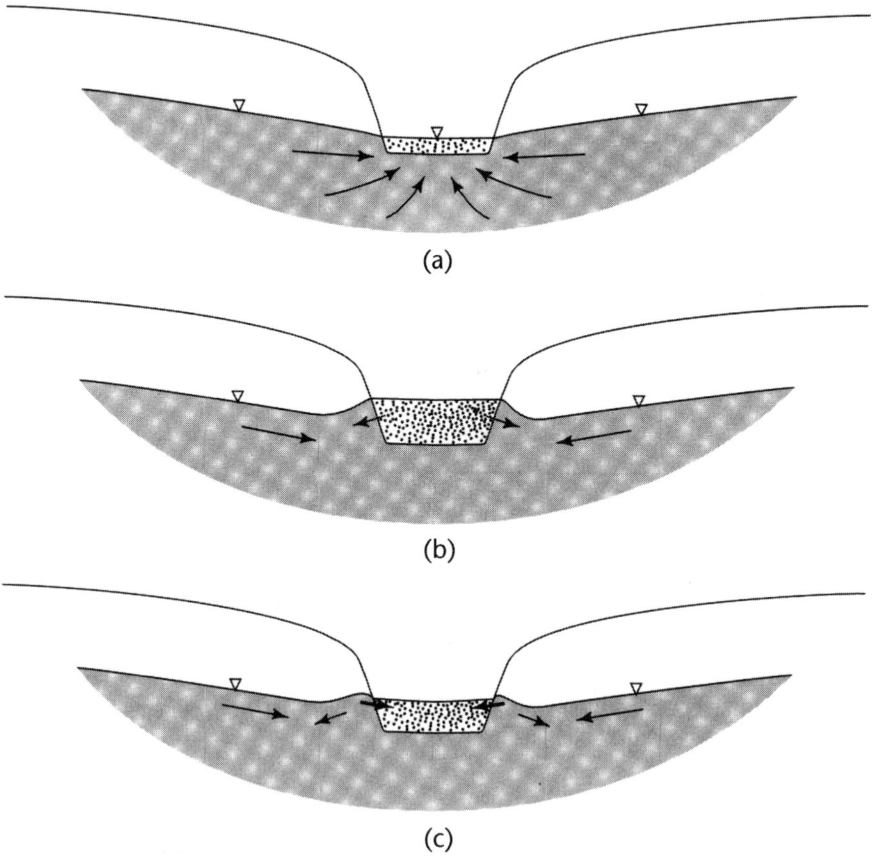

Figure 2.31 Diagram illustrating bank storage in a gaining stream. (a) Low flow with groundwater entering the stream (baseflow). (b) Peak flow passes, inducing flow from the stream into the bank. (c) After the peak of the wave passes, the bankward gradient declines. When the flood wave has passed, a streamward gradient is once again established.

stream network that has been affected by a given rainfall event, its shape is affected solely by channel hydraulics and bank-storage effects.

Figure 2.34 shows a typical example of how the effects of hillslope-response mechanisms are gradually superseded by channel-hydraulic effects through a stream network. The hydrograph shape for the smallest watershed is strongly influenced by the form of the hyetograph. Subsequently, the hydrograph is increasingly affected by tributary inputs and by the storage effects of the stream channels, and the net result is an increase in the lag time between the rainfall inputs and the peaks and a decrease in hydrograph ordinates (when scaled by drainage area). The hydrograph also becomes smoother, and at the lowest two gages, the formerly multiple-peaked hydrograph has become single-peaked.

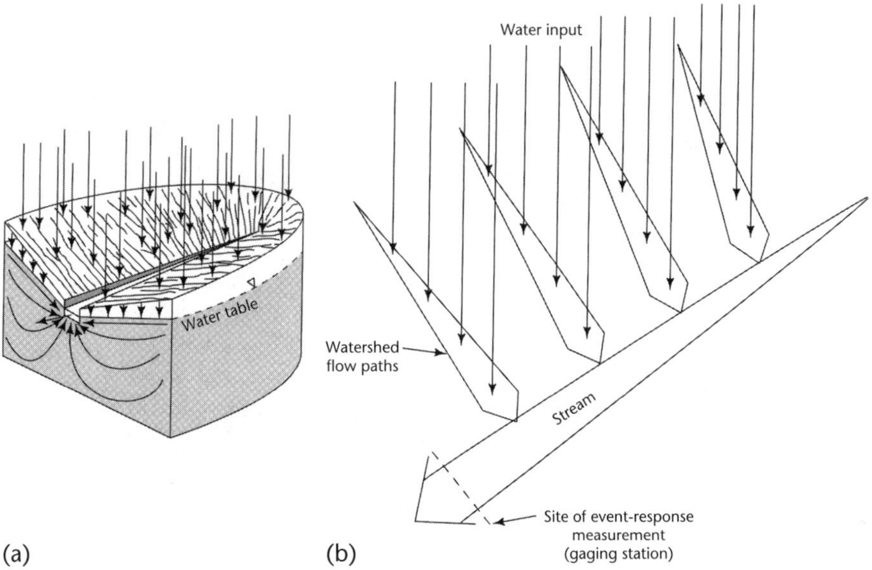

Figure 2.32 (a) Flow paths in a small upland watershed during a rain event. (b) The essence of watershed response as the spatially and temporally integrated result of accumulated lateral inflows.

2.5.6 Timing

2.5.6.1 Hydroclimatic Regimes

The **hydroclimatic regime** of a river reach is characterized by its typical seasonal (intra-annual) pattern of flow variability, its year-to-year (interannual) flow variability, and various quantitative and qualitative descriptors of the time series of low flows, average flows, and flood flows (Dingman 2002). The interannual flow regime can be summarized by the mean and standard deviation of annual streamflows. Vogel et al. (1999) gave equations that can be used to estimate those quantities in the water-resource regions of the conterminous United States based on drainage area, mean annual precipitation, and mean annual temperature.

Streams that flow all year are **perennial streams**, and those that flow only during wet seasons are **intermittent streams**; these stream types are almost always gaining streams that are sustained to varying degrees by groundwater flow between rain and snowmelt events. **Ephemeral streams** flow only in response to a water-input event; they are usually not connected to regional groundwater flows and are usually losing streams.

The seasonality of river flows was mapped globally by L'vovich (1974) and for North America by Riggs and Harvey (1990). More detailed examples of interannual and intra-annual variability are illustrated in figure 2.35.

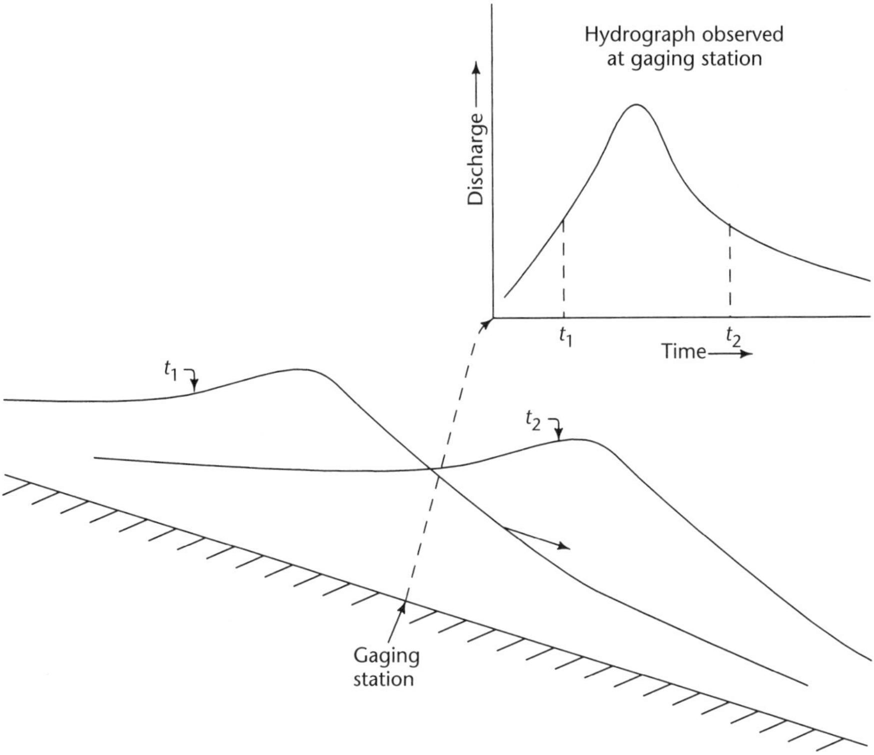

Figure 2.33 Streamflow in response to a rainfall or snowmelt event takes the form of a flood wave that moves downstream through the stream network. The observed hydrograph (inset) records the movement of the flood wave past the fixed point of measurement.

The following subsections introduce the two main statistical techniques used to summarize the time variability of streamflows in a particular reach: flow-duration curves and flood-frequency curves.

2.5.6.2 Flow-Duration Curves

The flow-duration curve is a conceptually simple but highly informative way to summarize the temporal variability of streamflow at a given location (cross section or reach):

> A **flow-duration curve** (FDC) is a cumulative-frequency curve that shows the fraction (percent) of days that the daily average discharge exceeded a specified value over a period of observation long enough to include a representative range of seasonal and interannual variability.

Dingman (2002) described how FDCs can be constructed for reaches that have long-term streamflow records and those that do not.

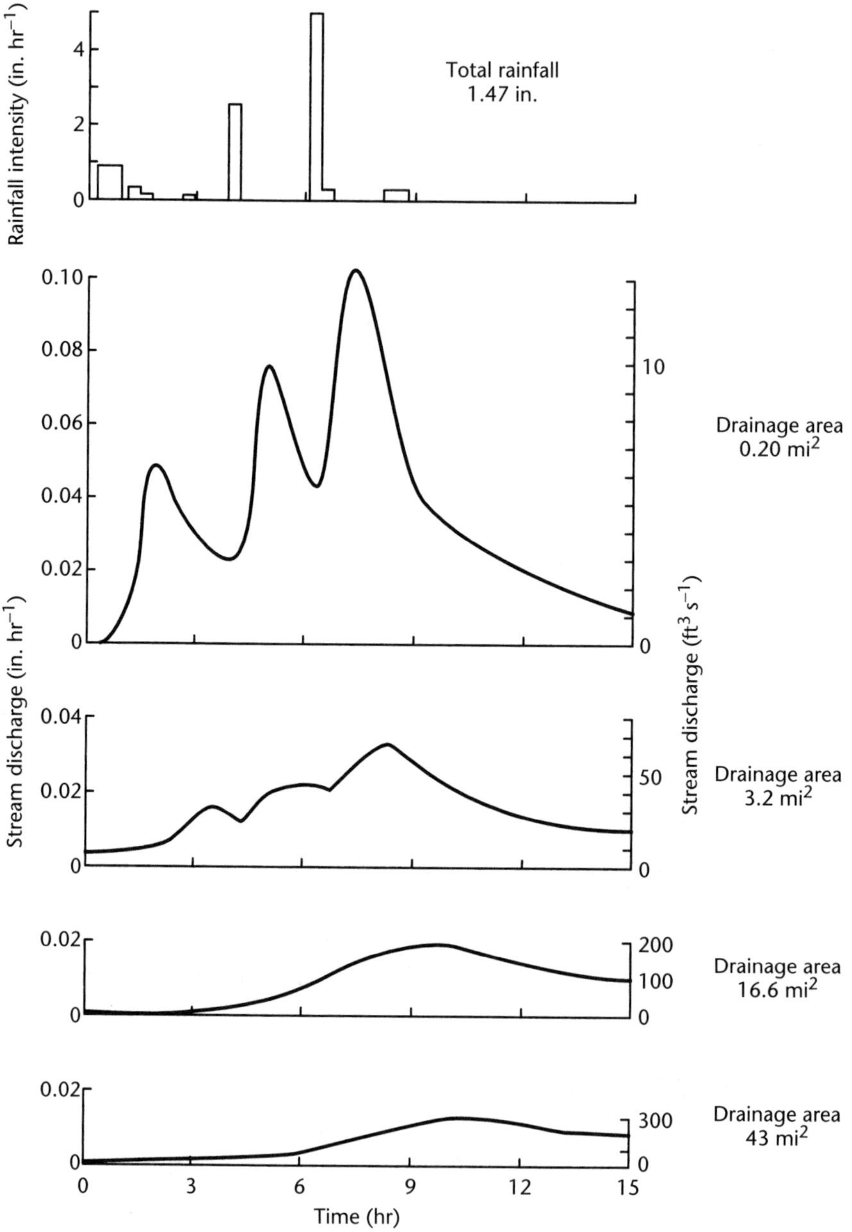

Figure 2.34 Evolution of a hydrograph in response to a rainfall event in the Sleepers River Research Watershed, Danville, Vermont (Dunne and Leopold 1978). The top graph is the rainfall hyetograph. The hydrograph on the smallest watershed closely resembles the hyetograph; on successively larger watersheds, the three peaks gradually merge into one, occur at increasingly later times, and have smaller ordinates on a per-unit-area basis. From *Environmental Planning* T. Dunne and Leopold (1978); reproduced with permission of W.H. Freeman and Company.

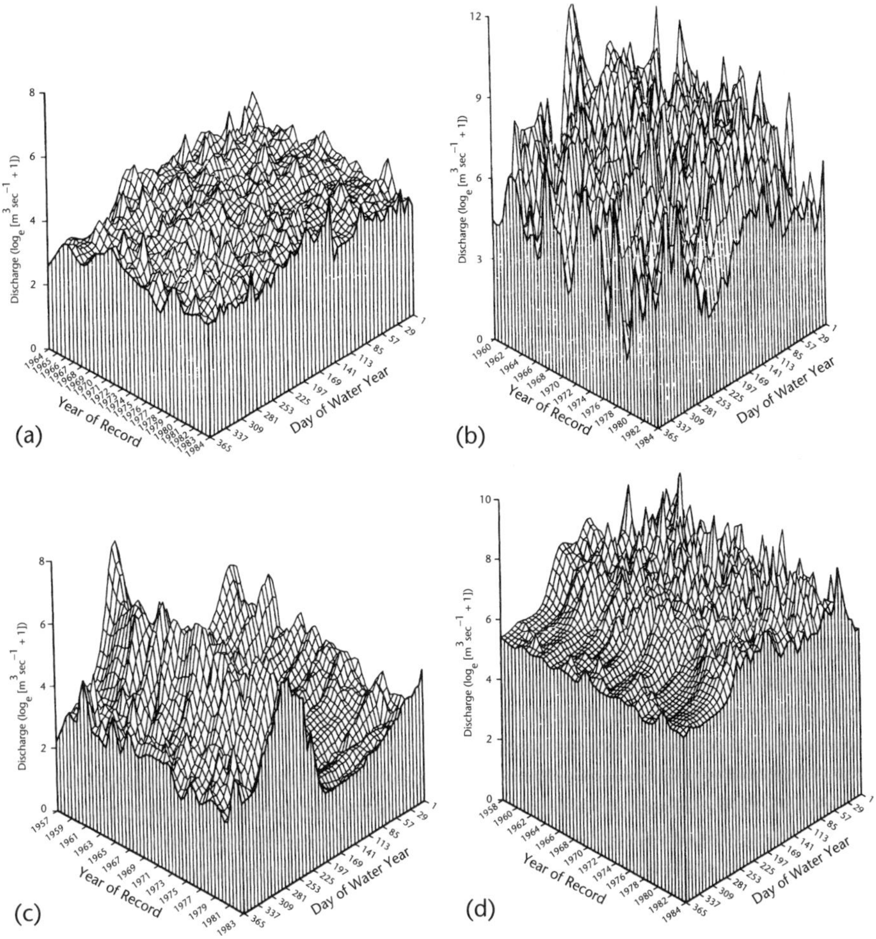

Figure 2.35 Examples of intra-annual flow-variability patterns. (a) Little variability due to relatively constant precipitation inputs and large groundwater contributions (Augusta Creek, MI). (b) High variability where snow is absent, groundwater contribution is small, and storms occur in all seasons (Satilla River, GA). (c) Relatively constant pattern of seasonal variability due to winter snow accumulation and spring snowmelt (upper Colorado River, CO). (d) Pronounced low-flow season due to high summer evapotranspiration, with random distribution of rain storms in other seasons (South Fork of MacKenzie River, OR). From Poff et al. (1997); reproduced with permission of the American Institute of Biological Science.

In statistical terms, the FDC is a graph plotting the daily average discharge (Q, y-axis) versus the fraction of time, or probability, that Q exceeds any specified value $Q = Q_{ep}$ (x-axis). This probability, designated $EP(Q_{ep})$, is called the **exceedence probability** (or **exceedence frequency**) and is defined in probability terms as

$$EP(Q_{ep}) \equiv Pr\{Q > Q_{ep}\} = ep, \tag{2.27}$$

where Pr{ } denotes the probability of the condition within the braces.

It is important to understand that, on FDCs, exceedence probability refers to the probability of exceedence on a day chosen randomly from a period of many years, rather than the probability of exceedence on any specific day or day of the year. Seasonal effects and hydrological persistence cause exceedence probabilities of daily flows to vary as a function of time of year and antecedent conditions, and the FDC does not account for those dependencies.

An example of an FDC is shown in figure 2.36. In figure 2.36a, discharge is plotted on a logarithmic scale and exceedence probability on a probability scale; this is the usual practice because it allows the curve to be more easily read at the high and low ends. This FDC shows that the discharge of the Boise River at the long-term measurement station at Twin Springs, Idaho, exceeded 9.2 m^3/s on 90% of the days; that is, $Q_{0.90} = 9.2\,m^3/s$, or $EP(9.2) = Pr\{Q > 9.2\} = 0.90$.

The integral of the FDC is equal to the long-term average flow for the period plotted. The flow exceeded on 50% of the days, $Q_{0.50}$, is the median flow; figure 2.36a shows that the median flow for the Boise River is 15.7 m^3/s. The long-term average flow for the Boise River is 34.0 m^3/s, which is exceeded only 31.6% of the time. The arithmetic plot of the Boise River FDC is shown in figure 2.36b; this emphasizes the virtually universal fact that river flows are well below the average flow most of the time. In less humid regions, the mean flow is exceeded even more rarely than is the case for the Boise River.

The steepness of the FDC is proportional to the variability of the daily flows. For streams unaffected by diversion, regulation, or land-use modification, the slope of the high-discharge end of the FDC is determined principally by the regional climate, and the slope of the low-discharge end by the geology and topography. The slope of the upper end of the FDC is usually relatively flat where snowmelt is a principal cause of floods and for large streams where floods are caused by storms that last several days. "Flashy" streams, where floods are typically generated by intense storms of short duration, have steep upper end slopes. At the lower end of the FDC, a flat slope usually indicates that flows come from significant storage in groundwater aquifers or in large lakes or wetlands; a steep slope indicates an absence of significant storage. The presence of reservoir regulation upstream of the point of measurement can greatly flatten the FDC by raising the low-discharge end and lowering the high-discharge end (Dingman 2002).

2.5.6.3 *Flood-Frequency Curves*

Definition and Properties In contrast to FDCs, exceedence probabilities for flood flows are calculated on an annual basis by statistical analysis of the highest instantaneous discharges in each year. Thus, in this context, Q designates the **annual peak discharge**. A **flood-frequency curve** is a cumulative-frequency curve that shows the fraction (percent) of years that the annual peak discharge exceeded a specified value over a period of observation long enough to be considered representative of the annual variability. Equation 2.27 applies for peak flows as well as daily flows, but the probability applies to years rather than days. Procedures for computing flood exceedence probabilities are described by Dingman (2002).

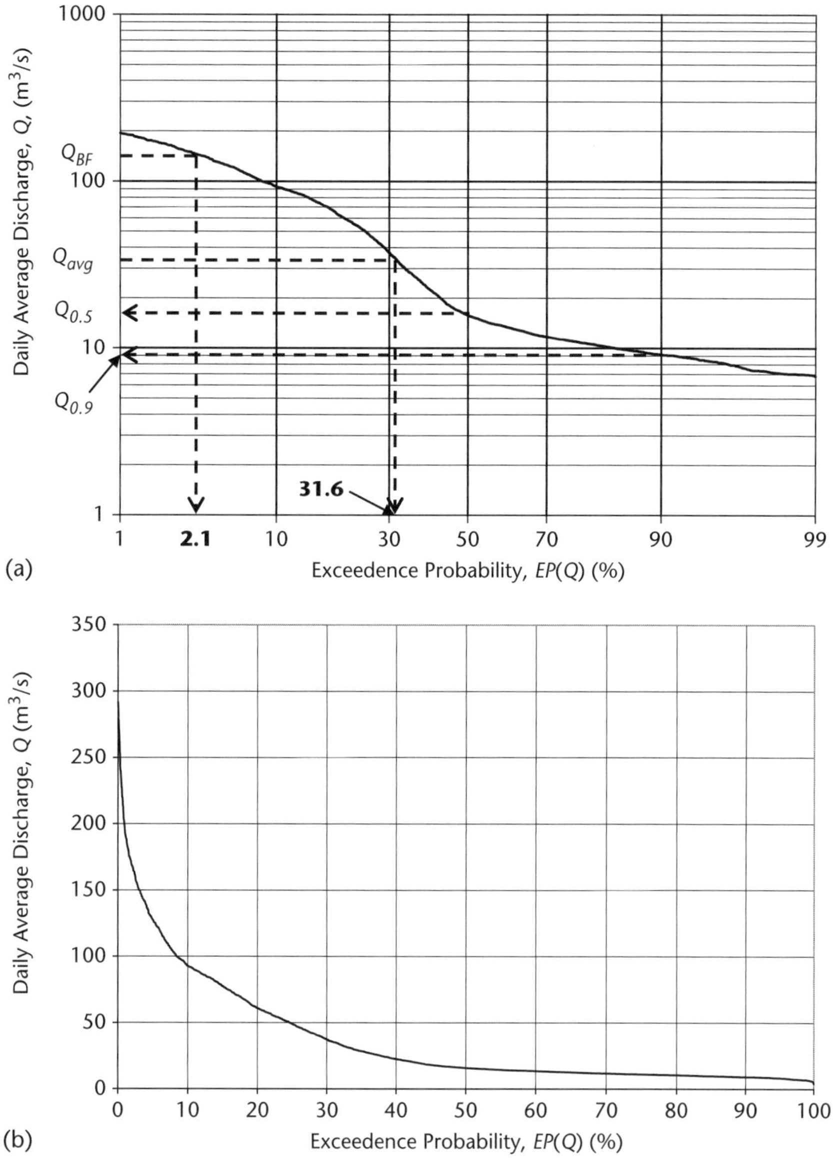

Figure 2.36 Flow-duration curve for the Boise River at Twin Springs, Idaho. (a) Log-probability plot. The average discharge exceeded on 90% of the days is $9.2\,m^3/s$ ($Q_{0.9} = 9.2\,m^3/s$); the median discharge is $Q_{0.5} = 15.7\,m^3/s$. The average discharge is $Q_{avg} = 34\,m^3/s$, which has an exceedence probability of 31.6%; the bankfull discharge is $Q_{BF} = 167\,m^3/s$, which has an exceedence probability of 2.1%. (b) Arithmetic plot.

Annual exceedence probability is often expressed in terms of the **recurrence interval** (also called **return period**), which is the average number of years between exceedences of the flood discharge with a given exceedence probability. The recurrence interval, $T_R(Q_{ep})$, of a flood peak, Q_{ep}, with annual exceedence probability ep [$= EP\ (Q_{ep})$], is simply the inverse of the exceedence probability:

$$T_R(Q_p) = \frac{1}{ep} = \frac{1}{EP(Q_{ep})}. \tag{2.28}$$

Thus, the "T_R-year flood" is the flood peak with an annual exceedence probability $= 1/T_R$.

Figure 2.37 shows the flood-frequency curve for the Boise River. It shows that a flood of 287 m^3/s has an exceedence probability of 0.10 ($Q_{0.10} = 287\ m^3/s$); that is, there is a 10% chance that the highest peak flow in any year will exceed 287 m^3/s. In terms of recurrence interval, 287 m^3/s is the "10-year flood," We can see that this is borne out by the historical record of annual peak flows shown in figure 2.38: there have been nine exceedences of 287 m^3/s in the 95-year record, and the average time between those exceedences is 8.75 years.

Relation to Bankfull Discharge Bankfull discharge in most regions has a recurrence interval of about 1.5 years (annual exceedence probability of $1/1.5 = 0.67$).

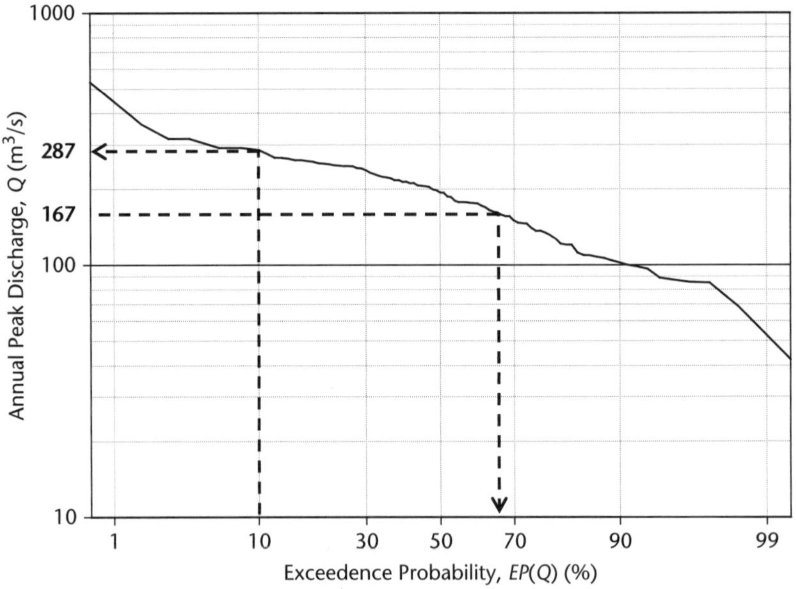

Figure 2.37 Flood-frequency curve for the Boise River at Twin Springs, Idaho. The flood peak with an annual exceedence probability of 10% (i.e., the 10-year flood) is 287 m^3/s. The bankfull discharge $Q_{BF} = 167\ m^3/s$ has an annual exceedence probability of 63%, so this discharge is the $1/0.63 = 1.6$-year flood.

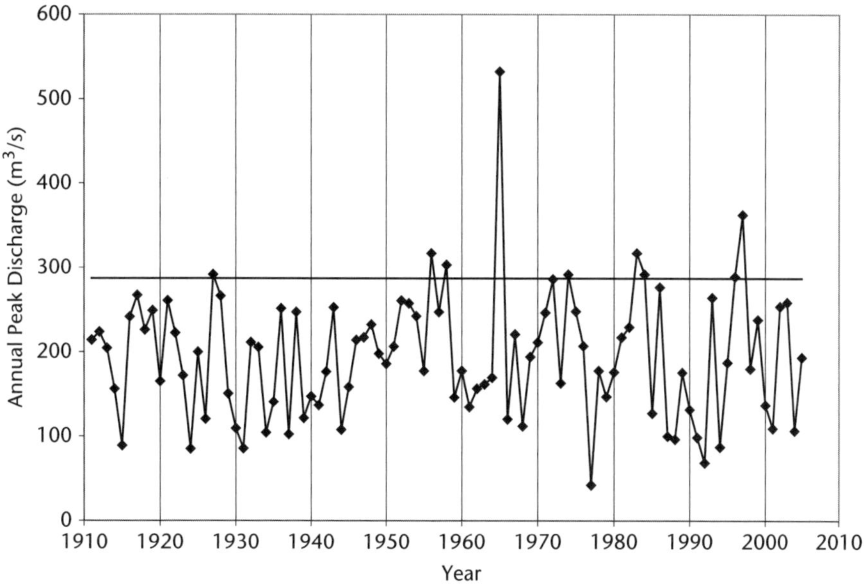

Figure 2.38 Time series of annual peak discharges of the Boise River, 1911–2005. The horizontal line represents a peak of 287 m^3/s, which is the 10-year flood. There have been nine exceedences of this flow, with an average of 8.75 years between exceedences.

This means that most streams experience overbank flooding in about two out of every three years. However, there is considerable regional and even local variability, and field studies such as those described in box 2.1 should be carried out to establish the relation for a particular stream reach: Williams (1978) found that, although 62% of most of the streams he studied had a bankfull recurrence interval between one and two years, the interval was as high as 32 years.

Field studies indicate that the bankfull discharge for the Boise River at this location is 167 m^3/s (Boise Adjudication Team 2004). We see from figure 2.37 that that discharge has an exceedence probability of 63%; this is equivalent to a recurrence interval of 1.6 years, close to the typical value. Note from figure 2.36 that this flow is exceeded on 2.1% of the days, or about 7.7 days per year on average.

2.6 Variables and Their Spatial and Temporal Variability

2.6.1 Principal Variables and Time and Space Scales

The principal variables discussed in this chapter, and in subsequent chapters, are summarized in table 2.10. Table 2.11 categorizes these variables as either measurable or derived. All these quantities vary on a range of spatial and temporal scales, and there is a general correlation between the size of a fluvial feature and the time scale at which it varies (table 2.12, figure 2.39).

Table 2.10 Measurable and derived variables characterizing stream morphology, materials, and flows.

Symbol	Variable	Dimensions
A	Cross-sectional area of flow	[L^2]
A_{BF}	Bankfull cross-sectional area of flow	[L^2]
A_D	Drainage area	[L^2]
$A_D(\omega)$	Average drainage area of streams of order ω	[L^2]
d_p	Particle diameter greater than $p\%$ of particles	[L]
D_D	Drainage density	[1]
K	Reach hydraulic conductance (equation 2.22)	[1]
L	Discharge of particulate sediment	[$F\ T^{-1}$]
N(braids)	Average number of braids in a cross section	[1]
$N(\omega)$	Number of streams of order ω	[1]
Q	Discharge	[$L^3\ T^{-1}$]
Q_{BF}	Bankfull discharge	[$L^3\ T^{-1}$]
r	Cross-section shape exponent (equation 2.20)	[1]
r_m	Radius of curvature of meanders	[L]
R_A	Area ratio (table 2.2)	[1]
R_B	Bifurcation ratio (table 2.2)	[1]
R_L	Length ratio (table 2.2)	[1]
S_0	Channel slope	[1]
S_s	Water-surface slope	[1]
S_v	Valley slope	[1]
u	Point velocity	[$L\ T^{-1}$]
U	Cross-section or reach average velocity	[$L\ T^{-1}$]
U_{BF}	Bankfull cross-section or reach average velocity	[$L\ T^{-1}$]
W	Cross-section or reach average water-surface width	[L]
W_{BF}	Cross-section or reach average bankfull water-surface width	[L]
X	Streamwise distance	[L]
$X(\omega)$	Average length of streams of order ω	[L]
Y	Cross-section or reach average depth	[L]
Y_{BF}	Bankfull cross-section or reach average depth	[L]
ΔX	Increment of streamwise distance	[L]
ΔX_v	Increment of valley distance	[L]
ΔZ_0	Difference in channel-bed elevation	[L]
ΔZ_s	Difference in water-surface elevation	[L]
ΔZ_{sBF}	Difference in water-surface elevation at bankfull	[L]
ζ	Sinuosity	[1]
λ_m	Meander wavelength	[L]
ΣX	Total stream length	[L]
ϕ	Angle of bank slope	[1]
Φ	Angle of repose of bank material	[1]
Ψ	Maximum depth in cross section	[L]
Ψ_{BF}	Maximum depth in cross section at bankfull	[L]
ω	Stream order	[1]

Table 2.11 Classification of measurable and derived variables characterizing stream morphology, materials, and flows.[a]

Domain	Extent	Measurable variables	Derived variables
Stream network	Area or watershed	$N(\omega)$, $X(\omega)$, $A_D(\omega)$, ΣX, A_D	R_B, R_L, R_A, D_D
Profile	Reach to entire stream	X, ΔX, ΔX_v, ΔZ_0, ΔZ_v	S_0, S_v
Planform	Reach to entire stream	λ_m, r_m, N(braids), ΔX, ΔX_v, ΔZ_0, ΔZ_v	ζ, S_0, S_v
Cross section	Cross section to reach	Q_{BF}, W_{BF}, Ψ_{BF}, A_{BF}, d_p, Φ, ΔZ_{sBF}, ΔZ_v, ΔX_v	Y_{BF}, U_{BF}, r, K_{BF}, S_0, S_{sBF}
Flow	Cross section to reach	Q, W, Ψ, A, L, u, ΔZ_S, ΔZ_v, ΔX, ΔX_v	Y, U, K, S_S

[a] See table 2.10 for symbol definitions.

Table 2.12 Space and time scales of fluvial features.

Spatial scale	Dimensions (km, km^2)	Feature	Major controlling factors	Time scale
Mega	$> 10^3$, $> 10^6$	Major watersheds, stream networks	Major climate zones, very long-term climate change, large-scale tectonic processes	10^6–10^7 years
Macro	10–10^3, 10^2–10^6	Large watersheds, major floodplains	Regional climate zones, long-term climate change, regional tectonic processes	10^3–10^6 years
Meso	0.5–10, 0.25–10^2	Meanders, changes in planform, channel shifts	Local climate, short-term climate change, local and regional tectonic processes, land-use change, engineering structures	10^2–10^3 years, "graded time"
Micro	0.1–0.5, 0.01–0.25	Local erosion and deposition, channel shifts	Major storms, engineering structures	1–10 years, "steady time"
Reach	0.01–0.1, <0.01	Local erosion and deposition	Major storms, engineering structures	< 1 year

Modified from Summerfield (1991).

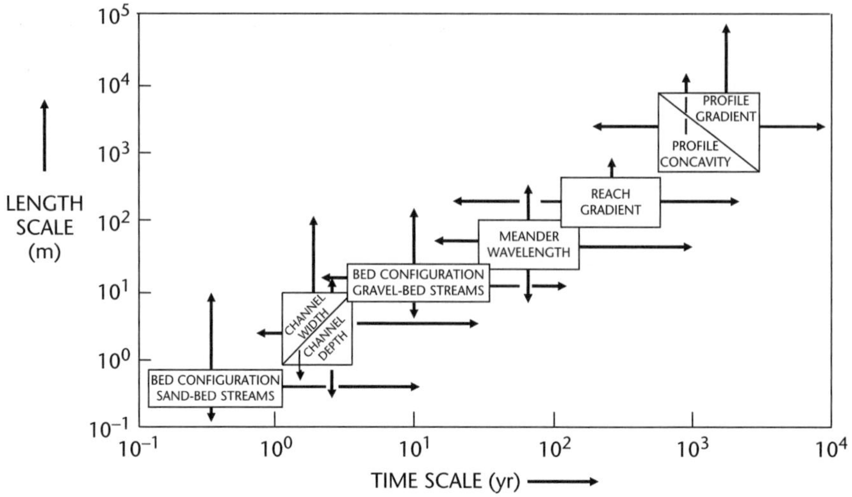

Figure 2.39 Relation of length scale of various aspects of channel form to time scale of adjustment. From *Fluvial Forms and Processes* (Knighton 1998); reproduced with permission of Edward Arnold Ltd.

A sense of the complex network of interrelationships among these variables is conveyed in figure 2.40. This text is mostly concerned with phenomena at the reach scale, which would typically be in the range of a few meters to a few kilometers, and thus with typical time scales of up to 1 year. At this scale, we can characterize the variables of interest as follows:

Fixed quantities: Average discharge, bankfull discharge, timing of flows, discharge associated with various flood frequencies, bed-material size (d), and channel slope (S_0) are determined by watershed size and regional geology, topography, and climate. The planform (ζ), reach and cross-section dimensions (W_{BF}, Y_{BF}), and shape (Φ, r) are determined by complex interactions among those factors. In general, channel dimensions increase downstream in a given watershed, and channel slope and bed-material size decrease.

Independent variables: Discharge (Q) and sediment discharge (L) are delivered to a reach from upstream. These vary with time but, at the reach scale and for short time periods, can be considered essentially constant, specified independent variables.

Dependent variables: Width (W), average depth (Y), average velocity (U), and sediment transport (L) out of the reach are the principal dependent variables. Their values are determined by the imposed discharge and the geometric and material properties of the reach and change as discharge changes spatially and temporally.

As discussed in detail in chapter 6, local conductance (K) in general changes with discharge but may be considered an independent variable to the extent the $Q-K$ relation is known. Local water-surface slope (S_s) may also change with discharge (discussed in chapter 11) but may often be considered a constant equal to the channel slope.

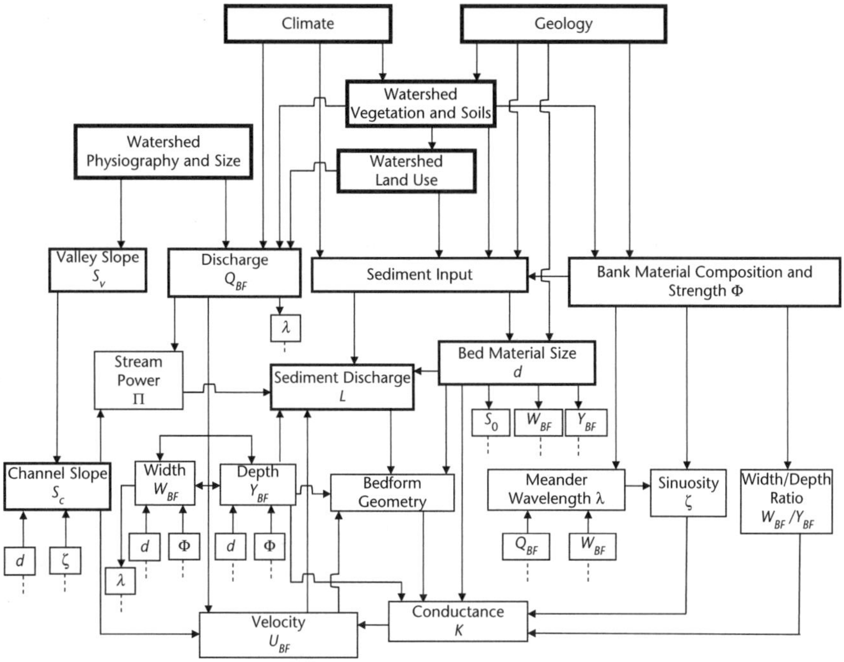

Figure 2.40 Interrelations among variables in the fluvial system. Arrows indicate direction of influence. Dashed lines indicate interrelations that are not fully diagrammed. Note that the figure contains some variables that have not yet been discussed (e.g., bedforms, stream power, and frictional resistance); these will be introduced in later chapters. Modified from Knighton (1998).

2.6.2 Channel Adjustment, Equilibrium, and the Graded Stream

It has been recognized at least since the writings of James Hutton in the late eighteenth century that the elements of the landscape are in a quasi-equilibrium state, implying relatively rapid mutual adjustment to changing conditions. John Playfair clearly articulated Hutton's observations as applied to streams and their valleys in 1802:

> Every river appears to consist of a main trunk, fed from a variety of branches, each running in a valley proportioned to its size, and all of them together forming a system of valleys, communicating with one another, and having such a nice adjustment of their declivities, that none of them joins the principal valley, either on too high or too low a level, a circumstance which would be infinitely improbable if each of these valleys were not the work of the stream which flows in it. (quoted in Summerfield 1991, p. 4)

In the fluvial geomorphological literature, this observation evolved into the concept of the **graded stream**, which was most notably articulated by J. Hoover Mackin:

> A graded river is one in which, over a period of years, slope and channel characteristics are delicately adjusted to provide, with available discharge, just the velocity required

for the transportation of the load supplied from the drainage basin. The graded stream is a system in equilibrium; its diagnostic characteristic is that any change in any of the controlling factors will cause a displacement of the equilibrium in a direction that will tend to absorb the effect of the change. (Mackin 1948, p. 471)

Figure 2.40 gives a sense of the complicated interactions that are involved in responding to changes in the driving variables of climate, geological processes, and human activities. Until the middle of the twentieth century, geomorphologists tended to emphasize mutual adjustments among only three of these variables: sediment load, channel slope, and velocity, such that an increase in sediment delivery from upstream causes deposition, which causes local slope to increase, which causes velocity to increase, which increases sediment transport out of the reach, which reduces slope and velocity back toward the original conditions.

It has since become recognized that changes in slope usually occur only very slowly, and that the mutual adjustments that tend to maintain an equilibrium form involve other aspects of flow and channel geometry that respond more rapidly to change. Thus, Leopold and Bull (1979) suggested that the concept of the graded stream be restated to be more consistent with this recognition and with figure 2.40: "A graded river is one in which, over a period of years, slope, velocity, depth, width, roughness, (planform) pattern and channel morphology ... mutually adjust to provide the power and the efficiency necessary to provide the load supplied by the drainage basin without aggradation or degradation of the channel" (p. 195).

The following section describes hydraulic geometry, which is the general term for the quantitative description of the adjustment of hydraulic variables to temporal and spatial changes in discharge.

2.6.3 Hydraulic Geometry

Leopold and Maddock (1953) coined the term "hydraulic geometry" to refer collectively to the quantitative relations between various hydraulic variables and discharge:

At-a-station hydraulic geometry refers to the changes of hydraulic variables as discharge changes with time in a given reach.

Downstream hydraulic geometry refers to the changes of hydraulic variables as discharge changes with space in a given stream or stream network.

Leopold and Maddock (1953) and subsequent researchers have focused on the hydraulic geometry relations for the components of discharge and postulated that these could be quantitatively represented by simple power-law equations:

Width versus discharge:

$$W = a \cdot Q^b, \tag{2.29}$$

Average depth versus discharge:

$$Y = c \cdot Q^f, \tag{2.30}$$

Average velocity versus discharge:

$$U = k \cdot Q^m, \tag{2.31}$$

Because $Q = W \cdot Y \cdot U$, it must be true that

$$b + f + m = 1 \tag{2.32}$$

and

$$a \cdot c \cdot k = 1. \tag{2.33}$$

The coefficients and exponents in equations 2.29–2.31 vary from reach to reach and differ for at-a-station and downstream relations in a given region.[3] Leopold and Maddock (1953) and most subsequent writers have determined the values of these coefficients and exponents empirically (by regression analysis; see section 4.8.3.1) and have identified tendencies for the exponents to center around particular values (different for at-a-station and downstream relations). As discussed in the following subsections, many researchers have attempted to find physical reasons for these tendencies.

2.6.3.1 Temporal Changes: At-a-Station Hydraulic Geometry

Dependence on Cross-Section Geometry and Hydraulics In at-a-station hydraulic geometry, the symbols Q, W, Y, and U refer to instantaneous values of those quantities at a given cross section or reach. Figure 2.41 shows the ranges of values of the exponents b, f, and m reported in a number of field studies summarized by Rhodes (1977). Although there is wide variation, there is a tendency for at-a-station values to center on $b \approx 0.11, f \approx 0.44, m \approx 0.45$; as an example, figure 2.42 shows the at-a-station hydraulic geometry relations for the Boise River, for which $b = 0.19$, $f = 0.45$, and $m = 0.35$.

There have been several attempts to understand the factors that determine the exponent values, as reviewed by Ferguson (1986) and Knighton (1998). Ferguson (1986) showed conceptually that the exponents and coefficients for a given reach are determined by the channel cross-section geometry and hydraulic relations. Following this reasoning, Dingman (2007a) used equation 2.20 along with generalized hydraulic relations to derive the relations shown in box 2.4. His analysis showed that the exponents depend only on the exponent r in the general equation for cross-section shape (equations 2.20 and 2B4.2) and the depth exponent p in the general hydraulic relation (equation 2B4.3). As shown in figure 2.41, the theoretical exponent values coincide with the central tendencies of the observed values. The effects of channel shape ($r = 1$, triangle; to $r \rightarrow \infty$, rectangle) and different values of p on the exponents can be clearly seen in figure 2.41. Box 2.4 also shows the theoretical relations for the coefficients, which can take on a wide range of values depending on the channel dimensions, conductance, and slope as well as on r and p. (Note that the coefficient values also depend on the units of measurement; the exponents do not.)

Application to Characterizing Stream Hydraulics It can be shown from equation 2.29 that $dW/W = b \cdot (dQ/Q)$, and analogously for equations 2.30 and 2.31; thus, the at-a-station hydraulic geometry relations give information on how small changes

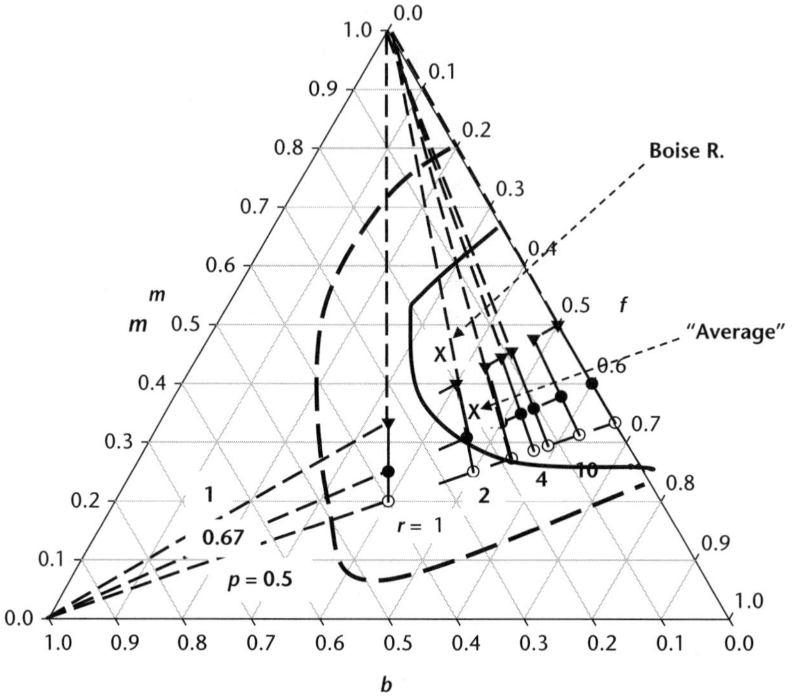

Figure 2.41 Tri-axial diagram showing values of exponents b (width), f (depth), and m (velocity) in at-a-station hydraulic geometry relations (equations 2.29–2.31). The inner (solid) curve encloses most of the empirical values reported by Rhodes (1977); virtually all the values he reported are enclosed by the outer (dashed) curve. The lines radiating from the lower left vertex show the loci of points dictated by the value of the depth exponent p in the generalized hydraulic relation (equation 2B4.3). The lines radiating from the upper vertex show the loci of points dictated by the value of the exponent r in the generalized cross-section relation (equation 2.20).

in discharge are allocated among changes in width, depth, and velocity in a reach. For example, if $b = 0.23$, $f = 0.46$, and $m = 0.31$, a 10% increase in discharge is accommodated by a 2.3% increase in width, a 4.6% increase in depth, and a 3.1% increase in velocity.

The hydraulic geometry relations, in conjunction with the flow-duration curve, can also be used to construct curves that show the time variability of width, depth, velocity, or any other quantity that depends on discharge, using the method described in box 2.5 and figure 2.43. The information presented in such curves is invaluable for such water resource management concerns as characterizing the suitability of the reach as habitat for aquatic organisms, which typically depend on velocity and depth; determining the frequency of overbank flooding, which is a function of depth; and evaluating the potential for stream-bed erosion at a bridge site, which is a function of velocity and depth (Dingman 2002).

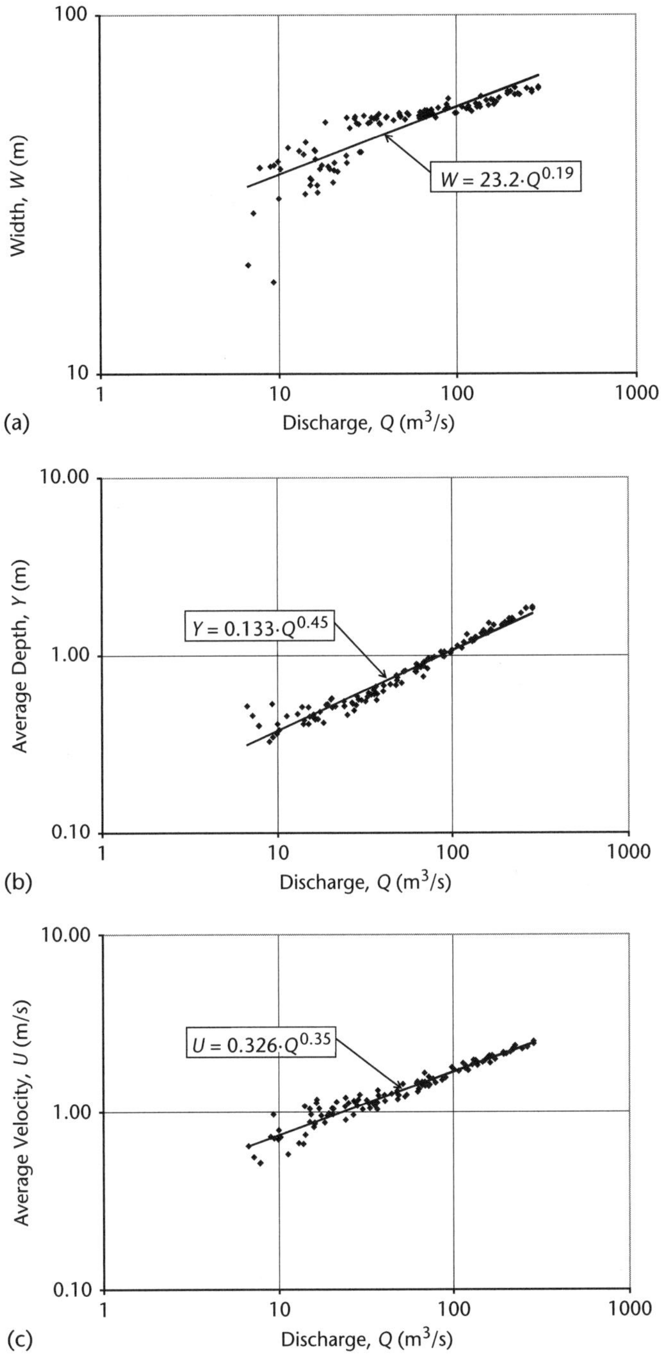

Figure 2.42 Log-log plots of (a) width, (b) average depth, and (c) velocity versus discharge for the Boise River at Twin Springs, ID, showing empirical at-a-station hydraulic-geometry relations established by regression analysis. Note that the fits are stronger at the higher discharges, and there is considerable scatter at lower flows, especially for the width relation.

BOX 2.4 Relations between the Exponents and Coefficients in At-a-Station Hydraulic Geometry Relations and Reach Properties

Starting with the basic continuity relation of equation 2.21,

$$Q = W \cdot Y \cdot U, \tag{2B4.1}$$

Dingman (2007a) used the general cross-section geometry model of equation 2.20,

$$z(w) = \Psi_{BF} \cdot \left(\frac{2}{W_{BF}}\right)^r \cdot w^r, \tag{2B4.2}$$

and a generalization of the hydraulic relation of equation 2.23,

$$Q = g^{1/2} \cdot K \cdot \left(\frac{r+1}{r}\right)^{1/r} \cdot \left(\frac{W_{BF}}{\Psi_{BF}^{1/r}}\right) \cdot Y^p \cdot S^q, \tag{2B4.3}$$

to derive the following relations:

Width exponent b:

$$b = \frac{1}{1+r+r\cdot p} = \frac{1}{\delta}.$$

Depth exponent f:

$$f = \frac{r}{1+r+r\cdot p} = \frac{r}{\delta}.$$

Velocity exponent m:

$$m = \frac{r\cdot p}{1+r+r\cdot p} = \frac{r\cdot p}{\delta}.$$

Width coefficient a:

$$a = W_{BF}^{(r+r\cdot p)/\delta} \cdot \left(\frac{1}{\Psi_{BF}}\right)^{(1+p)/\delta} \cdot \left(\frac{r+1}{r}\right)^{(1+p)/\delta} \cdot \left(\frac{1}{g^{1/2}\cdot K}\right)^{1/\delta} \cdot \left(\frac{1}{S^q}\right)^{1/\delta}.$$

Depth coefficient c:

$$c = \left(\frac{1}{W_{BF}}\right)^{r/\delta} \cdot \Psi_{BF}^{1/\delta} \cdot \left(\frac{r}{r+1}\right)^{1/\delta} \cdot \left(\frac{1}{g^{1/2}\cdot K}\right)^{r/\delta} \cdot \left(\frac{1}{S^q}\right)^{r/\delta}.$$

Velocity coefficient k:

$$k = \left(\frac{1}{W_{BF}}\right)^{r \cdot p/\delta} \cdot \Psi_{BF}^{p/\delta} \cdot \left(\frac{r}{r+1}\right)^{p/\delta} \cdot (g^{1/2} \cdot K)^{(1+r)/\delta} \cdot S^{q \cdot (1+r)/\delta}.$$

Symbols

g gravitational acceleration
K generalized conductance coefficient
p depth exponent in generalized hydraulic relation
q slope exponent in generalized hydraulic relation
r exponent in cross-section geometry relation
S energy or surface slope
U average cross-sectional velocity
w cross-channel distance from center
W water-surface width
W_{BF} bankfull water-surface width
Y cross-sectional average water depth
$\delta \equiv 1 + r + r \cdot p.$
Ψ_{BF} bankfull maximum water depth in cross section

BOX 2.5 Construction of Duration Curves for Quantities That Are Functions of Discharge

In figure 2.43 the graph in the upper right quadrant is the flow-duration curve (FDC), established using methods described by Dingman (2002). The curve in the upper left-hand quadrant is the relation between width, depth, or velocity (or any other quantity that depends on discharge) and discharge. The lower left quadrant is simply a 45°, or 1:1, line.

The duration curve for width, depth, or velocity is constructed in the lower right quadrant by first selecting a number of points on the FDC covering the entire curve. From each point, a vertical line is then projected into the lower right quadrant, and a horizontal line is projected into the upper left quadrant to its intersection with the relation plotted there. A vertical line is projected from each intersection to intersect with the 1:1 line in the lower left quadrant. Finally, horizontal lines are extended from those points to intersect with the vertical lines in the lower right quadrant. Those intersections define the relation between values of width, depth, or velocity and the corresponding exceedence probability, which defines the desired duration curve for width, depth, or velocity.

(*Continued*)

BOX 2.5 *Continued*

As noted in the text, the long-term average discharge, $\overline{Q}$, is equal to the integral of the flow-duration curve:

$$\overline{Q} = \int_0^1 Q(EP) \cdot \mathrm{d}\,EP. \qquad (2B5.1)$$

The curve constructed in the lower right quadrant of figure 2.43 is the duration curve for a quantity that is a function of Q. The long-term average value $\overline{X}$ of a quantity X that depends on discharge, $X(Q)$, is likewise found by integrating its duration curve:

$$\overline{X} = \int_0^1 X[\,Q(EP)] \cdot \mathrm{d}\,EP. \qquad (2B5.2)$$

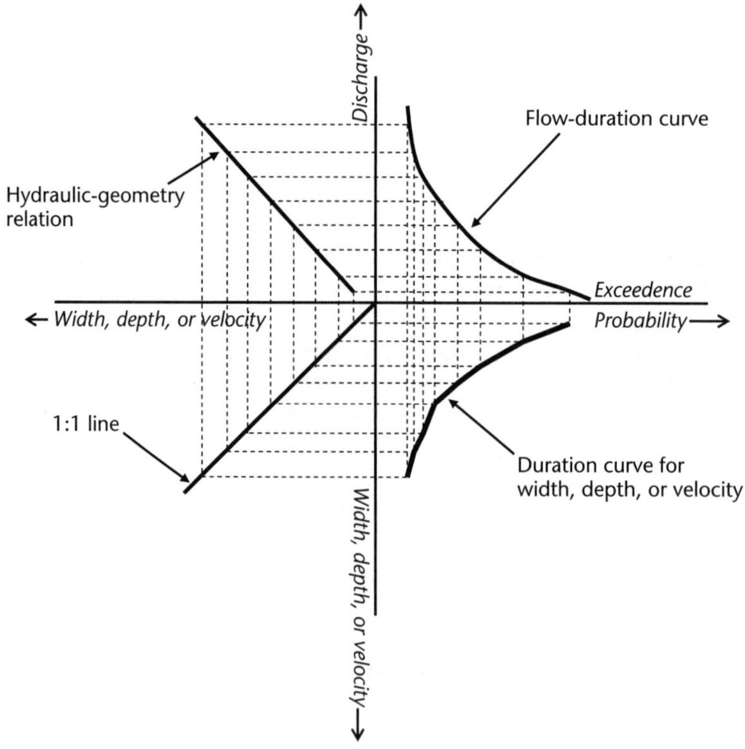

Figure 2.43 Diagram demonstrating construction of duration curves for width, depth, or velocity from the flow-duration curve and at-a-station hydraulic-geometry relations, as described in box 2.5.

2.6.3.2 Spatial Changes: Downstream Hydraulic Geometry

In downstream hydraulic geometry, the hydraulic geometry relations of equations 2.29–2.31 characterize spatial changes in width, depth, and velocity through a river system at a given reference discharge, which is usually taken to be the bankfull discharge, Q_{BF}. The values of the exponents for the downstream relations determined empirically for many regions of the world have been found to vary less than those for the at-a-station relations and are typically near $b = 0.5$, $f = 0.4$, and $m = 0.1$. The coefficients depend on the reference discharge used (as well as the units of measurement) and vary widely depending largely on climate. Again, many attempts have been made to derive these values theoretically, mostly based on considerations similar to the stable-channel approach described in section 2.4.3.1, but there is no generally accepted explanation (for reviews, see Ferguson 1986; Knighton 1998).

One practical application of the downstream relations is in estimating bankfull discharge, depth, and velocity using measurements of bankfull width remotely observed via satellite or air photographs (Bjerklie et al. 2003).

3

Structure and Properties of Water

3.0 Introduction and Overview

Water moves in response to forces acting on it, and its physical properties determine the qualitative and quantitative relations between those forces and the resulting motion. Thus, it is important for the student of hydraulics to have an understanding of these properties. This chapter begins with a description of the atomic and molecular structure of water that give rise to its unique properties, including the fact that it occurs in the gaseous, liquid, and solid phases at the earth's surface. The nature of water in its three phases and the phenomena that accompany phase transitions in nature are briefly described.

The last portion of the chapter uses a series of thought experiments to elucidate the properties of liquid water that are crucial to understanding its behavior in open-channel flows. This section emphasizes the dimensional nature of the various properties, and you may want to refresh your understanding of physical dimensions by reviewing appendix A.

3.1 Structure of Water

3.1.1 Atomic, Molecular, and Intermolecular Structures

The water molecule is formed by the combination of two hydrogen atoms (group Ia, with a nucleus consisting of one proton, and one electron in the outer shell) and one oxygen atom (group VIa, with a nucleus consisting of eight protons and eight

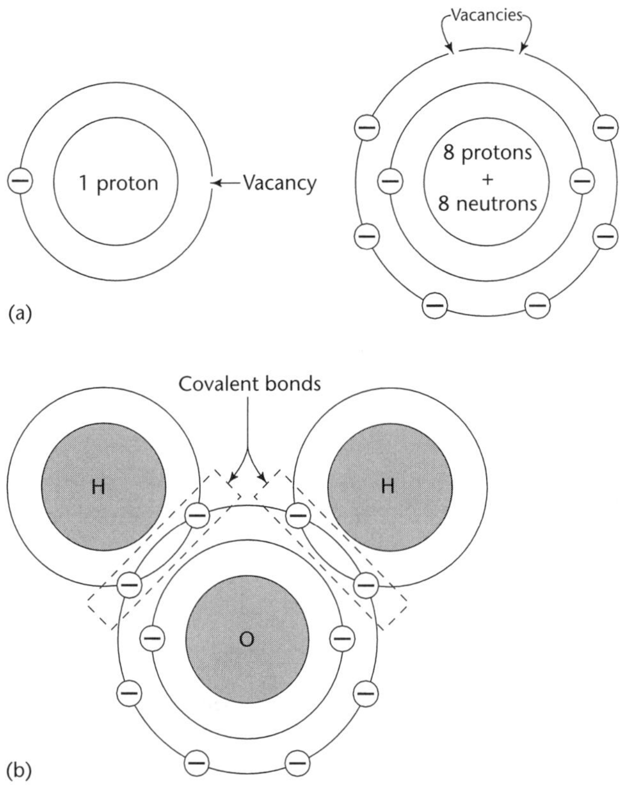

Figure 3.1 (a) Schematic diagram of a hydrogen atom (left) and an oxygen atom (right). (b) Schematic diagram of a water molecule showing sharing of electrons in covalent bonding.

neutrons, two electrons in its inner shell, and six electrons in the outer shell), so it has the chemical formula H_2O. As shown in figure 3.1a, the outer shell of oxygen can accommodate eight electrons, so it has two vacancies. The outer (and only) shell of hydrogen can hold two electrons, so it has one vacancy. The electron vacancies of two hydrogen atoms and one oxygen atom can be mutually filled by sharing outer-shell electrons, as shown schematically in figure 3.1b. This sharing is known as a **covalent bond**.

The two most important features of the water molecule are that 1) its covalent bonds are very strong (i.e., much energy is needed to break them) and 2) the molecular structure is asymmetric, with the hydrogen atoms attached on one "side" of the oxygen atom with an angle of about 105° between them (figure 3.2).

The asymmetry of the water molecule causes it to have a positively charged end (where the hydrogens are attached) and a negatively charged end (opposite the hydrogens), much like the poles of a magnet. Thus, H_2O molecules are **polar**, and the polarity produces an attractive force between the positively charged end of one molecule and the negatively charged end of another, so that liquid water has a cagelike structure (Liu et al. 1996), as shown in figure 3.3. The intermolecular force due to the

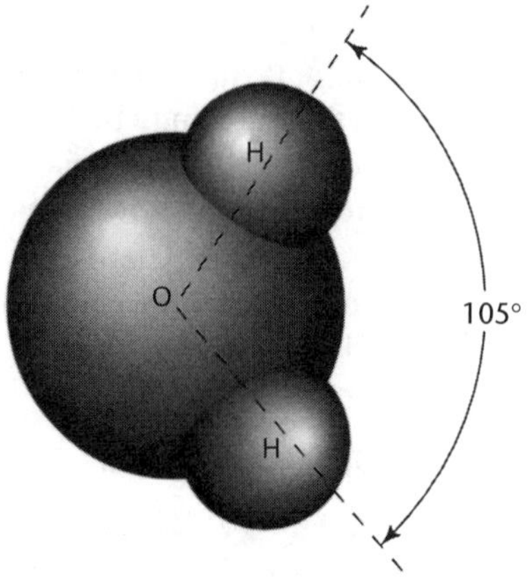

Figure 3.2 Diagram of a water molecule, showing the angle between the hydrogen atoms. After Davis and Day (1961).

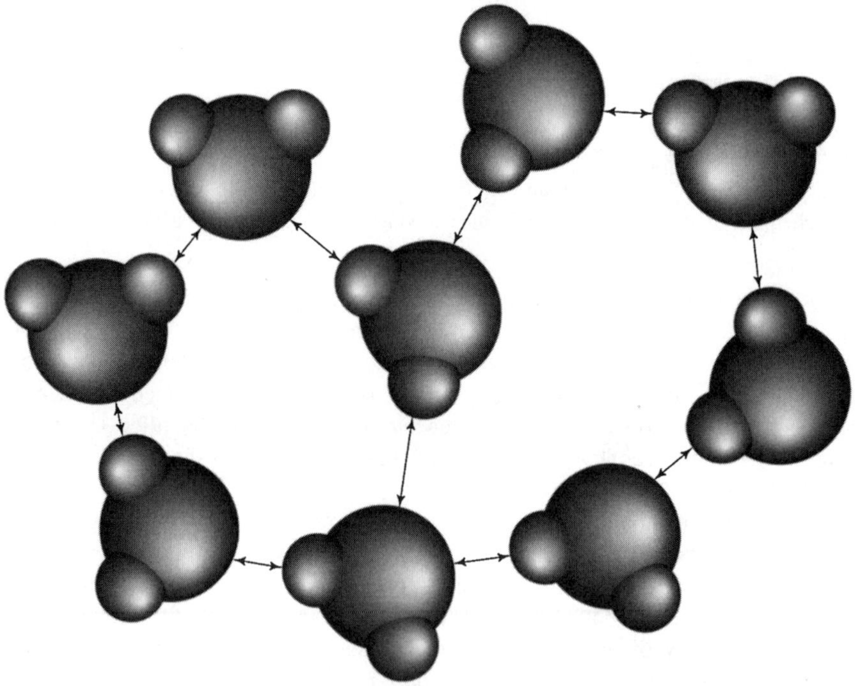

Figure 3.3 The cagelike arrangement of water molecules that characterizes liquid water. The arrows represent hydrogen bonds.

polarity, called a **hydrogen bond**, is absent in most other liquids. As we will see in section 3.3, liquid water has very unusual physical and chemical properties, most of which are due to its hydrogen bonds.

3.1.2 Dissociation

An **ion** is an elemental or molecular species with a net positive or negative electrical charge. At any given instant, a fraction of the molecules of liquid water are dissociated into positively charged hydrogen ions (**protons**), designated H^{+1}, and negatively charged hydroxide ions, designated OH^{-1}. Despite their generally very low concentrations, these ions participate in many important chemical reactions.

Hydrogen ions are responsible for the acidity of water, and acidity is usually measured in terms of **pH**, which is defined as

$$pH \equiv -\log_{10}[H^{+1}], \tag{3.1}$$

where $[H^{+1}]$ designates the concentration of hydrogen ions in mg L^{-1}. The concentration of hydrogen ions in pure water at 25°C is 10^{-7} mg L^{-1} (pH = 7). As $[H^{+1}]$ increases above this value (pH decreases below 7), water becomes more acid; as $[H^{+1}]$ decreases (pH > 7), it becomes more basic.

Certain chemical reactions change the concentration of hydrogen ions, causing the water to become more or less acid. The degree of acidity, in turn, determines the propensity of the water to dissolve many elements and compounds. The pH of cloud water droplets in equilibrium with the carbon dioxide in the atmosphere is about 5.7, and chemical reactions with pollutants reduce the pH of rainwater to the range of 4.0–5.6, depending on location (Turk 1983; see the maps published by the National Atmospheric Deposition Program [2008] at http://nadp.sws.uiuc.edu/isopleths/annualmaps.asp). Once rainwater reaches the ground, reactions with organic material and soil remove H^{+1} ions to increase the pH, so river water pH is typically in the range of pH 5.7–7.7.

3.1.3 Isotopes

Isotopes of an element have the same number of protons and electrons, but differing numbers of neutrons; thus, they have similar chemical behavior but differ in atomic weight. Some isotopes are radioactive and decay naturally to other atomic forms at a characteristic rate, whereas others are stable. Table 3.1 gives the properties and abundances of the isotopes of hydrogen and oxygen, from which it can be calculated that 99.73% of all water consists of "normal" ${}^{1}H_2{}^{16}O$.[1]

The various isotopes are involved in differing proportions in phase changes and chemical and biological reactions, so they are fractionated as water moves through the hydrological cycle (Fritz and Fontes 1980; Drever 1982). Thus, the relative concentrations of these isotopes can be used in some hydrological situations to identify the sources of water in aquifers or streams (see Dingman 2002).

The isotope ${}^{3}H$, called **tritium**, is radioactive and decays to ${}^{3}He$ (helium), with a half life of 12.5 years. It is produced in very small concentrations by natural processes

Table 3.1 Abundances of isotopes of hydrogen and oxygen.

Isotope	Natural abundance (%)
^{1}H	99.985
^{2}H (deuterium)	0.015
^{3}H (tritium)	Trace
^{16}O	99.76
^{17}O	0.04
^{18}O	0.20

and in larger concentrations by nuclear reactions; the increased atmospheric tritium created by atomic testing in the 1950s can be used to date water in aquifers and glaciers (e.g., Davis and Murphy 1987).

3.2 Phase Changes

3.2.1 Freezing/Melting and Condensation/Boiling Temperatures

Although the hydrogen bond is only about one-twentieth the strength of the covalent bond (Stillinger 1980), it is far stronger than the intermolecular bonds that are present in liquids with symmetrical, nonpolar molecules. We get an idea of this strength when we compare the freezing/melting temperature and the condensation/boiling temperature of the hydrides of the group VIa elements: oxygen (O), sulfur (S), selenium (Se), and tellurium (Te). These elements are all characterized by an outer electron shell that can hold eight electrons but has two vacancies. Thus, they all form covalent bonds with two hydrogens. However, except for water, the resulting molecules are nearly symmetrical and therefore nonpolar. In the absence of strong intermolecular forces that result from polar molecules, the melting/freezing and boiling/condensation temperatures of these compounds would be expected to rise as their atomic weights increase.

As shown in figure 3.4, these expectations are fulfilled, except—strikingly—in the case of H_2O. The reason for this anomaly is the hydrogen bonds, which attract one molecule to another and which can only be loosened (as in melting) or broken (as in evaporation) when the vibratory energy of the molecules is large—that is, when the temperature is high. Because of its high melting and boiling temperatures, water is one of the very few substances that exists in all three physical states—solid, liquid, and gas—at earth-surface temperatures (figure 3.5).

The abundance of water, and its existence in all three phases, makes our planet unique and makes the sciences of hydrology and hydraulics vital to understanding and managing the environment and our relation to it. Sections 3.2.2–3.2.3 describe the basic physics of phase changes and how they typically occur in the natural environment.

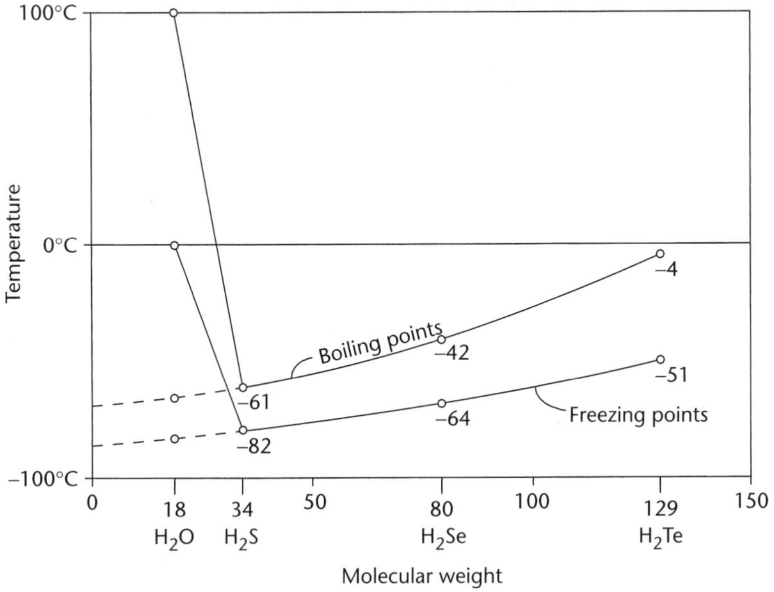

Figure 3.4 Melting/freezing (lower line) and boiling/condensation (upper line) temperatures of group VIa hydrides. In the absence of hydrogen bonds, water would have much lower melting/freezing and boiling/condensation points (dashed lines). After Davis and Day (1961).

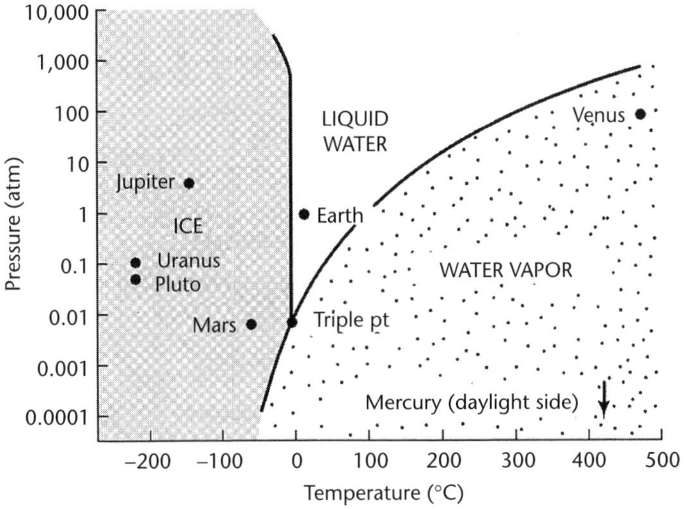

Figure 3.5 Surface temperatures and pressures (y-axis, in atmospheres) of the planets plotted on the phase diagram for water. From *Opportunities in the Hydrologic Sciences* (Eagleson et al. 1991). Reprinted with permission of National Academies Press.

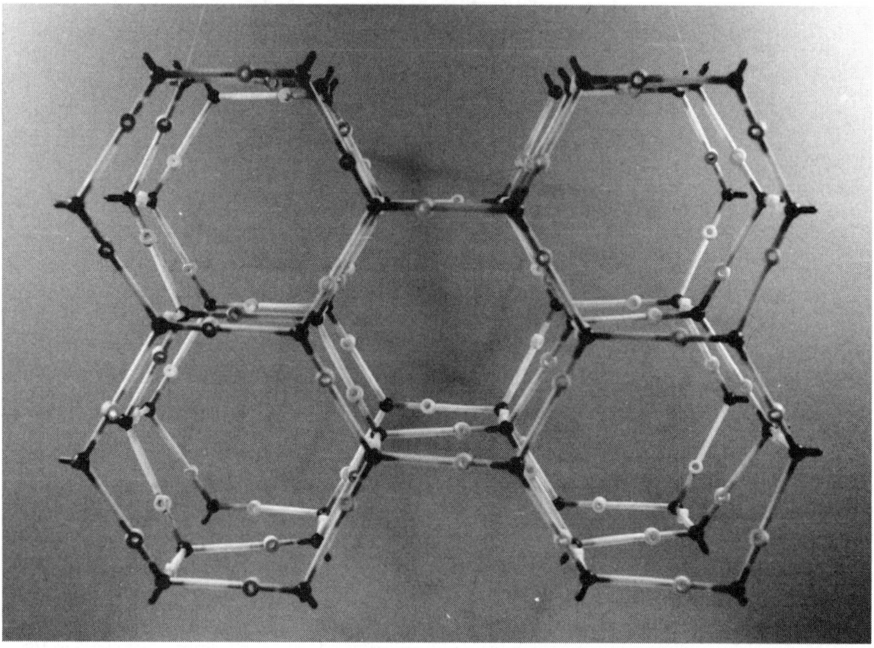

Figure 3.6 A model of the crystal lattice of ice, showing its hexagonal structure. White circles are hydrogen atoms, and dark circles are oxygen atoms; longer white lines are hydrogen bonds, darker shorter lines are covalent bonds. The crystallographic c-axis is perpendicular to the page through the centers of the hexagons; the three a-axes are in the plane of the page connecting the vertices. Photo by the author.

3.2.2 Freezing and Melting

3.2.2.1 Physics of Freezing and Melting

At temperatures below 0°C, the vibratory energy of water molecules is sufficiently low that the hydrogen bonds can lock the molecules into the regular three-dimensional crystal lattice of ice (figure 3.6). In the rigid ice lattice, a given number of molecules take up more space than in the liquid phase, and the density of ice is 91.7% of the density of liquid water at 0°C. Very few substances have a lower density in the solid state than in the liquid, and the fact that ice floats is of immense practical importance.

In the ice lattice, each molecule is hydrogen-bonded to four adjacent molecules. The angle between the hydrogen atoms in each molecule remains at 105°, but each molecule is oriented so that a puckered honeycomb of perfect hexagons is visible when the lattice is viewed from one direction. Thus, ice is a hexagonal crystal, and snowflakes show infinite variation on a theme of sixfold symmetry. The crystallographic c-axis passes through the center of the hexagons, and three a-axes are perpendicular to this, separated by angles of 120°. Interestingly, the layer of molecules at the surface of ice crystals appears to be liquid (i.e., more like figure 3.3) even at very

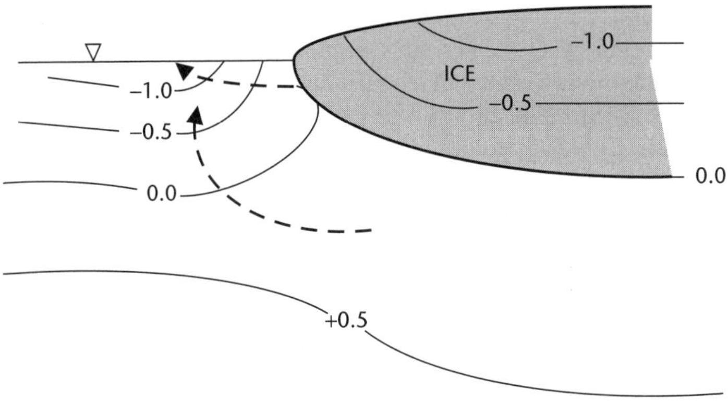

Figure 3.7 Freezing at the edge of an ice sheet or a frazil disk requires a temperature gradient away from the freezing location, and hence supercooling. Contours give temperature in °C; arrows show direction of heat flow. The inverted triangular hydrat symbol, ∇, designates a "free surface," that is, a surface of liquid water at atmospheric pressure. After Meier (1964).

low temperatures, and this layer is responsible for the low friction that makes skating and skiing possible (Seife 1996).

Although the ice lattice is the thermodynamically stable form of water substance at temperatures below 0°C, freezing does not usually take place exactly at the freezing point. Supercooling is required because freezing produces a large quantity of heat, the latent heat of fusion, that must be removed by conduction, and conduction can take place only if there is a temperature gradient directed away from the locus of freezing (figure 3.7). The value of the latent heat of fusion, λ_f, in the various unit systems is

$$\lambda_f = 3.34 \times 10^5 \text{ J kg}^{-1} = 79.7 \text{ cal g}^{-1} = 4620 \text{ Btu slug}^{-1} (= 144 \text{ Btu lb}^{-1}).$$

Once ice is warmed to 0°C, further additions of heat cause melting without a change in temperature. The heat required to melt a given mass of ice is identical to the amount liberated on freezing, that is, the latent heat of fusion, λ_f. Melting involves the rupturing of about 15% of the hydrogen bonds (Stillinger 1980), and the ice lattice consequently collapses into the denser but less rigid liquid structure of figure 3.3.

3.2.2.2 Freezing and Melting of Lakes and Ponds

Freezing In the relatively still water of lakes and ponds, the freezing process begins with cooling at the surface as the lake loses heat to the atmosphere. If the initial surface temperature is above 4°C, the temperature of maximum density (see section 3.3.1), the cooled surface water is denser than that below the surface and sinks. This process, called the **fall turnover**, continues until the entire water body is at 4°C (if there is strong mixing by wind, the entire lake may be cooled to a lower temperature). Further cooling produces a surface layer that is less dense than the water below, and this layer remains at the surface and continues to cool to just below the freezing point. Ice-cover growth usually begins when seed crystals are introduced into water that is supercooled by a few hundredths of a Celsius degree.[2] These seed crystals are usually snowflakes, or ice crystals formed in the air when tiny droplets produced by breaking waves or bubbles freeze (Daly 2004). However, bacteria, organic molecules, and clay minerals can also act as seeds for ice nucleation. If wind action is negligible, the seeds provide nuclei around which freezing occurs rapidly to form an ice skim.

In quiescent water, the initial ice skim thickens downward as latent heat is conducted upward through the ice to the subfreezing air. Under steady-state conditions (i.e., a constant subfreezing air temperature), the thickness of an ice sheet, $h_{\text{ice}}(t)$, increases in proportion to the square root of time, t:

$$h_{\text{ice}}(t) = \left[\frac{2 \cdot K_{\text{ice}} \cdot (T_f - T_a) \cdot t}{\rho_{\text{ice}} \cdot \lambda_f}\right]^{1/2}, \tag{3.2}$$

where K_{ice} is the thermal conductivity of ice, T_f is the freezing temperature of ice, T_a is the air temperature, ρ_{ice} is the mass density of ice, and λ_f is the latent heat of fusion (Stefan 1889). The thermal conductivity of pure ice is

$$K_{\text{ice}} = 2.24 \text{ J m}^{-1}\text{s}^{-1}\text{K}^{-1} = 5.35 \times 10^{-3}\text{cal cm}^{-1}\text{s}^{-1}\text{ C}^{\circ -1}$$

$$= 3.58 \times 10^{-4} \text{ Btu ft}^{-1}\text{s}^{-1}\text{F}^{\circ -1}.$$

The following empirical equation for predicting lake-ice thickness is based on equation 3.2 (Michel 1971):

$$h_{\text{ice}}(n) = \alpha_f \cdot D(n)^{1/2}, \tag{3.3}$$

where $h_{\text{ice}}(n)$ is ice thickness (units of meters, m) n days after the start of freezing, α_f is a coefficient that depends on the rate of heat transfer through the ice surface (see table 3.2), and $D(n)$ is accumulated **freezing-degree days** from the start of freezing, computed as

$$D(n) \equiv \sum_{j=1}^{n} (T_f - T_{aj}), \tag{3.4}$$

where T_f is the freezing temperature (0°C), and T_{aj} is the average air temperature on the jth day after freezing begins (°C).

Table 3.2 Values of coefficient α_f in empirical ice-thickness-prediction equation (equation 3.3).

Environment and condition	α_f
Lake: windy, no snow	0.027
Lake: average with snow	0.017–0.024
River: average with snow	0.014–0.017
Small river, rapid flow	0.007–0.014

From Michel (1971).

Melting Lakes begin to melt along the shore due to the absorption of thermal radiation from the land and vegetation, and the ice cover typically becomes free-floating. Further melting occurs at the surface due to absorption of solar radiation and contact with warmer air, and the meltwater drains to the margin or vertically through holes and cracks. If a snow cover existed, a lake usually develops a several-centimeter-thick porous layer underlain by a layer of water-logged ice above a still-solid layer (Williams 1966). When the upper, relatively light-colored layer is gone, the darker underlying ice rapidly absorbs solar heat and melts quickly. Wind usually assists by breaking up the ice cover, allowing warmer subsurface water to contact the ice, and the melting accelerates. The resulting rapid disappearance of the ice cover has led some observers to believe that the ice actually sank (Birge 1910), but this is impossible because of its lower density.

3.2.2.3 Freezing and Melting of Streams

Freezing Ice covers in streams begin forming along the banks where velocities are low, by the same process that operates in lakes. In faster flowing regions, however, ice initially forms in small disks called **frazil** that form around nuclei in water that is supercooled by a few hundredths of a degree. (Again, the supercooling illustrated in figure 3.7 is required to remove the latent heat, which is transported to the surface by the turbulence and lost to the air.) As in lakes, snowflakes or small ice crystals that form in the air provide the initial seeds, but the frazil disks themselves provide a rapid increase in nuclei through a process called **secondary nucleation** (Daly 2004). Frazil disks are typically less than a millimeter in diameter and 0.05–0.5 mm thick, and become distributed through the flow by turbulent eddies (see section 3.3.4) in concentrations up to 10^6 m^{-3}.

The evolution of a river-ice cover is shown in figure 3.8. Frazil disks are extremely "sticky," and as the frazil concentration grows, the disks collide and stick together (agglomerate) into **flocs**. Some agglomerated frazil flocs float to the surface, where they accumulate as slush pans and ultimately become **floes** (large essentially flat floating ice masses). Other flocs that contact the bottom become attached to bottom particles as **anchor ice**. Anchor ice can build up to the extent that its buoyancy plucks particles from the bottom and brings them to the surface.

A complete river-ice cover typically forms by growth of surface ice outward from slow-flowing near-shore areas (**border ice**) plus the coalescing of floes formed from frazil ice. This coalescing begins in relatively slow-flowing reaches, where floes arriving from upstream collect and merge with border ice in a process called **bridging**. The ice cover builds upstream as more floes arrive until it connects with the next upstream accumulation.

River ice covers are of great scientific and engineering interest. In addition to interfering with navigation, they cause significant increases in frictional resistance to flow (discussed in chapter 6). In fact, frazil ice can form almost complete flow obstructions by accumulating between an existing ice cover and the bottom (figure 3.8) and can also cause significant problems by collecting on and blocking flow through flow-intake structures. River freezing represents the temporary storage of water,

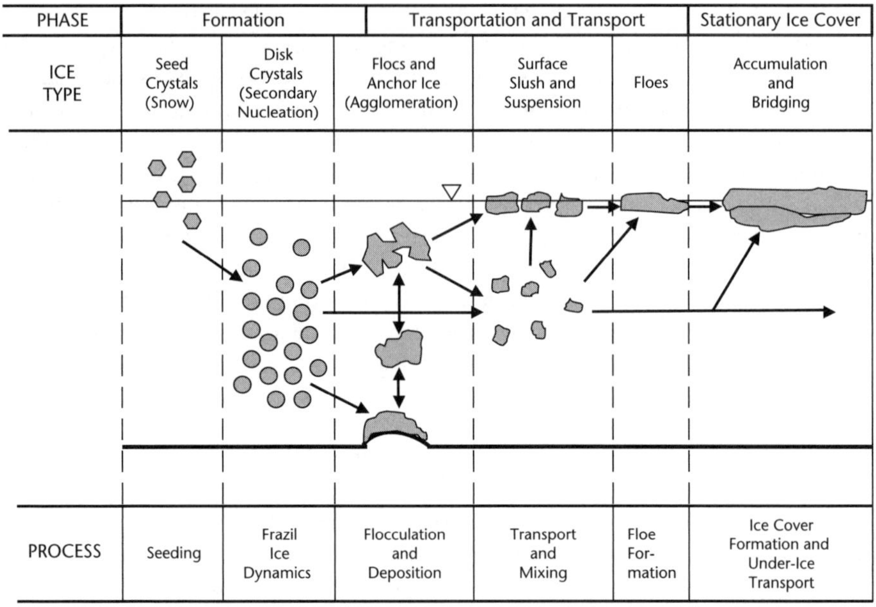

Figure 3.8 Processes involved in river ice-cover formation. After Daly (2004).

reducing streamflow quantities available for water supply, waste dilution, and power generation, and the ice cover reduces the dissolution of oxygen that is essential to aquatic life and to the oxygenation of wastewater.

Melting Michel (1971; see also Beltaos 2000) describes the typical river-ice breakup process as consisting of three phases (figure 3.9). The **prebreakup phase** usually begins with an increase in streamflow due to snowmelt in the drainage basin. The additional water tends to lift the ice cover, separating it from the shore and causing fractures that result in flooding over the ice surface. Further snowmelt, often produced in daily flood waves, ultimately removes the ice from areas of rapids; this ice is carried downstream to accumulate in **ice jams** at the upstream ends of the ice covers that remain in low-velocity reaches (figure 3.9a).

Continuing snowmelt runoff, accompanied by higher air temperatures and sometimes by rain, initiates the **breakup phase** in which the ice covers in various ice reaches are transported to an ice jam farther downstream. Depending on local conditions, this ice may cause further accumulation there, or may dislodge the cover in that reach and move it to form a larger jam at a downstream ice reach (figure 3.9b). Ultimately, if streamflow and warming continue, one of the larger ice jams gives way, and its momentum sweeps all downstream jams away in the **final drive,** typically freeing the river of ice in a few hours (figure 3.9c).

The temporary damming caused by ice jams exacerbates flooding and flood damages annually in large portions of the northern hemisphere, and the forces associated with the final drive can wreak tremendous damage on bridges and river-bank structures.

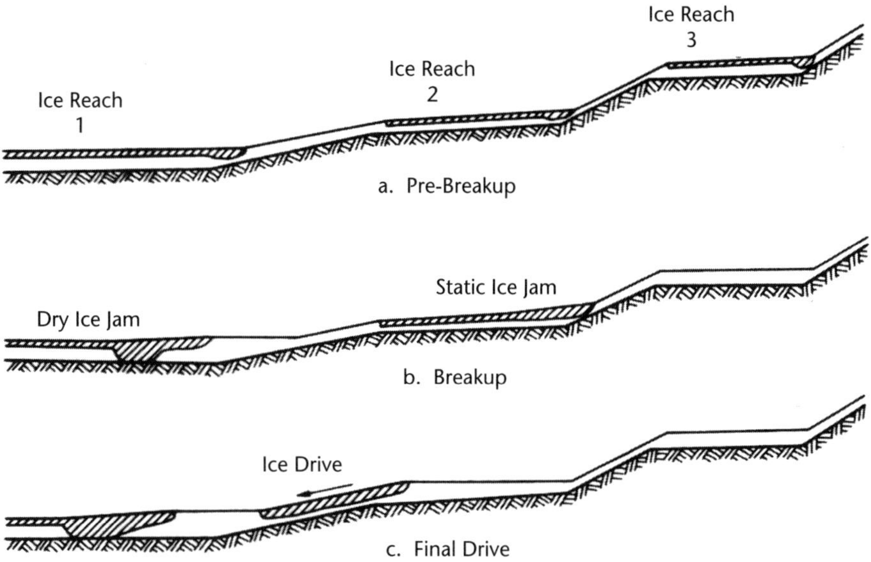

Figure 3.9 The stages of river-ice breakup. (a) In the prebreakup phase, snowmelt in the drainage basin increases river flow, which lifts the ice cover, separating it from the shore and ultimately removing the ice from steep reaches; this ice is carried downstream to accumulate in ice jams at the upstream ends of the ice covers that remain in low-velocity ice reaches. (b) In the breakup phase, continuing snowmelt runoff transports the ice covers in various ice reaches to an ice jam farther downstream. (c) As streamflow and warming continue, one of the larger ice jams gives way, and its momentum sweeps all downstream jams away in the final drive. From Michel (1971).

3.2.3 Evaporation, Condensation, and Sublimation

At temperatures less than 100°C, some molecules at the liquid–air or solid–air interface that have greater than average energy sever all hydrogen bonds with their neighbors and fly off to become **water vapor**, which consists of relatively widely spaced individual H_2O molecules; these are mixed with the other molecular species that constitute the atmosphere. Each constituent atmospheric gas exerts a partial pressure, and the atmospheric pressure is the sum of the partial pressures of all the constituents. For each constituent, the partial pressure is given by the ideal gas law:

$$e_i = R_i \cdot T_a \cdot \rho_i, \tag{3.5}$$

where e_i is the partial pressure of constituent i, R_i is the **gas constant** for constituent i, T_a is the air temperature, and ρ_i is the **vapor density** of constituent i (mass of constituent i per unit volume of atmosphere).

For water vapor,

$$e_v = 0.461 \cdot T_a \cdot \rho_v, \tag{3.6}$$

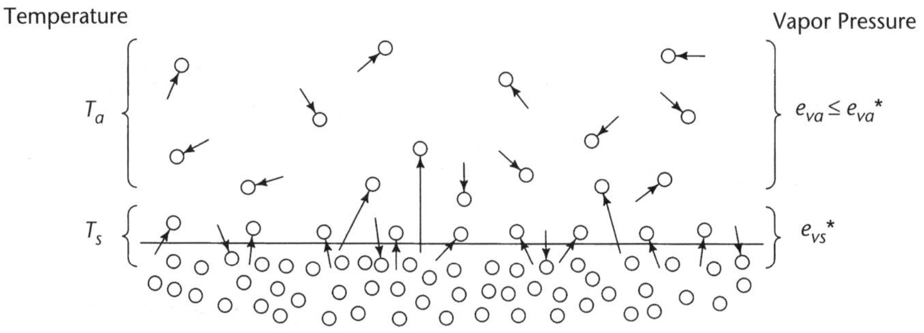

Figure 3.10 Schematic diagram of water-vapor flux near a water surface. Circles represent water molecules; arrows show paths of motion. T_a is air temperature, T_s is surface temperature, e_{va} is air vapor pressure, e_{va}^* is saturation vapor pressure at air temperature T_a, and e_{vs}^* is saturation vapor pressure at surface temperature T_s.

where e_v is water-vapor pressure (kPa), T_a is in K, and ρ_v is in kg m^{-3}. There is a thermodynamic maximum concentration of water vapor that the air can hold at a given temperature, which can be expressed as the **saturation vapor density**, ρ_v*, or the **saturation vapor pressure**, e_v*. This maximum corresponds to 100% relative humidity, and it is related to T_a approximately as

$$e_v* = 0.611 \cdot \exp\left(\frac{17.3 \cdot T_a}{T_a + 237.3}\right), \tag{3.7}$$

where e_v* is in kPa and T_a is in °C. The value of ρ_v* can be computed from equations 3.6 and 3.7.

Figure 3.10 schematically illustrates the movement of water vapor near a water or ice surface. Water molecules are continually entering and leaving the surface, and evaporation/condensation occurs if the amount leaving (per unit area per unit time) is greater/less than the amount entering. These amounts, in turn, are determined by 1) the difference in vapor pressure between the water surface and the overlying air and 2) the efficacy of air currents in removing/supplying vapor from/to the surface. For a liquid–water surface, the rate of evaporation/condensation, E (mm day^{-1}), can be estimated as

$$E = [0.95 \cdot (T_s - T_a)^{1/3} + 1.10 \cdot v_a] \cdot (e_{vs}^* - e_{va}), \text{ for } T_s > T_a, \tag{3.8a}$$

$$E = 1.10 \cdot v_a \cdot (e_{vs}^* - e_{va}), \text{ for } T_s \leq T_a, \tag{3.8b}$$

where T_s is surface temperature (°C), T_a is air temperature (°C), v_a is wind speed (m s^{-1}), $e_{vs}*$ is saturation vapor pressure of the surface (kPa), e_{va} is vapor pressure of the air (which may be less than or equal to the saturation value, e_{va}^*; kPa), and atmospheric variables are measured at a height of 2 m above the ground (Dingman 2002). Equation 3.8a accounts for situations in which vapor exchange is enhanced by convection that is induced when the surface is warmer than the air.

The breaking/forming of hydrogen bonds that accompanies evaporation/condensation results in an absorption/release of heat energy: the **latent heat**

of vaporization. Water has one of the largest latent heats of vaporization, λ_v of any substance; its value at 0°C is

$$\lambda_v = 2.495 \text{ MJ kg}^{-1} = 595.9 \text{ cal g}^{-1} = 3.457 \times 10^4 \text{ Btu slug}^{-1} (= 1{,}074 \text{ Btu lb}^{-1}).$$

This quantity, λ_v, decreases as the temperature of the evaporating surface increases approximately as

$$\lambda_v = 2.495 - (2.36 \times 10^{-3}) \cdot T_s, \tag{3.9}$$

where T_s is temperature in °C and λ_v is in MJ kg^{-1}. When liquid water is heated to 100°C, further additions of energy cause the eventual breaking of all the remaining hydrogen bonds, and the liquid is entirely transformed into a gas. At 100°C the latent heat of vaporization is 2.261 MJ kg^{-1}, more than six times the latent heat of fusion and more than five times the amount of energy it takes to warm the water from the melting point to the boiling point.

Note that the latent heat involved in the direct phase change between ice and water, without an intermediate liquid state (**sublimation**), is the sum of the latent heat of vaporization plus the latent heat of fusion.

Water's enormous latent heat of vaporization plays a critical role in global climate processes. It accounts for almost one-half the heat transfer from the earth's surface to the atmosphere, is a major component of meridional heat transport, and is a source of energy that drives the precipitation-forming process.

3.3 Properties of Liquid Water

The physical properties of water are determined by its atomic and molecular structures. As we have already seen, water is a very unusual substance with anomalous properties, and its strangeness is the reason it is so common at the earth's surface (figures 3.4 and 3.5). This section describes the basic physical properties of bulk liquid water that influence its movement through the hydrological cycle and its physical interactions with the terrestrial environment. More detailed discussions of these properties can be found in Dorsey (1940) and Davis and Day (1961), and they are very entertainingly described by van Hylckama (1979) and Ball (1999). Table 3.3 summarizes water's unique properties and their importance in earth-surface processes.

The variation of water's properties with temperature is important in many hydrological contexts. Thus, in the following discussion, the values of each property at 0°C are given in the three unit systems, and their relative variations with temperature are shown in table 3.4. Empirical equations for computing the values of the properties as functions of temperature are also given. Of course, water in the natural environment is never pure H_2O; it always contains dissolved solids and gases and often contains suspended organic and/or inorganic solids. Dissolved constituents are seldom present in high enough concentrations in streams and rivers to warrant accounting for those effects, but suspended sediment can affect water properties such as density and viscosity, and some information describing these effects is given.

Table 3.3 Physical and chemical properties of liquid water.

Property	Comparison with other substances	Importance to environment
Density	Maximum density at 4°C, not at freezing point; expands upon freezing	Prevents rivers and lakes from freezing solid; causes stratification in lakes
Melting and boiling points	Abnormally high (figure 3.4)	Permits water to exist at earth's surface (figure 3.5)
Heat capacity	Highest of any liquid except ammonia	Moderates temperatures
Latent heat of vaporization	One of the highest of any substance	Important to atmospheric heat transfer; moderates temperatures
Surface tension	Very high	Regulates cloud-drop and raindrop formation and water storage in soils
Absorption of electromagnetic radiation	Large in infrared and ultraviolet wavelengths; lower in visible wavelengths	Major control on atmospheric temperature (greenhouse gas); controls distribution of photosynthesis in lakes and oceans
Solvent properties	Strong solvent for ionic salts and polar molecules	Important in transfer of dissolved substances in hydrological cycle and biological systems

After Berner and Berner (1987).

Table 3.4 Properties of pure liquid water as functions of temperature.[a]

Temperature (°C)	Density (ρ, γ)	Surface tension (σ)	Dynamic viscosity (μ)	Kinematic viscosity (ν)
0	1.00000	1.0000	1.0000	1.0000
3.98	1.00013			
5	1.00012	0.9907	0.8500	0.8500
10	0.99986	0.9815	0.7314	0.7315
15	0.99926	0.9722	0.6374	0.6379
20	0.99836	0.9630	0.5637	0.5616
25	0.99720	0.9524	0.4983	0.4997
30	0.99580	0.9418	0.4463	0.4482

[a] Numbers are ratios of values at given temperature to value at 0°C.

3.3.1 Density

3.3.1.1 Definitions

Mass density, ρ, is the mass per unit volume $[\mathrm{M\,L^{-3}}]$ of a substance, whereas **weight density**, γ, is the weight per unit volume $[\mathrm{F\,L^{-3}}]$. These are related by Newton's second law (i.e., force equals mass times acceleration):

$$\gamma = \rho \cdot g, \tag{3.10}$$

where g is the acceleration due to gravity [LT^{-2}] ($g = 9.81$ m s^{-2} = 32.2 ft s^{-2}). Because gravitational force (= mass times gravitational acceleration) and momentum (= mass times velocity) are proportional to mass, and pressure depends on weight (see section 4.2.2.2), either ρ or γ appears in most equations describing the motion of fluids.

The **specific gravity**, G, of a substance is the ratio of its density to the density of pure water at 3.98°C; thus, it is dimensionless.

3.3.1.2 Magnitude

In the Système Internationale, or SI, system of units the kilogram is defined as the mass of 1 m^3 of pure water at its temperature of maximum density, 3.98°C. At 0°C,

$$\rho = 999.87 \text{ kg m}^{-3} = 0.99987 \text{ g cm}^{-3} = 1.9397 \text{ slug ft}^{-3},$$

$$\gamma = 9799 \text{ N m}^{-3} = 979.9 \text{ dyn cm}^{-3} = 62.46 \text{ lb ft}^{-3}.$$

Note that the kilogram and gram are commonly used as units of force as well as of mass: 1 kg of force (kgf) is the weight of a mass of 1 kg at the earth's surface, where $g = 9.81$ m s^{-2} (981 cm s^{-2}). Thus, 1 kg of force = 9.81 N; 1 g of force = 981 dyne, and at 0°C,

$$\gamma = 998.9 \text{ kgf m}^{-3} = 0.9989 \text{ gf cm}^{-3}.$$

As noted, water is anomalous in that the liquid at 0°C is denser than ice. The change in density of water with temperature is unusual (see tables 3.3 and 3.4) and environmentally significant. As liquid water is warmed from 0°C, its density initially increases, whereas most other substances become less dense as they warm. This anomalous increase continues until density reaches a maximum value of 1,000 kg m^{-3} at 3.98°C; beyond this, the density decreases with temperature as in most other substances. These density variations can be approximated as

$$\rho = 1000 - 0.019549 \cdot |T - 3.98|^{1.68}, \tag{3.11}$$

where T is temperature in °C and ρ is in kg m^{-3} (Heggen 1983). The variation of γ with temperature can be approximated via equations 3.10 and 3.11.

As noted in section 3.2.2, in lakes where temperatures reach 3.98°C, the density maximum controls the vertical distribution of temperature and causes an annual or semiannual overturn of water that has a major influence on biological and physical processes. However, except for lakes, the variation of density with temperature is small enough that it can usually be neglected in hydraulic calculations.

The addition of dissolved or suspended solids to water increases the density of the water–sediment mixture, γ_m, in proportion to the density of the solids, γ_s, and their volumetric concentration (volume of sediment per volume of water–sediment mixture), C_{vv}:

$$\gamma_m = \gamma_s \cdot C_{vv} + \gamma \cdot (1 - C_{vv}). \tag{3.12a}$$

Suspended sediment is usually assumed to have the specific gravity of quartz, $G_s = 2.65$, so $\gamma_s = 25{,}967$ N m^{-3}. Sediment concentrations are usually given in units

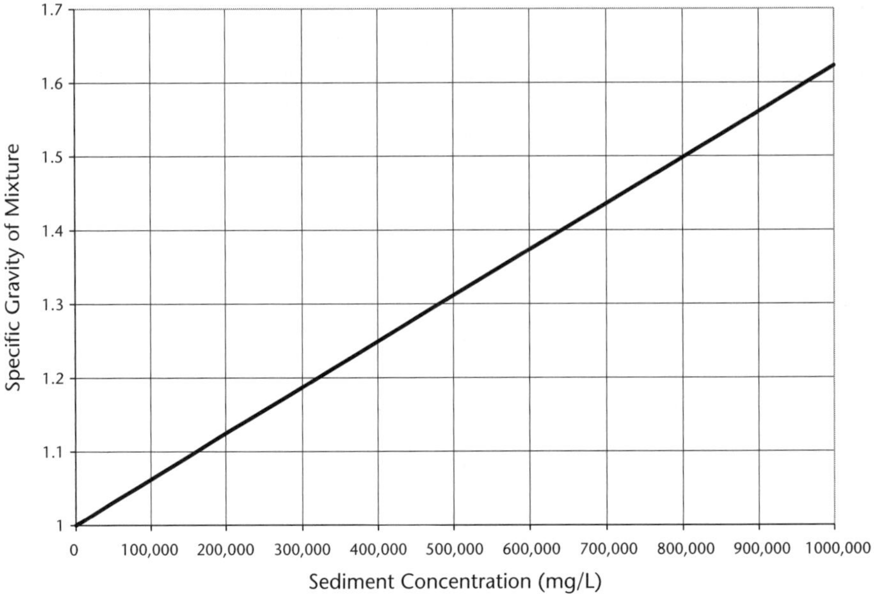

Figure 3.11 Effects of sediment concentration on the relative density (specific gravity) of water–sediment mixtures (equation 3.12).

of milligrams of sediment per liter of mixture, $C_{mg/L}$; using these units, equation 3.12a becomes

$$\gamma_m = \gamma_s \cdot \frac{C_{\mathrm{mg/L}}}{2.65 \times 10^6} + \gamma \cdot \left(1 - \frac{C_{\mathrm{mg/L}}}{2.65 \times 10^6}\right). \quad (3.12b)$$

Again, the effects of dissolved materials can be important in lakes, but are not usually significant in rivers. However, high concentrations of suspended matter can significantly increase the effective density of water in rivers, as shown in figure 3.11.

Water, like most liquids, has a very small compressibility, so changes of density with pressure can be neglected.

3.3.2 Surface Tension and Capillarity

Molecules in the surface of liquid water are subjected to a net inward force due to hydrogen bonding with the molecules below the surface (figure 3.12). This force tends to minimize the surface area of a given volume of water and produces surface tension and the phenomenon of capillarity.

3.3.2.1 Surface Tension

Surface tension is best understood by visualizing a thought experiment (figure 3.13). Consider a device consisting of an inverted U-shaped wire defining three sides of a rectangular area, with the fourth side formed by a straight wire that can slide along

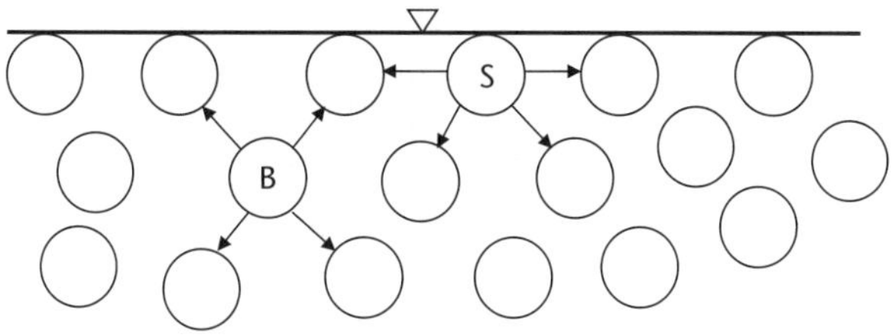

Figure 3.12 Intermolecular (hydrogen-bond) forces acting on typical surface (S) and nonsurface (B) molecules. The unbalanced forces on surface molecules produce the phenomenon of surface tension.

the arms of the U. The size of the area is a few square millimeters. When the device is dipped into water and removed, a film of water is retained in the opening. If the sliding wire can move without friction, it will be pulled toward the top of the inverted U (figure 3.13a). The force causing this movement is due to the intermolecular hydrogen bonds.

We can measure the magnitude of this force by suspending from the slide wire a small weight wt_s that just balances the upward force (figure 3.13b). The surface tension, σ, is equal to this weight divided by the distance over which the force acts, which is twice (because the film has two surfaces) the length, x_w, of the slide wire:

$$\sigma = \frac{wt_s}{2 \cdot x_w}. \tag{3.13}$$

The dimensions of σ are therefore $[\mathrm{F\,L}^{-1}]$.

Surface tension can also be thought of as the work required to increase the surface area of a liquid by a unit amount. If we add an increment of weight dwt to wt_s, the slide wire will be pulled down a distance dy_s, causing molecules within the film to move to the surface and increasing the surface area by $dA_s = 2 \cdot x_s \cdot dy_s$. The ratio of the increment of work dwt_s/dy_s to the increment of area dA_s is the surface tension:

$$\sigma \equiv \frac{dwt_s \cdot dy_s}{dA_s} = \frac{dwt_s \cdot dy_s}{2 \cdot x_s \cdot dy_s} = \frac{dwt_s}{2 \cdot x_s}. \tag{3.14}$$

3.3.2.2 Magnitude of Surface Tension

As might be expected from its strong intermolecular forces, water has a surface tension higher than most other liquids; its value at 0°C is

$$\sigma = 0.0756\ \mathrm{N\,m^{-1}} = 75.6\ \mathrm{dyn\,cm^{-1}} = 0.00518\ \mathrm{lb\,ft^{-1}}.$$

Surface tension decreases rapidly as temperature increases (table 3.4); the temperature effect can be approximated as

$$\sigma = 0.001 \cdot (20987 - 92.613 \cdot T)^{0.4348}, \tag{3.15}$$

where T is in °C and σ is in $\mathrm{N\,m^{-1}}$ (Heggen 1983). Dissolved substances can also increase or decrease surface tension, and certain organic compounds have a major effect on its value.

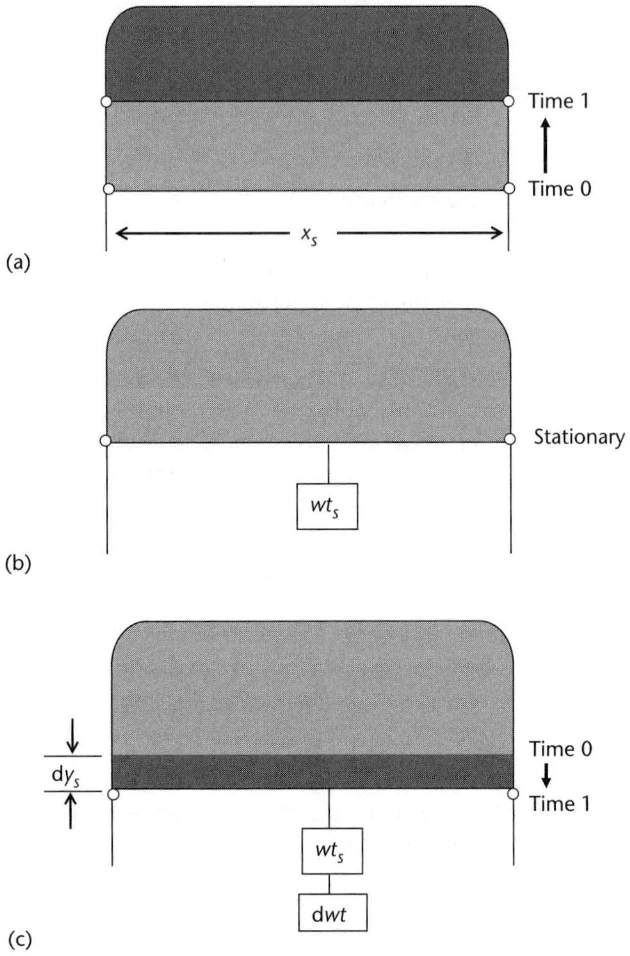

Figure 3.13 Thought experiment for surface tension, showing (a) the motion of a slide wire between time 0 and time 1 due to surface tension force (a). In (b) a weight wt_s has been attached to the slide wire to balance the upward surface-tension force. In (c) an increment of weight, dwt, has been added to the slide wire to pull it down a distance dy_s and increase the water-surface area by $2 \cdot x_s \cdot \mathrm{d}y_s$.

3.3.2.3 Capillarity

Interactions between water molecules and solid materials in combination with surface tension distort the water-surface configuration at the intersection of a water surface and a solid boundary. This phenomenon, called **capillarity**, can be understood by considering the small (diameter of a few millimeters or less) cylindrical tube immersed in a body of water with a free surface[3] shown in figure 3.14. If the material of the tube is such that the hydrogen bonds of the water are attracted

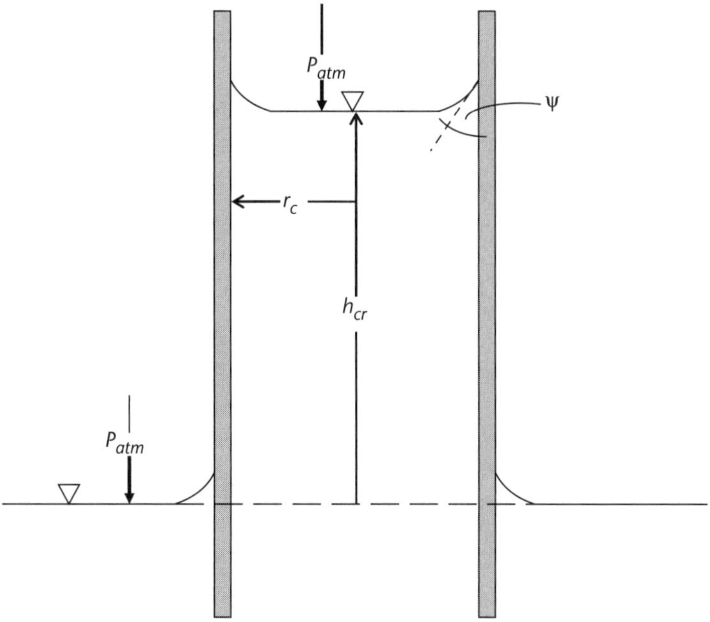

Figure 3.14 Definition sketch for computation of the height of capillary rise, h_{cr}, in a circular tube of radius r_c. Ψ is the contact angle between the meniscus and the tube wall, and P_{atm} is atmospheric pressure.

to it (called a **hydrophilic** material), the molecules in contact with the tube are drawn upward. The degree of attraction between the water and the tube is reflected in the contact angle, Ψ, between the water surface, or **meniscus**, and the tube: the stronger the attraction, the smaller the angle. Because of the intermolecular hydrogen bonds, the entire mass of water within the tube will be also drawn upward until the adhesive force between the molecules of the tube and those of the water is balanced by the downward force due to the weight of the water suspended within the tube.

The height to which the water will rise in the tube can thus be calculated by equating the upward and downward forces. The upward force, F_{st}, equals the vertical component of the surface tension times the distance over which that force acts:

$$F_{\text{st}} = \sigma \cdot \cos(\Psi) \cdot 2 \cdot \pi \cdot r_c, \tag{3.16}$$

where r_c is the radius of the tube. The downward force due to the weight of the column of water, F_g, is

$$F_g = \gamma \cdot \pi \cdot r_c^{\,2} \cdot h_{\text{cr}}, \tag{3.17}$$

where γ is the weight density of water, and h_{cr} is the height of the column. Equating F_{st} and F_g and solving for h_{cr} yields

$$h_{\text{cr}} = \frac{2 \cdot \sigma \cdot \cos(\Psi)}{\gamma \cdot r_c}. \tag{3.18}$$

Table 3.5 Surface-tension contact angles Ψ for water–air interfaces and various solids.

Solid	Contact angle, Ψ (°)	cos Ψ
Glass	0	1.0000
Most silicate minerals	0	1.0000
Ice	20	0.9397
Platinum	63	0.4540
Gold	68	0.3746
Talc	86	0.0698
Paraffin	105–110	−0.2588 to −0.3420
Shellac	107	−0.2924
Carnauba wax	107	−0.2924

Data from Dorsey (1940) and Jellinek (1972).

Thus, the height of capillary rise is inversely proportional to the radius of the tube and directly proportional to the surface tension and the cosine of the contact angle. Table 3.5 gives the contact angle for water in contact with air and selected solids; note that the value for most earth materials is close to 0° [$\cos(\Psi) = 1$]. Materials with contact angles greater than 180° are **hydrophobic** and repel rather than attract water molecules; in these materials, the meniscus curves downward.

We can construct a table showing the height of capillary rise as a function of tube radius for typical earth material for water at a temperature of 10°C. From equation 3.11, the value of ρ at 10°C is

$$\rho = 1000 - 0.019549 \times |10 - 3.98|^{1.68} = 999.60 \text{ kg m}^{-3}.$$

From equation 3.10,

$$\gamma = 9.81 \text{ m s}^{-2} \times 999.60 \text{ kg m}^{-3} = 9806.1 \text{ N m}^{-3}.$$

From equation 3.15, the value of σ at 10°C is

$$\sigma = 0.001 \times (20987 - 92.613 \times 10)^{0.4348} = 7.424 \times 10^{-2} \text{ N m}^{-1}.$$

Substituting these values into equation 3.18, assuming $\cos(\Psi) = 1$, and entering a range of values for r_c yields the values of h_{cr} shown in table 3.6.

These results show that capillary rise is significant only for tubes of very small radius. Because equation 3.18 applies also to vertical parallel plates if r_c represents the separation between the plates, we can also conclude that surface tension affects the water surface only in extremely small channels.

Other open-channel-flow situations in which surface-tension effects are appreciable include 1) the trickles of water that occur when rain collects on a window, whose approximately semicircular cross-sectional boundaries are formed by surface tension; and 2) capillary waves with wavelengths of a millimeter or so that occur near solid boundaries in open-channel flows (section 11.3.2). Although these phenomena

Table 3.6 Height of capillary rise, h_{cr}, as a function of tube diameter, r_c (equation 3.18).

r_c (mm)	1	2	5	10	20	50	100
h_{cr} (mm)	15.1	7.57	3.03	1.51	0.757	0.303	0.151

are not significant in the larger scale natural open-channel flows usually of interest to earth scientists, they may affect flows in physical models sometimes used in engineering studies.

3.3.3 Viscosity

When water flows over a solid boundary, hydrogen bonds cause the fluid molecules adjacent to the boundary to adhere to the boundary, so that the water velocity at a boundary equals the velocity of the boundary. This phenomenon, present in all natural flows, is called the **no-slip condition**.

The no-slip condition produces a frictional retarding force (**drag**) that is transmitted through the fluid for considerable distances normal to the boundary as a velocity gradient. Close to a boundary, the frictional force is transmitted into the flow by intermolecular attractions that manifest as **viscosity**.

3.3.3.1 Viscosity, Shear Stress, and Velocity Gradients

Viscosity can be understood by considering the thought experiment illustrated in figure 3.15a: The annular space of thickness Y_{ann} between a stationary cylinder and an outer movable cylinder is filled with water. The value of Y_{ann} is on the order of a few centimeters, and the annular space extends a distance normal to the page that is much greater than Y_{ann}, so that the flow is two-dimensional. The inner boundary of the outer cylinder has an area A_{cyl}, and we have some means of measuring the water velocity at arbitrary locations between the two boundaries. (Devices similar to this are used to measure the viscosities of liquids.) The system is initially at rest, and we begin the experiment by applying a tangential force F_{app} to rotate the outer boundary at a slow, steady rate. After an initial acceleration, the motion becomes steady.

If we now "zoom in" on a portion of the annular space (figure 3.15b), we can consider that the boundaries are planar, and designate the "downstream" direction as the x-direction and the direction normal to the boundary as the y-direction. The outer boundary is moving at a velocity U_x, and our velocity meters would show a linear increase in velocity, $u_x(y)$, from $u_x(0) = 0$ at the lower boundary and $u_x(Y_{ann}) = U$ at the outer boundary, due to the no-slip condition. If we repeat this experiment several times, each time with a different value of F_{app} (but keeping F_{app} and hence U_x relatively small) and plot the resulting velocity gradient, $du_x(y)/dy$, against the applied force per unit area, F_{app}/A_{cyl}, we would find a linear relation (figure 3.16). The inverse of the slope of this relation is called the **dynamic viscosity**, μ, and is due to intermolecular attractions. The flow in this experiment can be thought of as the sliding of layers (laminae) over each other, as in a stack of cards (figure 3.17), and is therefore called **laminar flow**; dynamic viscosity can be thought of as the friction between adjacent layers in laminar flow.

We can summarize these results with the relation

$$\frac{du_x(y)}{dy} = \frac{1}{\mu} \cdot \tau_{yx}, \tag{3.19a}$$

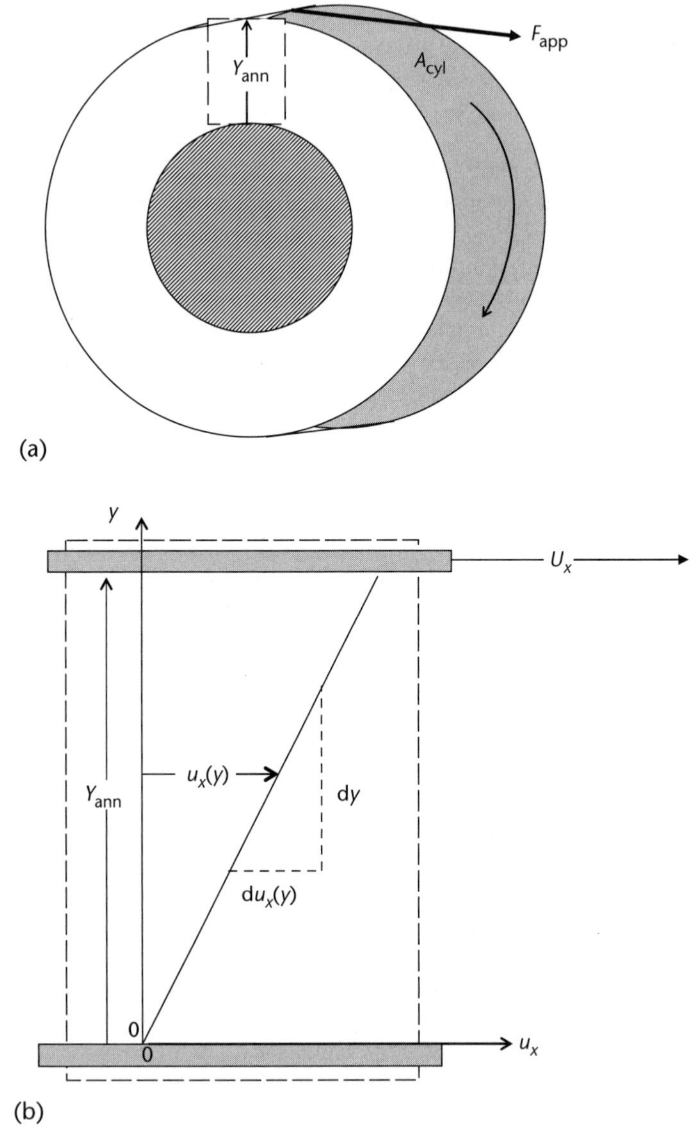

Figure 3.15 Thought experiment for viscosity. (a) The central cylinder is stationary; the outer cylinder of surface area A_{cyl} rotates when a tangential force F_{app} is applied. The cylinders are separated by a distance Y_{ann}, and the annular space is filled with water. (b) Enlarged area shown by the dashed rectangle in (a), where U_x is the velocity of the outer cylinder, $u_x(y)$ is the x-direction velocity at a distance y from the inner cylinder, and $\mathrm{d}u_x(y)/\mathrm{d}y$ is the linear velocity gradient that exists as long as Y_{ann} and U_x are not too large.

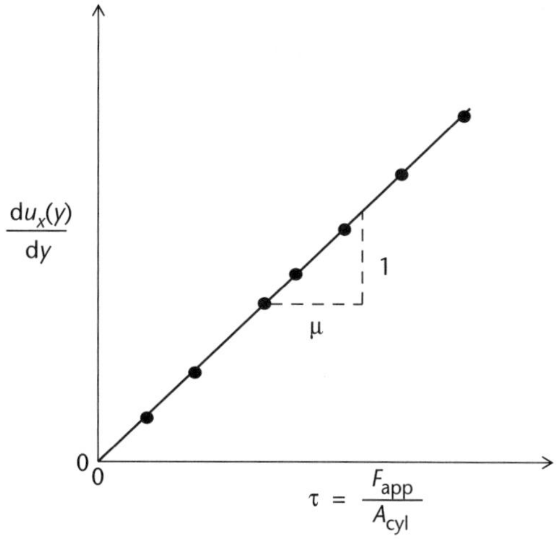

Figure 3.16 Graph of results of viscosity thought experiment (figure 3.15). As long as Y_{app} and U_x are not too large, there is a linear relation between the velocity gradient, $du_x(y)/dy$, induced by the applied shear stress, F_{app}/A_{cyl}. The slope of the relation = $1/\mu$, where μ is the dynamic (molecular) viscosity.

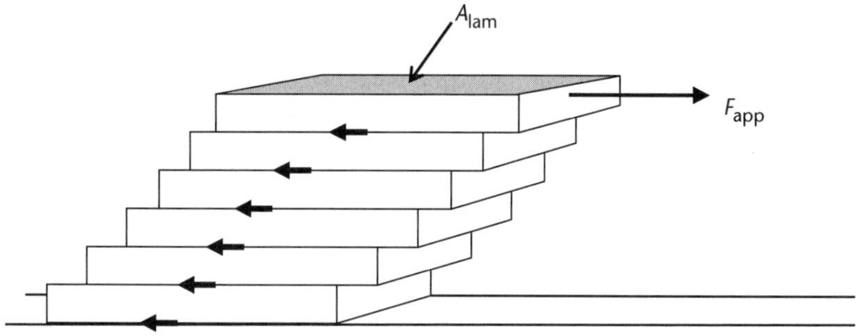

Figure 3.17 The viscous flow of figure 3.15 can be thought of as the sliding of layers (laminae) of water sliding over each other like a stack of cards; such flow is **laminar**. The dynamic viscosity μ is the friction between adjacent layers, represented by "upstream"-directed arrows.

where $\tau_{yx} = F_{app}/A_{cyl}$ (figure 3.16) or F_{app}/A_{lam} (figure 3.17). A force-per-unit-area is a **stress**, and a tangential stress such as τ_{yx} is a **shear stress**. The first subscript, y, indicates the direction normal to the stress, and the second, x, indicates the direction of the stress. Note that, since τ_{yx} has the dimensions [F L^{-2}], μ has the dimensions [F T L^{-2}] = [M L^{-1} T^{-1}].

The relation of equation 3.19a, usually written in the form

$$\tau_{yx} = \mu \cdot \frac{du_x(y)}{dy}, \tag{3.19b}$$

characterizes a **Newtonian fluid**. Water and air are Newtonian fluids, but in many substances (e.g., ice) the velocity gradient is nonlinearly related to the applied stress; and we will see in section 3.3.4 that, even for water, equation 3.19 applies only when the dimensions of the system are small and when the induced velocities remain small.

3.3.3.2 Magnitude of Dynamic Viscosity

Despite the strength of the hydrogen bonds, water's viscosity is relatively low because of the rapidity with which the hydrogen bonds break and reform (about once every 10^{-12} s). Dynamic viscosity at 0°C is

$$\begin{aligned}\mu &= 1.787 \times 10^{-3}\ \mathrm{N\ s\ m^{-2}\ (Pa\ s)} = 1.822 \times 10^{-4}\ \mathrm{kgf\ s\ m^{-2}} \\ &= 1.787 \times 10^{-2}\ \mathrm{dyn\ s\ cm^{-2}} \\ &= 3.735 \times 10^{-5}\ \mathrm{lb\ s\ ft^{-2}}.\end{aligned}$$

As shown in table 3.4, viscosity decreases rapidly as temperature increases. The temperature effect can be approximated as

$$\mu = 2.0319 \times 10^{-4} + 1.5883 \times 10^{-3} \cdot \exp\left[-\left(\frac{T^{0.9}}{22}\right)\right], \qquad (3.20)$$

where T is in °C and μ is in N s m^{-2} (Heggen 1983). Some dissolved constituents increase viscosity, whereas others decrease it, but these effects are usually negligible at the concentrations found in natural open-channel flows. However, moderate to high concentrations of suspended material can significantly increase the effective viscosity of the fluid; information about these effects is given in section 3.3.3.4.

3.3.3.3 Viscosity and Momentum Flux

The results of the thought experiment of figures 3.15 and 3.16 can be viewed in terms of momentum flux. Momentum, M, is mass times velocity [M L T^{-1}], so, assuming constant mass density, the existence of a velocity gradient implies the existence of a momentum gradient in the fluid. Analogously to the flow of heat from regions of high temperature (i.e., high concentration of heat) to those of lower temperature, there is a flow of momentum from regions of high velocity (i.e., high concentration of momentum) to regions of lower velocity.

We can show this more explicitly by noting that the dimensions of shear stress τ_{yx} [F L^{-2}] can be written as [M L^{-1} T^{-2}], which in turn is equivalent to [M L T^{-1}]/([L^2] · [T])—that is, momentum per unit area per unit time. And, just as heat flux is defined as the flow of heat energy per unit area per unit time, **momentum flux**, $\mathcal{F}_M$, is the flow of momentum per unit area per unit time. Note, however, that the direction of momentum flux is *down* the velocity gradient; thus, shear stress in the positive x-direction equals momentum flux in the negative y-direction:

$$\tau_{yx} = -\mathcal{F}_M. \qquad (3.21)$$

If we now modify equation 3.19b by multiplying and dividing by mass density and use equation 3.21, we can write

$$\mathcal{F}_{\mathrm{M}} = -\frac{\mu}{\rho} \cdot \frac{d[\rho \cdot u_x(y)]}{dy}. \tag{3.22}$$

The quantity $\rho \cdot u_x(y)$ has dimensions $[\mathrm{M\ L}^{-3}] \cdot [\mathrm{L\ T}^{-1}] = [\mathrm{M\ L\ T}^{-1}]/[\mathrm{L}^3]$ and represents the concentration of momentum (momentum per unit volume). Thus, we see that equation 3.19 also describes the momentum flux transverse to the flow and in the direction opposite to that of the velocity gradient (i.e., from regions of high velocity to regions of low velocity).

The ratio μ/ρ arises in many contexts; thus, it is convenient to define it as the **kinematic viscosity**, ν $[\mathrm{L}^2\ \mathrm{T}^{-1}]$,

$$\nu \equiv \frac{\mu}{\rho}, \tag{3.23}$$

and to write equation 3.22 as

$$\mathcal{F}_{\mathrm{M}} = -\nu \cdot \frac{d[\rho \cdot u_x(y)]}{dy}. \tag{3.24}$$

We will see in section 4.6 that equation 3.24 is Fick's law of diffusion written for momentum, and that the kinematic viscosity is the diffusivity of momentum in a viscous flow.

3.3.3.4 Magnitude of Kinematic Viscosity

Values of ν at 0°C are

$$\nu = 1.787 \times 10^{-6}\ \mathrm{m}^2\ \mathrm{s}^{-1} = 1.787 \times 10^{-2}\ \mathrm{cm}^2\ \mathrm{s}^{-1} = 1.926 \times 10^{-5}\ \mathrm{ft}^2\ \mathrm{s}^{-1}.$$

Changes of ν with temperature can be computed via equations 3.11 and 3.20. Simons et al. (1963) measured the effects of concentrations of two types of clay minerals on kinematic viscosity, and their results are summarized in figure 3.18. Clearly, the effects depend strongly on the nature of the suspended material; the suspensions of "rock flour" found in glacial streams are similar to kaolinite, and the effects of other typical clay mixtures probably lie between the two curves shown.

3.3.3.5 Summary

We can now summarize several important results from our thought experiment involving viscous flow:

- The frictional force exerted by the boundary due to the no-slip condition is transmitted into the fluid by viscosity and induces a linear velocity gradient (shear).
- For a Newtonian fluid, the velocity gradient induced by an applied shear stress is directly proportional to the stress, and as viscosity increases, a larger stress must be applied to induce a given gradient (equation 3.19a).
- Since the velocity gradient in viscous flow is linear, the shear stress (resistance) is proportional to the first power of the average velocity.

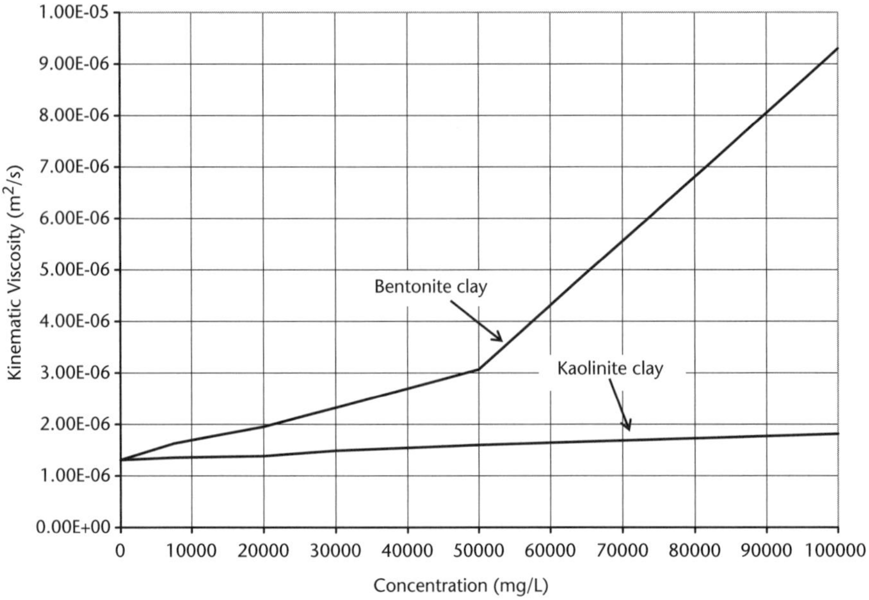

Figure 3.18 The effects of concentrations of two types of clay minerals on kinematic viscosity. Data from Simons et al. (1963).

- The viscous shear stress, τ_{yx}, is physically identical to the momentum flux perpendicular to the boundary due to viscosity.
- The relation between applied stress and shear (equation 3.19b) also describes the flux of momentum down the velocity gradient due to viscosity.
- The diffusivity of momentum due to viscosity is equal to the dynamic viscosity divided by the mass density and is called the kinematic viscosity.

3.3.4 Turbulence

If we were to expand the dimensions of the thought experiment of figure 3.15 beyond a few centimeters and/or apply a substantially larger force F_{app}, we would find that the velocity gradient $du_x(y)/dy$ is no longer linear and that the linear relationship between τ_{yx} and $du_x(y)/dy$ (figure 3.16) no longer holds. This is because, as distance from a boundary and velocity increase, the flow paths of individual water "particles" are increasingly likely to deviate from the parallel layers of laminar flow. At relatively modest distances and velocities, all semblance of parallel flow disappears, and the water moves in highly irregular eddies. This is the phenomenon of **turbulence**.

Turbulence is not a fluid property in the same sense as are density, surface tension, and molecular viscosity, because its magnitude is not directly determined by the atomic and molecular structure of water. However, it is appropriate to introduce the topic here because in most open-channel flows, turbulence, rather than molecular viscosity, is the principal means by which boundary friction is transmitted throughout the flow.

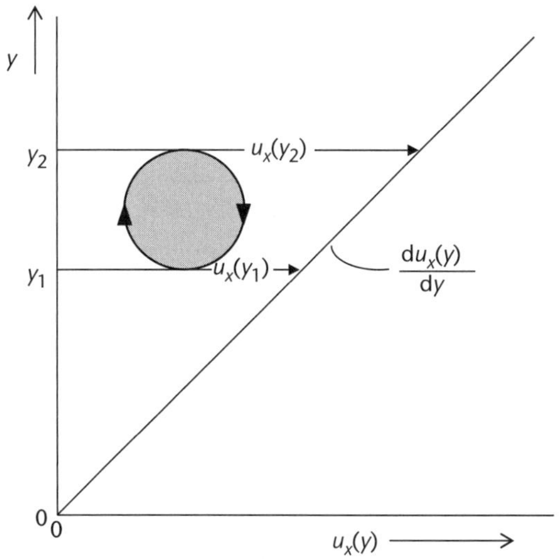

Figure 3.19 Velocity gradients, or shear, $du(y)/dy$, near a boundary tend to create quasi-circular eddies (shaded) that may be damped by viscous forces or grow and propagate through a flow as turbulence.

3.3.4.1 Qualitative Description

As the velocity of flow near a boundary increases, the no-slip condition necessitates an increase in the velocity gradient, or **shear**, normal to the boundary. As indicated in figure 3.19, this shear tends to generate quasi-circular eddies and wavelike fluctuations in flow paths. If the inertia of these fluctuations is small relative to the viscosity, the fluctuations will be damped and a laminar flow pattern reestablished. If the viscous forces are insufficient to damp the fluctuations, the induced velocity variations grow into vortices that induce additional fluctuations, and the instabilities grow and propagate through the flow as **turbulent eddies**. Individual fluid elements in such flows move in highly irregular flow paths (figures 3.20 and 3.21). figure 3.22 shows fully developed turbulence produced near flow boundaries, and figure 3.23 shows turbulent eddies in natural rivers.

Recent advances in instrumentation have revealed that the process of generating turbulence involves a quasi-repeating spatially complex pattern. In this process, known as **bursting**, rolling vortices are created by the near-boundary velocity gradients along low-velocity streaks. These vortices are ejected upward and then destroyed by sweeps of high-velocity eddies from above (Smith 1997). In rivers with large bed particles, the low-speed streaks are less conspicuous, and eddies that form on the lee side of the particles are ejected up into the flow (Bridge 2003). The bursting process repeats with a periodicity that is inversely related to the velocity gradient and ranges from a few seconds to several tens of seconds in natural rivers.

Thus, turbulence involves complex eddylike phenomena over a range of space and time scales. Based on observations on natural channels ranging from brooks to rivers

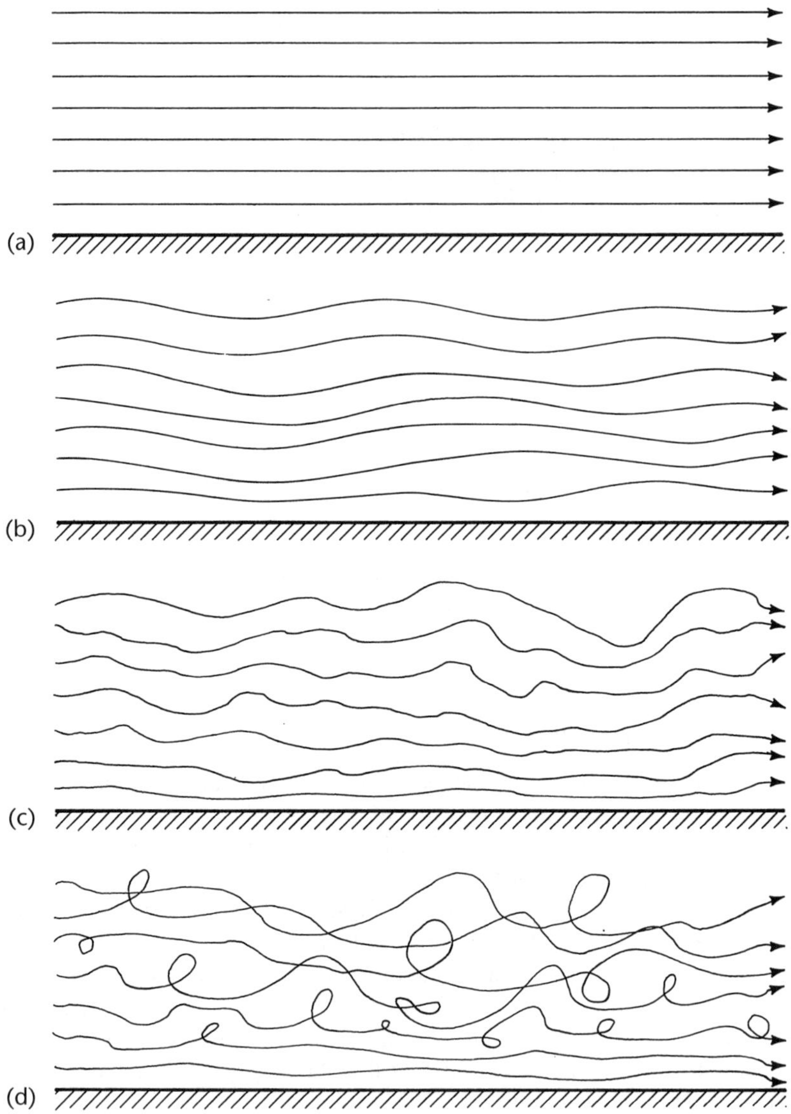

Figure 3.20 Schematic diagram showing the paths of individual fluid elements as flow changes from the laminar state in (a) to the fully turbulent state in (d). Flow in (b) and (c) is transitional.

the size of the Lower Mississippi, Matthes (1947) formulated the classification of "macroturbulence" phenomena that is summarized in box 3.1. As noted by Sundborg (1956), some of these phenomena are not true turbulence, but the classification and descriptions are very useful in conveying the spatial and temporal complexity of natural channel flows.

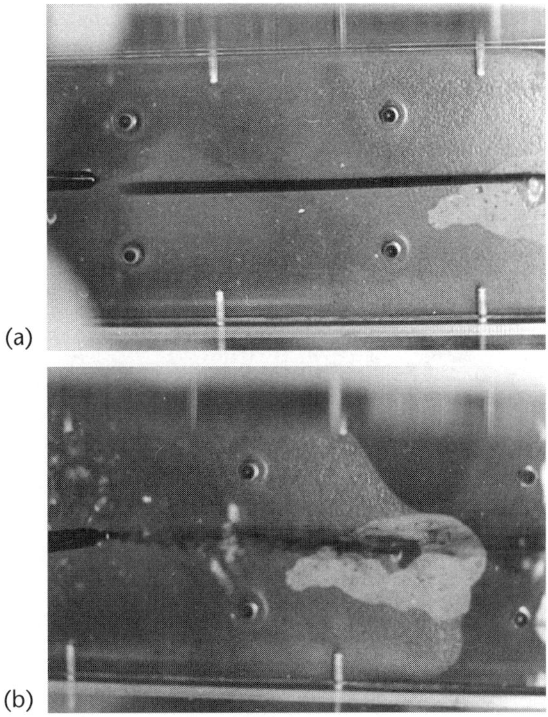

Figure 3.21 Dye injected into laboratory open-channel flows shows (a) laminar flow and (b) turbulent flow.

3.3.4.2 *Statistical Description*

The essentially random or chaotic nature of turbulence has resisted precise quantitative description and introduces an irreducible uncertainty into descriptions of river flow and sediment transport (it also limits accurate weather predictions to about 1 week). However, turbulence can be usefully characterized statistically, beginning with a thought experiment. Imagine that we could "tag" two adjacent fluid elements at an initial instant t_0. Richardson (1926) showed that the distance between these elements will increase in proportion to $(t - t_0)^{3/2}$ (figure 3.24).[4] It is this turbulent diffusion that disperses heat and dissolved and suspended sediment through a turbulent flow.

Another thought experiment leads to a statistical model of turbulence that, although crude, is a very useful approach to mathematical descriptions of turbulent flows. Consider a steady, two-dimensional turbulent flow, and superimpose a coordinate system with the x-direction downstream and the y-direction vertical. If we insert small, highly sensitive velocity sensors oriented in the x- and y-directions[5] into this flow (figure 3.25a), they will record rapid fluctuations of velocity (figure 3.25b,c). Focusing first on the downstream velocity, $u_x(t)$ (figure 3.25b), we can represent this instantaneous velocity as

$$u_x(t) = \bar{u}_x + u_x{}'(t), \tag{3.25}$$

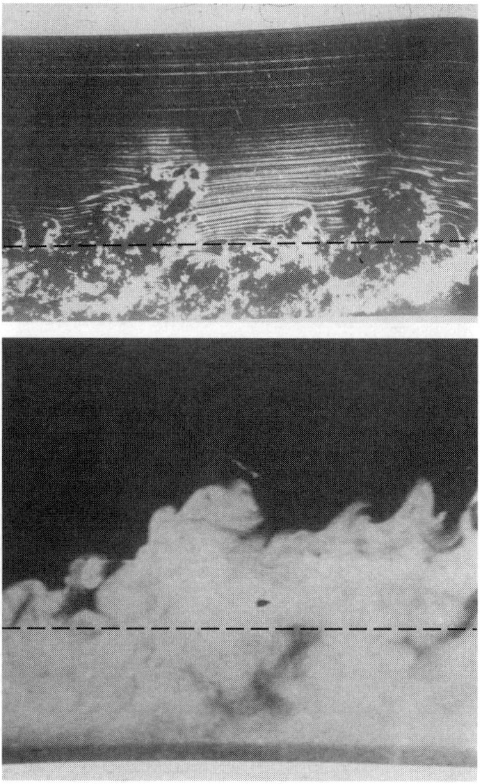

(a)

(b)

Figure 3.22 Turbulence generated by boundary friction in laboratory flows of air in wind tunnels (flow is from left to right). Turbulence in air is identical to turbulence in water, but in virtually all natural open-channel flows the turbulence extends all the way to the surface (simulated by dashed lines). (a) Turbulence made visible by smoke particles. From Van Dyke (1982). (b) Turbulence made visible by oil droplets. From Van Dyke (1982).

where $\bar{u}_x$ is the velocity averaged over a time period longer than the time scale of the velocity fluctuations, and $u_x{}'(t)$ is the deviation of the instantaneous velocity from the mean value. The value of $u_x{}'(t)$ can be positive or negative, and by definition, the time-average value of the deviations is zero, so

$$\overline{u_x{}'(t)} = 0 \tag{3.26}$$

and

$$\bar{u}_x(t) = \overline{u_x}. \tag{3.27}$$

We can similarly represent the instantaneous vertical velocity (figure 3.25c):

$$u_y(t) = \overline{u_y} + u_y{}'(t). \tag{3.28}$$

As with the downstream velocity fluctuations, $\overline{u_y{}'(t)} = 0$, but since the net flow is only in the x-direction, it is also true that $\overline{u_y(t)} = \overline{u_y} = 0$.

(a)

(b)

Figure 3.23 Turbulent eddies in natural river flows made visible at the interface between clear water and water containing suspended sediment. (a) The Yukon River in central Alaska; view upstream. A clear tributary enters on the river's right bank (left in photo). (b) A creek in southern Alaska; flow is from right to left. Note that the diameters of the largest eddies are proportional to the width of the streams.

Observations have shown that the average horizontal and vertical velocity fluctuations $\overline{u_x'(t)}$ and $\overline{u_y'(t)}$ decrease exponentially with distance from the boundary (Bridge 2003).

3.3.4.3 Eddy Viscosity

Because the water in turbulent eddies moves in directions other than the main flow direction, turbulence consumes some of the energy that would otherwise drive the main flow.

> Energy loss due to turbulence can be thought of as an addition to the internal friction of the fluid that operates exactly analogously to the molecular (dynamic) viscosity. Its effect is called the **eddy viscosity**, η.

BOX 3.1 Matthes's (1947) Classification of Macroturbulence Phenomena (from Sundborg 1956)

1. *Rhythmic and Cyclic Surges*

   - **Velocity pulsations:** ubiquitous; affect near-bottom velocities more than surface velocities
   - **Water-surface fluctuations:** periodic rise and fall of surface; more pronounced when flows are increasing
   - **Surge phenomena:** regular large-scale fluctuations in water-surface elevation; occur at local abrupt changes in flow direction, accompanied by eddying currents and sometimes reversals in flow direction

2. *Continuous Rotary Features*

   - **Slow bank eddies or rollers with quasi-vertical axes:** occur where channel has excessive width (side-channel bays or pockets); collect floating debris and deposit sediment
   - **Fast bank eddies or rollers with quasi-vertical axes (suction eddies):** occur at upstream and downstream ends of bridge abutments, bank-protection works, and projecting ledges; sites of concentrated erosion
   - **Slow bank rollers with quasi-horizontal axes:** occur during low flows where channel has excessive depth; promote deposition
   - **Fast bank rollers with quasi-horizontal axes:** occur at high stages downstream of natural bed sills or low obstructions; cause erosion and deepening

3. *Intermittent Upward Vortex Action*

   - **Nonrotating surface boils:** short-lived local upward displacements often carrying finer grained sediment; occur along main-current axis during increasing flows
   - **Vertical-axis vortices:** strong vortex action at stream bed; loses rotary motion while rising to surface, producing non-rotating boils; occurs at upstream or downstream edges of pronounced bottom obstructions; repeats at intervals; may carry sediment

4. *Sustained Downward Vortex Action*

Vortices with downward-trending axes inclined downstream occur during high-velocity flood flows. They are sustained but subject to interruption by temporary changes in current direction.

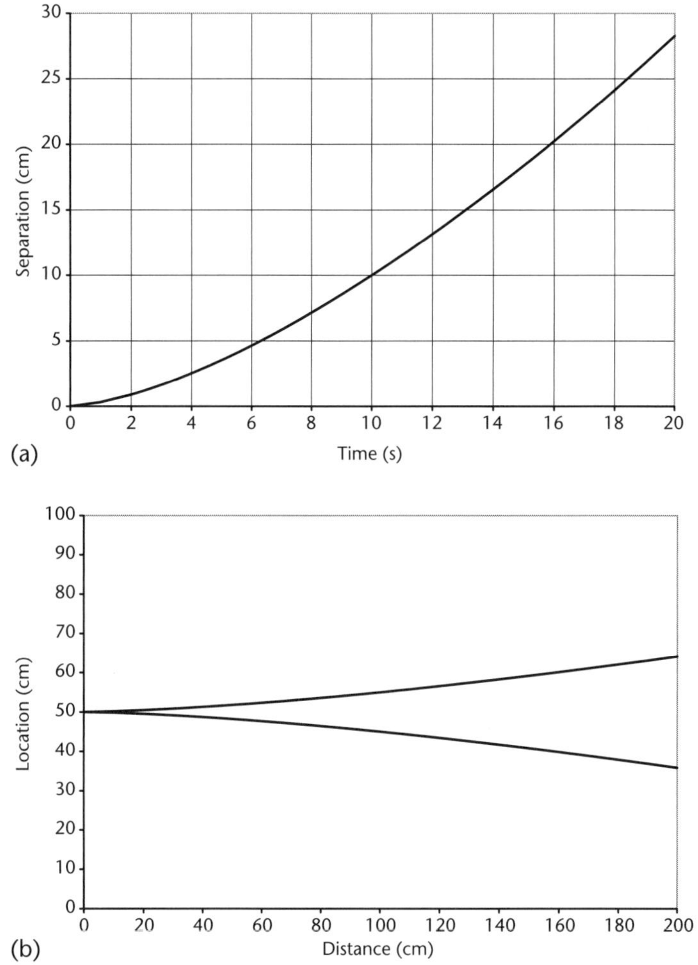

Figure 3.24 Richardson's (1926) 3/2-power law of turbulent diffusion proposes that the average separation between fluid particles increases in proportion to the 3/2 power of time. (a) Graph showing this relation, where the proportionality constant is arbitrarily set to 0.01. (b) Separation (horizontal or vertical) of two initially ($t = 0$) adjacent fluid elements as a function of distance in a flow with a uniform velocity of 10 cm/s.

Physically, the effect of molecular viscosity is always present and is the ultimate mechanism by which the retarding effect of a boundary is transmitted into the fluid. Thus, the flow resistances due to eddy viscosity and molecular viscosity are additive, and the general relation between total applied shear stress, τ_{yx}, and velocity gradient can be represented as

$$\tau_{yx} = \tau_{Vyx} + \tau_{Tyx} = \mu \cdot \frac{du_x(y)}{dy} + \eta \cdot \frac{du_x(y)}{dy} = (\mu + \eta) \cdot \frac{du_x(y)}{dy}, \tag{3.29}$$

where we now designate the viscous shear stress as τ_{Vyx}, and τ_{Tyx} is the shear stress due to turbulence.

Although eddy viscosity has the same dimensions as molecular viscosity, $[\mathrm{M\,L^{-1}\,T^{-1}}]$ or $[\mathrm{F\,T\,L^{-2}}]$, it depends not on the molecular structure of water, but on the characteristics of the flow, and varies from place to place in a given flow. In this section, we develop the relation between η and flow characteristics based on the statistical description of turbulent eddies developed in section 3.3.4.2.

3.3.4.4 Prandtl's Mixing-Length Hypothesis

Prandtl (1925) conceived a major breakthrough in quantifying the relation between turbulence and velocity gradient by introducing the concept of **mixing length,** l [L]. This quantity, which varies with location in a flow, can be thought of as "the average distance a small fluid mass will travel before it loses its increment of momentum to the region into which it comes" (Rouse 1938, p. 186) and can be taken to represent the average diameter of turbulent eddies (figure 3.25a).

Figure 3.26 shows a region of a two-dimensional steady turbulent flow with average vertical velocity gradient $d\overline{u_x(y)}/dy$, where y is distance from the flow boundary. Prandtl reasoned that a fluid element beginning at elevation y_1 and moving the distance l to y_2 before changing its momentum would cause a velocity fluctuation $u_x'(t)$ at y_2 with a magnitude proportional to the difference in average velocities at y_2 and y_1:

$$\overline{u_x'(t)} = l \cdot \frac{d\bar{u}_x}{dy}. \tag{3.30}$$

By this reasoning, a fluid element moving upward from y_1 will have a positive vertical velocity fluctuation $[u_y'(t) > 0]$ but, on arriving at y_2, will have a downstream velocity lower than the average there. This will therefore produce a negative fluctuation in the downstream velocity; that is, $u_x'(t) < 0$. Conversely, a fluid element moving downward a distance l to y_2 will have $u_y'(t) < 0$ and produce $u_x'(t) > 0$. Thus, Prandtl concluded that 1) vertical and horizontal velocity fluctuations are negatively correlated [i.e., a positive $u_y'(t)$ is associated with a negative $u_x'(t)$, and vice versa], and 2) considering that the mass of fluid at each level must be conserved, the magnitudes of co-occurring horizontal and vertical fluctuations are of similar magnitude. Subsequent studies indicate that

$$|\overline{u_y'(t)}| = k_{yx} \cdot |\overline{u_x'(t)}|, \tag{3.31}$$

where $k_{yx} \approx 0.55$ (Bridge 2003).

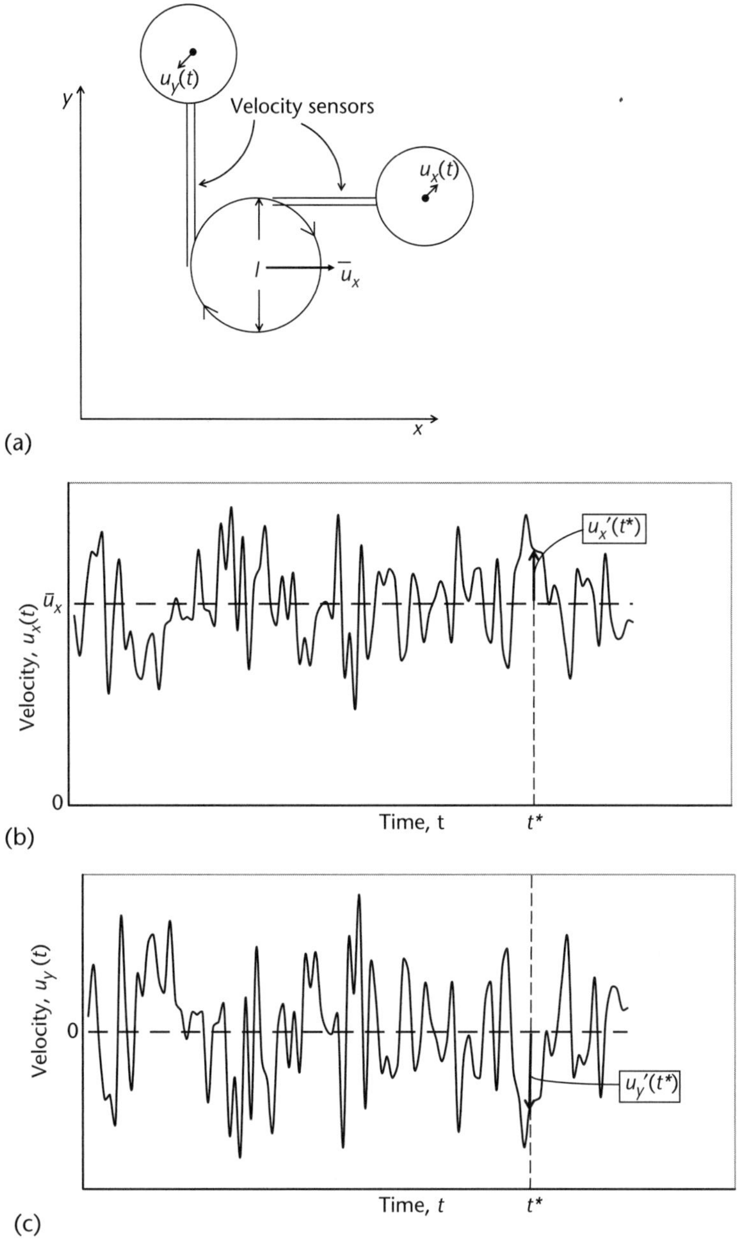

Figure 3.25 (a) Schematic diagram of a turbulent eddy with diameter l showing sensors for measuring and recording instantaneous velocities in the x- and y-directions, $u_x(t)$ and $u_y(t)$ respectively. $\bar{u}_x$ is the time-averaged velocity in the x-direction. (b) Hypothetical recording of horizontal-velocity fluctuations in a turbulent flow from experiment of (a); dashed horizontal line is time-averaged velocity $\bar{u}_x$ (>0); $u_x\prime(t^*)$ is horizontal-velocity fluctuation at arbitrary time t^*. (c) Hypothetical recording of vertical velocity u_y; horizontal dashed line is time-averaged velocity $\bar{u}_y$ ($= 0$); $u_y\prime(t^*)$ is vertical-velocity fluctuation at arbitrary time t^*.

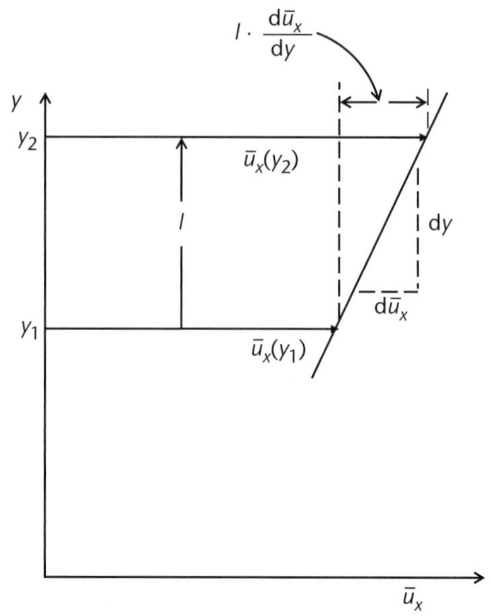

Figure 3.26 Diagram illustrating Prandtl's mixing-length hypothesis. See text for explanation.

These concepts can now be applied to show how turbulence affects momentum flux and produces an eddy viscosity. In figure 3.26, a fluid element moving from level y_1 to level y_2 transports an average increment of momentum (per unit volume) equal to $-\rho \cdot \overline{u_x'(t)}$ to y_2. The average rate of vertical movement (flux) of momentum involved in that motion is then $-\rho \cdot \overline{u_x'(t)} \cdot \overline{u_y'(t)}$. As in viscous flow, this flux has dimensions [M L^{-1} T^{-2}] or [F L^{-2}], the same as shear stress; thus, we can write the time-averaged shear stress due to turbulence, τ_{Tyx}, as

$$\tau_{Tyx} = -\rho \cdot \overline{u_x'(t)} \cdot \overline{u_y'(t)} = -\rho \cdot k \cdot \overline{u_x'(t)} \cdot |\overline{u_x'(t)}|. \tag{3.32}$$

This shear stress or momentum flux acting perpendicularly to the downstream flow direction has the same physical effect as viscous shear (equation 3.19) and represents a **frictional resistance** to the flow.

We can now combine equations 3.30 and 3.32 to write

$$\tau_{Tyx} = -\rho \cdot \overline{u_x'(t)} \cdot \overline{u_y'(t)} = \rho \cdot |\overline{u_y'(t)}| \cdot l \cdot \frac{d\bar{u}_x}{dy}. \tag{3.33}$$

Finally, making use of equations 3.31 and 3.30, we can write equation 3.33 as

$$\tau_{Tyx} = \rho \cdot l^2 \cdot \left|\frac{d\bar{u}_x}{dy}\right| \cdot \frac{d\bar{u}_x}{dy}, \tag{3.34}$$

where the constant k_{yx} has been absorbed into the definition of l.

Comparing equations 3.34 and 3.29, we see that

$$\eta = \rho \cdot l^2 \cdot \left| \frac{d\bar{u}_x}{dy} \right|, \tag{3.35}$$

and that the dimensions of η are [M L^{-1} T^{-1}], the same as for μ. We can also define a **kinematic eddy viscosity**, ε, with dimensions [L^2 T^{-1}], analogous to the kinematic viscosity ν (equation 3.23):

$$\varepsilon \equiv \frac{\eta}{\rho} = l^2 \cdot \left| \frac{d\bar{u}_x}{dy} \right|. \tag{3.36}$$

Thus, Prandtl's reasoning shows that the eddy viscosity depends essentially on two flow properties, the mixing length and the velocity gradient. We conclude this section by exploring how mixing length varies in such a flow, and we will use the relationships developed here to describe velocity gradients in turbulent flows in chapter 6.

Prandtl (1925) developed the relationship between mixing length and distance from a boundary by reasoning that the average eddy diameter (mixing length) must equal 0 at a fluid boundary and would increase in proportion to distance from the boundary:

$$l = \kappa \cdot y, \tag{3.37}$$

where κ is the proportionality constant, known as the **von Kármán constant**.[6] This seems logical, and experimental results for flows in pipes confirm this proportionality, with $\kappa \approx 0.4$ near the boundary (Schlichting 1979). This reasoning, though, breaks down when applied to open-channel flows, because it predicts that the largest eddies would be at the surface—that is, the surface of a river would be "boiling" with vertical eddies. It is more reasonable to assume that $l = 0$ at a water surface as well as at a solid boundary; thus, Henderson (1966) suggested an alternative model:

$$l = \kappa \cdot y \cdot \left(1 - \frac{y}{Y}\right)^{1/2}, \tag{3.38}$$

where Y is the total flow depth (i.e., $y = Y$ at the surface). This formulation is nearly identical to equation 3.37 for small y/Y, goes to 0 at $y = Y$ as well as $y = 0$ (figure 3.27), and is consistent with observed velocity distributions and other relations discussed later in this text. Thus, even though equation 3.38 is developed from purely conceptual reasoning rather than basic physics,[7] we will consider that it satisfactorily describes how mixing length depends on location in an essentially two-dimensional open-channel flow. Combining equations 3.36 and 3.38,

$$\eta = \rho \cdot \kappa^2 \cdot y^2 \cdot \left(1 - \frac{y}{Y}\right) \cdot \left| \frac{d\bar{u}_x}{dy} \right|, \tag{3.39}$$

we can write the relation between shear stress and velocity gradient for turbulent flow as

$$\tau_{Txy} = \rho \cdot \kappa^2 \cdot y^2 \cdot \left(1 - \frac{y}{Y}\right) \cdot \left| \frac{d\bar{u}_x}{dy} \right| \cdot \left(\frac{d\bar{u}_x}{dy} \right). \tag{3.40a}$$

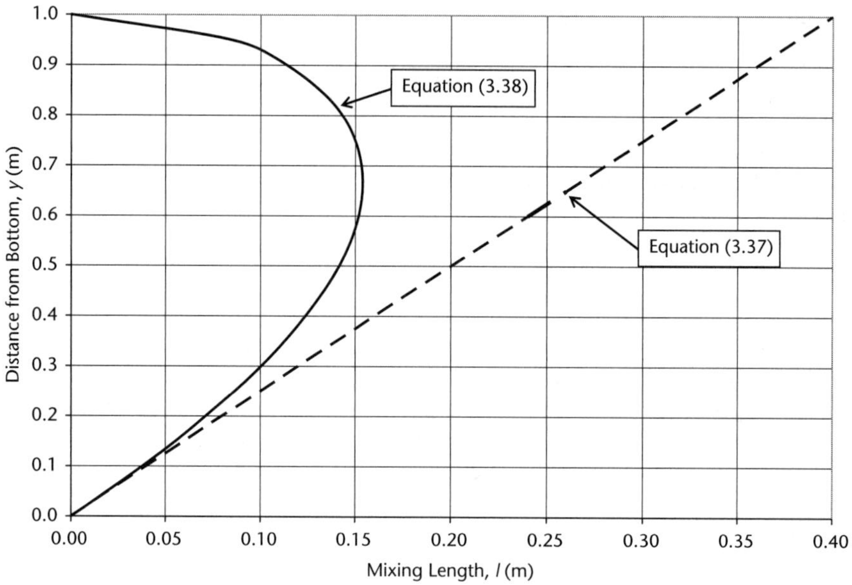

Figure 3.27 Mixing length, l, as a function of distance from the bottom. The linear relation of equation 3.37 is Prandtl's (1925) original hypothesis; equation 3.38 was suggested by Henderson (1961) and is more physically plausible. Flow depth arbitrarily chosen as $Y = 1$ m.

And, with equation 3.39, we can write an expression for turbulent flow that is exactly analogous to the basic relation for viscous flow of a Newtonian fluid (equation 3.19b):

$$\tau_{Txy} = \eta \cdot \left(\frac{d\bar{u}_x}{dy} \right) = \rho \cdot \varepsilon \cdot \left(\frac{d\bar{u}_x}{dy} \right). \quad (3.40b)$$

(Note that the dimensions of η are the same as those of dynamic viscosity.)

3.3.4.5 Summary

We can now summarize several important results concerning turbulent flow:

- The frictional force (resistance) exerted by the boundary due to the no-slip condition is transmitted into the fluid by viscosity and turbulence (equation 3.29) and induces a vertical velocity gradient (shear).
- The frictional resistance due to turbulence can be represented by the eddy viscosity, analogous to the dynamic viscosity.
- The eddy viscosity is not a fluid property, but depends on the location in the flow (distance from the boundary) and the local velocity gradient (equation 3.39).
- In turbulent flow, the velocity gradient induced by an applied shear stress is not linearly related to the stress.
- Since we can reason that vertical and horizontal velocity fluctuations will be proportional to the average velocity at any level, one important implication

of equations 3.32 and 3.40 is that resistance due to turbulence increases approximately as the square of the average velocity.

- Analogously to viscous shear stress, the turbulent shear stress, τ_{Tyx}, is physically identical to the momentum flux due to turbulence.
- The diffusivity of momentum due to turbulence is equal to the eddy viscosity divided by the mass density and is called the kinematic eddy viscosity.
- The relation between applied stress and shear (equation 3.40) also describes the flux of momentum down the velocity gradient due to turbulence.

3.4 Flow States, Boundary Layers, and the Reynolds Number

3.4.1 Flow States and Boundary Layers

Sections 3.3.3 and 3.3.4 have developed a basic understanding of two **flow states**, laminar (or viscous) and turbulent, with very different characteristics. In a final thought experiment, this section examines how laminar and turbulent flows develop in open-channel flows and develops a criterion for determining whether an open-channel flow is laminar or turbulent.

Consider the flow shown in figure 3.28. We focus only on the flow to the left of and above the boundary, and again orient the x-direction downstream along the boundary and the y-direction extending vertically from the boundary. At the left side of the diagram there is no solid boundary influencing the flow, so the velocity is equal everywhere at the value U_0, called the **free-stream velocity**.[8] The absence of a velocity gradient means that neither viscous nor turbulent shear stress is acting on the flow in this region (equation 3.29).

When the flow encounters the horizontal boundary, the no-slip condition induces a zero velocity adjacent to the boundary, and the effects of this retardation are transmitted into the flow by the dynamic viscosity. The vertical zone affected by the retardation is called the **boundary layer**. The top of this zone cannot be precisely located, so the **boundary layer thickness**, δ_{BL}, is defined as the distance above the boundary at which the velocity $u(y) = 0.99 \cdot U_0$ (i.e., $u(\delta_{BL}) = 0.99 \cdot U_0$). At the left edge of the boundary, the flow in the boundary layer is laminar, and the height δ_{BL} increases downstream in proportion to the square root of the downstream distance.

At a distance $x = X_1$ along the boundary, wavelike fluctuations develop in the formerly parallel laminae (figure 3.20b). (The location of X_1 would move upstream as U_0 increases, and downstream as viscosity increases.) These fluctuations increase rapidly downstream of X_1 and soon develop into turbulent eddies. The region occupied by these eddies grows vertically upward and downward; the upper boundary grows proportionally to the 0.8 power of distance from X_1 until it intersects the surface, while the lower region of laminar flow is increasingly suppressed. Downstream of the point $x = X_2$, a velocity gradient induced by turbulence extends throughout the flow except for a very thin layer of laminar flow adjacent to the boundary.

Flows in which the retarding effects of the boundary are present are called **boundary-layer flows**. To the left of X_1 in figure 3.28, δ_{BL} is the upper margin of a laminar boundary layer; to the right of X_2, a turbulent boundary layer extends

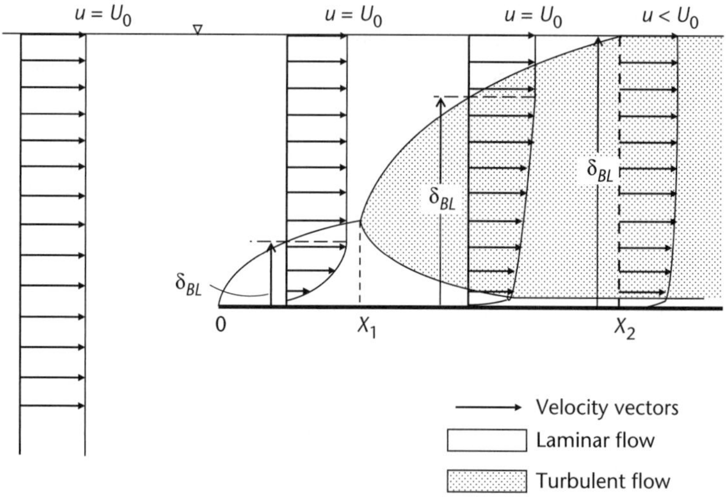

Figure 3.28 Growth of boundary layer thickness δ_{BL}. At the far left the flow is unaffected by a boundary and the velocity equals the free-stream velocity U_0 throughout. Where the flow encounters the boundary, friction retards the flow (velocity equals zero at the boundary) and frictional drag is transmitted into the flow, initially by molecular viscosity. Turbulence arises at distance X_1, and a turbulent boundary layer develops between X_1 and X_2. Downstream of X_2 the turbulent boundary layer is fully developed, and turbulence is present throughout the flow except for the very thin laminar sublayer adjacent to the boundary. Virtually all river flows are fully developed turbulent boundary-layer flows.

to the surface. (The region between X_1 and X_2 is a transitional zone.) Note that, because the velocity goes to zero at a smooth boundary, a **viscous sublayer** must always be present beneath a turbulent boundary layer. Thus, the effect of dynamic (molecular) viscosity is present in all flows, and it is the ultimate mechanism by which the retarding effect of a boundary is transmitted into the flow.

We will explore the velocity distributions in laminar and turbulent boundary layers and the thickness of the viscous sublayer in chapter 5; for now, note that virtually all open-channel flows of interest to hydrologists and engineers are turbulent boundary-layer flows.

3.4.2 The Reynolds Number

The criterion for determining whether a given open-channel flow is laminar or turbulent can be developed by writing the dimensionless ratio of eddy viscosity (equation 3.39) to dynamic viscosity,

$$\frac{\eta}{\mu} = \frac{\kappa^2 \cdot \rho \cdot y^2 \cdot \left(1 - \frac{y}{Y}\right) \cdot \left(\frac{du}{dy}\right)}{\mu}, \tag{3.41}$$

and reasoning that the larger this ratio, the more likely a flow is to be turbulent.

However, equation 3.41 is not useful for overall characterization of a flow, because η varies with location in the flow. To convert it to a form useful for categorizing entire flows, we can replace y with its "average" value, $Y/2$, and reason that the ratio U/Y, where U is the average flow velocity, characterizes the overall velocity gradient. With these substitutions, equation 3.41 becomes

$$\frac{\eta}{\mu} \approx \left(\frac{\kappa^2}{8}\right) \cdot \left(\frac{\rho \cdot Y \cdot U}{\mu}\right). \tag{3.42}$$

Finally, we absorb the proportionality constants into the definition of the **Reynolds number** for open-channel flows, Re:

$$Re \equiv \frac{\rho \cdot Y \cdot U}{\mu} = \frac{Y \cdot U}{\nu}. \tag{3.43}$$

The Reynolds number is named for Osborne Reynolds (1842–1912), an English hydraulician who first recognized the importance of this dimensionless ratio in determining the flow state. Reynolds found by experiment that when $Re < 500$, disturbances to the flow induced by vibration or obstructions (as in figure 3.20b,c) are damped out by viscous friction, and the flow reverts to the laminar state (figure 3.20a). When $Re > 2{,}000$, the inertia of water particles subject to even very small disturbance is sufficient to overcome the viscous damping, and the flow is almost always turbulent (figure 3.20d). When $500 < Re < 2{,}000$, small disturbances may persist, grow into full turbulence, or subside, depending on the frequency, amplitude, and persistence of the disturbance; the state of flows in this range is **transitional**.

As we will see in section 4.8.2.2, the Reynolds number also arises from dimensional analysis of open-channel flows. In fact, Reynolds numbers arise in analyses of many different types of flows and always have the form

$$Re \equiv \frac{\rho \cdot \mathcal{L} \cdot \mathcal{U}}{\mu} = \frac{\mathcal{L} \cdot \mathcal{U}}{\nu}, \tag{3.44}$$

where $\mathcal{L}$ is a "characteristic length" and $\mathcal{U}$ is a "characteristic velocity" that are defined differently in different flow situations (e.g., flow in pipes, settling of sediment particles, groundwater flows). Note that the Reynolds number defined in equation 3.43 is specifically applicable to open-channel flows, as are the numerical values that delimit the three flow states.

We can construct a graph showing the combinations of values of average depth, Y, and average velocity, U, that delimit flows in the laminar, transitional, and turbulent state. Assuming a water temperature of 10°C, we find from equation 3.11, that the value of ρ at 10°C is

$$\rho = 1000 - 0.019549 \cdot |10 - 3.98|^{1.68} = 999.60 \text{ kg m}^{-3}.$$

From equation 3.20, the value of μ at 10°C is

$$\mu = 2.0319 \times 10^{-4} + 1.5883 \times 10^{-3} \cdot \exp\left[-\frac{10^{0.9}}{22}\right] = 1.31 \times 10^{-3} \text{ N s m}^{-2}.$$

From equation 3.23, the value of ν at 10°C is therefore

$$\nu = \frac{1.31 \times 10^{-3} \text{ N s m}^{-2}}{999.60 \text{ kg m}^{-3}} = 1.31 \times 10^{-6} \text{ m}^2 \text{ s}^{-1}.$$

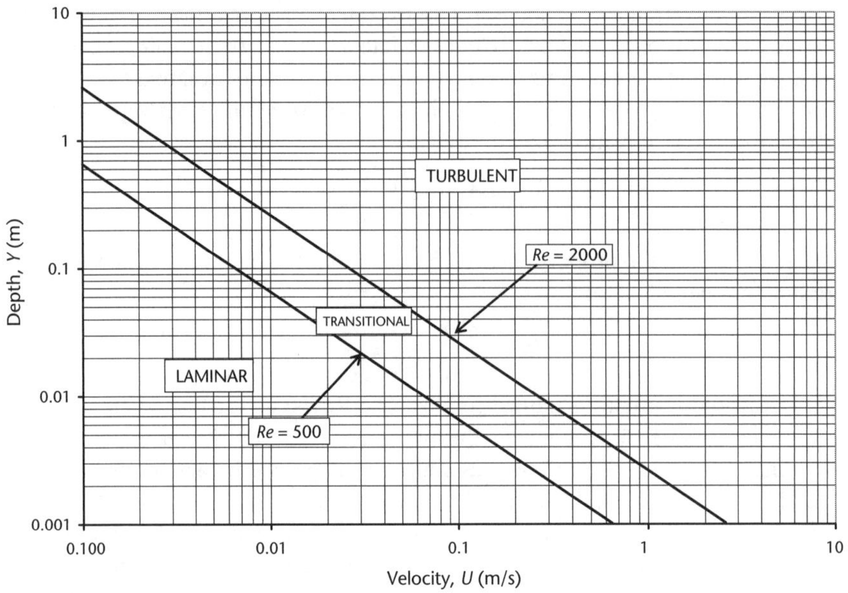

Figure 3.29 Laminar, transitional, and turbulent flow states as a function of flow depth, Y, and average velocity, U.

To find the boundary between laminar and transitional states, we can use this value of ν and solve equation 3.43 for Y (m) with $Re = 500$,

$$Y = \frac{500 \cdot (1.31 \times 10^{-6}\ \mathrm{m^2\ s^{-1}})}{U\ \mathrm{m\ s^{-1}}},$$

and substitute a range of values of U. To find the boundary between transitional and turbulent states, we repeat the calculations with $Re = 2{,}000$:

$$Y = \frac{2000 \cdot (1.31 \times 10^{-6}\ \mathrm{m^2\ s^{-1}})}{U\ \mathrm{m\ s^{-1}}} \tag{3.45}$$

The results are plotted in figure 3.29.

To summarize, the Reynolds number reflects the ratio of turbulent resistance to laminar resistance in a flow and therefore provides a fundamental characterization of a flow. And finally, it's clear from figure 3.29 that open-channel flows of even modest depths and velocities are turbulent.

4

Basic Concepts and Equations

4.0 Introduction and Overview

Chapter 2 developed an appreciation of the qualitative nature of natural rivers and river flows; the variables that characterize channels, flows, and sediment; and some of the quantitative relations among these variables. Chapter 3 described the properties of water that determine how it responds to forces acting on it. To complete the presentation of the foundations of the study of open-channel flows, this chapter focuses on the physical and mathematical concepts that underlie the basic equations relating fluid properties and hydraulic variables, with the objective of providing a deeper understanding of the origins, implications, and applicability of those equations. Most of these equations are based directly on the laws of classical (Newtonian) mechanics; however it is often useful or necessary to make use of equations that are not derived directly from basic physical laws, and these are introduced in the last section of the chapter.

The complete quantitative characterization of the behavior of natural rivers remains an elusive goal, largely due to 1) the infinite small-scale variability of the geological and biological environment, 2) the complications imposed by local climatic and geological history, and 3) the difficulty of completely describing turbulence. However, continuing improvements in instrumentation and computing power are making it possible for geomorphologists and hydrologists to move ever closer to that goal.

4.1 Basic Mathematical Concepts

The basic relations of open-channel flow and sediment transport are derived from the fundamental laws of classical physics, particularly the following:

Conservation of mass: Mass is neither created nor destroyed.

Newton's laws of motion: 1) The momentum of a body remains constant unless a net force acts upon the body (conservation of momentum). 2) The rate of change of momentum of a body is proportional to the net force acting on the body, and is in the same direction as the net force. (Force equals mass times acceleration.) 3) For every net force acting on a body, there is a corresponding force of the same magnitude exerted by the body in the opposite direction.

Laws of thermodynamics: 1) Energy is neither created nor destroyed (conservation of energy). 2) No process is possible in which the sole result is the absorption of heat and its complete conversion into work.

Fick's law of diffusion: A diffusing substance moves from where its concentration is larger to where its concentration is smaller at a rate that is proportional to the spatial gradient of concentration.

Equations based on these relations are developed by first stating the appropriate fundamental law(s) in mathematical form, incorporating the boundary and (if required) initial conditions appropriate to the situation, and then applying the principles of algebra and calculus. These mathematical formulations require two assumptions that are not physically realistic, but that fortunately lead to physically sound results: 1) the fluid continuum, and 2) the fluid element. Formal mathematical developments also require the specification of a formal system of spatial coordinates (usually the three mutually perpendicular Cartesian coordinates), and may also involve time as an additional dimension. These concepts are presented here.

4.1.1 Fluid Continuum

The techniques of calculus—taking derivatives and integrals—are essential tools for expressing basic physical principles in mathematical form. Underlying the application of these techniques to problems of fluid flow is the concept of the **fluid continuum**: To apply the mathematical concept of "taking limits," which underlies the definitions of derivatives and integrals, we must imagine that the bulk properties (density, pressure, viscosity, velocity, etc.) exist even as we consider infinitesimally small regions of the fluid. In reality, of course, fluids are made of discrete molecules, and the bulk properties are not defined at the molecular scale. Fortunately, the fiction of the fluid continuum serves us well for the purposes of earth sciences and engineering.

4.1.2 Fluid Element

Fluids are also continua in the sense that, in contrast to solids, there are no physical boundaries separating the elements of a flow. Thus, another useful fiction commonly invoked in analyzing fluid-flow situations is that of the **fluid element** or **fluid particle**: "Any fluid may be imagined to consist of innumerable small but finite particles, each having a volume so slight as to be negligible when compared with the total volume of the fluid, yet sufficiently large to be considered homogeneous in constitution" (Rouse 1938, p. 35). Each particle at any instant of time has its own particular velocity and other properties, which generally vary as it travels from point to point.

4.1.3 Coordinate Systems

Precise mathematical descriptions of objects in space require specification of a coordinate system. The two coordinate systems used in this text are illustrated in figure 4.1. We use the standard orthogonal Cartesian x-, y-, z-coordinate system when focusing on fluid elements and other phenomena at the "microscopic" scale (figure 4.1a). We will often restrict our interest to two dimensions, with the z-axis oriented vertically and the x-axis directed in the "downstream" direction.

When examining flows in channels at the more macroscopic scale, we will usually use a two-dimensional coordinate system, replacing the (x, y, z) coordinate directions with (X, y, z). We maintain the z-axis vertical and the X-axis downstream, but because the channel bottom will generally be sloping at an angle θ_0 (measured positive downward from the horizontal), the X-axis will make an angle of $\pi/4 + \theta_0$ $(90° + \theta_0)$ with the z-axis (figure 4.1b). The y-axis is oriented normal to the X-axis with $y = 0$ at the channel bottom, so distances in the y-direction are distances above the bottom. Distances measured along the y-axis are related to those measured along the z-axis as

$$y = (z - z_0) \cdot \cos \theta_0, \tag{4.1}$$

where z_0 is the elevation of the channel bottom above an arbitrary elevation datum. In a few instances, we define a "depth" (i.e., distance below the surface) direction as $h \equiv Y - y$, where Y is the height of the surface above the bottom.

For two-dimensional mathematical representations of channel cross sections (figure 4.1c), we use w for the cross-channel direction, generally taking $w = 0$ at the channel center. The vertical direction is represented by z.

In this text, we will assume that coordinate systems are fixed relative to points on the earth's surface, and that those points are stationary. In reality, points on the earth are moving through space and, more significantly, rotating due to the earth's rotation around its axis. This rotation gives rise to the Coriolis effect, which introduces accelerations to objects moving with respect to a fixed coordinate system. These accelerations increase from zero at the equator to a maximum at the poles. However, as we will show in chapter 7, the Coriolis effect becomes significant only for very large-scale flows such as ocean currents, and it is safe to ignore the effect at the scale of river flows.

Accelerations are also induced due to momentum when fluid elements follow curved paths in a fixed coordinate system. These accelerations are usually treated as centrifugal force and can be important in river flows, as discussed in chapters 6 and 7.

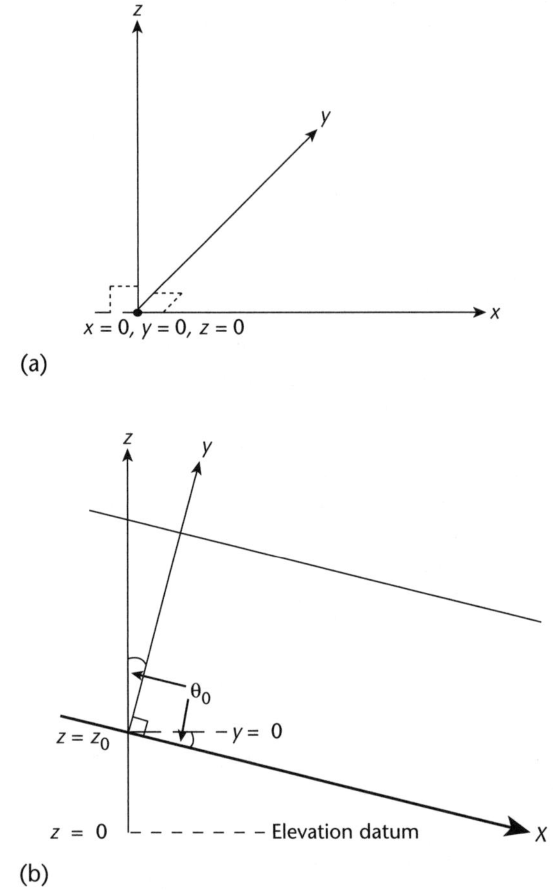

Figure 4.1 Coordinate systems used in this book. (a) The standard Cartesian coordinate system with x-, y-, z-axes orthogonal. The z-axis is usually oriented vertically, and the x-axis is usually directed in the principal flow direction (downstream). (b) The coordinate system used for two-dimensional flow macroscopic flow descriptions. The z-axis is oriented vertically with its 0-point the elevation of an arbitrary datum. The X-axis is directed in the principal flow direction (downstream). The y-axis represents distance above the bottom. It is oriented normal to the X-axis and makes an angle θ_0 with the z-axis; $y = 0$ at the channel bottom. (c) For channel cross sections, w represents the horizontal cross-channel direction, with $w = 0$ usually at the channel center. The z-axis is oriented vertically with its 0-point usually at the elevation of the deepest point of the channel.

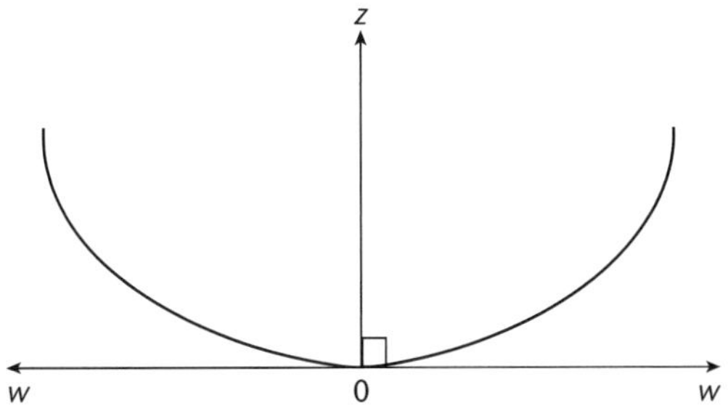

(c)

Figure 4.1 *Continued*

4.1.4 The Lagrangian and Eulerian Viewpoints

Problems of fluid flow can be analyzed in two formal viewpoints: In the **Lagrangian**[1] viewpoint, we follow the path of a fluid particle as it moves through space. In the Lagrangian approach the location of an individual fluid element is a function of time. Thus, for an element that is at location x_0, y_0, z_0 at time t_0, its subsequent locations are functions of its original location and time, t:

$$x = f_1(x_0, y_0, z_0, t), \quad y = f_2(x_0, y_0, z_0, t), \quad z = f_3(x_0, y_0, z_0, t). \tag{4.2}$$

In the **Eulerian** viewpoint, we observe the behavior of fluid elements as they pass fixed points. Thus, in the Eulerian approach the fluid properties are functions of fixed location coordinates and time:

$$q_x = f_1(x, y, z, t), \quad q_y = f_2(x, y, z, t), \quad q_z = f_3(x, y, z, t), \tag{4.3}$$

where q_x, q_y, q_z represent fluid properties (e.g., velocity, acceleration, density) that may vary in the three coordinate directions.

Comparing equations 4.2 and 4.3, we see that in the Eulerian approach the spatial coordinates, along with time, are independent rather than dependent variables. This is usually the simpler way of analyzing a flow problem and is the one we will most often use herein. However, it is sometimes possible to convert time-varying flows to simpler time-invariant flows by switching from a Eulerian to a Lagrangian viewpoint (e.g., in considering the settling of sediment particles, or the passage of a wave along a channel).

4.2 Kinematics and Dynamics

Relations that involve only velocities and/or accelerations (i.e., quantities involving only the dimensions length [L] and time [T]) are **kinematic** relations; those that involve quantities with the dimension of force [F] or mass [M] are **dynamic** relations. Newton's second law of motion, "force (F) equals mass (M) times acceleration (a)," provides the basic link between kinematics and dynamics:

$$F = M \cdot a, \tag{4.4a}$$

which also expresses the relation between the basic physical dimensions of force, [F], and mass, [M]:

$$[\mathrm{F}] = [\mathrm{M}] \cdot [\mathrm{L\,T}^{-2}], \tag{4.4b}$$

(see appendix A for a review of dimensions of physical quantities).

4.2.1 Kinematics

4.2.1.1 Velocity

The **velocity** in an arbitrary s-direction, u_s, is the time rate of change of the location of a fluid element:

$$u_s \equiv \frac{\mathrm{d}s}{\mathrm{d}t}, \tag{4.5}$$

where ds is the distance moved in the time increment dt. Thus, velocity is a vector quantity with dimension $[\mathrm{L\,T}^{-1}]$ that has direction as well as magnitude.

In the Eulerian viewpoint the direction can be specified by resolving the actual velocity into its components in the orthogonal coordinate directions (illustrated for two dimensions in figure 4.2) such that

$$\frac{\mathrm{d}s}{\mathrm{d}t} = \frac{1}{\cos\Psi_x}\cdot\frac{\mathrm{d}x}{\mathrm{d}t} = \frac{1}{\cos\Psi_y}\cdot\frac{\mathrm{d}y}{\mathrm{d}t} = \frac{1}{\cos\Psi_z}\cdot\frac{\mathrm{d}z}{\mathrm{d}t}, \tag{4.6}$$

where Ψ_x, Ψ_y, Ψ_z are the angles between the s-direction and the x-, y-, and z-directions, respectively. Defining the components of velocity in the three coordinate directions as

$$u_x \equiv \frac{\mathrm{d}x}{\mathrm{d}t}, \quad u_y \equiv \frac{\mathrm{d}y}{\mathrm{d}t}, \quad u_z \equiv \frac{\mathrm{d}z}{\mathrm{d}t}, \tag{4.7}$$

the magnitude of the velocity is

$$u_s = (u_x{}^2 + u_y{}^2 + u_z{}^2)^{1/2}. \tag{4.8}$$

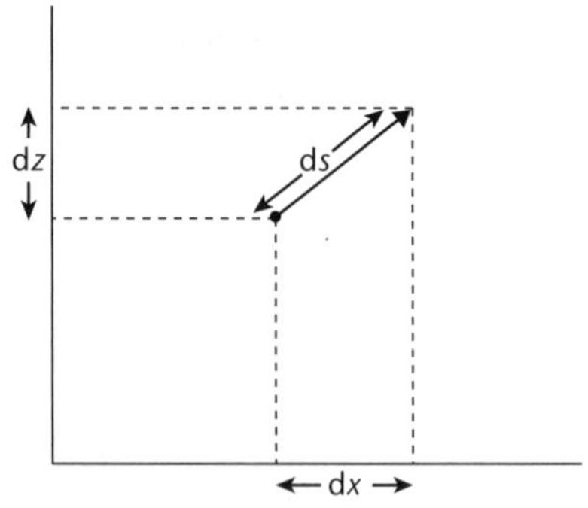

Figure 4.2 The distance ds traveled by a fluid element in an arbitrary direction in time dt can be resolved into distances parallel to the orthogonal x- and z-axes, dx and dz.

Recall from section 3.3.4 that most open-channel flows are turbulent, and the velocities of fluid elements change from instant to instant and have chaotic paths (see figures 3.20 and 3.21). Thus, to be useful in describing the overall flow, the velocities discussed in this chapter—and in most of this text—are time-averaged to eliminate the fluctuations due to turbulent eddies; that is, they are the $\bar{u}_i$ quantities defined in figure 3.25.

Velocity is, of course, a central concern in fluid physics, and although it is a vector quantity, "knowledge of vector analysis is not essential to the study of fluid motion, for the variation of a vector may be fully described by the changes in magnitude of its three components" (Rouse 1938, p. 35). These changes—accelerations—are discussed in the following section.

4.2.1.2 Acceleration

Acceleration is the time rate of change of velocity, with dimension $[\mathrm{L\,T}^{-2}]$. Acceleration is also a vector quantity, and in the Eulerian viewpoint we write the accelerations for each directional velocity component separately. A change in the component of velocity in the i-direction, $\mathrm{d}u_i$, where $i = x, y, z$ is the sum of its rate of change in time at a point $\partial u_i/\partial t$ times a small time increment $\mathrm{d}t$, plus its rates of change in each of the three coordinate directions times short spatial increments in each direction, $\mathrm{d}x$, $\mathrm{d}y$, $\mathrm{d}z$:

$$\mathrm{d}u_i = \frac{\partial u_i}{\partial t} \cdot \mathrm{d}t + \frac{\partial u_i}{\partial x} \cdot \mathrm{d}x + \frac{\partial u_i}{\partial y} \cdot \mathrm{d}y + \frac{\partial u_i}{\partial z} \cdot \mathrm{d}z \tag{4.9}$$

Acceleration in the i-direction is $\mathrm{d}u_i/\mathrm{d}t$, so from equation 4.9,

$$\frac{\mathrm{d}u_i}{\mathrm{d}t} = \frac{\partial u_i}{\partial t} + \frac{\partial u_i}{\partial x} \cdot \frac{\mathrm{d}x}{\mathrm{d}t} + \frac{\partial u_i}{\partial y} \cdot \frac{\mathrm{d}y}{\mathrm{d}t} + \frac{\partial u_i}{\partial z} \cdot \frac{\mathrm{d}z}{\mathrm{d}t}, \tag{4.10}$$

and using the definitions of equation 4.7, we can write the expression for acceleration in the i-direction as

$$\frac{\mathrm{d}u_i}{\mathrm{d}t} = \frac{\partial u_i}{\partial t} + \frac{\partial u_i}{\partial x} \cdot u_x + \frac{\partial u_i}{\partial y} \cdot u_y + \frac{\partial u_i}{\partial z} \cdot u_z. \tag{4.11}$$

Equation 4.11 gives the rates of change of velocity components u_x, u_y, u_z for a fluid element at a particular spatial location and instant of time. These accelerations are the sum of the local acceleration and the convective acceleration:

> **Local acceleration** is the time rate of change of velocity at a point, $\partial u_i/\partial t$. If the local acceleration in a flow is zero, the flow is **steady;** otherwise it is **unsteady.**
>
> **Convective acceleration** is the rate of change of velocity at a particular instant due to its motion in space, $(\partial u_i/\partial x) \cdot u_x + (\partial u_i/\partial y) \cdot u_y + (\partial u_i/\partial z) \cdot u_z$. If the convective acceleration in a flow is zero, the flow is **uniform;** otherwise it is **nonuniform.**

Flows may be steady and uniform (no acceleration), steady and nonuniform (convective acceleration only), or unsteady and nonuniform (both local and convective acceleration); unsteady uniform flows (those with local acceleration only) are virtually impossible. Again, these definitions refer to the time-averaged velocities neglecting the fluctuations due to turbulent eddies.

4.2.1.3 Streamlines and Pathlines

A **streamline** is an imaginary line drawn in a flow that is everywhere tangent to the local time-averaged velocity vector (figure 4.3). If a flow is either steady or uniform, the streamlines are also pathlines; that is, they represent the time-averaged paths of fluid elements, neglecting motion due to turbulent eddies. In uniform flow, the streamlines are parallel to each other (figure 4.3c). Many of the basic relationships of open-channel flows are developed first for "microscopic" fluid elements and streamlines, and then integrated to apply to macroscopic flows.

4.2.2 Dynamics

4.2.2.1 Forces in Fluid Flow

The forces involved in open-channel flows are as follows:

> Body forces: gravitational (directed downstream); Coriolis (apparent force perpendicular to flow); centrifugal (apparent force perpendicular to flow)
>
> Surface forces: pressure (directed downstream or upstream); shear (directed upstream)

Body forces act on all matter in each fluid element; surface forces can be thought of as acting only on the surfaces of elements, and are often expressed as stress— that is, force-per-unit area.

Gravitational and shear forces are important in all open-channel flows: Flow in open channels is induced by gravitational force due to the slope of the water surface. Shear forces arising from the frictional resistance of the solid boundary and the effects of viscosity and turbulence act to oppose the gravitationally induced flow. Pressure forces are present if there is a downstream gradient in depth, and may act in the upstream or downstream directions, depending on the direction of the gradient. As noted above, the Coriolis and centrifugal forces are apparent forces that arise from the earth's rotation and curvature of flow paths, respectively, when describing flows in a fixed coordinate system.

The nature of fluid pressure and shear are described further in the remainder of this section, and chapter 7 is devoted to a quantitative exploration of all forces in open-channel flows.

4.2.2.2 Fluid Pressure

Fluid pressure ($[F\,L^{-2}]$ or $[M\,L^{-1}\,T^{-2}]$), is the force normal to a surface due to the weight of the fluid above the surface, divided by the area of the surface. Like temperature, it is a state variable that may vary as a function of space and time. Pressure is a component of the potential energy of fluids (discussed more fully in section 4.5.1),

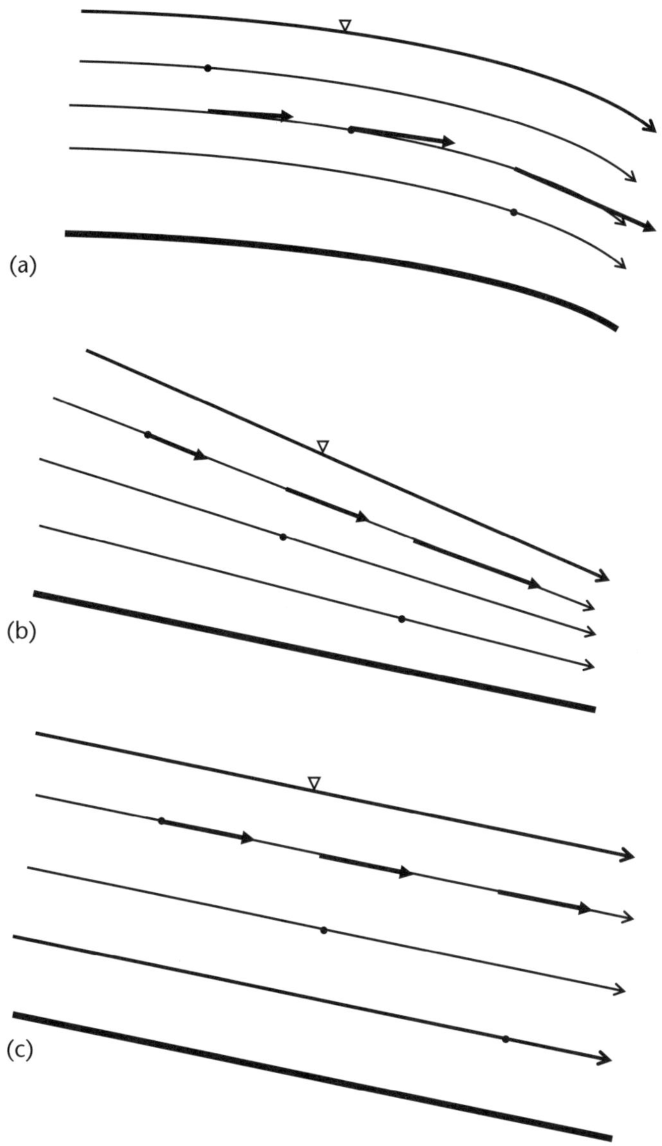

Figure 4.3 Streamlines in steady flows. The heavy arrows are velocity vectors at arbitrary points; streamlines are tangent to the time-averaged velocity vector at every point. Because the flows are steady, the streamlines are also time-averaged pathlines tracing the movement of fluid elements. (a) Steady nonuniform flow. Clearly, the direction and magnitude of velocity of fluid elements moving along the streamlines change spatially. (b) Steady nonuniform flow. Although the direction in which element is moving is constant, the magnitude of velocity changes spatially. (c) Steady uniform flow. The direction and magnitude of velocity of each fluid element remain constant.

and spatial differences in pressure create forces that cause accelerations and affect the movement of fluid elements. Here, we develop expressions for the magnitude of pressure in open-channel flows and show that the pressure at a point in a fluid is a scalar quantity that acts equally in all directions.

Magnitude To derive an expression for the magnitude of pressure, consider a horizontal plane of area A_h at a depth h in a static (nonflowing) body of water (figure 4.4a). The weight of the water column is $\gamma \cdot h \cdot A_h$, where γ is the weight density of water, so the total pressure on the plane, P_{abs}, is

$$P_{abs} = \gamma \cdot h \cdot A_h / A_h + P_{atm} = \gamma \cdot h + P_{atm}, \tag{4.12}$$

where P_{atm} is atmospheric pressure.

We shall see in section 4.5.1 that pressure is one component of potential energy, and in section 4.7 that flow is caused by spatial gradients in potential energy. Thus, we will almost always be concerned with pressure gradients rather than actual pressures,

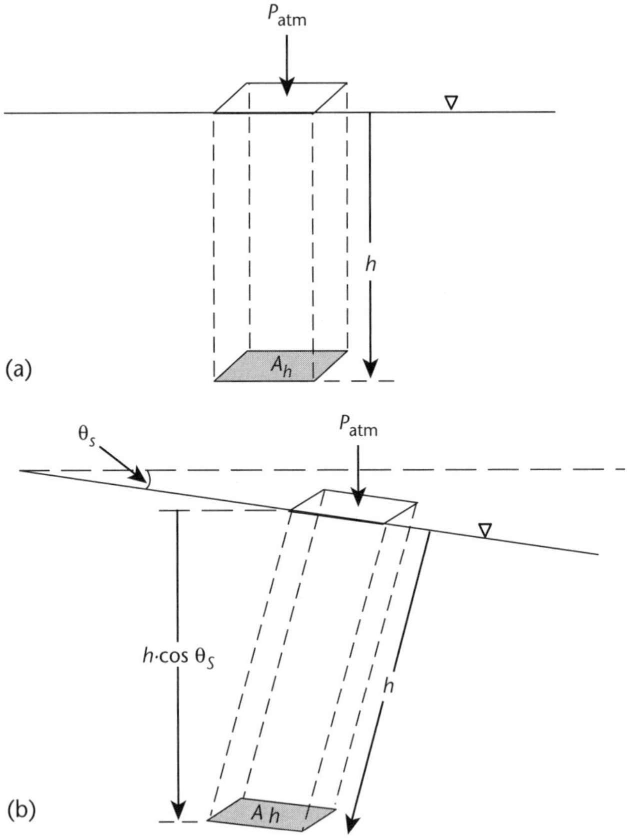

Figure 4.4 Definitions of terms for deriving the expression for pressure in (a) a water body at rest and (b) an open-channel flow (equation 4.13). See text.

and since atmospheric pressure is essentially constant for a given situation, we can neglect P_{atm} and be concerned only with the **gage pressure**, P:

$$P = \gamma \cdot h = \rho \cdot g \cdot h, \tag{4.13a}$$

where ρ is the mass density of water, and g is the gravitational acceleration. Because the situation in figure 4.4a is static, the pressure given by equation 4.13a is the **hydrostatic pressure**.

When water is flowing, the water surface is no longer horizontal but slopes at an angle θ_S (figure 4.4b) in the direction of flow. The force of gravity acts vertically, but since the depth is measured normal to the surface, the pressure in this situation is given by

$$P = \gamma \cdot h \cdot \cos\,\theta_S = \rho \cdot g \cdot h \cdot \cos\,\theta_S. \tag{4.13b}$$

However, since natural stream slopes almost never exceed 0.1 rad (5.7°), $\cos\,\theta_S$ is almost always greater than 0.995, and can usually be assumed = 1.

Equations 4.13a and 4.13b, represent the **hydrostatic pressure distribution** and applies to open-channel flows unless the water surface curves very sharply in the vertical plane (figure 4.5a). Such sharp curvature may occur, for example, near a free overfall or at the base of very steep rapids or artificial spillway; in these cases, centrifugal force increases or reduces pressure as shown in figure 4.5, b and c. With these exceptions, the hydrostatic pressure distribution given by equation 4.13 can be assumed to apply in open-channel flows, and because water is incompressible (section 3.3.1) and its mass density changes only very little with temperature, pressure is a linear function of depth as given by equation 4.13.

Direction If the fluid pressure at a point varied with direction, it would be possible to construct a perpetual motion machine like that shown in figure 4.6, in which the pressure difference induces a flow that drives a turbine. Because such a machine does not produce motion, this simple thought experiment shows that the magnitude of fluid pressure is equal in all directions. Note that this conclusion does not preclude the point-to-point variation of pressure.

4.2.2.3 Fluid Shear

We saw in sections 3.3.3 and 3.3.4 that the presence of a velocity gradient in a fluid implies a tangential force per unit area, called a **shear stress**, between adjacent fluid layers due to fluid viscosity and, usually, turbulence. As expressed in equation 3.29, the general relation is

$$\tau_{yx} = (\mu + \eta) \cdot \frac{\mathrm{d}u_x(y)}{\mathrm{d}y}, \tag{4.14}$$

where x is the direction of the flow, y is the direction of the velocity gradient (normal to x), τ_{yx} is the shear stress, μ is the dynamic viscosity, η is the eddy viscosity due to turbulence (if present), and u_x is the velocity in the x-direction.

The shear stress is directed upstream, that is, in the negative x-direction, and can be thought of as a force that tends to retard the flow. Recall also that the shear

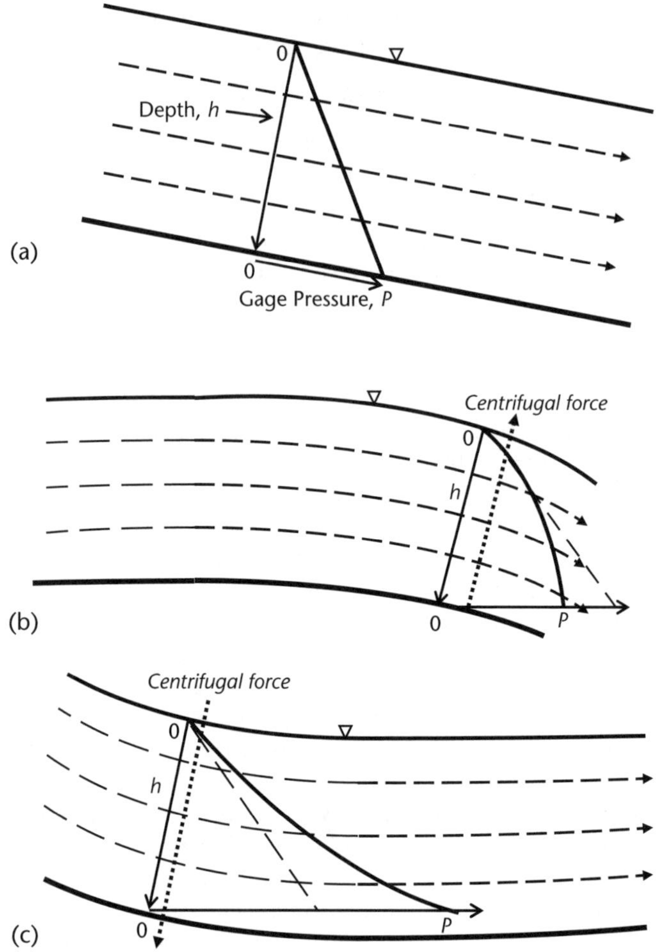

Figure 4.5 Pressure, P, as a function of depth, h, in open-channel flows (solid lines). Long-dashed arrows represent streamlines. (a) The linear hydrostatic pressure distribution (equation 4.13) applies unless distorted by centrifugal force (dotted arrows) where the water surface is strongly curved in the vertical plane, as in an overfall (b); and at the base of a steep rapids or artificial spillway, as in (c). The dashed lines in (b) and (c) show the hydrostatic distribution.

stress is physically equivalent to a momentum flux in the direction of the velocity gradient (y-direction) from regions of higher velocity to regions of lower velocity (section 3.3.3.3).

Equation 4.14 provides a link between the kinematics (the velocity gradient) and the dynamics (the shear force or momentum flux) of a flow. Velocity gradients are induced in open-channel flows by the solid boundaries and, as discussed in section 3.4.1, are present throughout most natural open-channel flows.

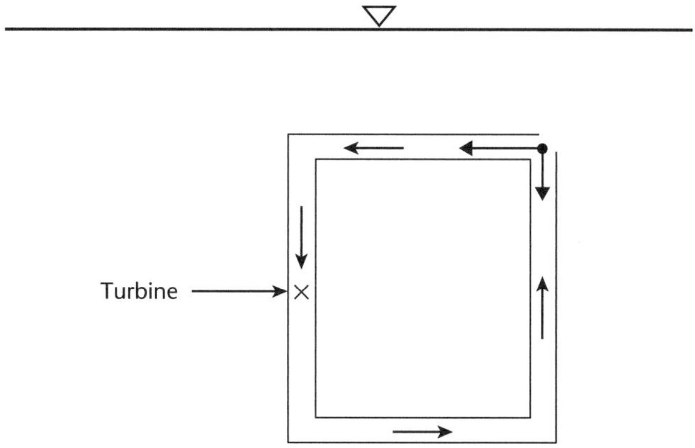

Figure 4.6 Thought experiment showing that, if the magnitude of fluid pressure at a point (•) were greater in one direction (e.g., to the left) than in another (downward), it would be possible to create a perpetual motion machine using pipes with a turbine.

4.3 Equations Based on Conservation of Mass (Continuity)

The **conservation-of-mass equation**, or **continuity equation**, applies to a conservative substance (i.e., a substance that is not produced or depleted by chemical reaction or radioactivity) entering and/or leaving a fixed region of space, called a **control volume**, during a defined period of time. It can be stated in words as follows:

> The quantity of mass of a conservative substance entering a control volume during a defined time period, minus the quantity leaving the volume during the time period, equals the change in the quantity stored in the volume during the time period.

In condensed form, we can state the conservation equation as

$$\text{Mass In} - \text{Mass Out} = \text{Change in Mass Stored}, \tag{4.15}$$

but we must remember that the equation is strictly true *only* for

- Conservative substances
- A defined control volume
- A defined time period

4.3.1 "Microscopic" Continuity Relation

The most general version of this equation is developed for a "microscopic" elemental control volume with infinitesimal dimensions dx, dy, dz aligned with the Cartesian coordinate axes and an infinitesimal time period dt (figure 4.7). Applying equation 4.15 to this situation leads to the expression

$$\frac{\partial(\rho \cdot u_x)}{\partial x} + \frac{\partial(\rho \cdot u_y)}{\partial y} + \frac{\partial(\rho \cdot u_z)}{\partial z} = -\frac{\partial \rho}{\partial t}, \tag{4.16a}$$

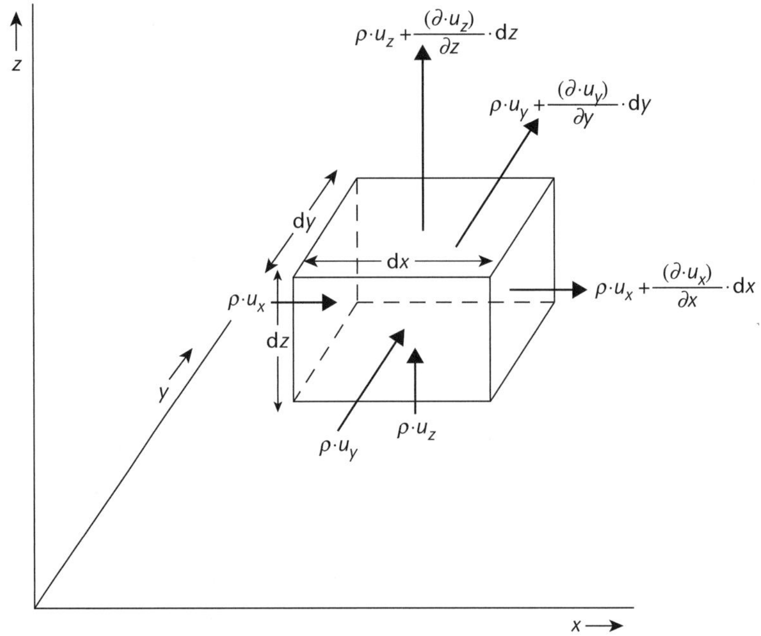

Figure 4.7 Definition diagram for derivation of the "microscopic" continuity equation 4.16. The control volume is the infinitesimal parallelepiped $dx \cdot dy \cdot dz$. The mass fluxes (flows of mass per unit area per unit time) into the control volume are the $\rho \cdot u_i$ terms; the mass fluxes out of the volume are the $\rho \cdot u_i + [\partial(\rho u_i)/\partial i] \cdot di$ terms, where $i = x, y, z$.

where ρ is the mass density of the water (for detailed development see, e.g., Daily and Harleman 1966; Furbish 1997). As noted in section 3.3.1, water is effectively incompressible, and its density changes only slightly with temperature, so we can usually assume that ρ will be constant in time and space. With that assumption, equation 4.16a reduces to

$$\frac{\partial u_x}{\partial x} + \frac{\partial u_y}{\partial y} + \frac{\partial u_z}{\partial z} = 0. \tag{4.16b}$$

Equation 4.16 is applicable to microscopic regions of open-channel flows with low sediment concentrations. It is used as the basis for detailed computer modeling of open-channel flows (e.g., Olsen 2004).

4.3.2 Macroscopic Continuity Relations

In the present text, we will usually be concerned with macroscopic open-channel flow in one direction only and so can develop the continuity equation for control volumes that have finite dimensions equal to the channel width and depth and are infinitesimal only in the flow direction. Referring to the idealized channel segment in figure 4.8 and applying equation 4.15 for flow only in the X-direction, the mass entering the

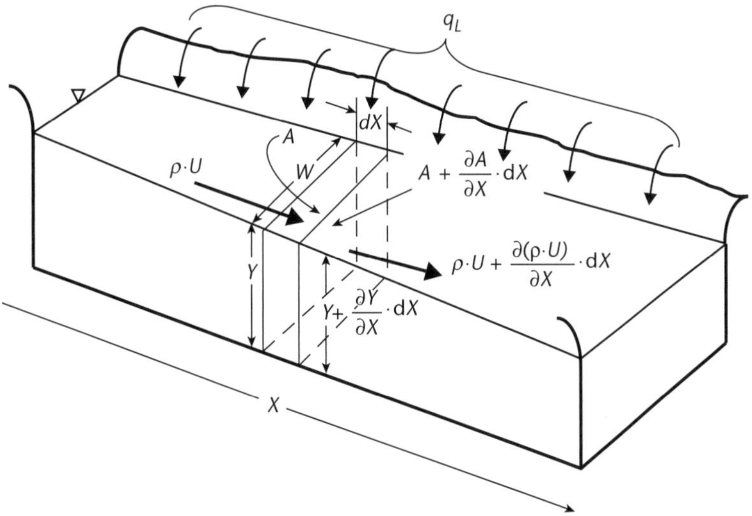

Figure 4.8 Definition diagram for derivation of macroscopic continuity equation 4.18 and macroscopic conservation-of-momentum equation 4.26. The areas of the upstream and downstream faces of the control volume are A and $A + \partial A/\partial X$, respectively.

control volume in dt is

$$\text{Mass In} = \rho \cdot U \cdot A \cdot \mathrm{d}t + \rho \cdot q_L \cdot \mathrm{d}X \cdot \mathrm{d}t, \tag{4.17a}$$

where U is cross-sectional average velocity [L T^{-1}], A is cross-sectional area [L^2], and q_L is the net rate of **lateral inflow** (which might include rainfall and seepage into and out of the channel) per unit channel distance [L^2 T^{-1}]. The mass leaving the control volume in dt is

$$\begin{aligned}\text{Mass Out} &= \left(\rho \cdot U + \frac{\partial(\rho \cdot U)}{\partial X} \cdot \mathrm{d}X\right) \cdot \left(A + \frac{\partial A}{\partial X} \cdot \mathrm{d}X\right) \cdot \mathrm{d}t \\ &= \left[\rho \cdot U \cdot A + \rho \cdot U \cdot \frac{\partial A}{\partial X} \cdot \mathrm{d}X + A \cdot \frac{\partial(\rho \cdot U)}{\partial X} \cdot \mathrm{d}X + \frac{\partial A}{\partial X} \cdot \frac{\partial(\rho \cdot U)}{\partial X} \cdot (\mathrm{d}X)^2\right] \cdot \mathrm{d}t,\end{aligned} \tag{4.17b}$$

and the change in mass occupying the control volume during dt is

$$\text{Change in Mass Stored} = \frac{\partial(\rho \cdot A)}{\partial t} \cdot \mathrm{d}X \cdot \mathrm{d}t. \tag{4.17c}$$

The macroscopic continuity equation is obtained by substituting equation 4.17a–c into 4.15. If we assume spatially and temporally constant density and neglect the term with $(\mathrm{d}X)^2$,[2] this substitution leads to

$$q_L - U \cdot \frac{\partial A}{\partial X} - A \cdot \frac{\partial U}{\partial X} = \frac{\partial A}{\partial t}. \tag{4.18a}$$

Since the discharge $Q = U \cdot A$, we can use the rules of derivatives to note that $U \cdot (\partial A/\partial X) + A \cdot (\partial U/\partial X) = \partial Q/\partial X$ and write equation 4.18a more compactly as

$$q_L - \frac{\partial Q}{\partial X} = \frac{\partial A}{\partial t} \tag{4.18b}$$

or, in the absence of lateral inflow,

$$-\frac{\partial Q}{\partial X} = \frac{\partial A}{\partial t}. \tag{4.18c}$$

As we will see in chapter 11, equation 4.18c is used to predict the passage of a flood wave through a channel reach.

In many of the developments in this text, we will be considering reaches with fixed geometry and specified constant discharge, Q. In these cases, the mass flow rate [M T $^{-1}$] through a channel cross section is given by $\rho \cdot Q$, where

$$W \cdot Y \cdot U = Q, \tag{4.19}$$

and W is the local water-surface width, Y is the local average depth, and U is the local average velocity. Thus, for constant discharge and constant mass density, we can write an even simpler macroscopic continuity relation as

$$U = \frac{Q}{W \cdot Y}. \tag{4.20}$$

4.4 Equations Based on Conservation of Momentum

Momentum is mass times velocity [M L T^{-1}]. The time rate-of-change of momentum has dimensions [M L T^{-2}] = [F], so the principle of **conservation of momentum** can be stated as follows:

> The time rate-of-change of momentum of a fluid element is equal to the net force applied to the element.

Mathematically, we can express it for a fluid element as

$$\frac{d\mathsf{M}}{dt} = \Sigma F, \tag{4.21}$$

where M is momentum, t is time, and ΣF is the net force acting on the element. Equation 4.21 is simply another way of stating Newton's second law.

The conservation-of-momentum principle is applied in various forms to solve fluid-flow problems, often in conjunction with the conservation of mass. A microscopic conservation-of-momentum equation can be derived for a fluid element in Cartesian coordinates, as shown in many fluid mechanics texts (e.g., Daily and Harleman 1966; Julien 2002), and the resulting three-dimensional relation can be simplified to apply to typical one-dimensional macroscopic open-channel flow situations.

Alternatively, we can apply the principle directly to the macroscopic channel shown in figure 4.8 to derive an expression for one-dimensional (downstream X-direction) momentum changes. In this case, we will assume that the discharge, Q, through the reach is spatially and temporally constant, that the channel width, W, and mass density ρ, are constant and that there is no lateral inflow. The time rate-of-change of momentum for an element passing through the channel segment is due only to its downstream change in velocity:

$$\frac{\mathrm{dM}}{\mathrm{d}t} = \rho \cdot Q \cdot \frac{\partial U}{\partial X} \cdot \mathrm{d}X, \tag{4.22}$$

where U is the cross-sectional average velocity at the upstream face.[3]

In general, as we shall see in chapter 7, the forces that are included in ΣF are those due to gravity, pressure, and friction. However, because the downstream dimension of the fluid element in figure 4.8 is infinitesimally short, we can ignore the gravitational force due the downstream component of the element's weight and the frictional force due to the channel bed. This leaves only the pressure force, which we can evaluate using the relations developed in section 4.2.2.2.

Assuming a hydrostatic pressure distribution, we can apply equation 4.13. The average pressure on the upstream face is then $\gamma \cdot Y/2$, where γ is the weight density of water; and the pressure force on the upstream face, F_{up}, is the product of the average pressure and the area of the face, $W \cdot Y$:

$$F_{\mathrm{up}} = \frac{\gamma \cdot W \cdot Y^2}{2} \tag{4.23}$$

Using similar reasoning for the downstream face (and neglecting terms with powers of dX) yields

$$F_{\mathrm{down}} = \frac{\gamma \cdot W}{2} \cdot \left(Y + \frac{\partial Y}{\partial X} \cdot \mathrm{d}X\right)^2 = \left(\frac{\gamma \cdot W}{2}\right) \cdot \left(Y^2 + 2 \cdot Y \cdot \frac{\partial Y}{\partial X} \cdot \mathrm{d}X\right). \tag{4.24}$$

Thus, the net downstream-directed pressure force on the element is

$$\Sigma F = F_{\mathrm{up}} - F_{\mathrm{down}} = -\gamma \cdot W \cdot Y \cdot \frac{\partial Y}{\partial X} \cdot \mathrm{d}X. \tag{4.25}$$

Note that if depth increases downstream ($\partial Y/\partial X > 0$), then $\Sigma F < 0$ and the net pressure force is directed upstream, and vise versa.

Substituting equations 4.25 and 4.22 into 4.21 and simplifying yields

$$\rho \cdot Q \cdot \frac{\partial U}{\partial X} = -\gamma \cdot W \cdot Y \cdot \frac{\partial Y}{\partial X}, \tag{4.26a}$$

and further noting that $\gamma = \rho \cdot g$, where g is gravitational acceleration, and $Q = W \cdot Y \cdot U$, we have

$$U \cdot \frac{\partial U}{\partial X} = -g \cdot \frac{\partial Y}{\partial X}. \tag{4.26b}$$

Note in equation 4.26 that if $\partial U/\partial X > 0$ (i.e., velocity increases downstream), then $\partial Y/\partial X < 0$ (depth decreases downstream). Given that discharge and width are constant, this is consistent with the conservation of mass (equation 4.20).

Equation 4.26 is the mathematical expression of the conservation-of-momentum principle for one-dimensional flow in an open channel. Note that it is a purely kinematic relation, although momentum is a dynamic quantity. We will encounter other forms of the conservation-of-momentum relation in chapters 8, 10, and 11.

4.5 Equations Based on Conservation of Energy

In this text, we will be much concerned with mechanical energy in its two forms, **potential energy** (*PE*) and **kinetic energy** (*KE*). Here we develop general expressions for these quantities in open-channel flows and show how the first and second laws of thermodynamics apply in such flows. Specific applications of these concepts to solve open-channel flow problems are described in chapters 9 and10.

4.5.1 Mechanical Potential Energy

Mechanical potential energy is a central concept because fluids flow in response to spatial gradients in mechanical potential energy of fluid elements, and the direction of the flow is from regions with higher potential energy to regions of lower potential energy (section 4.7).

To develop expressions for potential energy, we focus on two fluid elements with mass density ρ and volume V at different elevations within a static (nonflowing) body of water (figure 4.9). The **gravitational potential energy** of each element (PE_{gA}, PE_{gB}) is due to its mass ($\rho \cdot V$) and its elevation (z_A, z_B) above a datum (z_0) in a gravitational field of strength (acceleration) g, so

$$PE_{gA} = \rho \cdot V \cdot g \cdot (z_A - z_0); \tag{4.27a}$$

$$PE_{gB} = \rho \cdot V \cdot g \cdot (z_B - z_0). \tag{4.27b}$$

Clearly, the gravitational potential energies of the two elements differ. However, since there is no motion, the total potential energy of the two elements must be equal.

The total potential energy of the two elements can be made equal if we postulate that each element has an additional component of potential energy, PE_{pA} and PE_{pB}, respectively, and write

$$PE_{gA} + PE_{pA} = PE_{gB} + PE_{pB}. \tag{4.28}$$

Substituting equation 4.27 into 4.28 and using the facts that $h_A = z_S - z_A$ and $h_B = z_S - z_B$, where z_S is the surface elevation, leads to

$$PE_{pA} - PE_{pB} = \rho \cdot g \cdot V \cdot (z_B - z_A) = \rho \cdot g \cdot V \cdot h_A - \rho \cdot g \cdot V \cdot h_B. \tag{4.29}$$

Thus, we conclude that the general expression for the additional component of potential energy is

$$PE_p = \rho \cdot g \cdot V \cdot h = \gamma \cdot V \cdot h. \tag{4.30}$$

Comparing equations 4.30 and 4.13, we see that the second component of potential energy is due to pressure, and is called the **pressure potential energy**.

Thus, we conclude that the total potential energy, *PE*, of a fluid element is the sum of its gravitational and pressure potential energies:

$$PE = PE_g + PE_p = \rho \cdot g \cdot V \cdot [(z - z_0) + h] = \gamma \cdot V \cdot [(z - z_0) + h]. \tag{4.31}$$

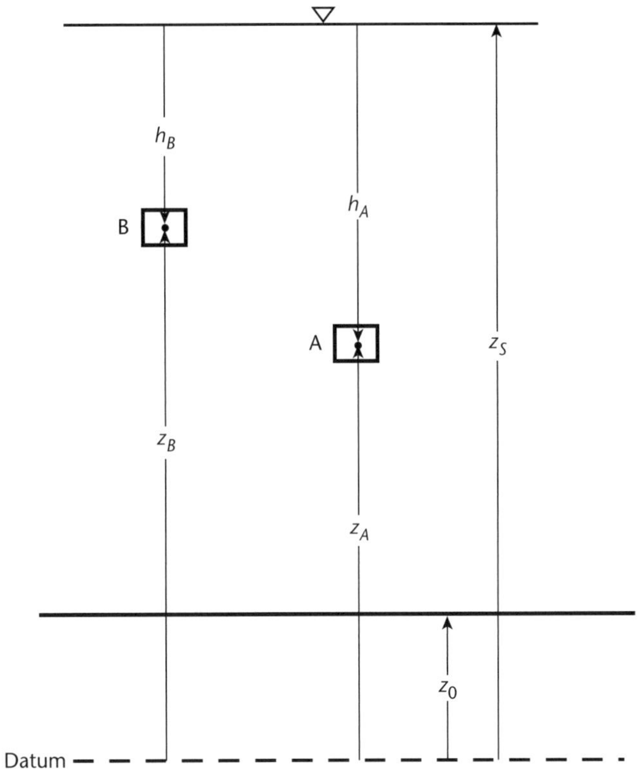

Figure 4.9 Definitions of terms for determining the magnitude of total potential energy in a stationary water body (equations 4.30–4.35). A and B are fluid elements of equal volume and density.

Equation 4.31 can be generalized by defining a quantity called head:

Head [L] is the energy [F L] of a fluid element divided by its weight [F].

Dividing 4.31 by the weight of the fluid element, $\gamma \cdot V$, yields

$$\hbar_{\mathrm{PE}} = (z - z_0) + h, \tag{4.32}$$

where $\hbar_{PE}$ is called the **potential head**. We can similarly divide the expressions for PE_{g} and PE_{p} by $\gamma \cdot V$ and define the **gravitational head** (or **elevation head**), $\hbar_g$, as the elevation above a datum,

$$\hbar_g = z - z_0, \tag{4.33}$$

and the pressure head, $\hbar_p$, as the distance below a water surface,

$$\hbar_p = h = z_S - z. \tag{4.34}$$

Obviously,

$$\hbar_{\mathrm{PE}} = \hbar_g + \hbar_p. \tag{4.35}$$

We can summarize the preceding by stating that, if the pressure distribution is hydrostatic,

> The potential energy of a fluid element an open channel is determined by its location in 1) a gravitational field and 2) a pressure field.

> The potential energy per unit weight of a fluid element in an open channel can be directly measured as the sum of 1) its elevation above a datum (gravitational potential) and 2) its distance below the water surface (pressure potential).

In a body of water with a horizontal surface, $\hbar_{PE} = z_S - z_0$ at all points, and there is no flow. If the surface is sloping, $\hbar_g$ and $\hbar_{PE}$ at a given depth will be lower where the surface is at a lower elevation, and flow will occur in response to this gradient. We will explore this further in chapter 7.

4.5.2 Mechanical Kinetic Energy

Mechanical kinetic energy is energy due to the motion of a fluid element. Consider a fluid element of mass M moving along a streamline in an arbitrary x-direction from point x_1, where it has velocity u_1, to point x_2, where it has velocity u_2. The difference in velocities represents an acceleration [L T^{-2}], and the force [F] applied, integrated over the distance traveled, represents the work [F L] done, or energy expended, to produce that acceleration. Thus, integrating Newton's second law over the distance traveled,

$$\int_{x_1}^{x_2} F \cdot \mathrm{d}x = M \cdot \int_{x_1}^{x_2} \frac{\mathrm{d}u}{\mathrm{d}t} \cdot \mathrm{d}x. \tag{4.36}$$

However, velocity is defined as $u \equiv \mathrm{d}x/\mathrm{d}t$, so we can write equation 4.36 as

$$\int_{x_1}^{x_2} F \cdot \mathrm{d}x = M \cdot \int_{u_1}^{u_2} u \cdot \mathrm{d}u, \tag{4.37}$$

from which we find

$$\int_{x_1}^{x_2} F \cdot \mathrm{d}x = \left(\frac{1}{2}\right) \cdot M \cdot ({u_2}^2 - {u_1}^2). \tag{4.38}$$

Note that the quantity $(1/2) \cdot M \cdot u^2$ has the dimensions of energy [M L^2 T^{-2}] = [F L]; this is the energy associated with the motion of the element and is called the **kinetic energy**. Thus, the kinetic energy, KE, of a fluid element of mass M moving with velocity u is

$$KE = \left(\frac{1}{2}\right) \cdot M \cdot u^2. \tag{4.39}$$

From these developments, we can state that

> The work done in accelerating a fluid element as it moves a given distance in a flow is equal to 1) the kinetic energy acquired by the element over that distance, and 2) the net force applied to the element in the direction of motion, times the distance.

As with potential energy, we can define the **kinetic-energy head** (or **velocity head**), $\hbar_{KE}$, by dividing KE by the weight of the element $\gamma \cdot V$ (and noting that $\rho \equiv M/V$ and $\gamma = \rho \cdot g$):

$$\hbar_{\mathrm{KE}} \equiv \frac{KE}{\gamma \cdot V} = \left(\frac{1}{2}\right) \cdot M \cdot \frac{u^2}{\gamma \cdot V} = \frac{u^2}{2 \cdot g}. \tag{4.40}$$

Thus, the kinetic energy per unit weight of a fluid element is proportional to the square of its velocity.

4.5.3 Total Mechanical Energy and the Laws of Thermodynamics

The total mechanical energy of a fluid element, $\hbar$, is the sum of its potential and kinetic energies, expressed most generally in terms of heads:

$$\hbar = \hbar_g + \hbar_p + \hbar_{\mathrm{KE}}. \tag{4.41}$$

Consider the movement of a fluid element along a streamline from point x_1 to point x_2 in an open-channel flow (figure 4.10). (As noted above, the water surface must be sloping if flow is occurring.) The difference in total mechanical energy at the two points is the following equation:[4]

$$\hbar_2 - \hbar_1 = \hbar_{g2} - \hbar_{g1} + \hbar_{p2} - \hbar_{p1} + \hbar_{\mathrm{KE}2} - \hbar_{\mathrm{KE}1}. \tag{4.42}$$

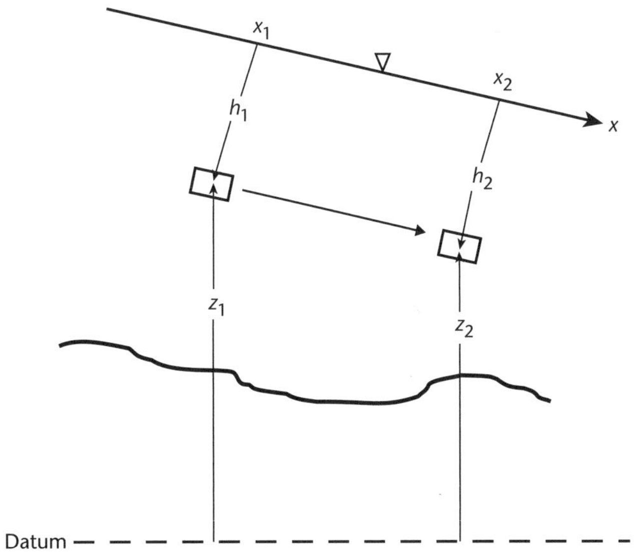

Figure 4.10 Movement of a fluid element along a streamline in an open-channel flow, defining terms for its total mechanical energy (equations 4.41–4.45).

To simplify the discussion, assume that the element remains at the same distance below the surface, so that $h_2 = h_1$ and $\hbar_{p2} = \hbar_{p1}$. Then equation 4.42 becomes

$$\hbar_2 - \hbar_1 = \hbar_{g2} - \hbar_{g1} + \hbar_{KE2} - \hbar_{KE1}. \tag{4.43}$$

The first law of thermodynamics may be stated as, "Energy is neither created nor destroyed." If we consider mechanical energy only, this principle would suggest that $\hbar$ for a given element does not change as it moves in an open-channel flow. Because the water surface in figure 4.10 slopes, $z_2 < z_1$ and $\hbar_g$ decreases in the direction of flow. The first law and equation 4.43 would then suggest that $\hbar_{KE}$ must increase by the same amount that $\hbar_g$ decreases, that is, that

$$\hbar_{g1} - \hbar_{g2} = \hbar_{KE2} - \hbar_{KE1} = \left(\frac{1}{2 \cdot g}\right) \cdot (u_2{}^2 - u_1{}^2). \tag{4.44}$$

Equation 4.44, which was derived by considering mechanical energy only, implies that an open-channel flow must continually accelerate in the direction of movement, like a free-falling body in a vacuum. However, real open-channel flows do not continually accelerate, so there is something missing from this analysis—namely, the effect of friction in converting mechanical (kinetic) energy to heat energy and the dissipation of the heat into the environment. The irreversible conversion of mechanical kinetic energy into heat is a manifestation of the second law of thermodynamics.

To incorporate this law into the statement of conservation of energy for open-channel flows, we must add to equation 4.44 a term representing the energy per unit weight that is converted to heat, called the **head loss** or **energy loss**, $\hbar_e$, and write the conservation-of-energy equation for a fluid element as

$$\hbar_2 - \hbar_1 = \hbar_{g2} - \hbar_{g1} + \hbar_{p2} - \hbar_{p1} + \hbar_{KE2} - \hbar_{KE1} + \hbar_e, \tag{4.45}$$

where $\hbar_e$ is always a positive number. $\hbar_e$ is the consequence of the friction induced by the presence of a flow boundary (as described in section 3.4) and transmitted into the fluid by viscosity and, in most flows, by turbulence.

We will see that some of the most important problems in open-channel hydraulics are approached by applying the energy equation, including predicting the response of flow configuration and velocity to changes in channel geometry (chapters 9 and 10). Addressing these problems requires integrating the elemental energy equation (equation 4.45) over a cross section; this development is the subject of section 8.1.1.

Meanwhile, we can summarize the energy laws for open-channel flow as follows:

- A gradient of gravitational potential energy (and of the water surface) is required to cause flow.
- The flow process involves the continuous conversion of potential energy into kinetic energy.
- A portion of the kinetic energy of a flow is continuously converted into heat due to friction that originates at the boundary and lost by dissipation into the environment.

And we should also note that if sediment transport occurs, some of the kinetic energy is transmitted from the water to the sediment.

4.6 Equations Based on Diffusion

The **flux** of a "substance," which may be material (e.g., sediment or dissolved constituents), momentum, or energy, is its rate of movement across a plane per unit area of the plane and per unit time. The dimensions of a flux are $[\mathrm{S\,L^{-2}\,T^{-1}}]$, where [S] represents the dimensions of the diffusing substance $\mathcal{S}$ (for matter, [S] = [M]; for momentum, $[\mathrm{S}] = [\mathrm{M\,L\,T^{-1}}]$; for energy, $\mathcal{S} = [\mathrm{M\,L^2\,T^{-2}}]$). In the phenomenon of **diffusion**, the flux of a "substance" through a medium occurs in response to a spatial gradient of the concentration of the "substance" (figure 4.11).

The physiologist Adolf Fick (1829–1901) determined the law governing this process, which is known as **Fick's law**:

$$\mathcal{F}_x(\mathcal{S}) = -D_S \cdot \frac{\mathrm{d}C(\mathcal{S})}{\mathrm{d}x}. \tag{4.46}$$

In words, this law states that

> The **flux** (flow per unit area per unit time), $\mathcal{F}_x(\mathcal{S})$, of substance $\mathcal{S}$ (matter, momentum, or energy) in the x-direction through a medium is proportional to the product of 1) the gradient of the concentration of $\mathcal{S}$, $C(\mathcal{S})$, in the x-direction, and 2) the **diffusivity** of $\mathcal{S}$ in the medium, $D_{\mathcal{S}}$.

The negative sign specifies that the flux is "down-gradient," that is, from a region where the concentration of $\mathcal{S}$ is larger to where it is smaller.

Fick's law governs the diffusion of tea from a tea bag in hot water, the movement of heat from the hotter to the colder end of a metal rod, the dispersion of sediment or pollutants in river flows and groundwater, and many other phenomena. Obviously, the substance involved and the mechanism causing the diffusion, and hence the numerical value of the diffusivity, differ in these various contexts, but the dimensions of diffusivity are always $[\mathrm{L^2\,T^{-1}}]$, regardless of whether $\mathcal{S}$ represents matter, momentum, or energy and regardless of the nature of the medium. And, since the concentration of $\mathcal{S}$ has dimensions $[\mathrm{S\,L^{-3}}]$, we can write Fick's law dimensionally as

$$[\mathrm{S\,L^{-2}\,T^{-1}}] = [\mathrm{L^2\,T^{-1}}] \cdot [\mathrm{S\,L^{-3}}]/[\mathrm{L}]. \tag{4.47}$$

In section 3.3.3, we saw that the relation between applied shear stress and velocity gradient for a Newtonian fluid also described the flux of momentum, M, down the

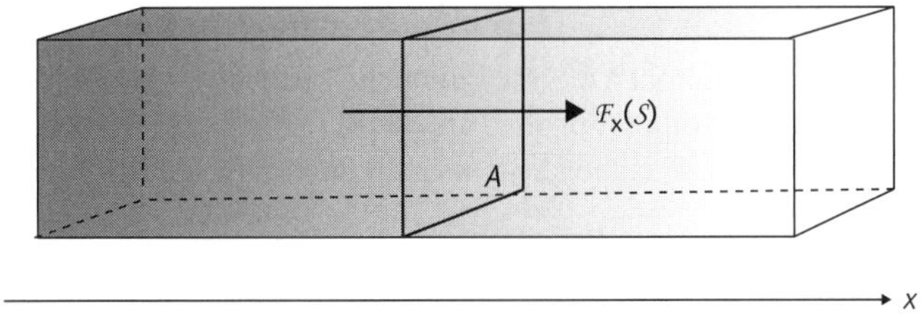

Figure 4.11 Conceptual diagram of the diffusion process (equation 4.46). The gray scale depicts the concentration of substance $\mathcal{S}$, $C(\mathcal{S})$, in the x-direction; $\mathcal{F}_x(\mathcal{S})$ is the **flux** of $\mathcal{S}$, that is, the amount of $\mathcal{S}$ flowing per unit area, A, per unit time.

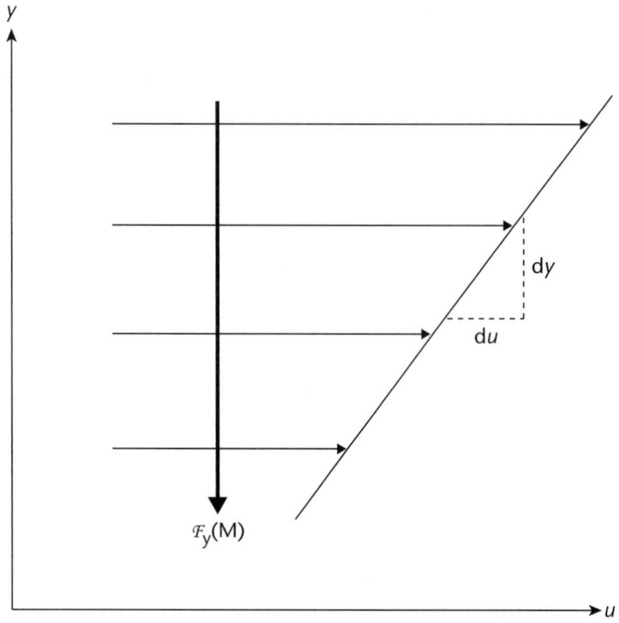

Figure 4.12 Diffusion of momentum in an open-channel flow. The horizontal arrows are vectors of the downstream velocity u; the velocity gradient is du/dy. The vertical arrow represents the flux of momentum, $\mathcal{F}_y(\mathrm{M})$, down the velocity gradient.

velocity gradient, $\mathrm{d}u_x(y)/\mathrm{d}y$ (equations 3.21 and 3.24). This flux is illustrated in figure 4.12. We can now show that this phenomenon is a manifestation of Fick's law describing the diffusion of momentum.

The concentration of momentum at any level, $C(\mathrm{M})$, is $\rho \cdot u$ ($[\mathrm{M\,L\,T}^{-1}]/[\mathrm{L}^3] = [\mathrm{M\,L}^{-2}\,\mathrm{T}^{-1}]$).[5] Writing equation 4.46 for this situation gives

$$\mathcal{F}_y(\mathrm{M}) = -D_\mathrm{M} \cdot \frac{\mathrm{d}[C(\mathrm{M})]}{\mathrm{d}y} = -D_\mathrm{M} \cdot \frac{\mathrm{d}(\rho \cdot u)}{\mathrm{d}y}. \tag{4.48}$$

Because we can almost always assume that ρ is constant,

$$\mathcal{F}_z(\mathrm{M}) = -D_\mathrm{M} \cdot \rho \cdot \frac{\mathrm{d}u}{\mathrm{d}y}. \tag{4.49a}$$

The diffusivity of momentum, D_M, is the kinematic viscosity, $\nu \equiv \mu/\rho$ (equation 3.23), and the dimensions of momentum flux are $[\mathrm{M\,L\,T}^{-1}]/[\mathrm{L}^2\,\mathrm{T}] = [\mathrm{M\,L}^{-1}\,\mathrm{T}^{-2}]$, or, equivalently, $[\mathrm{F\,L}^{-2}]$, which is the shear stress induced by viscosity, $-\tau_{yx}$. Thus, we can write

$$\mathcal{F}_y(\mathrm{M}) = -\tau_{yx} = -\nu \cdot \rho \cdot \frac{\mathrm{d}u}{\mathrm{d}y},$$

$$\tau_{yx} = \mu \cdot \frac{\mathrm{d}u}{\mathrm{d}y}, \tag{4.49b}$$

and see that equation 4.49b is identical to equation 3.19.

We will invoke Fick's law in several other contexts later in this text, including the movement of a flood wave along a river (section 11.5) and the vertical concentration of sediment (section 12.5.2).

4.7 Force-Balance and Conductance Equations

Many of the basic relations for fluid flow are derived by assuming steady uniform flow; that is, that the fluid elements are experiencing no convective or local accelerations and are therefore moving with constant velocity. From Newton's second law, this implies that there are no net forces acting on the fluid. Stated simply,

$$F_D = F_R, \tag{4.50}$$

where F_D represents the net forces tending to cause motion, and F_R represents the net forces tending to resist motion.

If we consider a fluid element of volume V within an open-channel flow with a water surface sloping at angle θ_S (figure 4.13), the force tending to cause motion of a fluid element in an open channel is the downslope component of its weight, given by

$$F_D = \gamma \cdot V \cdot \sin\,\theta_S. \tag{4.51}$$

Note that $\sin\theta_S = -dz/dx$ and expresses the gradient of gravitational potential energy. (There is no net pressure force on the element because its upstream and downstream ends are the same distance below the surface; thus, the pressure-potential-energy gradient is zero.)

As we will see, the forces resisting flow are due to the frictional resistance provided by the flow boundary, and are functions of the flow velocity, u. We will postpone

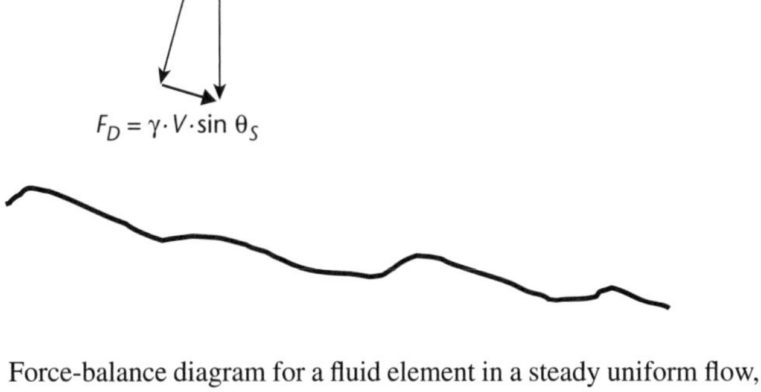

Figure 4.13 Force-balance diagram for a fluid element in a steady uniform flow, the basis for developing a generalized conductance equation (equations 4.51–4.54).

examining the exact forms of these functions and for now write

$$F_R = f_\Omega^*(u), \tag{4.52}$$

where $f_\Omega^*(u)$ represents an unspecified function of velocity. Combining equations 4.50–4.52,

$$f_\Omega^*(u) = \gamma \cdot V \cdot \sin\,\theta_S. \tag{4.53}$$

Solving (4.53) for u,

$$u = f_\Omega(\gamma \cdot V \cdot \sin\,\theta_S), \tag{4.54}$$

where $f_\Omega(.) = f_\Omega^{*-1}(.)$.

As we will see later, the function f_Ω reflects the **conductance** (inverse of the **resistance**) of the flow path, which depends on the water properties, the geometry of the boundary, and the flow state. Equation 4.54 is a generalized **conductance equation** for open-channel flow, and we summarize its development by stating that

> Conductance equations, which relate velocity to the gradient of gravitational potential energy, can be derived for various flow states and configurations by balancing the forces inducing flow with those resisting flow.

And, although derived under the assumption of steady uniform flow, conductance equations are usually assumed to apply to open-channel flows generally.

4.8 Other Bases for Equations

This section introduces the bases for equations that are not derived directly from the basic laws of physics but that are useful and, because of the limitations of our ability to measure and understand of all the factors that affect open-channel flows, often necessary for quantitative analysis.

4.8.1 Equations of Definition

It is often convenient to give a single name and symbol to the relation between two or more physical quantities. For example, as noted in section 3.3.3, the ratio of the dynamic viscosity, μ, to the mass density, ρ, often arises in the quantitative description of flow phenomena, and the kinematic viscosity, ν, is the name given to that ratio—that is,

$$\nu \equiv \frac{\mu}{\rho}. \tag{4.55}$$

Similarly, the ratio of the cross-sectional area, A, to the wetted perimeter, P_w, of a flow often arises (chapter 6), and that ratio is called the hydraulic radius, R:

$$R \equiv \frac{A}{P_w} \tag{4.56}$$

These equations are read as, "Kinematic viscosity is defined as the ratio of dynamic viscosity to mass density," and "Hydraulic radius is defined as the ratio of cross-sectional area to wetted perimeter," respectively.

Equations such as 4.55 and 4.56 are **equations of definition**. It is important to recognize these and to understand that the essential difference between an equation of definition and other types of equations is that equations of definition present no new information—they simply specify a convenient symbolic and nomenclatural shorthand. The use of the **identity sign** ($\equiv$), rather than the equal sign makes clear the distinction. However, many writers do not use the identity sign, so often you will have to study the text in order to identify equations of definition.

4.8.2 Equations Based on Dimensional Analysis

4.8.2.1 Theory of Dimensional Analysis

An equation that completely and correctly describes a physical relation has the same dimensions on both sides of the equal sign, and is thus **dimensionally homogeneous**. This truth is emphasized in the developments of sections 4.2–4.7; these developments begin with basic laws of physics, and subsequent mathematical operations preserve dimensional homogeneity. (Appendix A summarizes the rules for dealing with the dimensions of physical quantities.)

There are many fluid-flow problems for which we can identify the variables involved with reasonable confidence but, because of complicated boundary geometry and/or the random nature of turbulence, for which we cannot derive the relevant equations from the basic laws of physics. Because several variables are usually involved, it would be at best inefficient to try to determine the relations among all the variables by experiment. **Dimensional analysis** simplifies the analysis of such problems by incorporating the basic variables into a smaller number of dimensionless variables. Once this smaller number of variables is identified, one can conduct experiments to determine the relationships among them. As we will see in later chapters, this process has been frequently applied to fluid-flow problems and has led to theoretical insights as to the basic relations among variables and practical simplifications in the design of experiments.

This section describes the theory of dimensional analysis, presents a strategy for formulating physically sound universal relationships for such problems, and illustrates the types of insight that can be obtained from the procedure by applying it to an important problem of open-channel flow.

Dimensional analysis was introduced to English-speaking scientists and engineers by Edgar Buckingham (1915) and is based on the **Buckingham pi theorem**. Here we outline the basic approach; further description can be found in Rouse (1938), Middleton and Southard (1984), Middleton and Wilcock (1994), and Furbish (1997).

Buckingham's pi theorem can be summarized succinctly:

1. If a fluid-flow situation is completely characterized by N variables X_i, $i = 1, 2, \ldots, N$, then

$$0 = f(X_1, X_2, \ldots, X_N) \tag{4.57}$$

where "f" signifies some function.[6]

2. If these N variables have a total of n fundamental dimensions, they can be arranged into $N - n$ dimensionless pi terms, Π_j, $j = 1, 2, \ldots, N - n$, and the

relation can be characterized in the form

$$0=f(\Pi_1, \Pi_2, \ldots, \Pi_{N-n}). \tag{4.58}$$

For most problems of fluid flow, $n = 3$, that is, [M] or [F], [L], and [T]. If temperature [Θ] is also involved, $n = 4$.

3. Each pi term contains $n+1$ of the N variables.
4. Each pi term contains n common variables and one variable that is unique to it.

The steps for constructing pi terms are described in box 4.1.

The results and these steps are applied to a central problem of open-channel flow in the following subsection.

4.8.2.2 An Application of Dimensional Analysis to Open-Channel Flow

Equation 4.54 is a generalized relation between the velocity of a fluid element and the gradient of gravitational potential energy. We can use the combination of dimensional analysis and empirical observation to obtain further information about the specific relation between the average velocity of an open-channel flow (U) and the gradient of gravitational potential energy ($g \cdot \sin\theta_S$), where g is gravitational acceleration and θ_S is the water-surface slope angle. As indicated in step 1 of box 4.1, the process begins by identifying the variables thought to be relevant to the problem. For this case, we will assume that the relation between U and $g \cdot \sin\theta_S$ involves the geometry of the flow (width, W; depth, Y; and a quantity proportional to the height of roughness elements on the channel boundary, y_r) and the fluid properties mass density, ρ; surface tension, σ, and viscosity, μ.

Box 4.2 applies the steps of the Buckingham pi theorem to this problem. Substituting the results into equation 4.57, we have condensed the original problem with eight variables into one with five dimensionless variables:

$$0=f(\Pi_1, \Pi_2, \Pi_3, \Pi_4, \Pi_5) \tag{4.59a}$$

$$0=f\left(\frac{Y}{W}, \frac{Y}{y_r}, \frac{U^2}{Y \cdot (g \cdot \sin\theta_S)}, \frac{Y \cdot U^2 \cdot \rho}{\sigma}, \frac{Y \cdot U \cdot \rho}{\mu}\right). \tag{4.59b}$$

Since we are focusing on the relation between U and $g \cdot \sin\theta$, we can separate out the term containing those quantities and write

$$\Pi_3 = f_\Omega(\Pi_1, \Pi_2, \Pi_4, \Pi_5), \tag{4.60a}$$

$$\frac{U^2}{Y \cdot (g \cdot \sin\theta_S)} = f_\Omega\left(\frac{Y}{W}, \frac{Y}{y_r}, \frac{Y \cdot U^2 \cdot \rho}{\sigma}, \frac{Y \cdot U \cdot \rho}{\mu}\right), \tag{4.60b}$$

BOX 4.1 Construction of Buckingham Pi Terms

1. Identify all variables $X_1, X_2, \ldots, X_N$ required to describe the flow situation.
2. Assign each variable to one of the following categories (see Appendix A, table A.1): (a) *Geometric*—variables describing the boundaries and dimensions of the situation whose dimensions include [L] only: lengths, areas, volumes. (b) *Kinematic/dynamic*—variables whose dimensions include [M] or [F] and/or [T]: velocities, discharges, forces, stresses, accelerations, energies, momentums. (c) *Fluid properties*—for open-channel flow problems; these may include viscosity, density, surface tension.
3. Indicate the dimensions of each variable in the form $[\mathrm{L}^a\ \mathrm{M}^b\ \mathrm{T}^c]$.
4. Select n common variables, $X_{c1}, X_{c2}, \ldots, X_{cn}$, which have the following properties: (a) none can be dimensionless; (b) no two can have the same dimensions; (c) none can be expressible as the product of others (or as the product raised to a power); (d) collectively, the common variables must include all the n fundamental dimensions. One way to achieve these properties is to select one variable from each category of step 2 to be common.
5. Each pi term then includes the n common variables and one of the unique variables and takes the form
$$\Pi_j = X_{c1}{}^{xj} \cdot X_{c2}{}^{yj} \cdot X_{c3}{}^{xj} \cdot X_j^{\pm 1},\ j = 1, 2, \ldots, N - n, \qquad (4B1.1)$$
where X_j are successively chosen from the unique variables. [In equation 4B1.1 and subsequently we assume $n = 3$ (M, L, and T).] The exponent assigned to each noncommon variable is chosen arbitrarily as either $+1$ or -1.
6. Because each pi term must be dimensionless, its dimensions must satisfy
$$[\mathrm{L}^a\mathrm{M}^b\mathrm{T}^c]^{xj} \cdot [\mathrm{L}^d\mathrm{M}^e\mathrm{T}^f]^{yj} \cdot [\mathrm{L}^g\mathrm{M}^h\mathrm{T}^k]^{zj} \cdot [\mathrm{L}^p\mathrm{M}^q\mathrm{T}^r]^{\pm 1} = [\mathrm{L}^0\mathrm{M}^0\mathrm{T}^0], \qquad (4B1.2)$$
where the exponents $a, b, \ldots, r$ are those appropriate to each variable.
7. For each pi term, use equation 4B1.2 to write n simultaneous equations, one for each dimension:
$$[\mathrm{L}]: a \cdot xj + d \cdot yj + g \cdot zj + p \cdot (\pm 1) = 0 \qquad (4B1.3L)$$
$$[\mathrm{M}]: b \cdot xj + e \cdot yj + h \cdot zj + q \cdot (\pm 1) = 0 \qquad (4B1.3M)$$
$$[\mathrm{T}]: c \cdot xj + f \cdot yj + k \cdot zj + r \cdot (\pm 1) = 0. \qquad (4B1.3T)$$
8. Solve equations (4B1.3) to find the values of xj, yj, zj for the jth pi term.
9. Conduct experiments to determine the relations among the dimensionless pi terms.

BOX 4.2 Derivation of Pi Terms for Open-Channel Flow

Following the steps of box 4.1:

1. Geometric variables: W[L], Y[L], y_r[L] (y_r is the average height of roughness elements such as sand grains on the channel bed and banks)
2. Kinematic/dynamic variables: U [L T^{-1}], $g \cdot \sin \theta_S$ [L T^{-2}].
3. Fluid properties: ρ [M L^{-3}], σ [M T^{-2}], μ [M L^{-1} T^{-1}] (For this problem, $N = 8$ and $n = 3$. Thus, there will be $8 - 3 = 5$ pi terms.)
4. Select as common variables one from each category: Y, U, ρ. (These collectively contain all three dimensions.)
5. Write the pi terms:

$$\Pi_1 = Y^{x1} \cdot U^{y1} \cdot \rho^{z1} \cdot W^{-1}$$

$$\Pi_2 = Y^{x2} \cdot U^{y2} \cdot \rho^{z2} \cdot y_r^{-1}$$

$$\Pi_3 = Y^{x3} \cdot U^{y3} \cdot \rho^{z3} \cdot (g \cdot \sin \theta_S)^{-1}$$

$$\Pi_4 = Y^{x4} \cdot U^{y4} \cdot \rho^{z4} \cdot \sigma^{-1}$$

$$\Pi_5 = Y^{x5} \cdot U^{y5} \cdot \rho^{z5} \cdot \mu^{-1}$$

6. Write the dimensional equations for the pi terms:

$$\Pi_1 : [\mathrm{L}]^{x1} \cdot [\mathrm{L\,T^{-1}}]^{y1} \cdot [\mathrm{M\,L^{-3}}]^{z1} \cdot [\mathrm{L}]^{-1} = [\mathrm{L^0 M^0 T^0}]$$

$$\Pi_2 : [\mathrm{L}]^{x2} \cdot [\mathrm{L\,T^{-1}}]^{y2} \cdot [\mathrm{M\,L^{-3}}]^{z2} \cdot [\mathrm{L}]^{-1} = [\mathrm{L^0 M^0 T^0}]$$

$$\Pi_3 : [\mathrm{L}]^{x3} \cdot [\mathrm{L\,T^{-1}}]^{y3} \cdot [\mathrm{M\,L^{-3}}]^{z3} \cdot [\mathrm{L\,T^{-2}}]^{-1} = [\mathrm{L^0 M^0 T^0}]$$

$$\Pi_4 : [\mathrm{L}]^{x4} \cdot [\mathrm{L\,T^{-1}}]^{y4} \cdot [\mathrm{M\,L^{-3}}]^{z4} \cdot [\mathrm{M\,T^{-2}}]^{-1} = [\mathrm{L^0 M^0 T^0}]$$

$$\Pi_5 : [\mathrm{L}]^{x5} \cdot [\mathrm{L\,T^{-1}}]^{y5} \cdot [\mathrm{M\,L^{-3}}]^{z5} \cdot [\mathrm{M\,L^{-1} T^{-1}}]^{-1} = [\mathrm{L^0 M^0 T^0}]$$

7. Write and solve the three simultaneous equations for each pi term.

Π_1 :

$$[\mathrm{L}] : 1 \cdot x1 + 1 \cdot y1 - 3 \cdot z1 - 1 = 0$$

$$[\mathrm{M}] : 0 \cdot x1 + 0 \cdot y1 - 1 \cdot z1 + 0 = 0$$

$$[\mathrm{T}] : 0 \cdot x1 - 1 \cdot y1 + 0 \cdot z1 + 0 = 0$$

Therefore, $z1 = 0$, $y1 = 0$, and $x1 = 1$, so that $\Pi_1 = Y/W$.

Π_2 :

$[L]: 1 \cdot x2 + 1 \cdot y2 - 3 \cdot z2 - 1 = 0$

$[M]: 0 \cdot x2 + 0 \cdot y2 - 1 \cdot z2 + 0 = 0$

$[T]: 0 \cdot x2 - 1 \cdot y2 + 0 \cdot z2 + 0 = 0$

Therefore, $z2 = 0$, $y2 = 0$, and $x2 = 1$, so that $\Pi_2 = Y/y_r$.

Π_3 :

$[L]: 1 \cdot x3 + 1 \cdot y3 - 3 \cdot z3 - 1 = 0$

$[M]: 0 \cdot x3 + 0 \cdot y3 - 1 \cdot z3 + 0 = 0$

$[T]: 0 \cdot x3 - 1 \cdot y3 + 0 \cdot z3 - 2 = 0$

Therefore, $z3 = 0$, $y3 = 2$, and $x3 = -1$,

so that $\Pi_3 = U^2/[Y \cdot (g \cdot \sin \theta_S)]$.

Π_4:

$[L]: 1 \cdot x4 + 1 \cdot y4 - 3 \cdot z4 + 0 = 0$

$[M]: 0 \cdot x4 + 0 \cdot y4 + 1 \cdot z4 - 1 = 0$

$[T]: 0 \cdot x4 - 1 \cdot y4 + 0 \cdot z4 + 2 = 0$

Therefore, $z4 = 1$, $y4 = 2$, and $x4 = 1$, so that $\Pi_4 = Y \cdot U^2 \cdot \rho/\sigma$.

Π_5 :

$[L]: 1 \cdot x5 + 1 \cdot y5 - 3 \cdot z5 + 1 = 0$

$[M]: 0 \cdot x5 + 0 \cdot y5 + 1 \cdot z5 - 1 = 0$

$[T]: 0 \cdot x5 - 1 \cdot y5 + 0 \cdot z5 + 1 = 0$

Therefore, $z5 = 1$, $y5 = 1$, and $x5 = 1$, so that $\Pi_5 = Y \cdot U \cdot \rho/\mu$.

where f_Ω is an unknown function to be determined by experiment. To put equation 4.60b in a form similar to that of equation 4.54, we can take the square root of Π_3 (the term remains dimensionless) and write it as

$$U = f_\Omega\left(\frac{Y}{W}, \frac{Y}{y_r}, \frac{Y \cdot U^2 \cdot \rho}{\sigma}, \frac{Y \cdot U \cdot \rho}{\mu}\right) \cdot (Y \cdot g \cdot \sin \theta_S)^{1/2}. \tag{4.60c}$$

Although we still have a fairly large number of variables to sort out experimentally, we can use some intuition based on our knowledge of fluid properties and flows (which will become clearer as we proceed in this text) to identify what are likely to be the most important terms on the right side of equation 4.60c:

The quantity $\Pi_1 = Y/W$ is the ratio of flow depth to flow width, sometimes called the **aspect ratio**; its inverse is the **width/depth ratio**, W/Y. It is a potentially useful predictor of U because it can be independently determined a priori. We saw in section 2.4.2 that this quantity has an influence on flows in streams, because it is a measure of the relative importance of bed friction and bank friction on the flow (see figure 2.23). However, we also saw that most natural streams are "wide," so the influence of the bank is usually minor; thus, we can conclude that Y/W is probably only a minor factor in f_Ω.

The quantity $\Pi_2 = Y/y_r$ is the ratio of flow depth to the height of roughness elements on the channel boundary and is called the **relative smoothness** (its inverse, y_r/Y, is the **relative roughness**). This is a potentially useful predictor because, like Y/W, the value of Y/y_r can be determined a priori. Relative smoothness varies over a considerable range in natural streams, from near 1 in small bouldery mountain streams to over 10^5 in large silt-bed rivers. Thus, it seems reasonable to consider this variable a potentially important determinant of f_Ω. (We will explore this more fully in chapter 6.)

Π_4, the term involving surface tension, is called the **Weber number**, *We*, which expresses (inversely) the relative importance of surface tension in a flow:

$$We \equiv \frac{Y \cdot U^2 \cdot \rho}{\sigma}. \tag{4.61}$$

As we will see in chapter 7, *We* is very large even in small streams, reflecting the negligible role of surface tension. Thus, we can assume that *We* is not an important component of equation 4.60c. Note also that computation of *We* requires information about U, so it cannot be determined a priori.

Π_5, the term involving viscosity, is called the **Reynolds number**, *Re*:

$$Re \equiv \frac{Y \cdot U \cdot \rho}{\mu} \tag{4.62}$$

As we saw in section 3.4.2, the Reynolds number provides important information about the fluid flow state, because it expresses the relative importance of turbulence versus viscosity. This would thus seem to be an important factor in determining flow resistance, and we will explore this relation further in chapter 6. Note though that sorting out its effect on Π_3 experimentally is complicated because U must be known to calculate *Re*.

Based on these considerations, we can simplify equation 4.60c somewhat by dropping *We*:

$$U = f_\Omega\left(\frac{Y}{W}, \frac{Y}{y_r}, \frac{Y \cdot U \cdot \rho}{\mu}\right) \cdot (Y \cdot g \cdot \sin\theta_S)^{1/2} \tag{4.63}$$

Because we have identified Y/y_r as the component of equation 4.63 likely to have the greatest influence on the ratio f_Ω, the next step in the analysis is to make use of empirical data to explore the relation between the two dimensionless variables $U/(Y \cdot g \sin\theta_S)^{1/2}$ and Y/y_r. Figure 4.14 shows this relation for 28 New Zealand

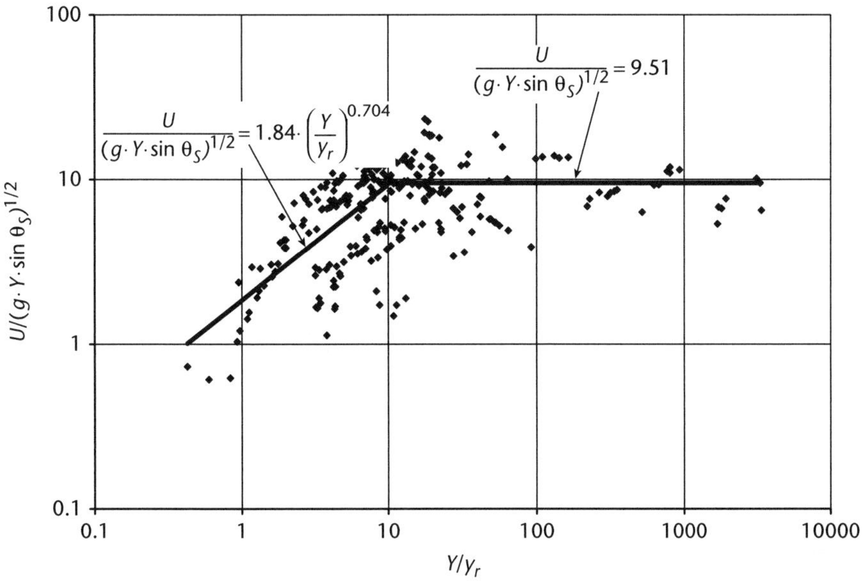

Figure 4.14 Combined plot of $U/(g \cdot Y \cdot \sin\theta_S)^{1/2}$ versus Y/y_r for 29 New Zealand stream reaches for which at least seven flows were measured and reported by Hicks and Mason (1991). The sloping line is equation 4.73.

stream reaches; for most reaches there is a strong dependence of $U/(Y \cdot g \sin \theta_S)^{1/2}$ on Y/y_r, as our analysis predicted. However, when all points are considered together, there is considerable scatter and a suggestion that the relationship is less important when Y/y_r exceeds about 50. The scatter is presumably due to the effects of the other dimensionless variables in equation 4.63, Y/W and Re, although it could also be due to important variables not considered in the problem formulation—for example, the effects of channel vegetation or channel curvature. However, dimensional analysis coupled with empirical observations allows us to state that, as a "first cut,"

$$U = f_\Omega\left(\frac{Y}{y_r}\right) \cdot (Y \cdot g \cdot \sin\,\theta_S)^{1/2}. \tag{4.64}$$

We will see in chapter 6 that the basic relation expressed in equation 4.64 is widely used for relating velocity to depth and slope in natural open-channel flows. Thus, we can conclude that dimensional analysis is a powerful tool for identifying dimensionless quantities characterizing flows and, when supplemented by observation, for revealing fundamental relations among flow variables. We will encounter other examples of the application of dimensional analysis throughout this text. The following section introduces approaches to identifying the mathematical form of empirical relations, such as that indicated for f_Ω in figure 4.14.

4.8.3 Empirical Equations

"Empirical" means "relying upon or derived from observation or experiment." **Empirical equations** are developed from measurements (observations), rather than from fundamental physical laws. Earth scientists frequently rely on empirical equations because earth processes are complex and distributed in space and/or time, and it is often not feasible to derive the applicable equations from the laws discussed in sections 4.2–4.7. However, it is important to understand that empirical relations differ fundamentally from those based on physical laws.

The next subsection outlines the most common approach to developing empirical equations and emphasizes the differences between such equations and those based on physical laws. The concluding subsection shows that one can often reduce some of the limitations associated with strictly empirical relations by combining empirical analysis with dimensional analysis as described in section 4.8.2.

4.8.3.1 Regression Equations

The standard approach to developing empirical equations is by regression analysis. Although the detailed methodology of regression analysis is beyond the scope of this text,[7] we will examine some of the basic characteristics of regression equations, beginning with the steps involved in developing them:

1. Select the variables of interest: Usually the objective is to develop an equation to estimate the value of a single dependent variable, Y, from measured values of one or more predictor (or independent) variables, $X_1, X_2, \ldots, X_P$.[8]
2. Formulate the regression model: The standard model is a **linear additive model**, which has the form

$$\hat{Y} = b_0 + b_1 \cdot X_1 + b_2 \cdot X_2 + \cdots + b_P \cdot X_P, \qquad (4.65)$$

   where $b_0 \ldots b_P$ are regression coefficients (b_0 is often called the **regression constant**). However, in hydraulics the most common model is the **linear multiplicative model**:

$$\hat{Y} = c_0 \cdot X_1{}^{c1} \cdot X_2{}^{c2} \cdot \ldots X_P{}^{cP} \qquad (4.66)$$

   Although the choice of additive or multiplicative model is up to the scientist, the regression process is identical in both, because equation 4.66 can be put in the form of 4.65 via a logarithmic transform:

$$\widehat{\log Y} = \log c_0 + c_1 \cdot \log X_1 + c_2 \cdot \log X_2 + \cdots + c_P \cdot \log X_P \qquad (4.67)$$

   Note that the "hat" notation in equations 4.65 and 4.67 denotes an *estimate* of the average value of the dependent variable Y or $\log Y$ associated with a particular set of x_{ji} values. This estimate is subject to uncertainty because 1) the model is always imperfect, and 2) the coefficients are derived for a specific set of data.
3. Collect the data: These are N measured values (observations) of the dependent and independent variables, $y_i, x_{1i}, x_{2i}, \ldots, x_{Pi}$, which must be associated in space or time.
4. Determine the values of the coefficients: The mathematics of ordinary regression analysis provide estimates of $b_0 \ldots b_p$ or $c_0 \ldots c_P$ that "best fit" the observations

in the sense that, for the data used, the coefficients minimize

$$\sum_{i=1}^{N}(y_i - \hat{y}_i)^2 \text{ or } \sum_{i=1}^{N}(\log y_i - \widehat{\log y_i})^2, \quad (4.68)$$

where the y_i are the actual measured values of the dependent variable and the $\hat{y}_i$ or $\widehat{\log y_i}$ are the values estimated by the regression equation (equation 4.65 or 4.67), and there are N sets of measured values ($i = 1, 2, \ldots, N$).

From these steps, it is clear that regression equations differ fundamentally from equations based on the laws of physics:

- The P variables included in an empirical equation are determined by the scientist, not by nature.
- The form of an empirical equation is determined by the scientist, not by nature.
- The numerical coefficients and exponents in an empirical equation are determined by the particular set of data analyzed (the N sets of y and x_j values) and, in general, are not universal.
- The relationships resulting from statistical analysis reflect association among variables, but not necessarily causation.

Because of these characteristics, uncertainty is an inherent aspect of regression analysis. There are some additional critical differences between regression equations and those derived from basic principles. One that is often overlooked is that ordinary regression equations are not invertible. To understand this, suppose we analyze a set of data and produce a regression equation

$$\hat{Y} = b_0 + b_1 \cdot X_1. \quad (4.69)$$

If this were a purely mathematical relation, we would consider that $\hat{Y} = Y$, and it would be true that

$$X_1 = -\frac{b_0}{b_1} + \frac{1}{b_1} \cdot Y. \quad (4.70)$$

However, if we use the same data to do an ordinary regression with X_1 as the dependent variable and Y as the predictor variable, the constant will not be equal to $(-b_0/b_1)$ and the coefficient will not be equal to $(1/b_1)$.[9]

A final fundamental difference between empirical equations and those derived from basic physics is that, in general, empirical equations are not dimensionally homogeneous. As explained in appendix A, this means that the coefficients estimated by the regression analysis must be changed for use in different measurement systems (e.g., British and SI).

4.8.3.2 Empirical Equations Based on Dimensional Analysis

The use of dimensional analysis to reduce a problem involving a large number of physical variables to one involving a smaller number of dimensionless quantities is described in section 4.8.2. Once the dimensional analysis is completed, the nature of the functional relationships among the dimensionless quantities is explored using observational data from laboratory experiments or field observations. Regression analysis can be a useful tool in this exploration.

Applying linear regression models to dimensionless quantities, we can write the analogs of equations 4.65 and 4.66, respectively, as

$$\hat{\Pi}_Y = b_0 + b_1 \cdot \Pi_1 + b_2 \cdot \Pi_2 + \cdots + b_P \cdot \Pi_P, \tag{4.71}$$

and

$$\hat{\Pi}_Y = c_0 \cdot {\Pi_1}^{c1} \cdot {\Pi_2}^{c2} \cdot \ldots {\Pi_P}^{cP}, \tag{4.72}$$

where one of the pi terms has been selected as the dependent variable and designated Π_Y.

Whichever model we choose, all the quantities are dimensionless, so in addition to simplifying the problem, we avoid having to worry about changing equations for use with different unit systems.

To illustrate this approach, we return to the dimensional analysis example in section 4.8.2.1. We focus on equation 4.64 and plot $U/(Y \cdot g \cdot \sin \theta_S)^{1/2}$ versus Y/y_r for 29 stream reaches in New Zealand in figure 4.14 using data provided by Hicks and Mason (1991). Note that both axes of that plot are logarithmic, and the distribution of plotted points suggests that one could approximate the relation by an upward-sloping straight line for $Y/y_r \leq 10$. Thus, we select the multiplicative (logarithmic) model (equation 4.66 with $P = 1$), and the regression analysis yields

$$\frac{U}{(Y \cdot g \cdot \sin \theta_S)^{1/2}} = 1.84 \cdot \left(\frac{Y}{y_r}\right)^{0.704} \tag{4.73}$$

as a first approximation of f_Ω; this line is plotted in figure 4.14. For $Y/y_r > 10$, the relationship can be approximated as simply the average value of $U/(Y \cdot g \cdot \sin \theta_S)^{1/2} =$ 9.51. Thus, the dimensional analysis combined with the measured data suggests the following model for predicting velocity:

$$U = 1.84 \cdot \left(\frac{Y}{y_r}\right)^{0.704} \cdot (Y \cdot g \cdot \sin \theta_S)^{1/2}, \; Y/y_r \leq 10; \tag{4.74a}$$

$$U = 9.51 \cdot (Y \cdot g \cdot \sin \theta_S)^{1/2}, \; Y/y_r > 10. \tag{4.74b}$$

Equation 4.74 is clearly an approximation, as there is much scatter about the line. Plotting the same data but identifying the points associated with each individual reach (figure 4.15) shows that the general form of the relation applies, but that the relationship is shifted from reach to reach. This pattern suggests that other factors that vary from reach to reach, perhaps including the pi terms W/Y and Re or other factors not included in the dimensional analysis, also affect velocity. Thus, we might conduct further analyses to explore approaches to reducing the scatter, focusing on 1) accounting for the effects of the other pi terms identified in the dimensional analysis, and 2) looking for factors not included in the original dimensional analysis that might affect the relationship, such as the presence of vegetation or channel curvature.

However, the dimensional analysis combined with measured data have clearly been a useful first step, and we can conclude that many important hydraulic relationships can be developed by empirical analysis of the relations between dimensionless variables identified via dimensional analysis. We will encounter several examples of this approach in subsequent chapters.

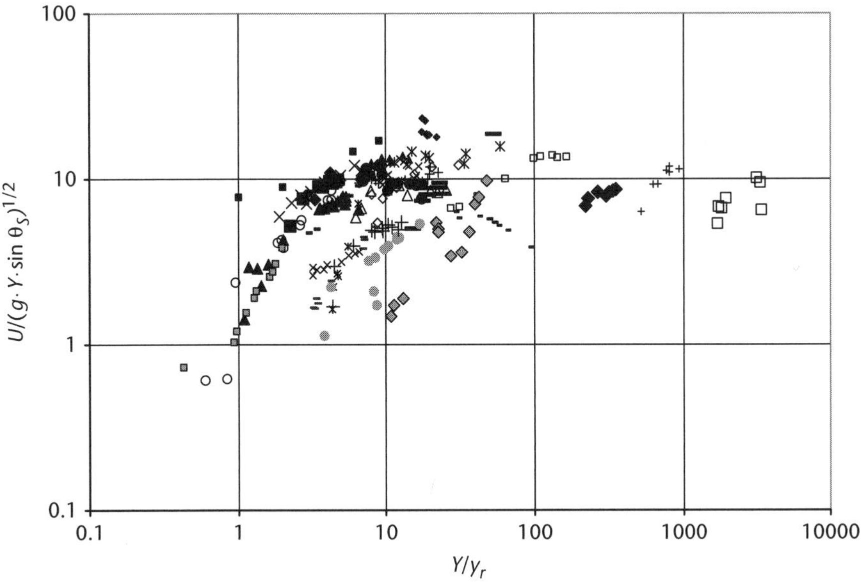

Figure 4.15 $U/(g \cdot Y \cdot \sin\theta_S)^{1/2}$ versus Y/y_r for 29 New Zealand stream reaches, where $y_r = d_{84}$. Flows from each reach are identified by a different symbol. Data from Hicks and Mason (1991).

4.8.4 Heuristic Equations

"Heuristic" means "helping to discover or learn; guiding or furthering investigation." A **heuristic equation** is one that, though not derived from basic physics or based on statistical analysis of observations, seems physically plausible and is generally consistent with observations. Hydrologists often invoke heuristic equations as conceptual models of complex processes when it is not practicable to develop detailed physically based representations or to collect all the data that would be necessary as input for such representations.

Probably the most common heuristic equation is the simple model of a hydrological or hydraulic reservoir as

$$Q = a_R \cdot V^{b_R}, \tag{4.75}$$

where Q is the rate of output $[\mathrm{L}^3\,\mathrm{T}^{-1}]$ from the reservoir (which might be a lake, a segment of a river channel, an aquifer, or a watershed), V is the volume of water $[\mathrm{L}^3]$ stored in the reservoir, and a_R and b_R are selected to best represent the particular situation.

In many situations, the exponent is assigned a value $b_R = 1$, and equation 4.75 then represents a linear reservoir. In this case, a_R has the dimensions $[\mathrm{T}^{-1}]$ and is equal to the inverse of the **residence time** of the reservoir, which is the average length of time an element of water spends in the reservoir (see Dingman 2002). Although the linear reservoir model does not strictly represent the way most natural hydraulic and hydrological reservoirs work, it does capture many of the essential aspects and is mathematically (and dimensionally) tractable.

We will incorporate the linear reservoir model in a simplified approach to predicting how flood waves move through stream channels in chapter11, and you will probably encounter heuristic equations in other hydrological and hydraulic contexts.

5

Velocity Distribution

5.0 Introduction and Overview

Previous chapters have discussed the velocity of individual fluid elements (point velocities), denoted as u_x, u_y, u_z, and the average velocity through a stream cross section, denoted as U. The main objective of the present chapter is to explore the connection between point velocities and cross-section average velocity by developing physically sound quantitative descriptions of the distribution of velocity in cross sections.

However, there has been little research on the distribution of velocities in entire cross section, so most of the discussion here will be devoted to velocity profiles:

> The **velocity profile** is the relation between downstream-directed velocity $u(y)$ and normal distance above the bottom, y.[1]

After an exploration of theoretical and actual velocity profiles, the last section of this chapter discusses the characterization of cross-sectional velocities.

Velocity profiles are the basis for formulating expressions for resistance, which can be viewed as the central problem of open-channel flow (chapter 6): The velocity profile is the consequence of the no-slip condition and the effects of viscosity and turbulence and thus is the manifestation of boundary friction, or resistance (see figure 3.28). Understanding velocity profiles is also critical for measuring streamflow and for understanding how sediment is entrained and transported (chapter 12).

Velocity profiles are developed from the force-balance concepts discussed in section 4.7, and the starting point is the balance of driving forces, F_D, and resisting forces, F_R, given by equation 4.50 for uniform flows:

$$F_D = F_R. \tag{5.1}$$

The other essential components of the derivations are 1) the relation between shear stress and velocity gradient given by equation 3.19 for laminar flow and equation 3.40 for turbulent flow; and 2) the relation between shear stress and distance above the bottom, which is derived in the following section. To simplify the profile derivations, we specify that the channel is "wide," that is, that we can neglect any frictional effects from the banks and assume that the flow is affected only by the friction arising from the channel bed (section 2.4.2).

The local average "vertical" velocity U_w is given by the integral of the velocity profile over the local depth, Y_w:

$$U_w = \frac{1}{Y_w} \cdot \int_0^{Y_w} u(y) \cdot \mathrm{d}y \tag{5.2a}$$

The average cross-section velocity, U, is given by

$$U = \frac{1}{A} \cdot \int_0^{W} U_w(w) \cdot \mathrm{d}A(\mathrm{w}), \tag{5.2b}$$

where A is cross-sectional area, W is water-surface width, and w is the distance from one bank measured at the water surface. For a wide rectangular channel, the local depth Y_w equals the average depth, Y, and equation 5.2a gives U directly. Chapter 6 explores how integrated velocity profiles provide the basis for fundamental flow-resistance relations for a cross section or channel reach.

As shown in chapter 3 (see figure 3.29), the great majority of natural open-channel flows are turbulent, so the turbulent velocity distribution is of primary interest. However, the laminar distribution does have relevance: Even in fully turbulent flows, the no-slip condition induces very low velocities and viscous flow near the flow boundary (figure 3.28), and the laminar distribution applies in that region if the boundary is smooth (discussed further in section 5.3.1.5). Furthermore, there are natural flows in which the Reynolds numbers are in the laminar or transitional range, including very thin "overland flows" that occur on slopes following rainstorms (Lawrence 2000) and some flows in wetlands. For example, the Florida Everglades "River of Grass," which is 10–15 km wide and 1–2 m deep, has a velocity on the order of 210 m day^{-1} (2.4×10^{-3} m s^{-1}) and a Reynolds number of about 1,000, well into the transitional range (Bolster and Saiers 2002).

As in most of this text, the term "velocity" in this chapter refers to the velocity time-averaged to eliminate the fluctuations due to turbulence.

5.1 "Vertical" Force Profile in Uniform Flows

The balance of forces expressed in equation 5.1 is the essential feature of uniform flow. As shown in figure 4.3c, uniform flows are characterized by parallel streamlines, which means that 1) average velocity and depth do not change in the downstream direction, and 2) water-surface slope is identical to the channel slope. Of course, in natural channels, flow can be assumed to be uniform only over a reach of limited downstream extent.

For both laminar and turbulent uniform flows, the velocity profiles normal to the channel bottom are developed by balancing the driving and resisting forces at each

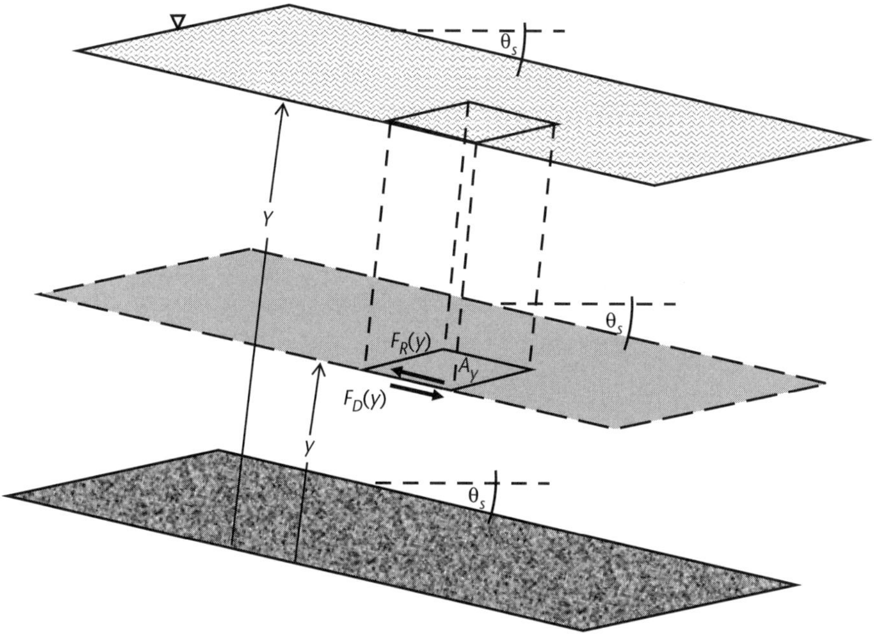

Figure 5.1 Definition diagram for deriving the relation between shear stress, τ, and distance above the bottom, y (equation 5.6).

level y within the flow, that is, by applying equation 5.1 in the form

$$F_D(y) = F_R(y), 0 \leq y \leq Y_w. \tag{5.3}$$

In this section we develop general expressions for $F_D(y)$ and $F_R(y)$ that we will use in deriving the velocity profiles for both flow states.

Figure 5.1 shows a plane parallel to the bottom and surface at an arbitrary height y above the bottom in a two-dimensional ("wide") uniform flow of depth Y. Because the depth does not vary along the channel, there is no pressure gradient (equation 4.25) and no pressure force to consider. Thus, the driving force in uniform flow is solely due to the downslope component of the weight of the water column. Isolating an area of size A_y on a plane at level y above the bottom, the downslope force on that area, $F_D(y)$, is thus

$$F_D(y) = \gamma \cdot (Y - y) \cdot A_y \cdot \sin\theta_S, \tag{5.4}$$

where γ is the weight density of water and θ_S is the slope.

In light of equation 5.3, it must also be true that

$$F_R(y) = \gamma \cdot (Y - y) \cdot A_y \cdot \sin\theta_S. \tag{5.5}$$

Dividing this force by the area A_y gives the shear stress $\tau(y)$:

$$\tau(y) \equiv \frac{F_R(y)}{A_y} = \gamma \cdot (Y - y) \cdot \sin\theta_S. \tag{5.6}$$

Recall from the discussions in sections 3.3.3 and 3.3.4 that this resisting force per unit area is the shear stress caused by molecular viscosity and, if the flow is turbulent, by the shear stress due to turbulent eddies. Thus, this simple derivation is independent of the flow state and leads to the important conclusion that, in a uniform flow, shear stress is a linear function of distance below the surface (figure 5.2).

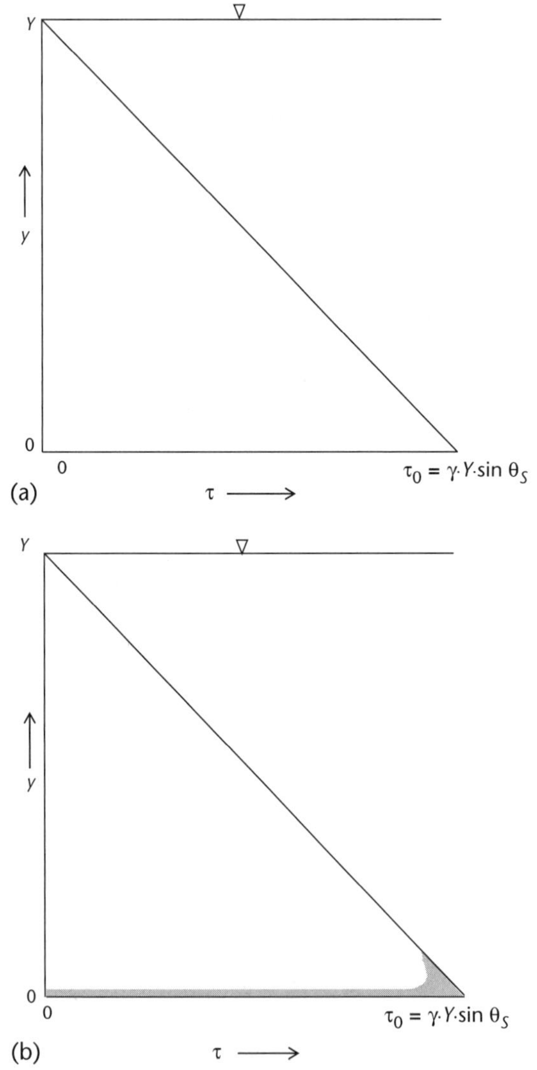

Figure 5.2 (a) The linear relation between shear stress, τ, and distance above the bottom, y, given by equation 5.6. This relation applies to both laminar and turbulent flow states. (b) Shear-stress distribution in a turbulent flow. The shaded area schematically represents the portion of total shear stress that is due to molecular viscosity. Total shear stress is the sum of that due to molecular viscosity and that due to eddy viscosity.

Where turbulence is fully developed, the eddy viscosity overwhelms the effects of molecular viscosity. However, even in turbulent flows, the velocity must go to zero at the bed due to the no-slip condition, so there is a region near the bed where turbulence is suppressed and molecular viscosity dominates. The relative importance of viscous and turbulent shear through a turbulent flow is schematically illustrated in figure 5.2b. This phenomenon is discussed more quantitatively in section 5.3.1.5.

Note that the derivation of the shear-stress profile in equation 5.6 is identical to the derivation of the hydrostatic pressure distribution in section 4.2.2.2, except that the shear stress depends on the sine of the slope (which gives the downslope component of the weight of overlying fluid) and the pressure on the cosine (which gives the component of the weight of overlying fluid that is normal to the bed). As with pressure, the profile of the downslope component of gravity and shear stress becomes significantly nonlinear in flows in which the streamlines are strongly curved (see figure 4.3). We will discuss such rapidly varied flows in chapter 11, but otherwise will assume that shear stress is a linear function of distance below the surface.

From equation 5.6, we see that the shear stress at the surface is zero and the shear stress at the bed, called the **boundary shear stress**, τ_0, is given by

$$\tau_0 = \gamma \cdot Y_w \cdot \sin\theta_S, \tag{5.7}$$

where Y_w is the local depth. The quantity τ_0 is a critically important quantity in open-channel flows because the boundary shear stress is the magnitude of the frictional force per unit area that the boundary exerts on the flow. And, following Newton's third law, the boundary shear stress is the magnitude of the erosive force per unit area that the flow exerts on the boundary. Chapter 6 will explore the role of τ_0 as a descriptor of boundary resistance; its role as a descriptor of erosive force plays a central role in the discussion of sediment transport in chapter 12.

5.2 Velocity Profile in Laminar Flows

5.2.1 Derivation

Equation 3.19b provides the relation between the shear stress and the "vertical" (i.e., y-direction, normal to the bottom) velocity gradient in laminar (viscous) flows:

$$\tau(y) = \mu \cdot \frac{du(y)}{dy}, \tag{5.8}$$

where μ is the dynamic viscosity. Equating 5.8 and 5.6, we have

$$\gamma \cdot (Y_w - y) \cdot \sin\theta_S = \mu \cdot \frac{du(y)}{dy};$$

$$du(y) = \frac{\gamma}{\mu} \cdot (Y_w - y) \cdot \sin\theta_S \cdot dy;$$

$$\int du(y) = \frac{\gamma \cdot \sin\theta_S}{\mu} \cdot \int (Y - y) \cdot dy. \tag{5.9}$$

Expression equation 5.9 is readily evaluated to give

$$u(y) = \frac{\gamma \cdot \sin \theta_S}{\mu} \cdot \left(Y_W \cdot y - \frac{y^2}{2} \right) + C_L, \tag{5.10}$$

where C_L is a constant of integration. The value of C_L is determined by noting the **boundary condition** dictated by the no-slip condition (section 3.3.3): $u(0) = 0$. Thus, $C_L = 0$, and the velocity profile for a wide laminar open-channel flow is given by

$$u(y) = \frac{\gamma}{\mu} \cdot \left(Y_W \cdot y - \frac{y^2}{2} \right) \cdot \sin \theta_S. \tag{5.11}$$

To visualize this distribution, we can first use equation 5.11 to calculate the velocity at the surface, $u(Y_W)$:

$$u(Y_W) = \frac{\gamma}{\mu} \cdot \left(\frac{Y_W{}^2}{2} \right) \cdot \sin \theta_S. \tag{5.12}$$

Then we can plot the dimensionless relative velocity $u(y)/u(Y_W)$ versus relative distance above the bottom, y/Y_W, in figure 5.3, where from equation 5.11 and 5.12,

$$\frac{u(y)}{u(Y_W)} = 2 \cdot \frac{y}{Y_W} - \left(\frac{y}{Y_W} \right)^2. \tag{5.13}$$

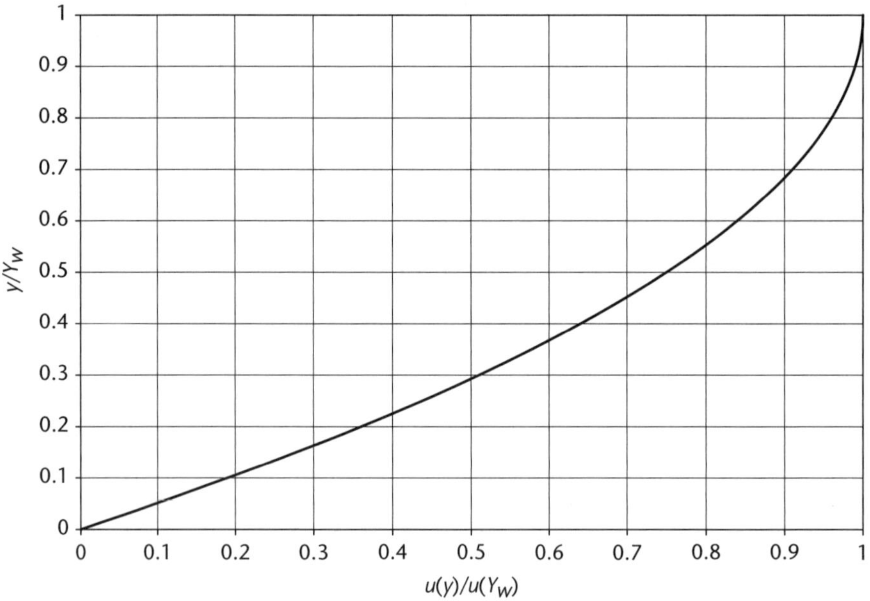

Figure 5.3 Relative velocity $u(y)/u(Y_W)$ as a function of relative distance above the bottom, y/Y_W, for laminar open-channel flows (equation 5.13).

From equation 5.13 and figure 5.3, we see that the velocity distribution in a laminar open-channel flow takes the form of a parabola, with the maximum velocity at the surface and, of course, zero velocity at the boundary.

5.2.2 Average "Vertical" Velocity

The average local "vertical" velocity of a wide laminar flow, U_W, is given by substituting equation 5.11 into 5.2 and integrating; evaluating that expression leads to

$$U_W = \left(\frac{\gamma}{3 \cdot \mu}\right) \cdot {Y_W}^2 \cdot \sin\theta_S. \tag{5.14}$$

Recall from section 3.4.2 that laminar flow only occurs when the Reynolds number, Re, is less than 500, where

$$Re \equiv \frac{U_W \cdot Y_W}{\nu}. \tag{5.15}$$

If we substitute 5.14 into 5.15 and recall that $\nu \equiv \mu/\rho$ and $\gamma = \rho \cdot g$, we arrive at

$$Re = \frac{g \cdot {Y_W}^3 \cdot \sin\,\theta_S}{3 \cdot \nu^2}, \tag{5.16}$$

and if $Re = 500$, the limiting value for laminar flow, we have

$$Y_W = \left(\frac{1500 \cdot \nu^2}{g \cdot \sin\,\theta_S}\right)^{1/3}. \tag{5.17}$$

We can use equation 5.17 to find the maximum depth for which a flow will be laminar at a specified slope; this relation is shown in figure 5.4. Note that even for surfaces with very low slopes (e.g., parking lots), this depth is in the centimeter range; for hillslopes, for which typically $\sin\theta_S > 0.01$, the maximum depth is in the millimeter range.

5.3 Velocity Profile in Turbulent Flows

5.3.1 The Prandtl-von Kármán Velocity Profile

5.3.1.1 Derivation

The "vertical" velocity distribution for wide turbulent flows can be derived using the same approach that was used for laminar flows. Note that equation 5.6 describes the distribution of shear stress for turbulent as well as laminar flows, but we now equate shear stress to equation 3.40a, which applies to turbulent flow:

$$\gamma \cdot (Y_W - y) \cdot \sin\theta_S = \rho \cdot \kappa^2 \cdot y^2 \cdot \left(1 - \frac{y}{Y_W}\right) \cdot \left(\frac{du(y)}{dy}\right)^2. \tag{5.18}$$

Recall from section 3.3.4.4 that κ is a proportionality factor known as von Kármán's constant. Noting that

$$\left(1 - \frac{y}{Y_W}\right) = \left(\frac{1}{Y_W}\right) \cdot (Y_W - y),$$

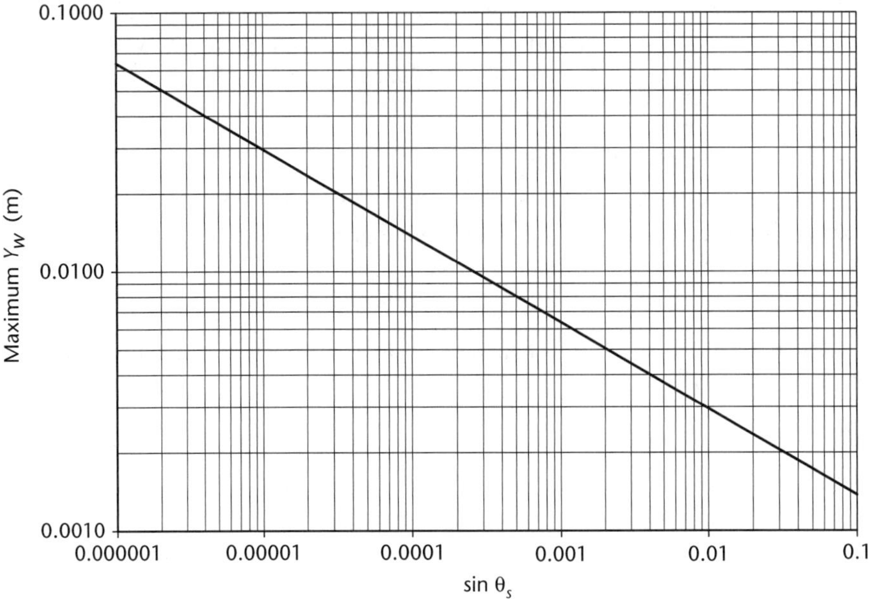

Figure 5.4 Maximum depth at which laminar flow occurs as a function of slope (equation 5.17). Kinematic viscosity is assuming a water temperature of 10°C.

equation 5.18 reduces to

$$\mathrm{d}u(y) = \left(\frac{1}{\kappa}\right) \cdot (g \cdot Y_w \cdot \sin\,\theta_S)^{1/2} \cdot \left(\frac{\mathrm{d}y}{y}\right), \tag{5.19a}$$

and

$$\int \mathrm{d}u(y) = \left(\frac{1}{\kappa}\right) \cdot (g \cdot Y_w \cdot \sin\,\theta_S)^{1/2} \cdot \int \frac{\mathrm{d}y}{y}. \tag{5.19b}$$

Carrying out the integration,

$$u(y) = \left(\frac{1}{\kappa}\right) \cdot (g \cdot Y_w \cdot \sin\,\theta_S)^{1/2} \cdot \ln(y) + C_T, \tag{5.20}$$

where C_T is once again a constant of integration. To evaluate this constant, we would like to again invoke the no-slip condition and specify $u(0) = 0$. This is mathematically precluded, however, because ln(0) is not defined. To get around this, we instead specify that $u(y_0) = 0$, where y_0 is a very small distance above the bottom. This allows us to evaluate C_T and arrive at

$$u(y) = \left(\frac{1}{\kappa}\right) \cdot (g \cdot Y_w \cdot \sin\,\theta_S)^{1/2} \cdot \ln\left(\frac{y}{y_0}\right). \tag{5.21}$$

Equation 5.21 is known as the **Prandtl-von Kármán universal velocity-distribution law**, and we will subsequently often refer to it as the "P-vK law."

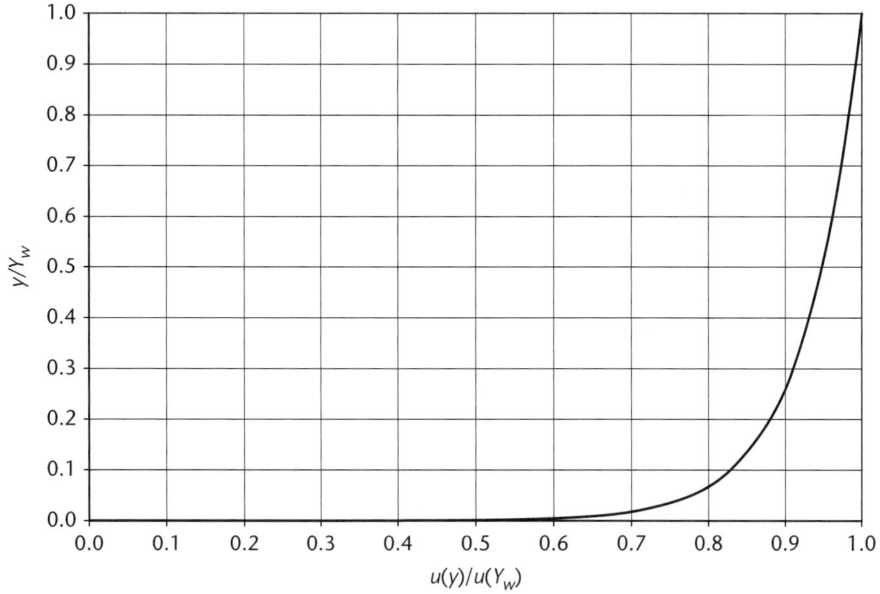

Figure 5.5 Relative velocity $u(y)/u(Y_w)$ as a function of relative distance above the bottom, y/Y_w, as given by the Prandtl-von Kármán universal velocity distribution (equation 5.21) for a turbulent open-channel flow with a depth $Y_w = 1\text{m}$ and a slope $\sin\theta = 0.001$.

The P-vK law allows $u(y)$ to be calculated when slope θ_S and flow depth Y_w are specified—provided that we can also determine y_0 as an independent parameter. We will see in section 5.3.1.6 that y_0 can be specified a priori, and figure 5.5 shows the form of the velocity distribution given by equation 5.21. Note that most of the change in velocity occurs very close to the bed, and the velocity gradient throughout most of the flow is much smaller than for laminar flow (figure 5.3). This is because turbulent eddies, which are present throughout most of the flow, are much more effective distributors of momentum than is molecular viscosity, which controls the momentum distribution very close to the bed.

Several aspects of the P-vK law require further exploration; these are discussed in the following subsections.

5.3.1.2 The P-vK Law and Shear Distribution

Section 5.1 showed that shear stress decreases linearly with distance below the surface (equation 5.6) in both laminar and turbulent flows. The development in section 3.3.4 used Prandtl's mixing-length hypothesis to arrive at the following expression for shear stress in a turbulent flow:

$$\tau = \rho \cdot \kappa^2 \cdot y^2 \cdot \left(1 - \frac{y}{Y_w}\right) \cdot \left(\frac{du}{dy}\right)^2. \tag{5.22}$$

This expression, which is identical to equation 3.40a, was incorporated in the derivation of the P-vK law (equation 5.18). To show that 5.22 is consistent with the linear shear-stress distribution, note from equation 5.19a that

$$\frac{du}{dy} = \frac{(g \cdot Y_w \cdot \sin \theta_S)^{1/2}}{\kappa \cdot y}. \tag{5.23}$$

Substituting equation 5.23 into 5.22 and noting that $\gamma = \rho \cdot g$ leads to equation 5.6, showing that the P-vK law is consistent with the linear shear-stress distribution.

5.3.1.3 Shear Velocity (Friction Velocity)

The quantity $(g \cdot Y_w \cdot \sin\theta_S)^{1/2}$ in equation 5.21 has the dimensions of a velocity. This quantity is called the **shear velocity**, or **friction velocity**, designated u_*:

$$u_* \equiv (g \cdot Y_w \cdot \sin\theta_S)^{1/2}. \tag{5.24}$$

The shear velocity is a measure of the intensity of turbulent velocity fluctuations. To see this, recall from equation 3.32 that the shear stress at a height y above the bed in a turbulent flow, $\tau(y)$, is related to the average turbulent velocity fluctuations as

$$\tau(y) = -\rho \cdot \bar{u}_x'(y) \cdot \bar{u}_y'(y), \tag{5.25}$$

where $\bar{u}_x'(y)$ and $\bar{u}_y'(y)$ are the average fluctuations in the x- and y-directions, respectively. We also saw from equation 3.31 that the magnitudes of these fluctuations are proportional, so we can write

$$\tau(y) = -k_{yx} \cdot \rho \cdot [\bar{u}_x'(y)]^2, \tag{5.26}$$

where k_{yx} is the proportionality constant. Now, noting equation 5.7, we see that

$$u* = \left(\frac{\tau_0}{\rho}\right)^{1/2} \tag{5.27a}$$

and

$$\tau_0 = \rho \cdot u_*^2. \tag{5.27b}$$

Comparing equation 5.27 with 5.26, we see that in turbulent flows u_* and τ_0 are alternate ways of expressing both the intensity of turbulence and the boundary shear stress. Shear velocity u_* expresses these physical quantities in kinematic (velocity) terms, whereas τ_0 expresses them in dynamic (force) terms. Also note that u_* can be thought of as a characteristic near-bed velocity in a turbulent flow.

5.3.1.4 Value of von Kármán's Constant, κ

Recall that von Kármán's constant, κ, is a proportionality factor in the heuristic relation between mixing length (i.e., the characteristic eddy diameter) and distance above a boundary (equation 3.38). The value of κ can only be determined by careful measurement of velocity distributions and thus is subject to uncertainty depending on experimental conditions and measurement accuracy. The most widely used value for this constant for clear water is $\kappa = 0.40$, although recent studies suggest $\kappa = 0.41$ (Bridge 2003), and many writers use that value.

However, some experimental data suggest that κ may not be a constant and may take on different values depending on location in a flow and on sediment concentration. Daily and Harleman (1966) suggest that $\kappa = 0.27$ away from the boundary. Some researchers have found that κ decreased with suspended-sediment concentration and reasoned that the intensity of turbulence is damped because the energy required to maintain the suspension comes from the turbulence. Einstein and Chien (1954) further developed this line of reasoning and presented data indicating values as low as $\kappa = 0.2$ at high sediment concentrations. (For reviews of these and other studies on this problem, see Middleton and Southard [1984] or Chang [1988].) In general, in this text, however, we will assume $\kappa = 0.4$ but will keep in mind that the value may be substantially lower for flows carrying high concentrations of suspended sediment.

5.3.1.5 Velocity Near the Boundary

The P-vK law is derived by assuming that the total shear stress throughout the flow (above y_0) is due to turbulence. However, as we have seen in figure 5.2b, this is not the case: Eddy viscosity decreases as one approaches the bed, so molecular viscosity becomes increasingly important near the bed and is the only source of shear stress in a region next to the boundary. To refine our understanding of the region over which the P-vK law describes the flow, we must look in more detail at the velocity structure of the near-bed region (figure 5.6).

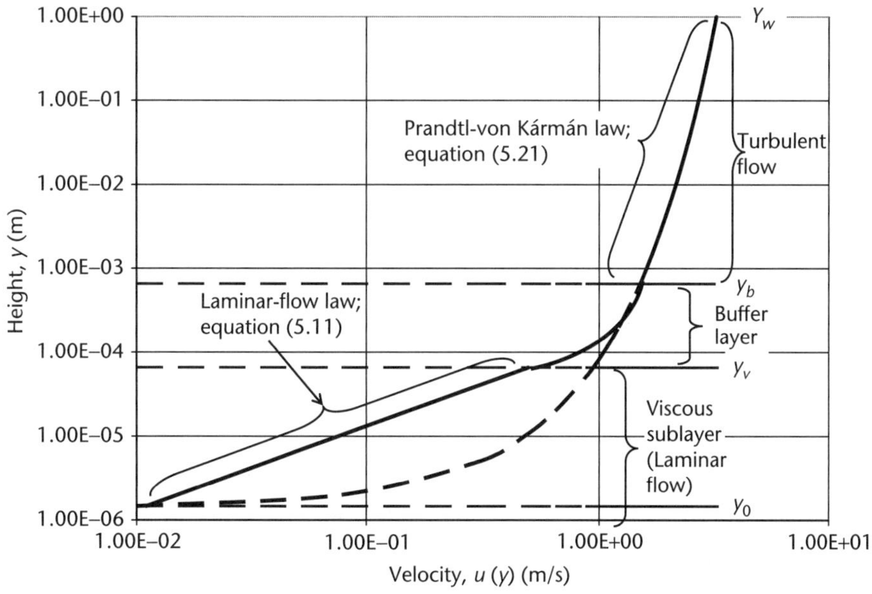

Figure 5.6 Velocity structure in a turbulent boundary-layer flow. The heavy line is the actual velocity profile. The P-vK profile applies from the top of the buffer layer y_b to the surface; the laminar profile applies from the bottom to y_v. See text for detailed explanation.

We know from the no-slip condition that the velocity at the bed is zero, that is, that $u(0) = 0$. Thus, if the bed is "smooth" (defined in the following subsection), there must be a zone extending some distance above the bed in which velocities and Reynolds numbers are low enough to be in the laminar range; this zone is called the **viscous sublayer**. The upper boundary of the viscous sublayer is indefinite and varies with time in a given flow as the turbulent bursts and sweeps described in section 3.3.4.1 impinge on it. By dimensional analysis and experiment, the average thickness of the viscous sublayer, y_v, has been found to be

$$y_v = \frac{5 \cdot \nu}{u_*}. \tag{5.28}$$

Using typical values for $\nu = 1.3 \times 10^{-6}$ m^2 s^{-1} and $u_* = 0.1$ m s^{-1}, we find $y_v \approx 6.5 \times 10^{-5}$ m or 6.5×10^{-2} mm—very small!

The velocity distribution within the viscous sublayer is given by the relation derived for laminar open-channel flows (equation 5.11). However, since y within the viscous sublayer is very small, the y^2 term in 5.11 is negligible, and the velocity gradient is effectively linear:

$$u(y) = \frac{\gamma \cdot Y_w \cdot \sin \theta_S}{\mu} \cdot y, \quad y \leq y_v; \tag{5.29a}$$

or

$$u(y) = \frac{u_*^2}{\nu} \cdot y, \quad y \leq y_v. \tag{5.29b}$$

As indicated in figure 5.6, the velocity gradient in the viscous sublayer is very steep.

Above the viscous sublayer is the **buffer layer**, where Reynolds numbers are in the transitional range and in which the transition to full turbulence occurs. In this zone the velocity gradient is still large and both viscous and turbulent shear stresses are important. As described by Middleton and Southard (1984, p. 104): "Very energetic small-scale turbulence is generated here by instability of the strongly sheared flow, and there is a sharp peak in the conversion of mean-flow kinetic energy to turbulent kinetic energy, and also in the dissipation of this turbulent energy; for this reason the buffer layer is often called the turbulence-generation layer."

As with the viscous sublayer, the upper boundary of the buffer layer fluctuates due the random nature of turbulence. Dimensional analysis and observations show that the average position of the upper boundary of the buffer layer is at a height y_b above the bottom, where

$$y_b = \frac{50 \cdot \nu}{u_*} \tag{5.30}$$

(Daily and Harleman 1966). Again using typical values for $\nu = 1.3 \times 10^{-6}$ m^2 s^{-1} and $u_* = 0.1$ m s^{-1}, we find $y_b \approx 6.4 \times 10^{-4}$ m, still less than 1 mm.

The velocity transitions smoothly from its value at the top of the viscous sublayer to its value at the top of the buffer layer, where full turbulence is present (on average). Above this point, the shear stress is essentially entirely due to turbulence, so the top of the buffer layer is the lowest elevation for which the P-vK law describes the velocity distribution.

Since shear in the buffer layer is due to both viscosity and turbulence, it is difficult to derive an equation for velocity distribution in this zone. Bridge and Bennett (1992) presented a semiempirical velocity profile for the buffer layer. However, this layer is so thin relative to typical flow depths that it can be neglected in integrating the "vertical" velocity profile.

5.3.1.6 Smooth and Rough Flow and the Determination of y_0

The practical application of the P-vK law requires some a priori way of determining the value of y_0. The approach to this determination depends whether the flow is hydraulically smooth or hydraulically rough. To understand the distinction, we consider the flow boundary (bed) to be covered with roughness elements of a typical height, y_r (figure 5.7). These roughness elements are usually thought of as sediment grains and y_r is generally taken to be proportional to the median (or other percentile) diameter of the bed material (see section 2.3.2.1; definitions of y_r are also discussed in chapter 6).

In hydraulically smooth flow, the height of the roughness elements is less than the thickness of the viscous sublayer (figure 5.7a). In rough flow, the element height is greater than the sublayer thickness, and the sublayer is not present as a continuous layer (figure 5.7b). Of course, the no-slip condition always requires a zero velocity at the boundary, but in rough flow eddies impinge on the bed and pressure forces due the irregularities of the bed particles exceed the viscous friction force (Middleton and Southard 1984).

Thus, the criterion for whether a flow is smooth or rough is simply to compare the thickness of the sublayer y_v given by equation 5.28 with y_r. This criterion is usually expressed by defining a **boundary Reynolds number** (also called the **roughness Reynolds number**), Re_b:[2]

$$Re_b \equiv \frac{u_* \cdot y_r}{\nu}. \tag{5.31}$$

Experiments have determined that the following numerical values of Re_b give the ranges of hydraulically smooth, transitionally rough, and fully rough flows:

Smooth	*Transitional*	*Rough*
>5	5–70	>70

It can easily be shown that the value of $Re_b = 5$ for the upper limit of hydraulically smooth flow corresponds to the situation when $y_r = y_v$ as given by equation 5.28.

Experiments have also shown that the value of y_0 in the P-vK law is as follows:

$$\text{Smooth flows } (Re_b < 5): \quad y_0 = \frac{\nu}{9 \cdot u_*}; \tag{5.32a}$$

$$\text{Transitional and fully rough flows } (Re_b \geq 5): \quad y_0 = \frac{y_r}{30}. \tag{5.32b}$$

It is important to note that, although the value of y_0 is determined by physical quantities and is an essential parameter of the P-vK law, the height y_0 is not a physically

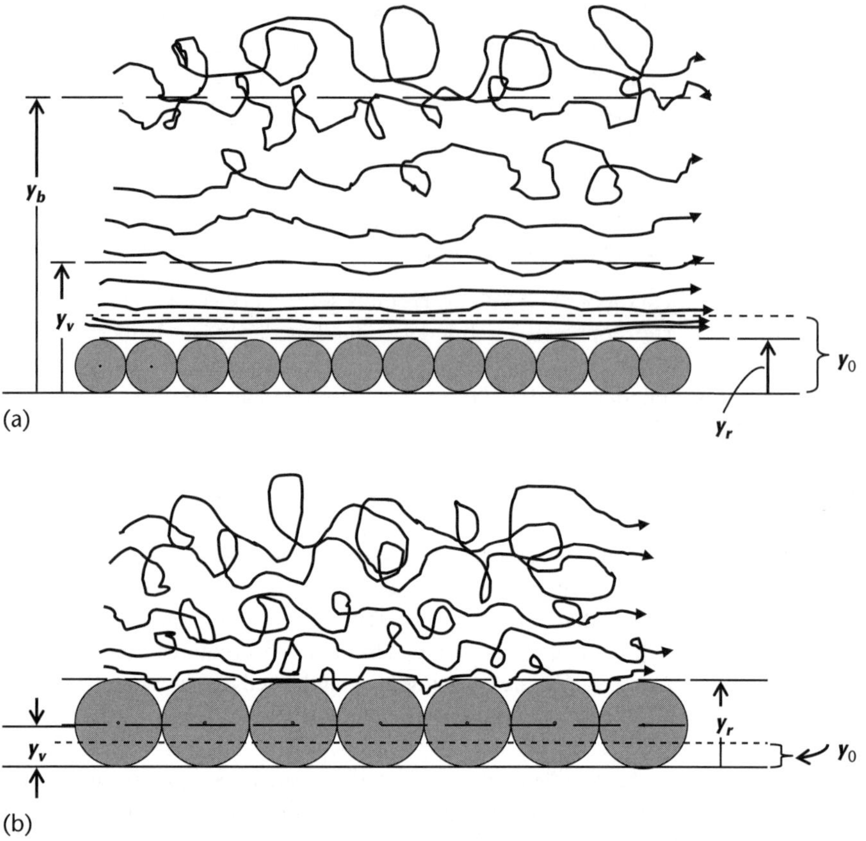

Figure 5.7 Schematic diagram of hydraulically (a) smooth and (b) rough turbulent flow. Arrows represent flow paths. In smooth flow, the viscous sublayer thickness y_v exceeds the height of the roughness elements y_r, and the viscous sublayer is present at the bed. In rough flow, the roughness height exceeds the viscous sublayer height, and no sublayer is present.

identifiable level in a flow. It is clear from figure 5.7 and equations 5.28 and 5.32 that y_0 is well within the viscous sublayer in smooth flows, well below the tops of the roughness elements in rough flows, and way below the level at which the P-vK law describes the velocity profile (i.e., the top of the buffer layer). Thus, y_0 should be thought of as an "adjustment factor" that depends on the boundary and flow characteristics (height of roughness elements, depth, and slope) and forces the P-vK law to fit the actual velocity profile above the buffer layer.

5.3.1.7 Zero-Plane Displacement Adjustment

In hydraulically smooth flows, fixing the origin of the y-axis height scale (i.e., the level at which $y = 0$) at the boundary is straightforward. However, in rough flows, it is not obvious where the origin should be placed (see figure 5.7b). A logical choice is to take $y = 0$ at the tops of the grains, because that is the surface on which a staff for depth measurement would be placed. However, where large bed particles are present, there are spaces between the particles in which flow occurs, and this causes deviations from the standard P-vK law in the region just above the grains. A common approach to accounting for these deviations is to modify the P-vK law by introducing a height, y_z, to give

$$u(y) = 2.5 \cdot u_* \cdot \ln\left(\frac{y - y_z}{y_0}\right), \quad y > y_0 + y_z. \tag{5.33}$$

Note that if y is measured from the tops of the particles, y_z is a negative number. You can see from equation 5.33 that velocity equals 0 when $y = y_0 + y_z = y_0 - |y_z|$; thus, y_z is called the **zero-plane displacement**. Including this term has the effect of lowering the effective "bottom," and for a given value of $y > y_0 + y_z$, the actual velocity is greater than that given by the original P-vK law.

Note that the effect of y_z on velocity at a given level is greatest for small y and decreases steadily as y increases to eventually become negligible. Thus, when the bed material is large, modifying the P-vK law by including the zero-plane displacement shifts the plotted velocities near the bed so that they form a straight line when plotted against height using a logarithmic axis. Figure 5.8 shows an example of this, with velocities measured at fixed levels in a steady flow in the Columbia River, where $y_r = 0.69$ m (boulders). In this case, a value of $y_z = -0.14$ m brings the points into a linear relation. This is consistent with Middleton and Southard's (1984) statement that, for a wide variety of roughness geometries, $|y_z|$ has been found to be between $0.2 \cdot y_r$ and $0.4 \cdot y_r$.

5.3.1.8 The P-vK Law: Summary

To summarize the discussions of sections 5.3.1.2–5.3.1.7, we use equations 5.24 and 5.32 to write the P-vK law in the forms that we will usually apply it:

Smooth flows, $Re_b \leq 5$:

$$u(y) = 2.50 \cdot u_* \cdot \ln\left(\frac{9 \cdot u_* \cdot y}{\nu}\right); \tag{5.34a}$$

Rough flows, $Re_b > 5$:

$$u(y) = 2.50 \cdot u_* \cdot \ln\left(\frac{30 \cdot y}{y_r}\right). \tag{5.34b}$$

Note that these are mathematically equivalent to Smooth flows, $Re_b \leq 5$:

$$\frac{u(y)}{u_*} = 2.50 \cdot \ln\left(\frac{u_* \cdot y}{\nu}\right) + 5.49 = 5.76 \cdot \log\left(\frac{u_* \cdot y}{\nu}\right) + 5.49; \tag{5.34c}$$

Rough flows, $Re_b > 5$:

$$\frac{u(y)}{u_*} = 2.50 \cdot \ln\left(\frac{y}{y_r}\right) + 8.50 = 5.76 \cdot \log\left(\frac{y}{y_r}\right) + 8.50; \tag{5.34d}$$

and the P-vK law may be written in any of these forms.

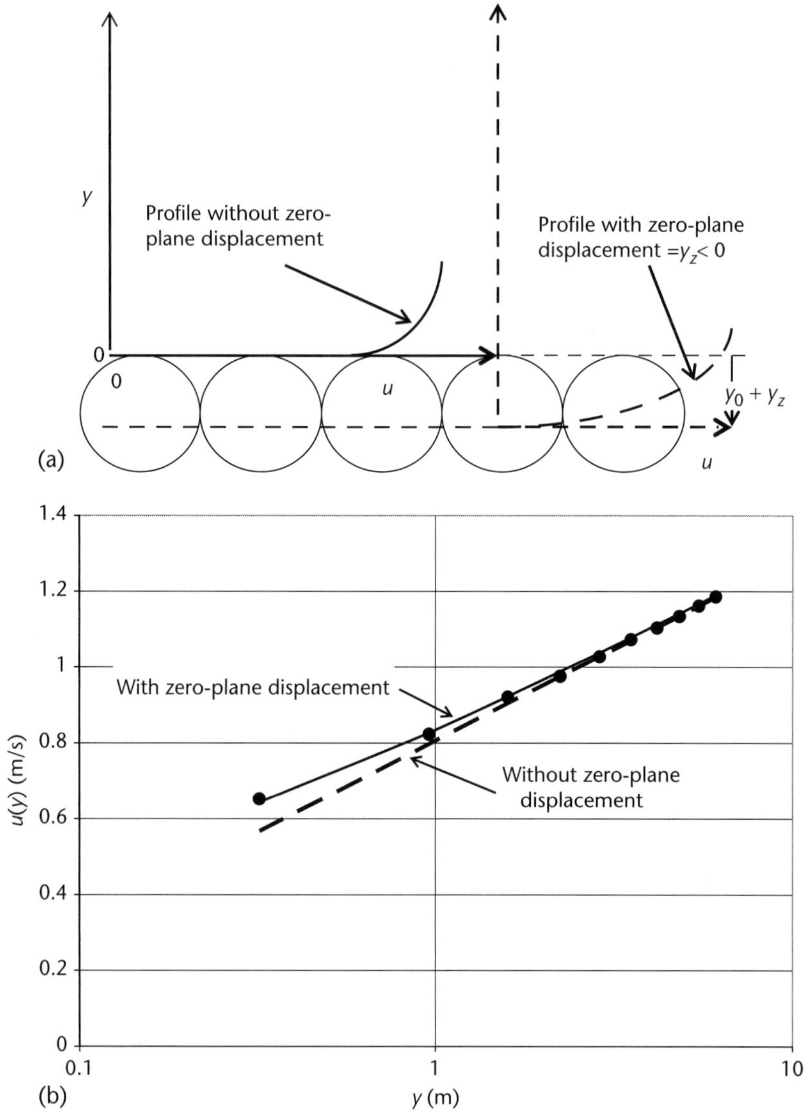

Figure 5.8 The zero-plane-displacement adjustment. (a) Velocity profiles are measured with respect to the normal y-direction with $y = 0$ at the tops of the roughness elements (solid axes and velocity profile). Using the zero-plane-displacement height y_z shifts the level of $u(y) = 0$ to $y = y_0 + y_z$, where $y_z < 0$ (dashed axes and profile). (b) The points are a velocity profile measured by Savini and Bodhaine (1971) in the Columbia River where the bed material consists of boulders averaging 0.69 m in diameter. The dashed line is a logarithmic velocity profile fit to the upper seven points; note that the actual velocities of the lower three points lie well above this line. A logarithmic profile including a zero-plane displacement value of $y_z = -0.14$ m (solid line, equation 5.33) fits the data over the entire profile.

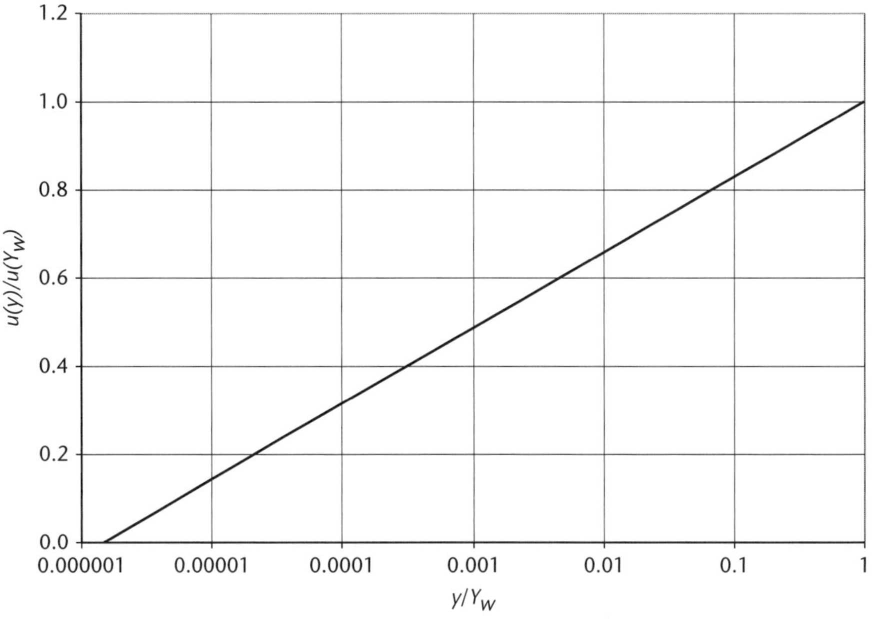

Figure 5.9 Relative velocity $u(y)/u(Y_w)$ as a function of relative distance above the bottom, y/Y_w as given by the Prandtl-von Kármán (P-vK) universal velocity distribution for turbulent open-channel flows (equation 5.21). The data plotted in this graph are identical to those in figure 5.5, but the axes have been reversed and the y-axis is logarithmic rather than arithmetic. Velocity profiles are commonly plotted in this way to check for conformance to the P-vK law.

Note also that, according to the P-vK law, a plot of velocity versus distance above the bottom will define a straight line when velocity $u(y)$ is plotted against an arithmetic axis and height-above bottom y is plotted against a logarithmic axis (figure 5.9). Measured velocity profiles are commonly plotted in this way to check for conformance to the P-vK law.

5.3.1.9 Average "Vertical" Velocity

As for laminar flow, the average "vertical" velocity U_w for turbulent flow can be derived by integration of the P-vK law (equation 5.34) over its range of validity above the top of the buffer zone, $y \geq y_b$:

$$U_w = \frac{1}{Y_w - y_b} \cdot \int_{y_b}^{Y_w} 2.50 \cdot u_* \cdot \ln\left(\frac{y}{y_0}\right) \cdot dy. \tag{5.35}$$

Using the facts that

$$\ln\left(\frac{y}{y_0}\right) = \ln(y) - \ln(y_0)$$

and

$$\int \ln(y) \cdot dy = y \cdot \ln(y) - y,$$

we can evaluate equation 5.35 as

$$U_w = \left(\frac{2.50 \cdot u_*}{Y_w - y_b} \right) \cdot \left[Y_w \cdot \ln\left(\frac{Y_w}{y_0} \right) - Y_w - y_b \cdot \ln\left(\frac{y_b}{y_0} \right) + y_b \right]. \tag{5.36a}$$

However, we have seen that y_b is generally very small relative to the depth Y_w (figure 5.6), and if this is true, then $y_b \approx 0$ and equation 5.36a can be simplified to

$$U_w = 2.50 \cdot u_* \cdot \left[\ln\left(\frac{Y_w}{y_0} \right) - 1 \right]. \tag{5.36b}$$

This expression for the local mean "vertical" velocity in a turbulent flow can be used to solve a practical problem—the measurement of discharge through a stream cross section. Recall that discharge Q is

$$Q = U \cdot Y \cdot W, \tag{5.37}$$

where U is average cross-section velocity, Y is average cross-section depth, and W is water-surface width. The velocity-area method of discharge measurement (described in section 2.5.3.1) involves dividing the cross section into I subsections and determining Q as

$$Q = \sum_{i=1}^{I} U_i \cdot Y_i \cdot W_i, \tag{5.38}$$

where U_i and Y_i are the local velocities and depths U_w and Y_w, respectively, at successive points i = 1, 2, ..., I, and W_i is the width of subsection i. Measurement of depth and width for each subsection is straightforward, but since velocity varies vertically, there is the problem of how to determine an average without measuring velocity at a large number of heights at each subsection.

This problem is solved by noting that the actual velocity $u(y)$ must equal the average value U_w at some height $y = y_U$. Taking $y_U = k_U \cdot Y_w$, the P-vK law gives

$$U_w = 2.50 \cdot u_* \cdot \ln\left(\frac{k_U \cdot Y_w}{y_0} \right), \tag{5.39}$$

and equating this to equation 5.36b gives

$$2.50 \cdot u_* \cdot \ln\left(\frac{k_U \cdot Y_w}{y_0} \right) = 2.50 \cdot u_* \cdot \left[\ln\left(\frac{Y_w}{y_0} \right) - 1 \right]. \tag{5.40}$$

The value of k_U can be found from equation 5.40 as

$$k_U = \frac{1}{e} = 0.368\ldots, \tag{5.41}$$

where $e = 2.718$ ... is the base of natural logarithms.

Thus, we see that, according to the P-vK law, the velocity measured at a distance $0.368 \cdot Y_w$ above the bottom equals the average value for the profile. This finding is

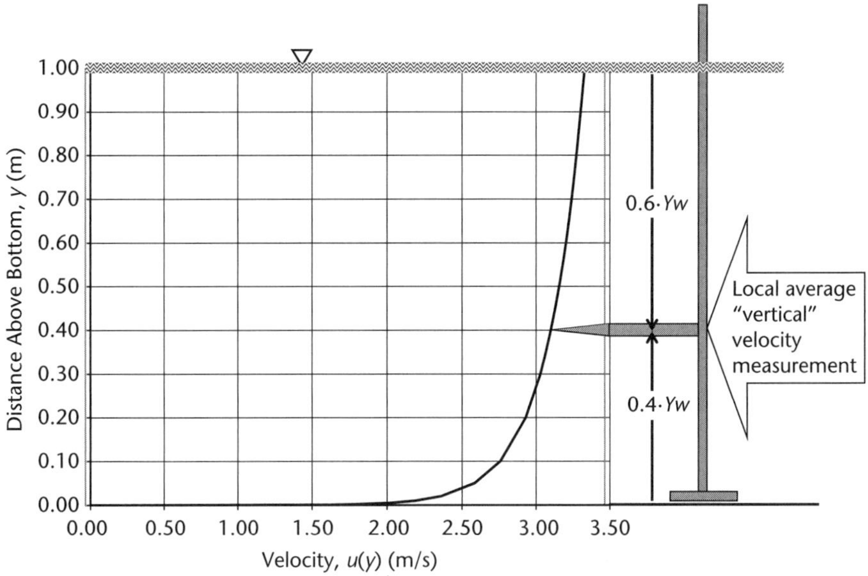

Figure 5.10 P-vK velocity profile for a turbulent flow with $Y_w = 1$ m, showing velocity measurement by current meter at six-tenths of the depth measured from the surface. According to the P-vK law, the actual velocity $u(y)$ equals the average velocity U_w at $y/Y_w = 0.368\ldots$. This is the basis for the "six-tenths-depth rule" for measuring local average "vertical" velocity.

the basis for the **six-tenths-depth rule** used by the U.S. Geological Survey and others for discharge measurement:

> If the P-vK law applies, the average velocity U_w at a point in a cross section is found by measuring the velocity six-tenths of the total depth downward from the surface, or four-tenths (≈ 0.368) of the depth above the bottom (figure 5.10).

It is also worth noting that the P-vK law also provides information about the relation between surface velocity and mean velocity that can be useful for measuring discharge. From the P-vK law and equation 5.36b,

$$\frac{U_w}{u(Y)} = \frac{\ln\left(\frac{Y_w}{y_0}\right) - 1}{\ln\left(\frac{Y_w}{y_0}\right)}, \tag{5.42}$$

and if we assume rough flow, we can use equation 5.34b and evaluate $U_w/u(Y_w)$ as a function of Y_w/y_r (figure 5.11). This information can be exploited to estimate mean velocity by measuring the surface velocity by means of floats. Note that for typical rivers, the mean velocity ranges from 0.82 to 0.92 of the surface velocity, and an approximate general value ≈ 0.87. (Note, however, that surface and mean velocity will vary across a stream.)

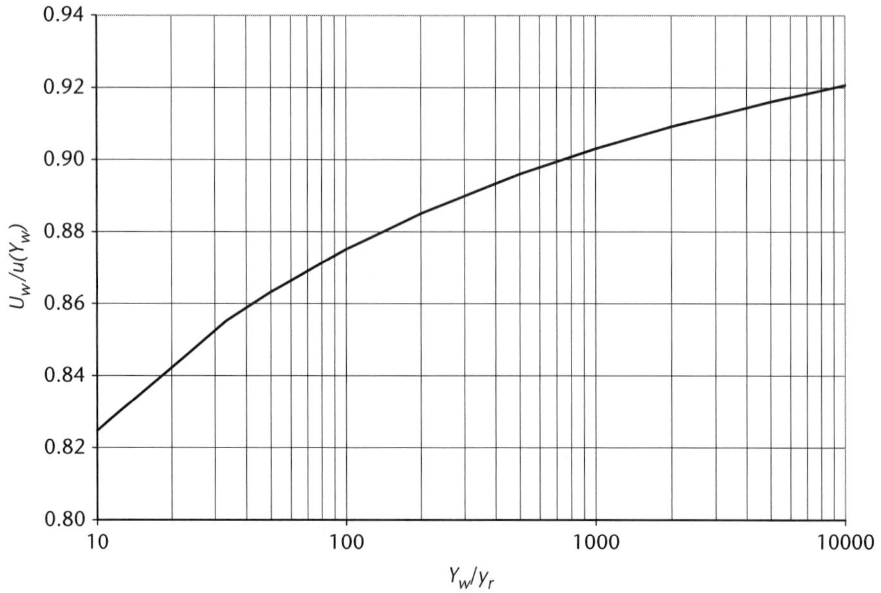

Figure 5.11 Ratio of local mean velocity U_w to local surface velocity $u(Y_w)$ as a function of the ratio of flow depth Y_w to bed-material size, y_r. For most large rivers, $0.85 \le U_w/u(Y_w) \le 0.90$.

Other approaches to estimating the average "vertical" velocity based on the P-vK law are presented in box 5.1.

5.3.2 The Velocity-Defect Law

In Prandtl's (1926) original development of the P-vK law, the shear stress was considered to be constant throughout the flow at the boundary (or "wall") value τ_0. Because of this, the P-vK law is also known as the **law of the wall**, and there has been considerable discussion about how far above the boundary the P-vK law applies.

It is widely accepted that "far" from the bed, the velocity gradient does not depend on viscosity (as in the P-vK law for smooth flows) or on bed roughness (as in the P-vK law for rough flows), but only on distance above the bed. In this region, the velocity profile is represented as a **velocity defect**, that is, as the difference between the velocity at the surface (or at the top of the turbulent boundary layer; see figure 3.28), $u(Y_w)$, and the velocity at an arbitrary level, $u(y)$, and is a function only of y/Y_w:

$$\frac{u(Y_w) - u(y)}{u_*} = f_{VD}\left(\frac{y}{Y_w}\right) \tag{5.43a}$$

or

$$u(y) = u(Y_w) - u_* \cdot f_{VD}\left(\frac{y}{Y_w}\right), \tag{5.43b}$$

BOX 5.1 Methods for Estimating Average "Vertical" Velocity from Velocity Profile Measurements

Average Velocity Accounting for Zero-Plane Displacement

The integration of the P-vK law including the zero-plane displacement (equation 5.33) gives the following relation for U_W:

$$U_W = \left(\frac{2.5 \cdot u_*}{Y_W - y_z - y_0}\right) \cdot [(Y_W - y_z) \cdot \ln(Y_W - y_z) - Y_W + y_z - y_0 \cdot \ln(y_0) + y_0], \tag{5B1.1a}$$

or, if y_0 is negligibly small,

$$U_W = 2.5 \cdot u_* \cdot [\ln(Y_W - y_z) - 1]. \tag{5B1.1b}$$

Two-Tenths/Eight-Tenths–Depth Method

If the velocity profile is given by the P-vK law, it can be shown that

$$u(0.4Y_W) = \frac{u(0.2 \cdot Y_W) + u(0.8 \cdot Y_W)}{2}. \tag{5B1.2}$$

Thus, average vertical velocity can be estimated as the average of the velocities at $0.2 \cdot Y_W$ and $0.8 \cdot Y_W$.

The two-tenths/eight-tenths–depth method has been found to give more accurate estimates of average velocity than does the six-tenths–depth method (Carter and Anderson 1963), and standard U.S. Geological Survey practice is to use the two-tenths/eight-tenths–depth method where $Y_W > 2.5$ ft (0.75 m).

General Two-Point Method

If velocity is measured at two points, each an arbitrary fixed distance above the bottom, the relative depths of those sensors will change as the discharge changes. Again assuming the P-vK law applies with $y_{0W} \ll Y_W$, Walker (1988) derived the following expression for calculating the average vertical velocity from two sensors fixed at arbitrary distances above the bottom, y_{W1} and y_{W2}, where $y_{W2} > y_{W1}$:

$$U_W = \frac{[1 + \ln(y_{W2})] \cdot u(y_{W1}) + [1 + \ln(y_{W1})] \cdot u(y_{W2})}{\ln(y_{W2}/y_{W1})} \tag{5B1.3}$$

Walker (1988) also calculated the error in estimating U_W for sensors located at various combinations of relative depths.

(Continued)

BOX 5.1 *Continued*

Multipoint Method

The assumption of the applicability of the P-vK law with $y_{0W} \ll Y_W$ is not valid in cross sections where there are roughness elements (boulders, weeds) with heights that are a significant fraction of depth, or where there are significant obstructions upstream and downstream of the measurement section. In these cases, Buchanan and Somers (1969) recommended estimating U_W as

$$U_W = 0.5 \cdot u(0.4 \cdot Y_W) + 0.25 \cdot [u(0.2 \cdot Y_W) + u(0.8 \cdot Y_W)] \qquad (5B1.4)$$

However, the highest accuracy in these situations is assured by measuring velocity several heights at each vertical, with averages found by numerical integration over each vertical or over the entire cross section. Alternatively, a statistical sampling approach over the cross section may be appropriate (Dingman 1989; see section 5.4.3).

where $f_{VD}(y/Y_w)$ is determined by experiment. Equation 5.43 is the general form of a **velocity-defect law**, which experiments have shown to be applicable in the region where $(y/Y_w) > 0.15$ for both smooth and rough boundaries.

Note that an a priori value of the surface velocity $u(Y_w)$ is required to apply this relation. To get this value, Daily and Harleman (1966) assume that the P-vK law for smooth boundaries can be applied, but that the value of κ and the constant determining y_0 may be different from 0.4 and 9, respectively. They used experimental data to arrive at two forms of the velocity-defect law, one of which applies for $y/Y_w < 0.15$ and the other for $y/Y_w > 0.15$. For the latter, they find

$$\frac{u(Y_w) - u(y)}{u_*} = -3.74 \cdot \ln\left(\frac{y}{Y_w}\right), y/Y_w > 0.15. \qquad (5.44)$$

The velocity-deflect law is extensively reviewed by Middleton and Southard (1984) and Bridge (2003), and both sources conclude that the P-vK law "fits the velocity profile without great error all the way to the free surface" in turbulent boundary-layer flows (Middleton and Southard 1984, p. 153). We can see this in figure 5.12, which compares the profile given by equation 5.44 with that given by the P-vK law for a smooth bed, where the average velocity over the profile is matched to that given by the P-vK law. Above a height of $y/Y_w = 0.15$, where the velocity-defect law is supposed to apply, there is less than 4% difference in the velocities predicted by the two relations.

The theoretical reason for introducing the velocity-defect law was that Prandtl's (1926) original derivation of the P-vK law was based on two assumptions that hold only near the boundary: 1) mixing length $l = \kappa \cdot y$ (equation 3.37), and 2) shear stress equals the boundary value τ_0 throughout the flow rather than decreasing with height above the bottom as given by equation 5.6. However, as shown in section 5.3.1.1,

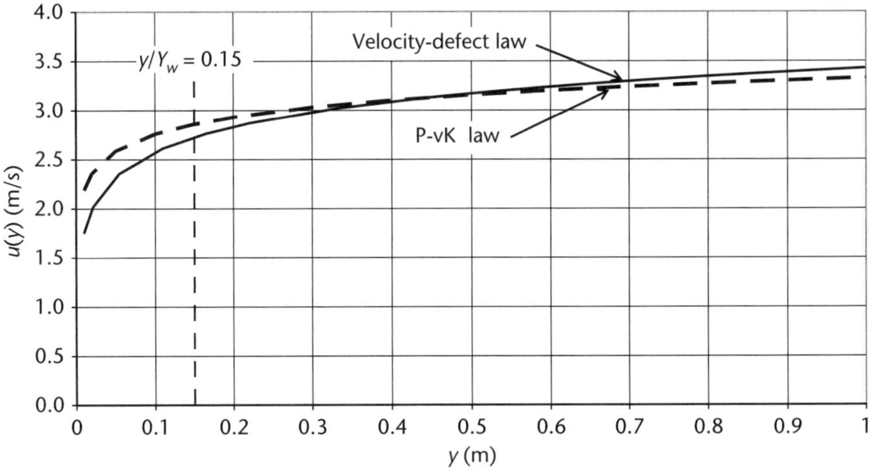

Figure 5.12 Comparison of velocity profiles given by the P-vK law (dashed line, equation 5.34) and the velocity-defect law (solid line, equation 5.44). The average velocities over the two profiles are identical. The difference between the velocities given by the two profiles differs by less than 4% for $y/Y_w > 0.15$.

the P-vK law can also be derived from the more realistic assumptions that mixing length is given by equation 3.38 and that the shear-stress distribution is linear with depth (equation 5.18). Thus, the theoretical justification for restricting the P-vK law to the region near the boundary is not compelling. Furthermore, we saw that velocities given by the velocity-defect law do not differ greatly from the P-vK law (figure 5.12). Therefore, we can conclude that there is usually no need to invoke the velocity-defect law in preference to the P-vK law.

5.3.3 Power-Law Profiles

Many observers have noted that turbulent velocity profiles can be represented by **power-law** (PL) relations of the form

$$u(y) = k_{PL} \cdot u_* \cdot \left(\frac{y}{y_0}\right)^{m_{PL}}, \tag{5.45}$$

where y_0 is defined separately for smooth and rough flow as in equation 5.32, and the values of the coefficient k_{PL} and the exponent m_{PL} are discussed below.

Power-law profiles have a mathematical advantage over the P-vK law in that they satisfy the no-slip condition that $u(0) = 0$. However, Chen (1991) showed that a universal power-law formulation cannot be derived from basic principles and found that 1) relations of this form are identical to the P-vK law only when $m_{PL} \cdot k_{PL} = 0.920$, and 2) different values of m_{PL} and k_{PL} are required to approximate the P-vK law for different ranges of y/y_0 (table 5.1). Note that this may mean that m_{PL} and k_{PL} may need to change in different depths for a given profile. Chen (1991) recommended using $m_{PL} = 1/7$ for hydraulically smooth flows and $m_{PL} = 1/6$ for

Table 5.1 Values of m_{PL} and k_{PL} required for power-law (equation 5.45) approximation of the P-vK law in various (overlapping) ranges of y/y_0.[a]

Lower limit of y/y_0	Upper limit of y/y_0	m_{PL}	k_{PL}
0.0737	86.8	1/2 = 0.500	1.40
0.759	232	1/3 = 0.333	2.44
3.07	591	1/4 = 0.250	3.45
10.9	1,450	1/5 = 0.200	4.43
36.8	**3,490**	**1/6 = 0.167**	**5.39**
123	**8,230**	**1/7 = 0.143**	**6.34**
409	19,200	1/8 = 0.125	7.29
1,360	44,200	1/9 = 0.111	8.23
4,500	101,000	1/10= 0.100	9.16
14,900	230,000	1/11 = 0.0909	10.1
49,300	521,000	1/12 = 0.0833	11.0

[a]Chen (1991) recommends using $m_{PL} = 1/7$ for hydraulically smooth flows and $m_{PL} = 1/6$ for hydraulically rough flows (shown in boldface in the table).
From Chen (1991).

hydraulically rough flows, grading to smaller m_{PL} values at larger y values in rough flows. Figure 5.13 compares a power-law profile with that given by the P-vK law.

When integrated per equation 5.2, equation 5.45 gives the average "vertical" velocity as

$$U_w = u_* \cdot \left(\frac{k_{PL}}{m_{PL} + 1}\right) \cdot \left(\frac{Y_w}{y_0}\right)^{m_{PL}}. \tag{5.46}$$

Note, however, that 5.46 only applies if a single pair of (m_{PL}, k_{PL}) values is used for the entire profile.

5.3.4 The Hyperbolic-Tangent Profile

In general, significant deviations from the P-vK law profile occur when the bed roughness is of the same order of magnitude as the flow depth (large relative roughness). As we have seen, one way to adjust for this is to use a zero-plane displacement adjustment (section 5.3.1.7). Recently, Katul et al. (2002) suggested a new form for the velocity profile in flows in which the bottom roughness is large relative to the depth:

$$u(y) = 4.5 \cdot u_* \cdot \left[1 + \tanh\left(\frac{y - y_r}{y_r}\right)\right], \tag{5.47}$$

where $\tanh(\xi)$ is the **hyperbolic tangent** of the quantity ξ, defined as

$$\tanh(\xi) \equiv \frac{e^{\xi} - e^{-\xi}}{e^{\xi} + e^{-\xi}}.$$

This profile is illustrated in figure 5.14 for a case where $Y_w = 2$ m and $y_r = 0.5$ m. Note that the profile has a point of inflection at $y = y_r$.

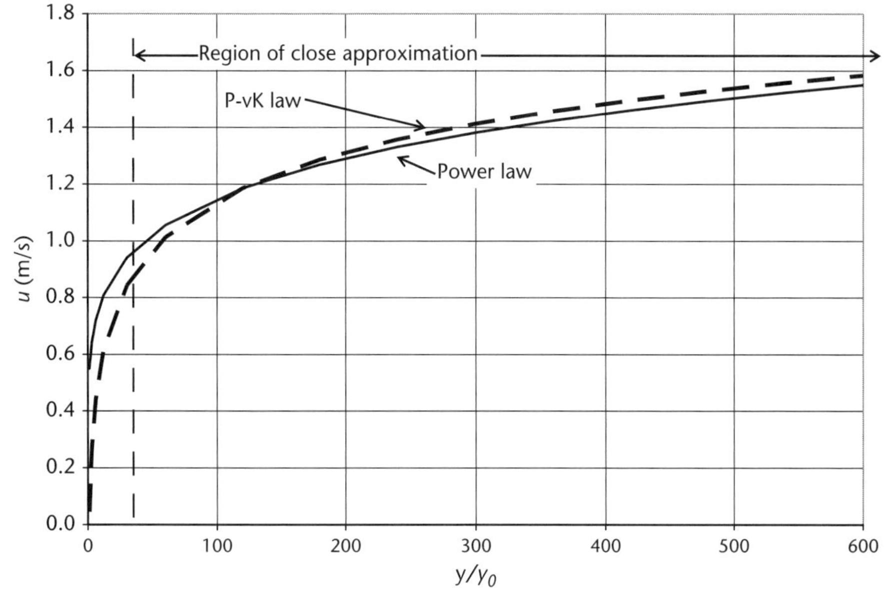

Figure 5.13 Velocity profiles for a flow with $Y_w = 1$ m, $\sin\theta = 0.001$, and $y_r = 50$ mm as given by the P-vK law (dashed) and the power-law (solid). The power-law profile is computed via equation 5.45 with $m_{PL} = 1/6, k_{PL} = 5.39$, and $y_0 = 1.67 \times 10^{-3}$ m and gives a good approximation only in the range $36.8 < y/y_0 < 3,490$ (table 5.1).

When integrated per equation 5.2, equation 5.47 gives the average "vertical" velocity as

$$U_w = 4.5 \cdot u_* \cdot \left[1 + \left(\frac{y_r}{Y_w} \right) \cdot \ln \left(\frac{\cosh\left(1 - \frac{Y_w}{y_r}\right)}{\cosh(1)} \right) \right], \tag{5.48}$$

where $\cosh(\xi) \equiv 0.5 \cdot (e^{\xi} - e^{-\xi})$ and $\cosh(1) = 1.543\ldots$

As discussed more fully in chapter 6, application of equation 5.48 to actual flows indicates that it gives useful results over a wide range of (Y_w/y_r) values and suggests that equation 5.47 may be a useful approach to modeling turbulent velocity profiles in flows with large relative roughness.

5.3.5 Other Theoretical Profiles

Here we briefly note some studies that explore velocity profiles under conditions that deviate markedly from those assumed in deriving the P-vK law: a smooth bed or a bed of similar-sized particles at low to moderate relative roughness.

Wiberg and Smith (1991) found that average velocity profiles in flows with highly variable bed-sediment size (including flows in which the surface is below the tops

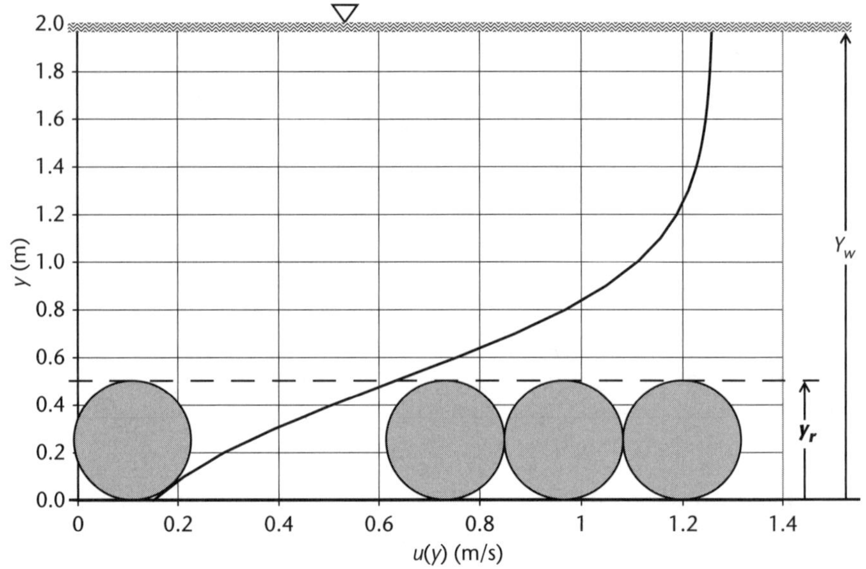

Figure 5.14 Velocity profile given by Katul et al. (2002) hyperbolic-tangent profile for shallow flows with large bed material (equation 5.47). Here, the flow depth $Y_w = 2$ m, slope $\sin\theta = 0.001$, and the particle size $y_r = 0.5$ m. The velocity profile has an inflection point at $y = y_r$, but it is not very apparent in this case.

of the largest bed particles) deviated significantly from the logarithmic profile. They applied force-balance concepts to develop expressions for the profiles in such flows and found they were similar to profiles measured in mountain streams.

Rowinski and Kubrak (2002) used similar concepts to deduce profiles for flows through trees (which are commonly present on floodplains) and confirmed their model experimentally.

5.3.6 Observed Velocity Profiles

Figure 5.15 shows a velocity profile measured in the central portion of a large river (width = 550 m, depth = 12 m), the Columbia. The smooth curve shows the logarithmic velocity profile that best fits the observed values; the good fit indicates that the velocity profile here is well modeled by the P-vK law [the curve would plot as a straight line on a graph of $u(y)$ vs. $\ln(y)$]. Figure 5.16 shows two profiles measured in a much smaller stream (width = 5.1 m, depth = 0.55 m). The profile measured near the center of the stream (2.9 m from the bank, triangular points), like that of the Columbia, has the maximum velocity at the surface and is well fit by the P-vK law (solid curve). However, in the profile measured nearer the bank (1.4 m out, square points) the maximum velocity is well below the surface and the profile is not well modeled by the P-vK law fitted to the lowest four points (dashed curve).

The depression of the maximum velocity below the surface, which is often observed in natural streams, is contrary to the prediction of the P-vK law and the

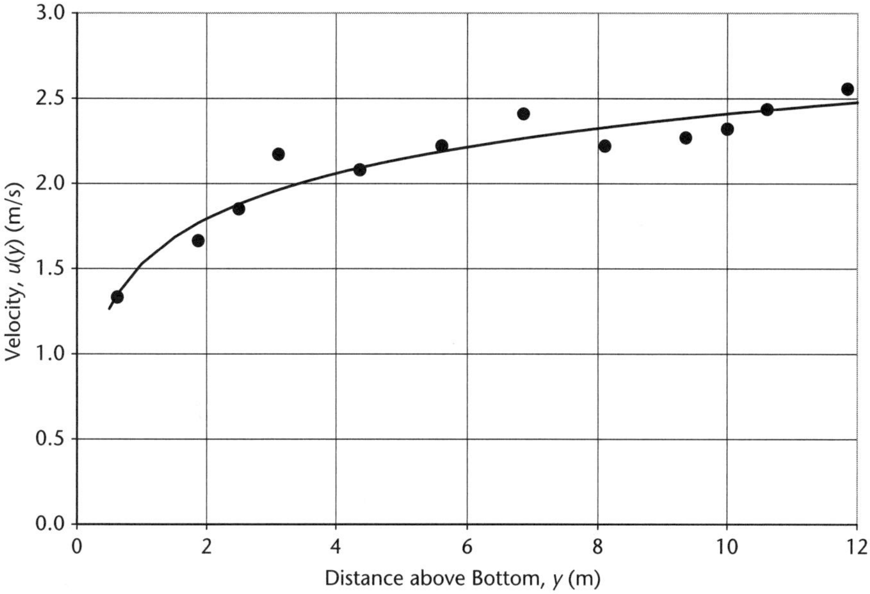

Figure 5.15 Velocity profile measured in central portion of the Columbia River, Washington (points), where the flow is 12 m deep and about 550 m wide. The smooth curve is a logarithmic fit to the measured points showing that the profile closely approximates the P-vK law.

other theoretical profiles discussed in sections 5.3.1–5.3.5. In profiles measured near the bank, or at any location in channels with relatively small width/depth ratios ($W/Y < \sim 10$), this depression is due to the effects of bank friction, which induces a spiral, or **helicoidal**, circulation (figure 5.17). Note that although the velocity profile is strongly affected, the magnitudes of the cross-channel (figure 5.17d) and vertical (figure 5.17e) velocities are less than 10% of the downstream velocity, and the average downstream velocity is little affected by the circulation (compare figure 5.17b,c).

In natural channels, helicoidal circulation and depression of the thread of maximum velocity may also be caused by 1) the proximity of significant irregularities of the bed, 2) downstream or upstream obstructions that create "threads" of high or low velocity that disrupt the theoretical patterns, or 3) centrifugal forces induced by channel curvature. We will examine these phenomena further in the exploration of velocity distributions in cross sections in section 5.4.

5.3.7 Summary: Velocity Profiles in Turbulent Flow

Because the original derivation of the P-vK law invoked conditions that are true only near the bed, theoretical justifications have been advanced for using the velocity-defect law at heights that exceed $y/Y_w = 0.15$. However, the P-vK law can also be derived from less restrictive conditions, and since the profiles given by the two laws do not differ greatly even far from the boundary, it does not seem necessary to invoke the velocity-defect law. Furthermore, as we see in figures 5.15 and 5.16, a single

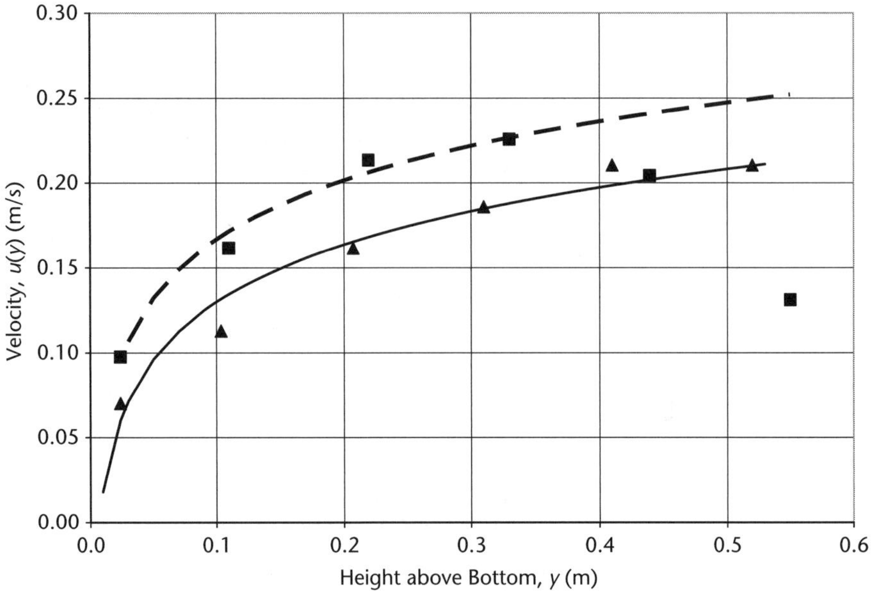

Figure 5.16 Two profiles measured in Casper Kill, New York, (width = 5.1 m, depth = 0.55 m). The profile measured near the center of the stream (2.9 m from the bank, triangular points), like that of the Columbia in figure 5.15, has the maximum velocity at the surface and is well fit by the P-vK law (solid curve). However, in the profile measured nearer the bank (1.4 m from the bank, square points), the maximum velocity is well below the surface and the overall profile is not well modeled by the P-vK law fitted to the lowest four points (dashed curve).

curve following the P-vK law often provides a good fit to measured velocities over the entire velocity profile.

Although there are sometimes mathematical advantages to power-law profiles, the law cannot be derived from basic principles. Furthermore, application of the power-law model is hindered because a given pair of coefficient and exponent values approximates the P-vK law only over a limited range of (y/Y_w) values.

Thus, we conclude that the P-vK law as given in equation 5.34 can be generally accepted as the theoretical local velocity profile in wide uniform turbulent flows, at least when Y_w/y_r is not too small (>10) and the bed is smooth or the bed roughness elements are uniformly distributed and of uniform size.

Profiles other than the P-vK law are appropriate for conditions that deviate markedly from those assumed in its derivation. When y_r is larger than gravel size (>50 mm), the zero-plane adjustment (equation 5.33) may be required to fit the profile near the bed. The alternative profile given by Katul et al. (2002) (equation 5.47) also appears to give good results for large relative roughness and may prove to be preferable under those conditions. The profile of Wiberg and Smith (1991) appears to fit conditions of highly nonuniform bed-particle sizes, and that of Rowinski and Kubrak (2002) can be used for flows through trees.

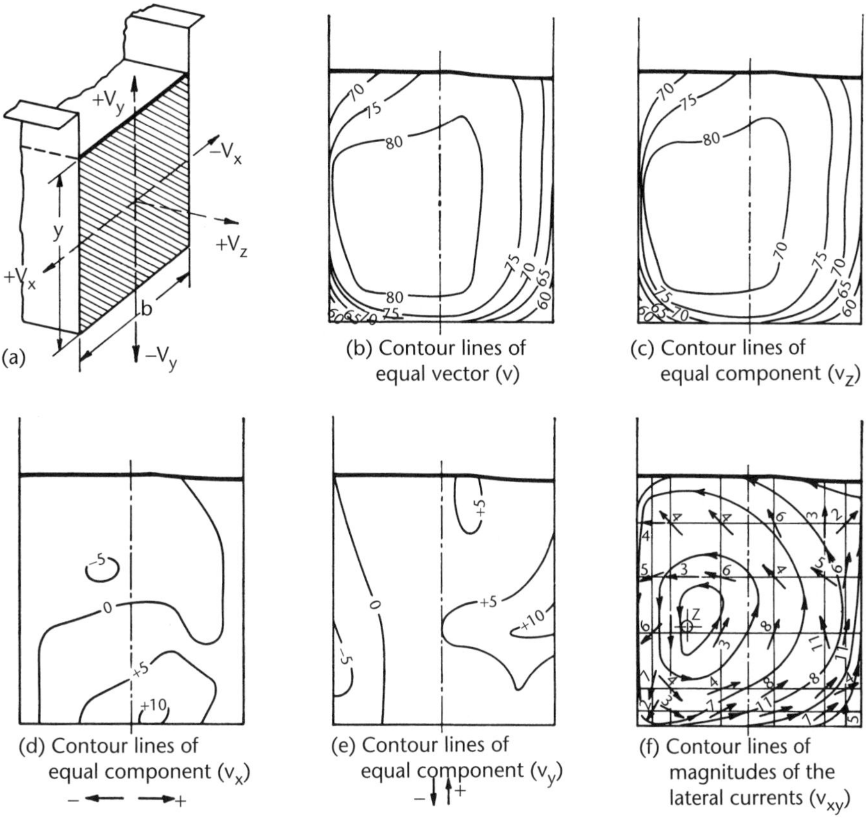

Figure 5.17 Velocity components in a rectangular flume with $W/Y \approx 1$, showing the presence of helicoidal flow. Isovels (velocity contours) labeled with velocities in cm/s. (a) Coordinate system. (b) Isovels of total velocity vector.(c) Isovels of downstream component. (d) Isovels of cross-stream component. (e) Isovels of vertical component. (f) Isovels and vectors of helicoidal currents. From Chow (1959).

Recall that the theoretical velocity profiles discussed in this chapter are local: They apply to the "vertical" distribution of velocity at a point in a cross section and were derived under the assumption of uniform flow in "wide" channels, where only the bed friction affects the flow. Because of these assumptions, all the theoretical profiles predict that the maximum velocity occurs at the surface. Bank friction and channel curvature can generate cross-channel **secondary currents**, which can suppress the maximum velocity some distance below the surface; this phenomenon is discussed further in section 5.4 and in chapter 6. However, as suggested by figure 5.17, these secondary currents generally have only a small effect on the average downstream velocity.

In most practical problems of fluvial hydraulics, we are interested in the cross-section average velocity and its relation to depth, slope, bed material, and other channel characteristics. The integrated forms of the appropriate theoretical profile equations give the local "vertically" averaged velocity U_w; this average may be a reasonable approximation of the cross-section average velocity U for wide channels with regular cross sections, but is not generally acceptable for natural streams.

The following section briefly explores the distribution of velocity in entire cross sections. The relation between cross-section average velocity, depth, slope, and bed material and other factors that affect flow resistance in natural channels is discussed in chapter 6.

5.4 Velocity Distributions in Cross Sections

5.4.1 Velocity Distribution in an Ideal Parabolic Channel

Interestingly, the theoretical distribution of velocity in cross sections has been little studied, and there are no generally accepted theoretical models. A starting point for formulating such models is to assume that vertical velocity profiles follow the P-vK law at each point in the cross section. This is done in the "synthetic channel model" described in appendix C. Figure 5.18a shows velocity contours (**isovels**) in a parabolic channel generated by this model: the channel shape, dimensions, slope, and roughness height are specified, and the P-vK law is applied at points along the cross section. The cross-channel distribution of surface velocity for this case is plotted in figures 5.18b (arithmetic plot) and 5.18c (semilogarithmic plot). Interestingly, the cross-channel distribution of surface velocity closely mimics the P-vK law for much of the distance, as evidenced by the straight-line fit in figure 5.18c.

However, the application of the P-vK law at each point in a cross section as in the synthetic channel model does not account for cross-channel shear, which distorts vertical profiles modeled as being affected only by bed shear. Thus, we would expect actual isovel patterns to differ somewhat from those shown in figure 5.18, even for prismatic parabolic channels.

5.4.2 Observed Velocity Distributions

5.4.2.1 Narrow Channels

As noted above, the effects of bank friction become significant in channels with small width/depth ratios, usually depressing the location of maximum velocity below the surface and generating helicoidal currents (figure 5.17). Figure 5.19 shows isovels in two small rectangular flumes and the velocity profile measured at the center. Note that the depression of the maximum velocity is greatest at the center and diminishes toward the boundary, and has only a minor effect on the form of the vertical profile, even in the center.

5.4.2.2 Bends

Figure 5.20 shows the typical strongly asymmetric cross section and pattern of isovels at the apex of a meander bend. The maximum velocity is fastest where the water is deepest, toward the outside of the bend. The asymmetry produces distortions

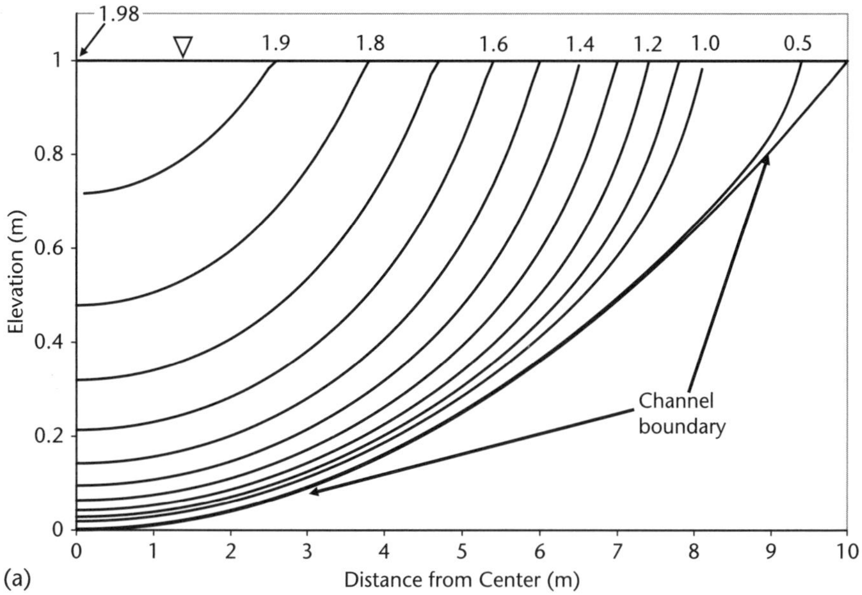

(a)

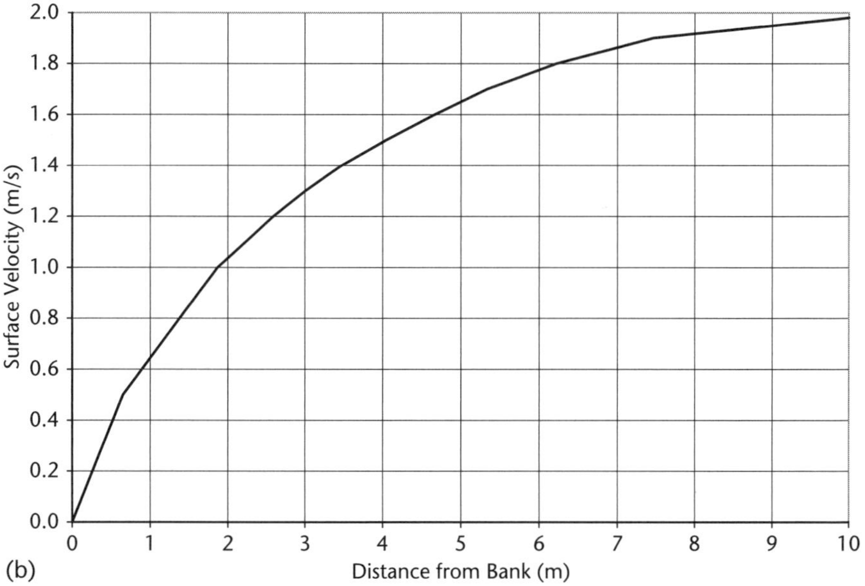

(b)

Figure 5.18 Velocities in one half of a parabolic channel as generated by the synthetic channel model (appendix C), which assumes that the P-vK law applies at each cross-channel location. (a) Isovels (m/s). Note vertical exaggeration. (b) Arithmetic plot of cross-channel distribution of surface velocity (c) Semilogarithmic plot of cross-channel distribution of surface velocity showing approximation to a P-vK-type law (dashed straight line). *(continued)*

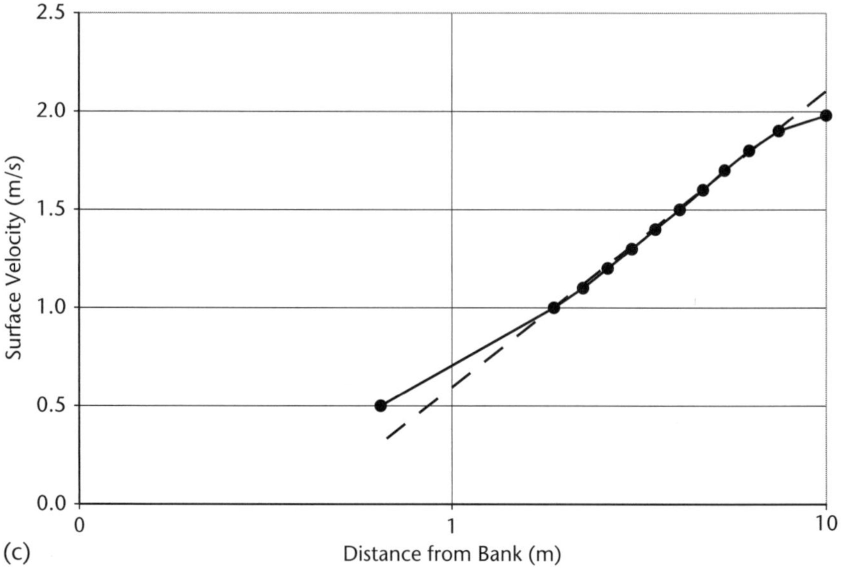

(c)

Figure 5.18 *Continued*

from the P-vK profile and the maximum velocity tends to be slightly below the surface (a distance too small to be seen in the bend shown here). Centrifugal force (which is proportional to u^2 and inversely proportional to the radius of curvature; see section 6.6.1.2) carries the faster surface velocity threads more strongly to the outside of the bend than the slower near-bed threads. Thus, helicoidal circulation is also a feature of river bends, with surface currents flowing toward the outside of the bend and near-bed currents flowing toward the convex bank (figure 5.21a); the outside concave bank therefore tends to be a site of erosion, and the inside bank a site of deposition that produces a **point bar**. In a meandering stream, the maximum-velocity thread follows the pattern shown in figure 5.21b.

The centrifugal force also produces a cross-channel tilting of the water surface (figure 5.21a); this phenomenon is called **superelevation**. The total difference in elevation, Δz, can be calculated as

$$\Delta z = \frac{U^2 \cdot W}{g \cdot r_c}, \tag{5.49}$$

where U is average velocity, W is width, g is gravitational acceleration, and r_c is the radius of curvature of the bend (Leliavsky 1955). (For the bend shown in figure 5.20, for which $r_c \approx 500$ m, Δz is only about 1 cm.)

5.4.2.3 Irregular Natural Channels

Figure 5.22 shows isovels in two natural-channel cross sections, one a bouldery mountain stream and the other a meandering sand-bed stream. Although velocities

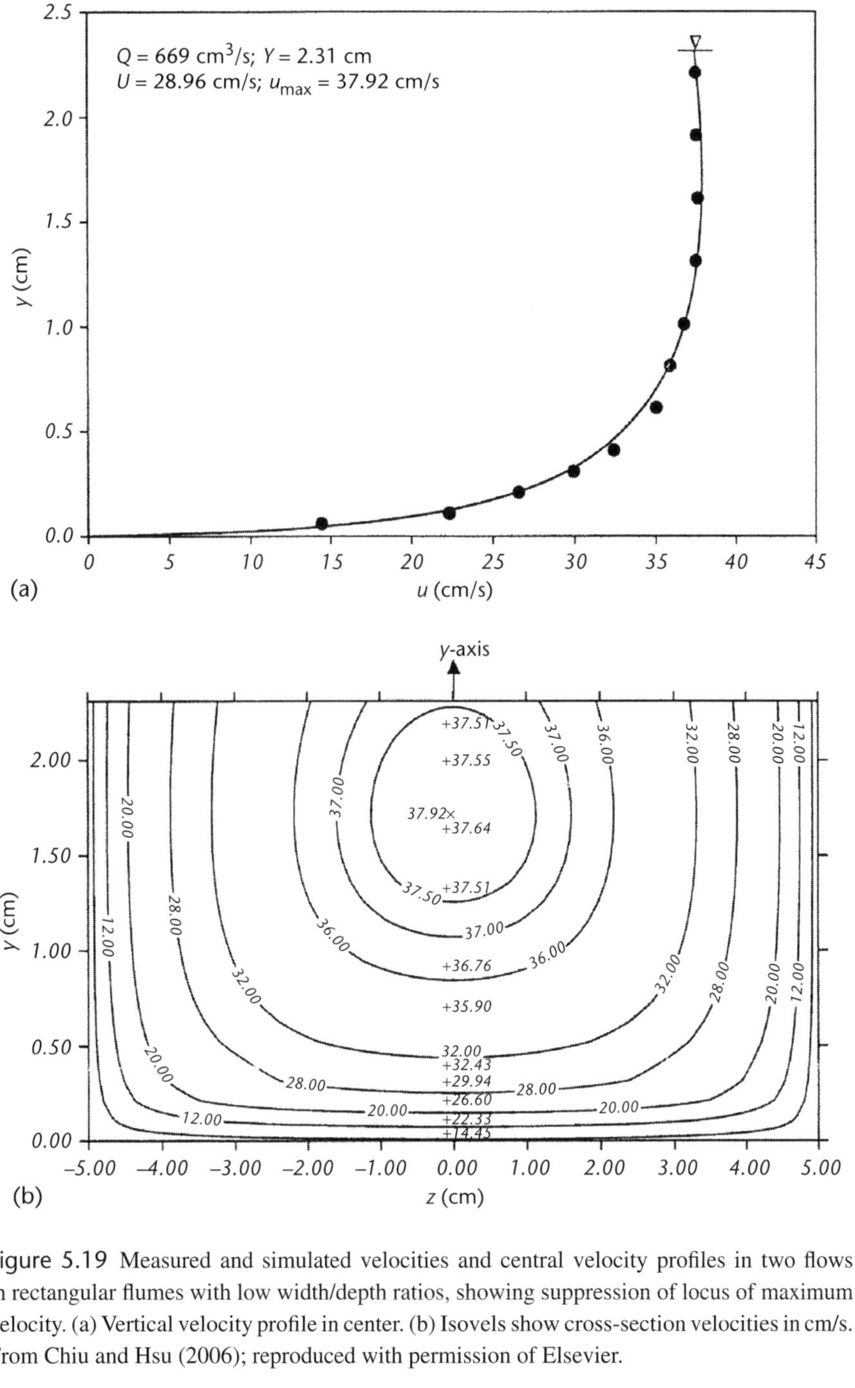

Figure 5.19 Measured and simulated velocities and central velocity profiles in two flows in rectangular flumes with low width/depth ratios, showing suppression of locus of maximum velocity. (a) Vertical velocity profile in center. (b) Isovels show cross-section velocities in cm/s. From Chiu and Hsu (2006); reproduced with permission of Elsevier.

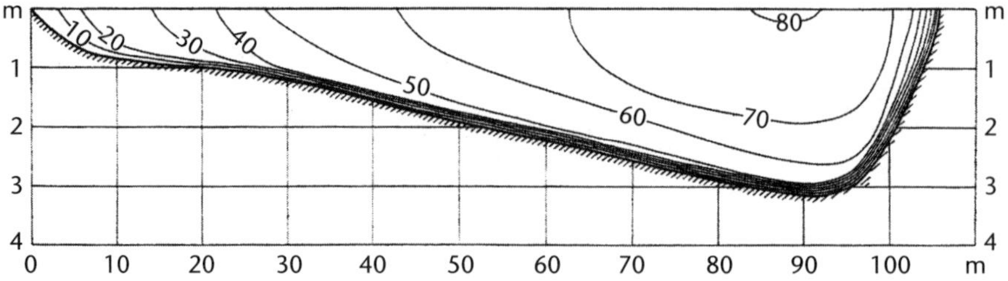

Figure 5.20 Isovels (cm/s) in a meander bend of the River Klarälven, Sweden, showing typical pattern of highest velocities in deepest portion of the cross section leading to helicoidal flow as shown in figure 5.21a. Note vertical exaggeration. From Sundborg (1956); reproduced with permission of Blackwell.

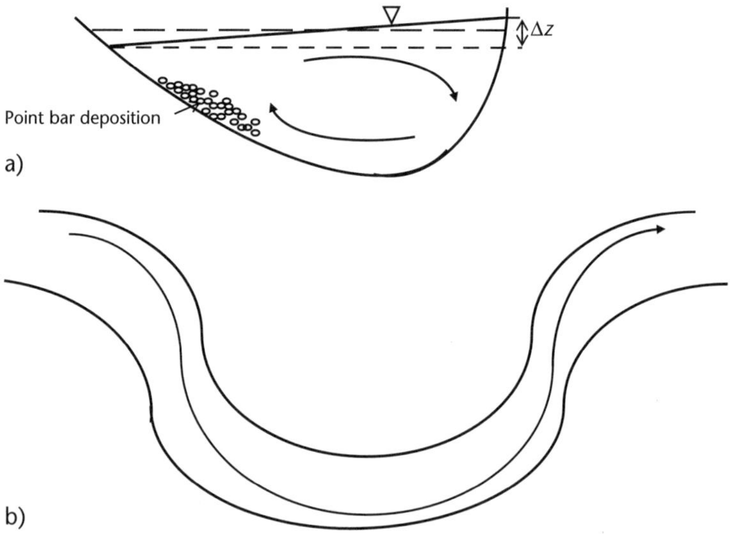

Figure 5.21 (a) Diagram of a meander bend (vertically exaggerated), showing typical asymmetry, helicoidal flow, point-bar deposition on inside of bend, and superelevation Δz. (b) Diagrammatic plan view of successive meander bends showing trace of thread of maximum velocity.

increase monotonically upward virtually everywhere in both, widely varying vertical profiles with clear deviations from the P-vK law are apparent throughout. Most of these deviations are caused by obstructions (large boulders and large woody debris) that are upstream and downstream of the measured sections. The effects of such obstructions change as the discharge changes and as the obstructions change over time. Clearly, it is impossible to predict such effects, and one should not expect the P-vK law or any other theoretical profile to be widely applicable in streams with irregularly distributed obstructions that are large relative to the depth.

One practical implication of this unpredictability is that measurements of velocity to determine discharge in such streams should not assume that the average velocity can be measured at "six-tenths depth," as described in section 5.3.1.9. Rather, one should determine average velocity at each vertical using the multipoint method described in box 5.1. As noted there, the highest accuracy in these situations is obtained by measuring velocity several heights at each vertical, with averages found by numerical integration over each vertical or over the entire cross section. Alternatively, a statistical sampling approach may be appropriate, as described in the next section.

5.4.3 Statistical Characterizations of Velocity Distribution

A promising approach to characterizing cross-section velocities in highly irregular channels such as those shown in figure 5.22 is to treat the problem statistically.

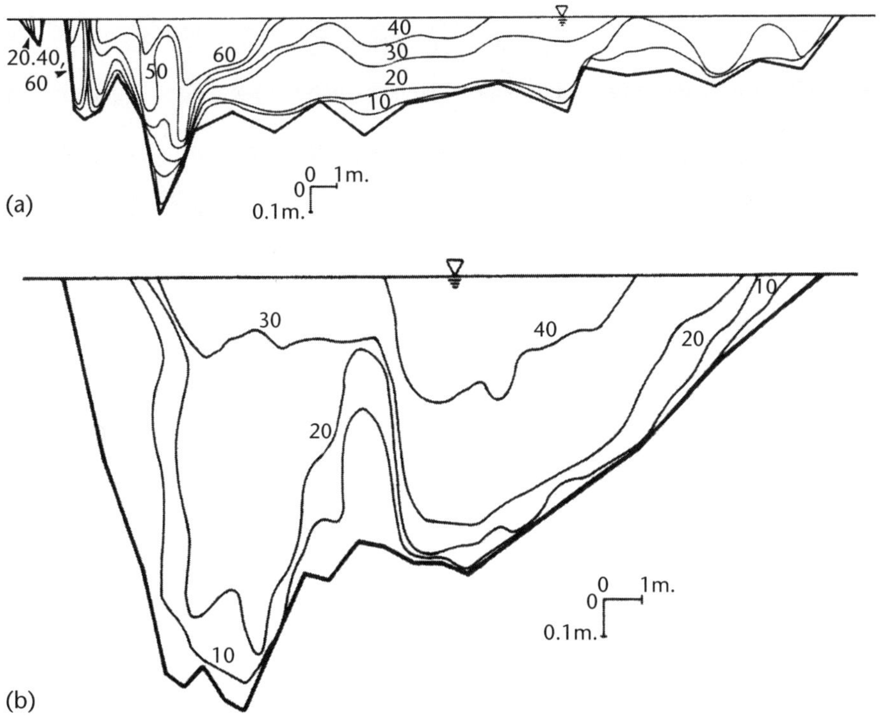

Figure 5.22 Isovels in two natural channels. (a) A wide, shallow, bouldery mountain stream (Mad River, Campton, NH). (b) a meandering sand-bed stream (Lovell River, Ossipee, NH).Velocities increase toward surface throughout both sections but do not generally follow the P-vK law largely due to disturbances by large boulders and woody debris upstream and downstream of measured sections. Note vertical exaggeration. From Dingman (1989).

Dingman (1989, 2007b) proposed that cross-section velocity followed a power-law distribution:

$$\Pr\left\{u \leq u'\right\} = \left(\frac{u'}{u_{\max}}\right)^{J}, \tag{5.50}$$

where $\Pr\{u \leq u'\}$ is the probability that a randomly chosen point velocity, u, is less than a particular value u', $u_{\max}$ is the maximum velocity in the section, and J is an exponent $(0 < J)$.[3] If velocities can be characterized by equation 5.50, the average cross-section velocity U is given by

$$U = \left(\frac{J}{J+1}\right) \cdot u_{\max}. \tag{5.51}$$

As we have seen, the maximum velocity will almost always be found near the surface at the deepest point in the channel and can be found relatively easily by trial measurements at likely locations. The value of J can be estimated by measuring velocity at a number of points over the entire section and computing

$$\hat{J} = \frac{1}{\ln(u_{\max}) - E[\ln(u)]}, \tag{5.52}$$

where $\hat{J}$ is the estimate of J, and $E[\ln(u)]$ is the average of the natural logarithms of the measured point velocities.

6

Uniform Flow and Flow Resistance

6.0 Introduction and Overview

The central problem of open-channel-flow hydraulics can be stated as follows: Given a channel reach with a specified geometry, material, and slope, what are the relations among flow depth, average velocity, width, and discharge? Solutions to this problem are essential for solving important practical problems, including 1) the design of channels and canals, 2) the areal extent of flooding that will result from a storm or snowmelt event, 3) the rate of travel of a flood wave through a channel network, and 4) the size and quantity of material that can be eroded or transported by various flows.

The characterization of flow resistance (defined precisely in section 6.4) is essential to the solutions of this central problem, because it provides the relation between velocity (usually considered the dependent variable) and 1) specified geometric and boundary characteristics of the channel, usually considered to be essentially constant; and 2) the flow magnitude expressed as discharge or depth, considered as the independent variable that may change with time in a given reach.

The definition of flow resistance is developed from the concepts of uniform flow (section 4.2.1.2) and force balance (section 4.7). Recall that in a steady uniform flow, there is no acceleration; thus, by Newton's second law of motion, there is no net force acting on the fluid. Although uniform flow is an ideal state seldom strictly achieved in natural flows, it is often a valid assumption because open-channel flows are self-adjusting dynamic systems (negative feedback loops) that are always tending toward a balance of driving and resisting forces: an increase (decrease) in velocity produces an increase (decrease) in resistance tending to decrease (increase) velocity.

To better appreciate the basic concepts underlying the definition and determination of resistance, this chapter begins by reviewing the basic geometric features of river reaches and reach boundaries presented in section 2.3. We then adapt the definition of uniform flow as applied to a fluid element to apply to a typical river reach and derive the Chézy equation, which is the basic equation for macroscopic uniform flows. This derivation allows us to formulate a simple definition of resistance. We then undertake an examination of the factors that determine flow resistance; this examination involves applying the principles of dimensional analysis developed in section 4.8.2 and the velocity-profile relations derived in chapter 5. The chapter concludes by exploring resistance in nonuniform flows and practical approaches to determining resistance in natural channels.

As we will see, there is still much research to be done to advance our understanding of resistance in natural rivers.

6.1 Boundary Characteristics

As noted above, the nature as well as the shape of the channel boundary affects flow resistance. The classification of boundary characteristics in figure 2.15 provides perspective for the discussion in the remainder of this chapter: Most of the analytical relations that have been developed and experimental results that have been obtained are for rigid, impervious, nonalluvial or plane-bed alluvial boundaries, while many, if not most, natural channels fall into other categories.

In this chapter, we consider cross-section-averaged or reach-averaged conditions rather than local "vertically" averaged velocities (U_w) and local depths (Y_w), and will designate these larger scale averages as U and Y, respectively. Figure 6.1 shows the spatial scales typically associated with these terms. Since our analytical reasoning will be based on the assumption of prismatic channels, there is no distinction between cross-section averaging and reach averaging. We will often invoke the wide

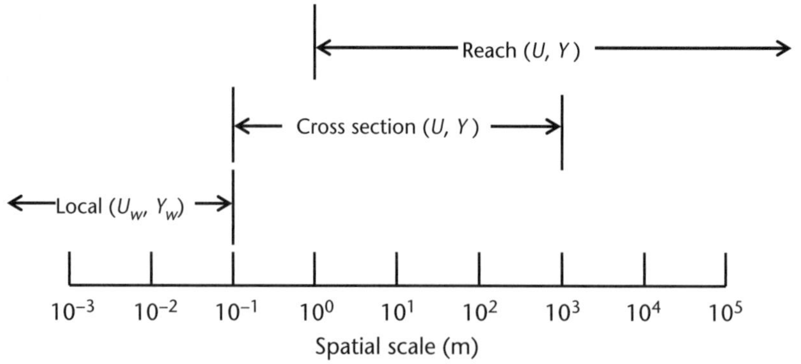

Figure 6.1 Spatial scales typically associated with local, cross-section-averaged, and reach-averaged velocities, depths, and resistance. After Yen (2002).

open-channel concept to justify applying the local, two-dimensional "vertical" velocity distributions discussed in chapter 5 [especially the Prandtl-von Kármán (P-vK) law] to entire cross sections.

We saw in section 5.3.1.6 that channel boundaries can be hydraulically "smooth" or "rough" depending on whether the boundary Reynolds number Re_b is greater or less than 5, where

$$Re_b \equiv \frac{u_* \cdot y_r}{\nu} \tag{6.1}$$

u_* is shear velocity, ν is kinematic viscosity, and y_r is the **roughness height**, that is, the characteristic height of roughness elements (projections) on the boundary (see figure 5.7). In natural alluvial channels, the bed material usually consists of sediment grains with a range of diameters (figure 2.17a). For a particular reach the characteristic height y_r is usually determined as shown in figure 2.17b:

$$y_r = k_r \cdot d_p, \tag{6.2}$$

where d_p is the diameter of particles larger than p percent of the particles on the boundary surface and k_r is a multiplier ≥ 1. Different investigators have used different values for p and k_r (see Chang 1988, p. 50); we will generally assume $k_r = 1$ and $p = 84$ so that $y_r = d_{84}$.

Of course, other aspects of the boundary affect the effective roughness height, especially the spacing and shape of particles. And, as suggested in figure 2.15, the appropriate value for y_r is affected by the presence of bedforms, growing and dead vegetation, and other factors.

6.2 Uniform Flow in Open Channels

6.2.1 Basic Definition

The concepts of steady flow and uniform flow were introduced in section 4.2.1.2 in the context of the movement of a fluid element in the x-direction along a streamline:

> If the element velocity u at a given point on a streamline does not change with time, the flow is **steady** (local acceleration $du/dt = 0$); otherwise, it is **unsteady**.
> If the element velocity at any instant is constant along a streamline, the flow is **uniform** (convective acceleration $du/dx = 0$); otherwise, it is **nonuniform**.

In the remainder of this text we will be concerned with the entirety of a flow within a reach of finite length rather than an individual fluid element flowing along a streamline. Furthermore, in turbulent flows, which include the great majority of natural open-channel flows, turbulent eddies preclude the existence of strictly steady or uniform flow. To account for these conditions we must modify the definition of "steady" and "uniform." To do this, we first designate the X-coordinate direction as the downstream direction for a reach and define U as the downstream-directed velocity,

1) time-averaged over a period longer than the time scale of turbulent fluctuations and 2) space-averaged over a cross section. Then,

- In **steady flow**, $dU/dt = 0$ at any cross section.
- In **uniform flow**, $dU/dX = 0$ at any instant.

As noted by Chow (1959, p. 89), unsteady uniform flow is virtually impossible of occurrence. Thus, henceforth, "uniform flow" implies "steady uniform flow." Note, however, that a nonuniform flow may be steady or unsteady.

We will usually assume that the discharge, Q, in a reach is constant in space and time, where

$$Q = W \cdot Y \cdot U, \tag{6.3}$$

W is the water-surface width, and Y is average depth.

In uniform flow with spatially constant Q, it must also be true that depth and width are constant, so "uniform flow" implies $dY/dX = 0$ and $dW/dX = 0$.[1] And, since the depth does not change, "uniform flow" implies that the water-surface slope is identical to the channel slope. Thus, it must also be true that for strictly uniform flow, cross-section shape is constant through a reach (i.e., the channel is prismatic).

Figure 6.2 further illustrates the concept of uniform flow. Here, a river or canal with constant channel slope θ_0, geometry, and bed and bank material, and no other inputs of water, connects two large reservoirs that maintain constant surface elevations. Under these conditions, the discharge will be constant along the entire channel. As the water leaves the upstream reservoir, it accelerates from zero velocity due to the downslope component of gravity, $g \cdot \sin\theta_S$, where θ_S is the local slope of the water surface. As it accelerates, the frictional resistance of the boundary is transmitted into the fluid by viscosity and turbulence (as in figure 3.28). This resistance increases as the velocity increases and soon balances the gravitational force,[2] at which point there is no further acceleration. Downstream of this point, the water-surface slope θ_S equals the channel slope θ_0, the cross-section-averaged velocity and depth become constant, and uniform flow is established. The velocity and depth remain constant

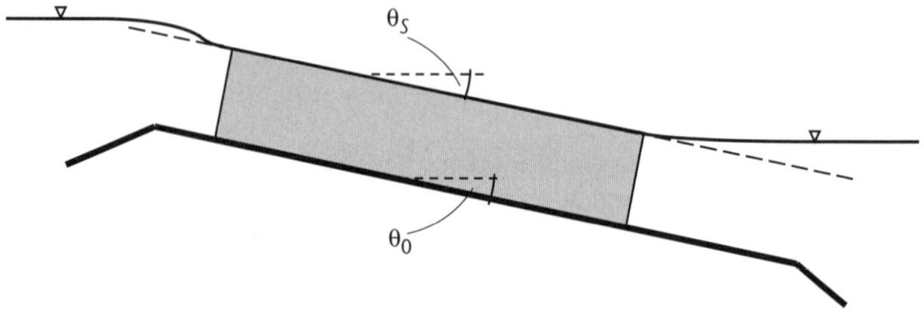

Figure 6.2 Idealized development of uniform flow in a channel of constant slope, θ_0, geometry, and bed material connecting two reservoirs. The shaded area is the region of uniform flow, where the downstream component of gravity is balanced by frictional resistance and the water-surface slope θ_S equals θ_0.

until the water-surface slope begins to decrease ($\theta_s < \theta_0$) to allow transi water level in the downstream reservoir, which is maintained at a level h that associated with uniform flow. This marks the beginning of negative ac and the downstream end of uniform flow.

6.2.2 Qualifications

Even with the above definitions, we see that strictly uniform flow is an idealization that cannot be attained in nonprismatic natural channels. And, even in prismatic channels there are hydraulic realities that usually prevent the attainment of truly uniform flow; these are described in the following subsections. Despite these realities, the concept of uniform flow is the starting point for describing resistance relations for all open-channel flows. If the deviations from strict uniform flow are not too great, the flow is quasi uniform, and the basic features of uniform flow will be assumed to apply.

6.2.2.1 Uniform Flow as an Asymptotic Condition

Although figure 6.2 depicts a long channel segment as having uniform flow, in fact uniform flow is approached asymptotically. As stated by Chow (1959, p. 91), "Theoretically speaking, the varied depth at each end approaches the uniform depth in the middle asymptotically and gradually. For practical purposes, however, the depth may be considered constant (and the flow uniform) if the variation in depth is within a certain margin, say, 1%, of the average uniform-flow depth." Thus, the shaded area in figure 6.2 is the portion of the flow that is within this 1% limit.

6.2.2.2 Water-Surface Stability

Under some conditions, wavelike fluctuations of the water surface prevent the attainment of truly uniform flow. As we will discuss more fully in chapter 11, a gravity wave in shallow water travels at a speed relative to the water, or **celerity**, C_{gw}, that is determined by the depth, Y:

$$C_{gw} = (g \cdot Y)^{1/2}, \tag{6.4}$$

where g is gravitational acceleration. ("Shallow" in this context means that the wavelength of the wave is much greater than the depth.) Note from figure 6.3 that this celerity is of the same order as typical river velocities. The **Froude number**, Fr, defined as

$$Fr \equiv \frac{U}{C_{gw}} = \frac{U}{(g \cdot Y)^{1/2}}, \tag{6.5}$$

is the ratio of flow velocity to wave celerity and defines the **flow regime**:[3]

> When $Fr = 1$, the flow regime is **critical**; when $Fr < 1$ it is **subcritical**, and when $Fr > 1$ it is **supercritical**.

Figure 6.4 shows the combinations of velocity and depth that define flows in the subcritical and supercritical regimes. Most natural river flows are subcritical

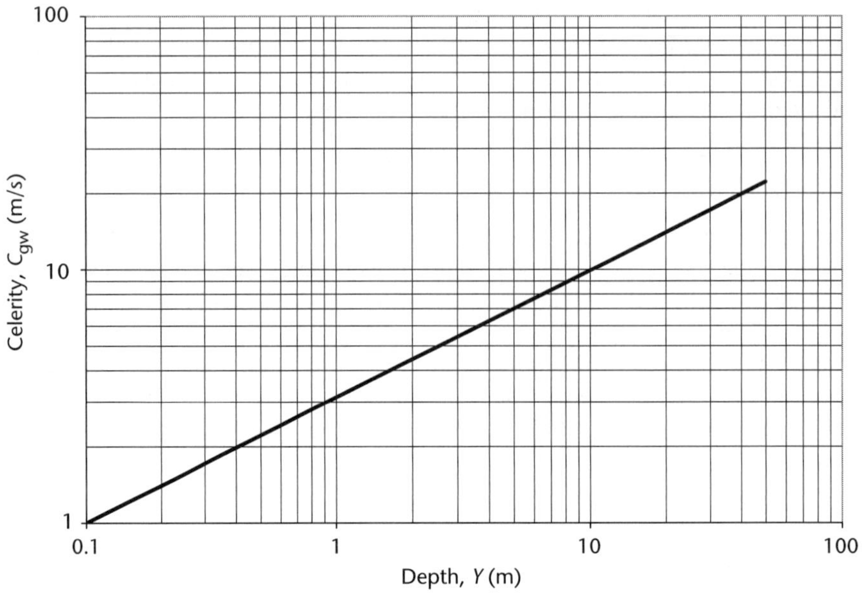

Figure 6.3 Celerity of shallow-water gravity waves, C_{gw}, as a function of flow depth, Y (equation 6.4). Note that C_{gw} is of the same order of magnitude as typical river velocities.

(Grant 1997), but when the slope is very steep and/or the channel material is very smooth (as in some bedrock channels and streams on glaciers, and at local steepenings in mountain streams), the Froude number may approach or exceed 1. When *Fr* approaches 1, waves begin to appear in the free surface, and strictly uniform flow is not possible. In channels with rigid boundaries, the amplitude of these waves increases approximately linearly with *Fr* (figure 6.5). When *Fr* approaches 2 (Koloseus and Davidian 1966), the flow will spontaneously form **roll waves**—the waves you often see on a steep roadway or driveway during a rainstorm (figure 6.6). However, this situation is unusual in natural channels. In channels with erodible boundaries (sand and gravel), wavelike bedforms called dunes or antidunes begin to form when *Fr* approaches 1. The water surface also becomes wavy, either out of phase (dunes) or in phase (antidunes) with the bedforms; these are discussed further in section 6.6.4 and in sections 10.2.1.5 and 12.5.4.

In situations where surface instabilities occur, it may be acceptable to relax the definition of "uniform" by averaging dU/dX and dY/dX over distances greater than the wavelength of the surface waves.

6.2.2.3 Secondary Currents

The concept of uniform flow as described in section 6.2.1 implicitly assumes that flow is the downstream direction only, and this assumption underlies most of the analyses in this text. However, as we saw in section 5.4.2, even in straight rectangular

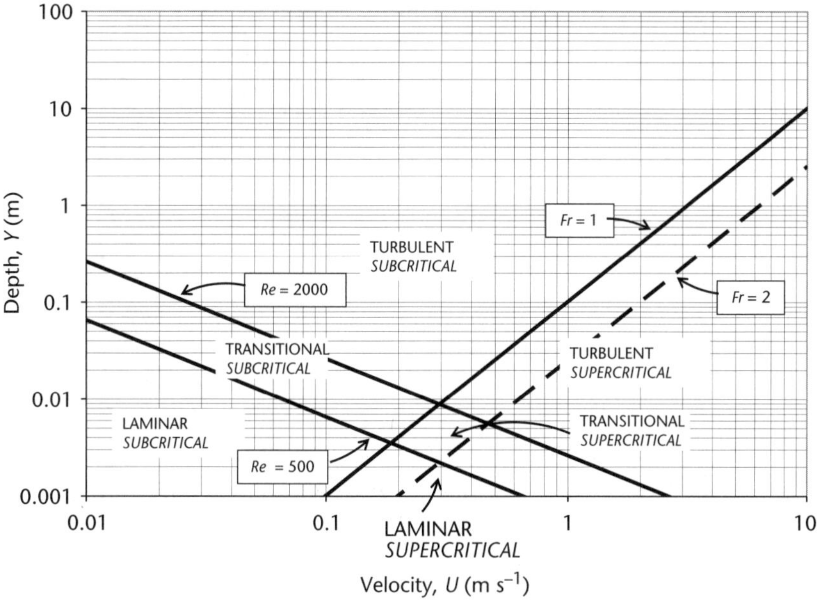

Figure 6.4 Flow states and flow regimes as a function of average velocity, U, and depth Y. The great majority of river flows are in the turbulent state ($Re > 2000$) and subcritical regime ($Fr < 1$). When the Froude number Fr (equation 6.5) approaches 1, the water surface becomes wavy, and strictly uniform flow cannot occur. When Fr approaches 2, pronounced waves are present. Note that some authors (e.g., Chow 1959) use the term "regime" to apply to one of the four fields shown on this diagram rather than to the subcritical/supercritical condition.

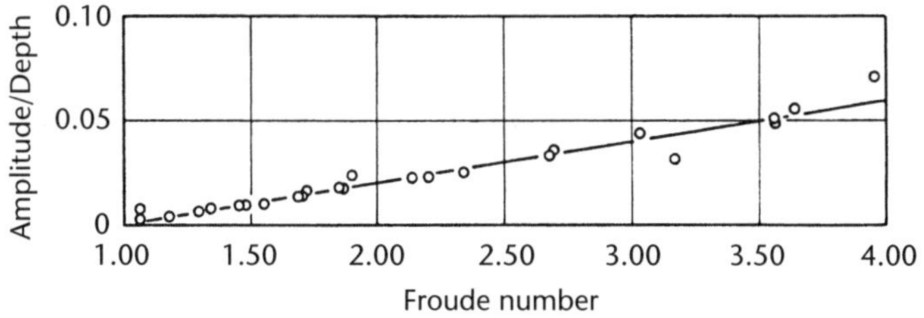

Figure 6.5 Ratio of wave amplitude to mean depth as a function of Froude number as observed in flume experiments by Tracy and Lester (1961, their figure 6).

channels spiral circulations are often present, making the velocity distribution three-dimensional and suppressing the level of maximum velocity below the surface. These secondary or helicoidal currents spiral downstream with velocities on the order of 5% of the downstream velocity and differ in direction by only a few degrees from the downstream direction (Bridge 2003). Thus, their effect on the assumptions of uniform flow is generally small.

Figure 6.6 Roll waves on a steep driveway during a rainstorm. These waves form when the Froude number approaches 2. Photo by the author.

6.3 Basic Equation of Uniform Flow: The Chézy Equation

In this section, we derive the basic equation for strictly uniform flow. This equation forms the basis for understanding fundamental resistance relations and other important aspects of flows in channel reaches.

Because there is no acceleration in a uniform flow, Newton's second law states that there are no net forces acting on the fluid and that

$$F_D = F_R, \tag{6.6}$$

where F_D represents the net forces tending to cause motion, and F_R represents the net forces tending to resist motion. The French engineer Antoine Chézy (1718–1798) was the first to develop a relation between flow velocity and channel characteristics from the fundamental force relation of equation 6.6.[4] Referring to the idealized rectangular channel reach of figure 6.7, Chézy expressed the downslope component of the gravitational force acting on the water in a channel reach, F_D, as

$$F_D = \gamma \cdot W \cdot Y \cdot X \cdot \sin\theta = \gamma \cdot A \cdot X \cdot \sin\theta, \tag{6.7}$$

where γ is the weight density of water, A is the cross-sectional area of the flow, and θ denotes the slope of the water surface and the channel, which are equal in uniform flow.

Chézy noted that the resistance forces are due to a boundary shear stress τ_0 [F L^{-2}] caused by boundary friction. This is the same quantity defined in equation 5.7, but

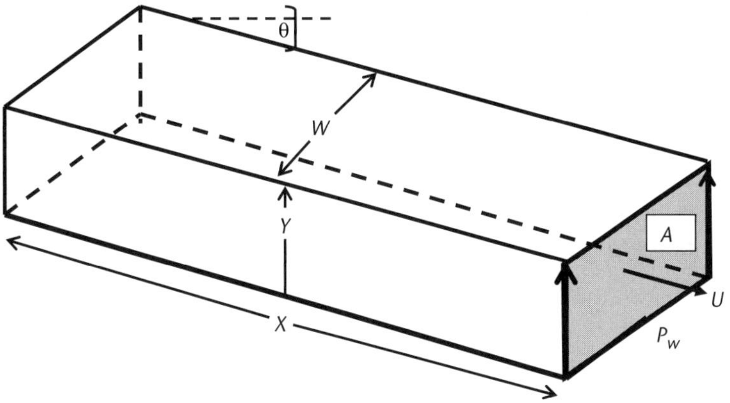

Figure 6.7 Definitions of terms for development of the Chézy relation (equation 6.15). The idealized channel reach has a rectangular cross-section of slope θ, width W, and depth Y. A is the wetted cross-sectional area (shaded), P_w is the wetted perimeter, and U is the reach-averaged velocity.

now applies to the entire cross section, not just the local channel bed. Chézy further reasoned that this stress is proportional to the square of the average velocity:

$$\tau_0 = K_T \cdot \rho \cdot U^2, \tag{6.8}$$

where K_T is a dimensionless proportionality factor. This expression is dimensionally correct and is physically justified by the model of turbulence developed in section 3.3.4, which shows that shear stress is proportional to the turbulent velocity fluctuations (equation 3.32; see also equation 5.27b) and that these fluctuations are proportional to the average velocity.[5]

This boundary shear stress acts over the area of the channel that is in contact with the water, A_B (the frictional resistance at the air-water interface is negligible), which in the rectangular channel shown in figure 6.7 is given by

$$A_B = (2Y + W) \cdot X = P_w \cdot X, \tag{6.9}$$

where P_w is the **wetted perimeter** of the flow. Thus,

$$F_R = \tau_0 \cdot A_B = K_T \cdot \rho \cdot U^2 \cdot P_w \cdot X, \tag{6.10}$$

where τ_0 designates the shear stress acting over the entire flow boundary.

Combining equations 6.6, 6.7, and 6.10 gives

$$\gamma \cdot A \cdot X \cdot \sin\theta = K_T \cdot \rho \cdot U^2 \cdot P_w \cdot X, \tag{6.11}$$

which (noting that $\gamma/\rho = g$) can be solved for U to give

$$U = \left(\frac{g}{K_T}\right)^{1/2} \cdot \left(\frac{A}{P_w}\right)^{1/2} \cdot (\sin\theta)^{1/2} \tag{6.12}$$

The ratio of cross-sectional area to wetted perimeter is called the **hydraulic radius**, R:

$$R \equiv \frac{A}{P_w}. \tag{6.13}$$

Incorporating equation 6.13 and defining

$$S \equiv \sin\theta, \quad (6.14)$$

We can write the Chézy equation as

$$U = \left(\frac{1}{K_T}\right)^{1/2} \cdot (g \cdot R \cdot S)^{1/2}. \quad (6.15a)$$

For "wide" channels we can approximate the hydraulic radius by the average depth. Thus, we can usually write the Chézy equation as

$$U = \left(\frac{1}{K_T}\right)^{1/2} \cdot (g \cdot Y \cdot S)^{1/2}. \quad (6.15b)$$

In engineering contexts, the Chézy equation is usually written as described in box 6.1.

The Chézy equation is the basic uniform-flow equation and is the basis for describing the relations among the cross-section or reach-averaged values of the fundamental hydraulic variables velocity, depth, slope, and channel characteristics. It provides a partial answer to the central question posed at the beginning of the chapter, as we have found that

> The average velocity of a uniform open-channel flow is proportional to the square root of the product of hydraulic radius (R) and the downslope component of gravitational acceleration ($g \cdot S$).

Also note that the Chézy equation was developed from force-balance considerations and is a macroscopic version of the general conductance relation (equation 4.54, section 4.7). The Chézy equation was derived by considering the water in the channel as a "block" interacting with the channel boundary; we did not consider phenomena within the "block" except to justify the relation between τ_0 and the square of the velocity (equation 6.8).

A more complete answer to the central question posed at the beginning of this chapter requires some way of determining the value of K_T. This quantity is the proportionality between the shear stress due to the boundary and the square of the velocity; thus, presumably it depends in some way on the nature of the boundary. Most of the rest of this chapter explores the relation between this proportionality and the nature of the boundary. We will see that the velocity profiles derived in chapter 5 along with experimental observations provide much of the basis for formulating this relation. But before proceeding to that exploration, we use the Chézy derivation to formulate the working definition of resistance.

6.4 Definition of Reach Resistance

By comparison with equation 5.24, the quantity $(g \cdot R \cdot S)^{1/2}$ can be considered to be the **reach-averaged shear velocity**, so henceforth

$$u_* \equiv (g \cdot R \cdot S)^{1/2}. \quad (6.16a)$$

Again, we have seen that we can usually approximate this definition as

$$u_* = (g \cdot Y \cdot S_0)^{1/2}. \quad (6.16b)$$

BOX 6.1 Chézy's *C*

In engineering texts, the Chézy equation is usually written as

$$U = C \cdot (R \cdot S)^{1/2}, \tag{6B1.1}$$

where *C* expresses the reach conductance and is known as "Chézy's *C*." Note from equation 6.15a that

$$C \equiv \left(\frac{g}{K_T}\right)^{1/2}, \tag{6B1.2}$$

and thus has dimensions [$L^{1/2}\ T^{-1}$].

In engineering practice, however, *C* is treated as a dimensionless quantity so that it has the same numerical value in all unit systems. This can be a dangerous practice: equation 6B1.1 is in fact correct only if the British (ft-s) unit system is used. If *C* is to have the same numerical value in all unit systems, the Chézy equation must be written as

$$U = u_C \cdot C \cdot (R \cdot S)^{1/2}, \tag{6B1.3}$$

where u_C is a unit-adjustment factor that takes the following values:

Unit system	u_C
Système Internationale	0.552
British	1.00
Centimeter-gram-second	5.52

No systematic method for estimating Chézy's *C* from channel characteristics has been published (Yen 2002). The following statistics from a database of 931 flows in New Zealand and the United States collated by the author give a sense of the range of *C* values in natural channels:

Statistic	**C value**
Mean	32.5
Median	29.3
Standard deviation	17.7
Maximum	86.6
Minimum	2.1

Using this definition, we define **reach resistance**, Ω, as the ratio of reach-averaged shear velocity to reach-averaged velocity:

$$\Omega \equiv \frac{u_*}{U}. \tag{6.17}$$

This definition simply provides us with a notation that will prove to be more convenient than using K_T: the relation between them is obviously

$$\Omega = K_T^{1/2}. \tag{6.18}$$

Box 6.2 defines the **Darcy-Weisbach friction factor**, a dimensionless resistance factor that is commonly used as an alternative to K_T and Ω.

Note that using equation 6.17, we can rewrite the Chézy equation as

$$U = \Omega^{-1} \cdot u_* . \quad (6.19)$$

BOX 6.2 The Darcy-Weisbach Friction Factor

In 1845 Julius Weisbach (1806–1871) published the results of pioneering experiments to determine frictional resistance in pipe flow (Rouse and Ince 1963) and formulated a dimensionless factor, f_{DW}, that expresses this resistance:

$$f_{DW} \equiv 2 \cdot \left(\frac{h_e}{X} \right) \cdot \left(\frac{D \cdot g}{U^2} \right), \quad (6B2.1)$$

where h_e (L) is the loss in mechanical energy per unit weight of water, or head (see equation 4.45) in distance X, D is the pipe diameter, U is the average flow velocity, and g is gravitational acceleration. In 1857, the same Henry Darcy (1803–1858) whose experiments led to Darcy's law, the central formula of groundwater hydraulics, published the results of similar pipe experiments, and f_{DW} is known as the **Darcy-Weisbach friction factor**.

The pipe diameter D equals four times the hydraulic radius, R, so

$$f_{DW} \equiv 8 \cdot \left(\frac{h_e}{X} \right) \cdot \left(\frac{R \cdot g}{U^2} \right). \quad (6B2.2)$$

The quantity h_e/X in pipe flow is physically identical to the channel and water-surface slope, $S \equiv \sin \theta$, in uniform open-channel flow, so the friction factor for open-channel flow is

$$f_{DW} \equiv 8 \cdot \frac{g \cdot R \cdot S}{U^2}. \quad (6B2.3a)$$

From the definition of shear velocity, u_* (equation 6.16a), 6B2.3a can also be written as

$$f_{DW} = 8 \cdot \frac{u_*^2}{U^2}, \quad (6B2.3b)$$

and from the definition of Ω (equation 6.17), we see that

$$f_{DW} = 8 \cdot \Omega^2; \quad (6B2.4a)$$

$$\Omega = \left(\frac{f_{DW}}{8} \right)^{1/2} = 0.354 \cdot f_{DW}^{1/2}. \quad (6B2.4b)$$

The Darcy-Weisbach friction factor is commonly used to express resistance in open channels as well as pipes. However, the Ω notation is used herein because it is simpler: It does not include the 8 multiplier and is written in terms of u_* and U rather than the squares of those quantities.

The inverse of a resistance is a conductance, so we can define Ω^{-1} as the **reach conductance**, and we can use the two concepts interchangeably. The central problem of open-channel flow can now be stated as, "What factors determine the value of Ω?"

6.5 Factors Affecting Reach Resistance in Uniform Flow

In section 4.8.2.2, we used dimensional analysis to derive equation 4.63:

$$U = f_\Omega\left(\frac{Y}{y_r}, \frac{Y}{W}, Re\right)\cdot(g\cdot Y\cdot S)^{1/2} = f_\Omega\left(\frac{Y}{y_r}, \frac{Y}{W}, Re\right)\cdot u_*, \tag{6.20}$$

where *Re* is the flow Reynolds number. Thus, we see that the Chézy equation is identical in form to the open-channel flow relation developed from dimensional analysis. And, comparing 6.19 and 6.20, we see that the dimensional analysis provided some clues to the factors affecting resistance/conductance:

$$\Omega = f_\Omega\left(\frac{Y}{y_r}, \frac{Y}{W}, Re\right), \tag{6.21}$$

where f_Ω denotes the resistance/conductance function. Thus, we have reason to believe that, in uniform turbulent flow, resistance depends on the **relative smoothness** Y/y_r (or its inverse, **relative roughness** y_r/Y),[6] the depth/width ratio Y/W (or W/Y), and the Reynolds number, *Re*. However, as we saw in section 2.4.2, most natural channels have small Y/W values, so the effects of Y/W should usually be minor; thus, we focus here on the effects of relative roughness and Reynolds number.

The nature of f_Ω has been explored experimentally in pipes and wide open channels and can be summarized as in figure 6.8. Here, Ω (y-axis) is shown as a function of *Re* (x-axis) and Y/y_r (separate curves at high *Re*) for wide open channels with rigid impervious boundaries. Graphs relating resistance to *Re* and Y/y_r are called **Moody diagrams** because they were first presented, for flow in pipes, by Moody (1944). The original Moody diagrams were based in part on experimental data of Johann Nikuradse (1894–1979), who measured resistance in pipes lined with sand particles of various diameters. These relations have been modified to apply to wide open channels (Brownlie 1981a; Chang 1988; Yen 2002).

Figure 6.8 reveals important aspects of the resistance relation for uniform flow. First, note that, overall, Ω tends to decrease with *Re* and that the $\Omega - Re$ relation f_Ω differs in different ranges of *Re*. For laminar flow and hydraulically smooth turbulent flow, Ω depends only on Reynolds number:

Laminar flow ($Re < 500$):

$$\Omega = \left(\frac{3}{Re}\right)^{1/2} = \frac{1.73}{Re^{1/2}}. \tag{6.22}$$

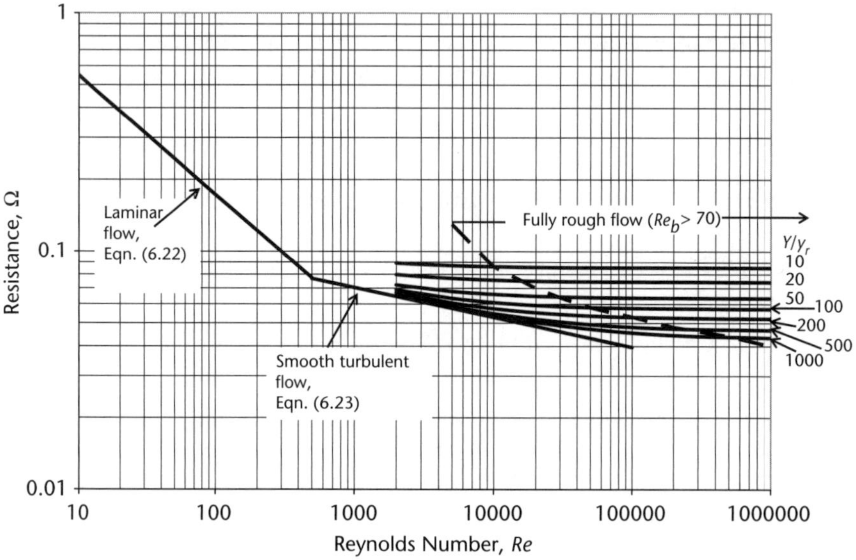

Figure 6.8 The Moody diagram: Relation between resistance, Ω; Reynolds number, *Re*; and relative smoothness, Y/y_r, for laminar, smooth turbulent, and rough turbulent flows in wide open channels. Y/y_r affects resistance only for rough turbulent flows ($Re > 2000$ and $Re_b > 5$). The effect of *Re* on resistance in rough turbulent flows decreases with *Re*; resistance becomes independent of *Re* for "fully rough" flows ($Re_b > 70$).

Smooth turbulent flow ($Re > 500$; $Re_b < 5$):

$$\Omega = \frac{0.167}{Re^{1/8}}. \tag{6.23}$$

For turbulent flow in hydraulically rough channels ($Re_b > 5$), the relation depends on both *Re* and Y/y_r and can be approximated by a semiempirical function proposed by Yen (2002):

$$\Omega = 0.400 \cdot \left[-\ln\left(\frac{y_r}{11 \cdot Y} + \frac{1.95}{Re^{0.9}} \right) \right]^{-1} \tag{6.24}$$

Note that at very high values of *Re*, the second term in 6.24 becomes very small and resistance depends only on Y/y_r (i.e., the curves become horizontal); this is the region of **fully rough flow**, $Re_b > 70$. The transition to fully rough flow occurs at lower *Re* values as the boundary gets relatively rougher (i.e., as Y/y_r decreases). Figure 6.9 shows the relation between Ω and Y/y_r given by 6.24 for fully rough flow, that is, where

$$\Omega_* = 0.400 \cdot \left[-\ln\left(\frac{y_r}{11 \cdot Y} \right) \right]^{-1} = 0.400 \cdot \left[\ln\left(\frac{11 \cdot Y}{y_r} \right) \right]^{-1}. \tag{6.25}$$

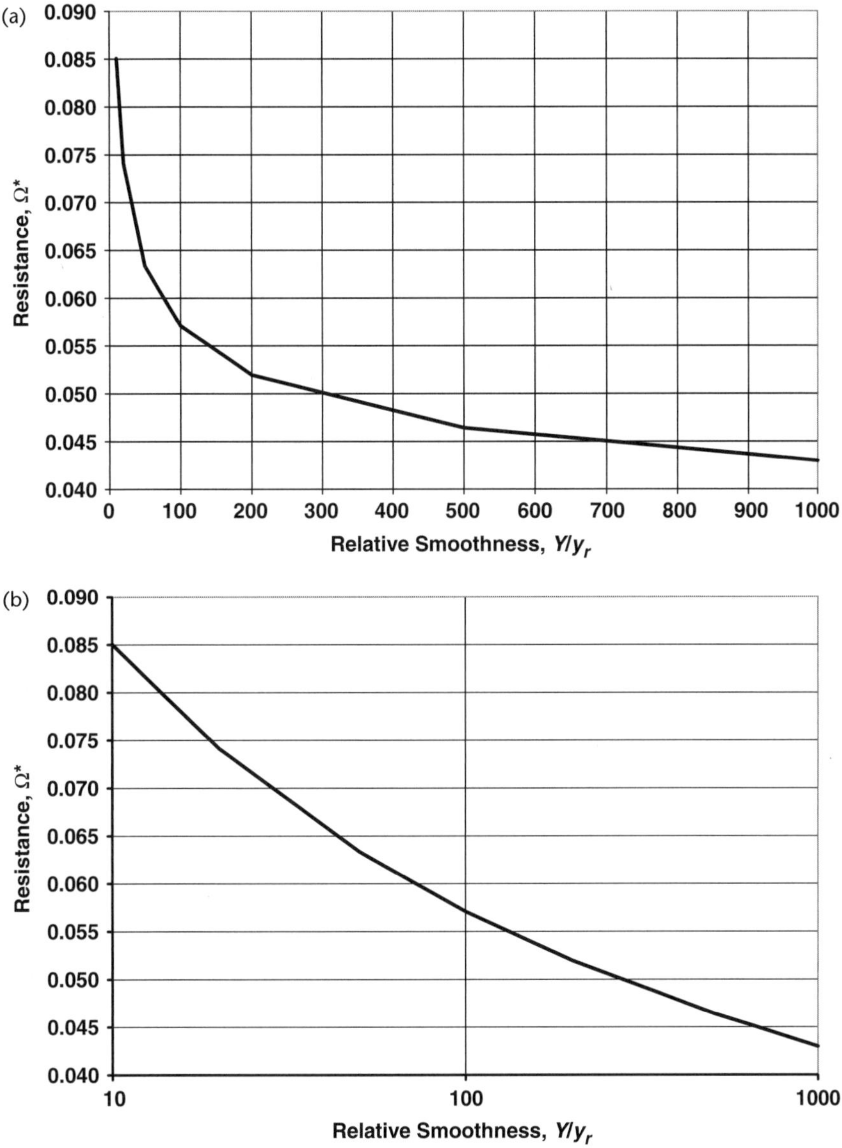

Figure 6.9 Baseline resistance, Ω_*, as a function of relative smoothness, Y/y_r, for fully rough turbulent flow in wide channels as given by equation 6.25. This is identical to the relation given by the integrated P-vK velocity profile (equation 6.26). (a) Arithmetic plot; (b) semilogarithmic plot.

In the reminder of this chapter, we designate the resistance given by 6.25 as Ω_* and use it to represent a baseline resistance value that applies to rough turbulent flow in wide channels.

In general, natural channels will have a resistance greater than Ω_* due to the complex effects of many factors that affect resistance in addition to Y/y_r and *Re*. These additional factors are explored in section 6.6.

For fully rough flow and very large values of *Re*, equation 6.25 can be inverted and written as

$$U = 2.50 \cdot u_* \cdot \ln\left(\frac{11 \cdot Y}{y_r}\right), \tag{6.26}$$

a form that looks similar to the vertically integrated P-vK velocity profile (equation 5.34a–d). In fact, if we combine equations 5.39–5.41 and recall from equation 5.32b that $y_0 = y_r/30$ for rough flow, the integrated P-vK law is identical to equation 6.26. This should not be surprising, given that the integrated P-vK profile gives the average velocity for a wide open channel. Equation 6.26 is often called the **Keulegan equation** (Keulegan 1938); we will refer to it as the **Chézy-Keulegan** or **C-K equation**.

We can summarize resistance relations for uniform turbulent flows in wide open channels with rigid impervious boundaries as follows:

- Although width/depth ratio potentially affects reach resistance, most natural flows have width/depth values so high that the effect is negligible.
- In smooth flows, resistance decreases as the Reynolds number increases.
- In rough flows with a given relative roughness, resistance decreases as the Reynolds number increases until the flow becomes fully rough, beyond which it ceases to depend on the Reynolds number.
- In rough flows at a given Reynolds number, resistance increases with relative roughness.
- In wide fully rough flows, resistance depends only on relative roughness and the relation between resistance and relative roughness is given by the integrated P-vK profile (C-K equation).

6.6 Factors Affecting Reach Resistance in Natural Channels

The analysis leading to equation 6.21 indicates that resistance in uniform flows in prismatic channels is a function of the relative smoothness, Y/y_r; the Reynolds number, *Re*; and the depth/width ratio, Y/W. Because flow resistance is determined by any feature that produces changes in the magnitude or direction of the velocity vectors, we can expect that resistance in natural channels is also affected by additional factors. We will use the quantity $(\Omega - \Omega_*)/\Omega_*$ to express the dimensionless "excess" resistance in a reach, that is, the difference between actual resistance Ω and the resistance computed via equation 6.25. Figure 6.10 shows this quantity plotted against Y/W for a database of 664 flows in natural channels. Although for many of these flows actual resistance is close to that given by 6.25 [i.e., $(\Omega - \Omega_*)/\Omega_* = 0$], a great majority (86%) have higher resistance, and some have resistances several times Ω_*. This plot

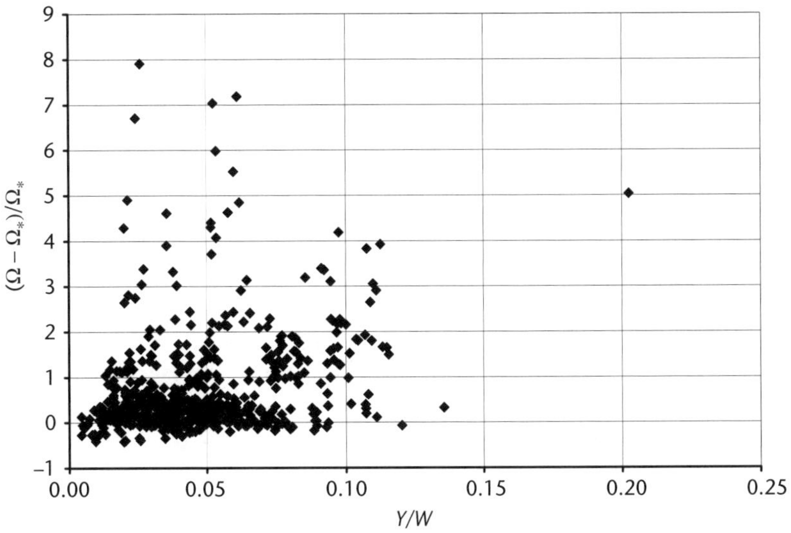

Figure 6.10 Ratio of "excess" resistance to baseline resistance computed from equation 6.25, $(\Omega - \Omega_*)/\Omega_*$, plotted against Y/W for a database of 664 flows in natural channels. Most (86%) of these flows have resistance greater than Ω_*. Clearly, the additional resistance is due to factors other than Y/W.

clearly indicates that, in general, factors other than Y/W cause "excess" resistance in natural channels.

The following subsections discuss, for each of four classes of factors that may produce this excess resistance, 1) approaches to quantifying its contribution, and 2) evidence from field and laboratory studies that gives an idea of the magnitude of the excess resistance produced. Keep in mind, however, that the variability of natural rivers makes this a very challenging area of research and that the approaches and results presented here are not completely definitive.

6.6.1 Effects of Channel Irregularities

Clearly, any irregularities in channel geometry will cause velocity vectors to deviate from direct downstream flow, producing accelerations and concomitant increases in resisting forces. Figure 6.11 shows three categories of geometrical irregularities: in cross section, in plan (map) view, and in reach-scale longitudinal profile (slope). These geometrical irregularities are usually the main sources of the excess resistance apparent in figure 6.10.

6.6.1.1 Cross-section Irregularities

Equation 6.25 gives resistance in hydraulically rough flows in wide open channels in which the depth is constant, the P-vK velocity profile applies at all locations in the

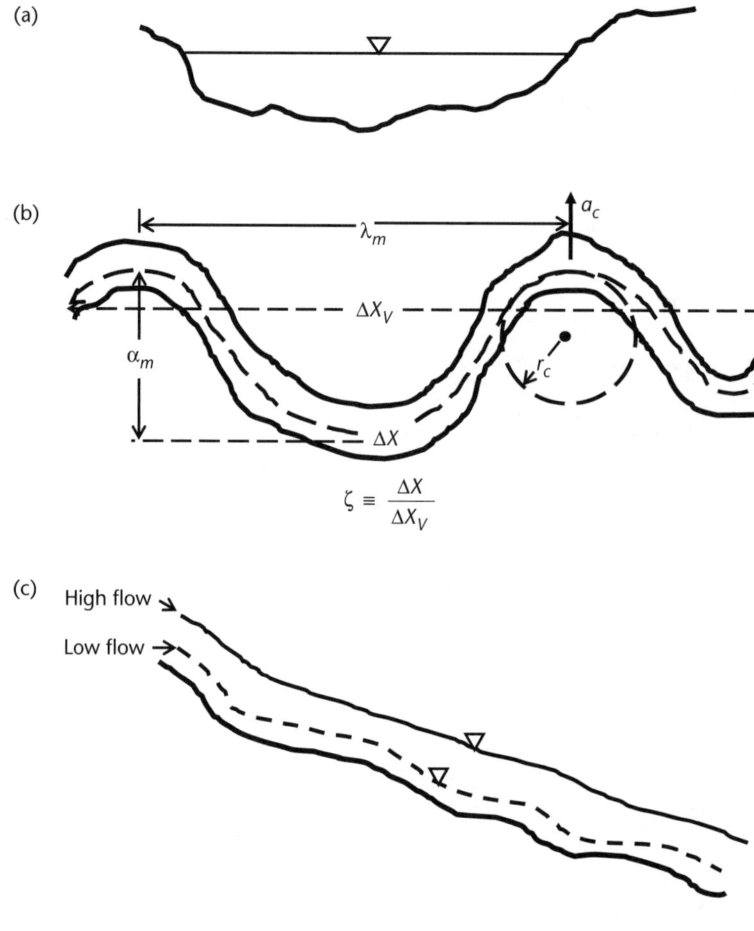

Figure 6.11 Three categories of channel irregularity that cause changes in the magnitude and/or direction of velocity vectors and hence increase flow resistance beyond that given by equation 6.25. (a) Irregularities in cross-section. (b) Irregularities in plan (map) view. ζ designates sinuosity, the streamwise distance ΔX divided by the valley distance ΔX_v; r_c is the radius of curvature of a river bend, λ_m is meander wavelength, a_m is meander amplitude, and a_c represents the centrifugal acceleration. (c) Reach-scale irregularities in longitudinal profile (channel slope); these are more pronounced at low flows and less pronounced at high flows.

cross section, and the only velocity gradients are "vertical." Under these conditions, the isovels (lines of equal velocity) are straight lines parallel to the bottom.

As shown in figure 6.12, irregularities in cross section (represented here by the sloping bank of a trapezoidal channel) cause deviations from this pattern and introduce horizontal velocity gradients that increase shear stress and produce excess resistance. These effects are also apparent in figure 5.22, which shows isovels in two natural channels, where bottom irregularities and other factors produce marked horizontal velocity gradients and significant excess resistance. The presence of obstructions also

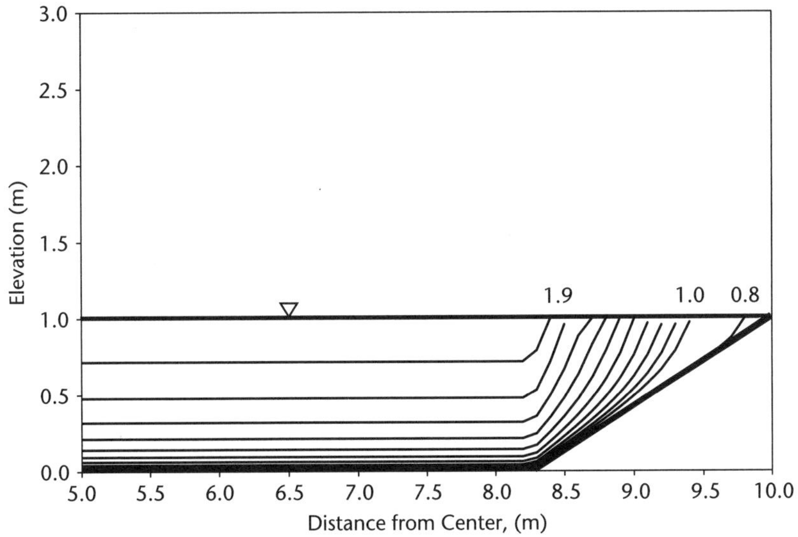

Figure 6.12 Isovels in the near-bank portion of an idealized flow in a trapezoidal channel. The P-vK vertical velocity distribution applies at all points; contours are in m/s. Cross-section irregularities, represented here by the sloping bank, induce horizontal velocity gradients that increase turbulent shear stress and therefore resistance.

induces secondary circulations and tends to suppress the maximum velocity below the surface (see figures 5.17, 5.19, and 5.20), further increasing resistance.

These effects are very difficult to quantify. However, the effects of cross-section irregularity should tend to diminish as depth increases in a particular reach, so at least to some extent these effects are accounted for by the inclusion of the relative smoothness Y/y_r in equation 6.25. Apparently, there been no systematic studies attempting to relate resistance to some measure of the variation of depth in a reach or cross section (e.g., the standard deviation of depth).

Bathurst (1993) reviewed resistance equations for natural streams in which gravel and boulders are a major source of cross-section irregularity. For approximately uniform flow in gravel-bed streams, he found that resistance could be estimated with $\pm 30\%$ error as

$$\Omega = 0.400 \cdot \left[-\ln\left(\frac{d_{84}}{3.60 \cdot R} \right) \right]^{-1}, \tag{6.27}$$

for reaches in which 39 mm $\leq d_{84} \leq$ 250 mm and $0.7 \leq R/d_{84} \leq 17$. For boulder-bed streams, Bathurst (1993) suggested the following equation, which is based on data from flume and field studies:

$$\Omega = 0.410 \cdot \left[-\ln\left(\frac{d_{84}}{5.15 \cdot R} \right) \right]^{-1}, \tag{6.28}$$

for reaches in which $0.004 \leq S \leq 0.04$ and $R/d_{84} \leq 10$. Note that the form of equations 6.27 and 6.28 is identical to that of equation 6.25, assuming $y_r = d_{84}$.

Figure 6.13 shows that excess resistance for gravel and boulder-bed streams given by equations 6.27 and 6.28 is typically in the range of 20% to well more than 50%. However, it seems surprising that resistance in gravel-bed streams is larger than in boulder-bed streams, and this result may reflect the very imperfect state of knowledge about resistance in natural streams, as Bathurst (1993) emphasizes. In some recent studies, Smart et al. (2002) developed similar relations for use in the relative-roughness range $5 \leq R/d_{84} \leq 20$, and Bathurst (2002) recommended computing resistance as a function of R/d_{84} via the formulas shown in table 6.1 as *minimum* values for resistance in mountain rivers with $R/d_{84} < 11$ and $0.002 \leq S_0 \leq 0.04$.

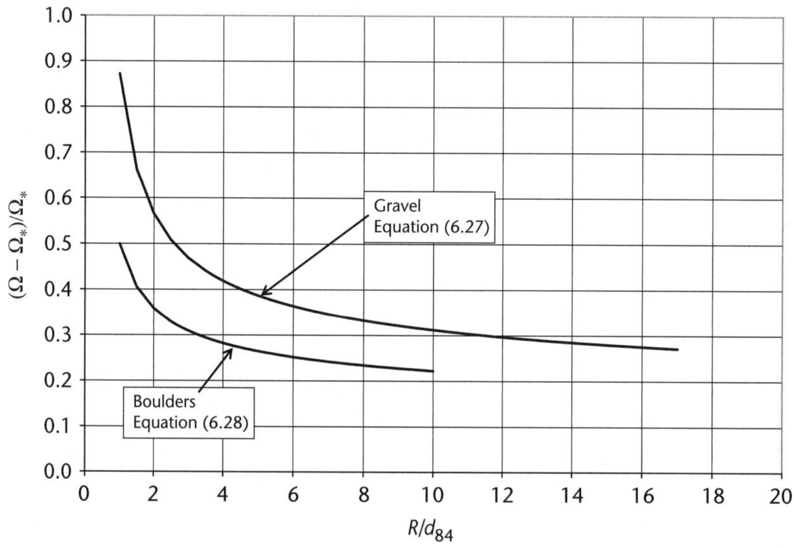

Figure 6.13 Ratio of excess resistance to baseline resistance for gravel and boulder-bed streams according to Bathurst (1993) (equations 6.27 and 6.28). Values are typically in the range of 20% to well more than 50%.

Table 6.1 Minimum values of resistance recommended by Bathurst (2002) for mountain rivers with $R/d_{84} < 11$ and $0.002 \leq S_0 \leq 0.04$.[a]

Slope range	Resistance (Ω)
$0.002 \leq S_0 \leq 0.008$	$3.84 \cdot \left(\frac{Y}{d_{84}}\right)^{0.547}$
$0.008 \leq S_0 \leq 0.04$	$3.10 \cdot \left(\frac{Y}{d_{84}}\right)^{0.93}$

[a]These values apply to situations in which resistance is primarily due to bed roughness; variations in planform, longitudinal profile, vegetation, and so forth, increase Ω beyond values given here.

6.6.1.2 Plan-View Irregularities

As we saw in section 2.2, few natural river reaches are straight, and there are several ways in which plan-view irregularities can be characterized. The overall degree of deviation from a straight-line path is the sinuosity, ζ, defined as the ratio of streamwise distance to straight-line distance (figure 6.11b). The local deviation from a straight-line path can be quantified as the radius of curvature, r_c (figure 6.11b).

From elementary physics, we know that motion with velocity U in a curved path with a radius of curvature r_c produces a **centrifugal acceleration** a_c where

$$a_c = \frac{U^2}{r_c}. \tag{6.29}$$

This acceleration multiplied by the mass of water flowing produces an apparent force, and because this force is directed at right angles to the downstream direction, it adds to the overall flow resistance.

Because velocity is highest near the surface, water near the surface accelerates more than that near the bottom; this produces secondary circulation in bends, with surface water flowing toward the outside of the bend and bottom water flowing in the opposite direction (see figure 5.21a). Thus, curvature enhances the secondary currents, increasing the resistance beyond that due to the curved flow path alone (Chang 1984).

The magnitude of the resistance due to curvature computed from a set of laboratory experiments (see box 6.3) is shown in figure 6.14. The data indicate that resistance can be increased by a factor of 2 or more when U^2/r_c exceeds 0.8 m/s^2 or sinuosity exceeds 1.04; as noted by Leopold (1994, p. 64), these experiments showed that "the frictional loss due to channel curvature is much larger than previously supposed." Sinuosities of typical meandering streams range from 1.1 to about 3.

6.6.1.3 Longitudinal-Profile Irregularities

At the reach scale, the longitudinal profiles of many streams have alternating steeper and flatter sections. In meandering streams (see section 2.2.3), the spacing of pools usually corresponds closely to the spacing of meander bends, so that pools tend to occur at spacings of about five times the bankfull width (equation 2.14). Steep mountain streams (see section 2.2.5, table 2.4) are characterized by relatively deep pools separated by steep rapids or cascades (step/pool reaches). On gentler slopes, the pools are shallower and separated by rapids (pool/riffle reaches).

The Chézy equation (equation 6.15) shows that velocity is proportional to the square root of slope. Thus, variations in slope produce accelerations and decelerations, vertical deflections of velocity vectors, and changes in depth along a river's course. Where longitudinal slope alterations are marked, they are typically a major component of overall resistance (Bathurst 1993). However, the effect in a given reach is dependent on discharge: At high flows, the water surface smoothes out and is less affected by alterations in the channel slope, whereas at low flows,

BOX 6.3 Flume Experiments on Resistance in Sinuous Channels

Leopold et al. (1960) conducted a series of experiments in a tiltable flume with a length of 15.9 m. Sand with a median diameter of 2 mm was placed in the flume, and a template was designed that could mold straight or curved trapezoidal channels in the sand. Once the channels were molded, they were coated with adhesive to prevent erosion. Plan-view geometries were as in table 6B3.1.

Table 6B3.1

Wavelength λ (m)	Radius of curvature r_c (m)	Sinuosity ζ
Straight	Straight	1.000
1.22	1.01	1.024
1.18	0.58	1.056
0.65	0.31	1.048
0.70	0.19	1.130

Flows were run at two depths; cross-section geometries were as in table 6B3.2.

Table 6B3.2

Maximum depth Y_m(m)	Bottom width W_b(m)	Water-surface width W (m)	Average depth Y (m)	Cross-sectional area A (m^2)	Wetted perimeter P_w (m)	Hydraulic radius R (m)
0.027	0.117	0.191	0.020	0.00418	0.209	0.020
0.041	0.117	0.224	0.027	0.00697	0.252	0.028

For each run, slope (S) and discharge (Q) could be set to obtain constant depth (uniform flow) throughout. The ranges of velocities (U), Reynolds numbers (Re) and Froude numbers (Fr) observed are listed in table 6B3.3.

Table 6B3.3

	S	Q (m^3/s)	U (m/s)	Re	Fr
Maximum	0.0118	0.00326	0.466	12100	0.970
Minimum	0.00033	0.00048	0.097	2130	0.187

The results of these experiments were used to plot figure 6.14 and gain quantitative insight on the effects of curvature on resistance.

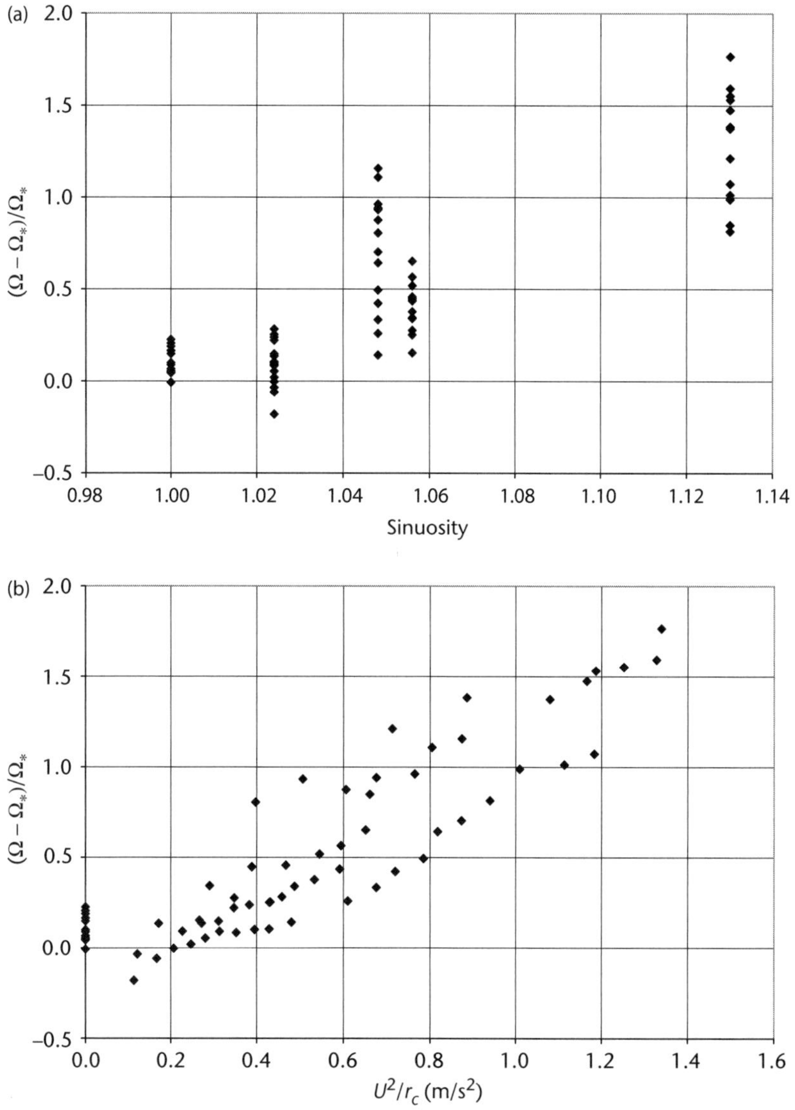

Figure 6.14 Effects of plan-view curvature on flow resistance from the experiments of Leopold et al. (1960) (see box 6.3). Excess resistance, $(\Omega - \Omega_*)/\Omega_*$, is plotted against (a) sinuosity, ζ and (b) centrifugal acceleration, $a_c = U^2/r_c$.

water-surface slope tends to parallel the local bottom slope and be more variable (figure 6.11c).

In one of the few detailed hydraulic studies of pool/fall streams, Bathurst (1993) measured resistance at three discharges in a gravel-bed river in Britain. As shown in figure 6.15, the effects of step/pool configuration are very pronounced at low discharges (low relative smoothness) and decline as discharge increases.

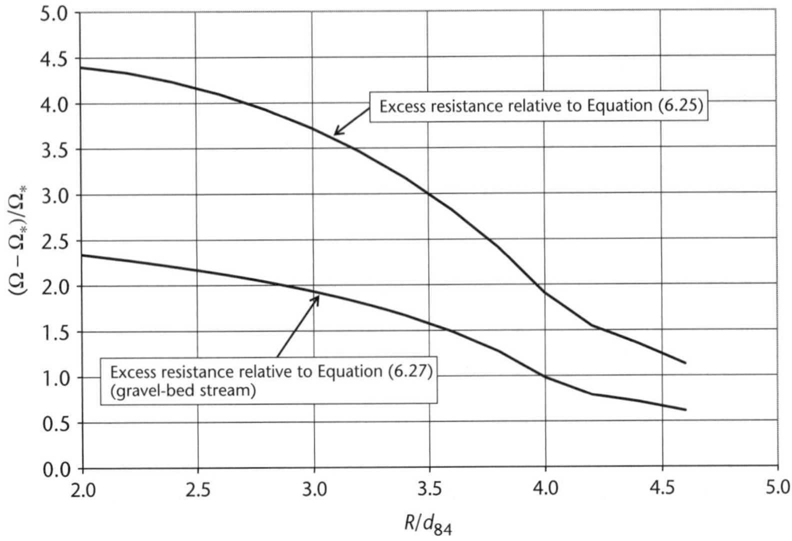

Figure 6.15 Excess resistance due to slope variations in a gravel-bed step-pool stream (River Swale, UK). The upper curve shows the excess resistance computed relative to the baseline relation (equation 6.25); the lower curve shows the excess relative to that of a uniform gravel stream (equation 6.27). The effect of the slope alterations decreases at higher discharges (higher relative smoothness). Data from Bathurst (1993).

6.6.2 Effects of Vegetation

Floodplains are commonly covered with brush or trees, and active channels can also contain living and dead plants. The effects of vegetation on resistance are complex and difficult to quantify; the major considerations are the size and shape of plants, their spacing, their heights, and their flexibility. The effects can change significantly during a particular flow event due to relative submergence and to the bending of flexible plants. Over longer time periods, the height and spacing of plants can vary seasonally and secularly due to, for example, anthropogenic increases in nutrients contained in runoff or simply to ecological processes (succession) or tree harvesting.

Kouwen and Li (1980) formulated an approach to estimating vegetative resistance that is conceptually similar to that of equations 6.27 and 6.28:

$$\Omega = k_{veg} \cdot \left[-\ln\left(\frac{y_{veg}}{K_{veg} \cdot Y} \right) \right]^{-1}, \tag{6.30}$$

where y_{veg} is the deflected vegetation height, and k_{veg} and K_{veg} are parameters. Approaches to determining values of y_{veg}, k_{veg}, and K_{veg} are given by Kouwen and Li (1980). Arcement and Schneider (1989) presented detailed field procedures

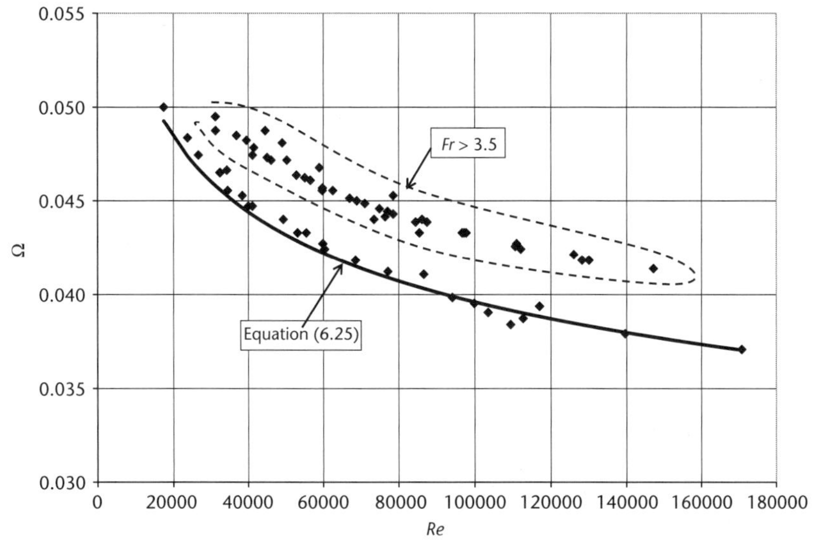

Figure 6.16 Plot of flow resistance, Ω, versus Reynolds number, *Re*, showing the effect of surface instability on flow resistance. The curve is the standard resistance relation for smooth channels given in equation 6.25; the points are resistance values measured in flume experiments of Sarma and Syala (1991). The points clustering close to the curve have $1 < Fr < 3.5$; those plotting substantially above the curve have $Fr > 3.5$.

for estimating resistance due to vegetation on floodplains. Recent analyses and experiments evaluating resistance due to vegetation are given by Wilson and Horritt (2002) and Rose et al. (2002) and summarized by Yen (2002).

6.6.3 Effects of Surface Instability

As noted in section 6.2.2.2, wavelike fluctuations begin to appear in the surfaces of open-channel flows as the Froude number *Fr* approaches 1. A few experimental studies in flumes have examined the effects of these instabilities on flow resistance.

Figure 6.16 summarizes measurements of supercritical flows in a straight, smooth, rectangular flume (Sarma and Syala 1991). It shows that for flows with $1 < Fr < 3.5$, flow resistance is essentially as predicted by the standard relation for smooth turbulent flows (equation 6.25). However, when *Fr* exceeds a threshold value of about 3.5, there is a discontinuity, and resistance jumps to a value about 10% larger than the standard value. Because Froude numbers in natural channels seldom exceed 1, Sarma and Syala's (1991) results suggest that one can usually safely ignore the effects of surface instabilities on resistance in straight channels.

However, the experiments of Leopold et al. (1960) described in box 6.3 indicate the existence of discontinuities in resistance that they attributed to surface instabilities at channel bends and called **spill resistance**. These sudden increases in resistance occurred at Froude numbers in the range of 0.4–0.55, much lower than found by Sarma and Syala (1991) in straight smooth flumes. Thus, spill resistance may be a significant contributor to excess resistance at high flows in channel bends.

6.6.4 Effects of Sediment

Sediment transport affects flow resistance in two principal ways: 1) the effects of suspended sediment on turbulence characteristics, and 2) the effects of bedforms that accompany sediment transport on channel-bed configuration.

6.6.4.1 Effects of Sediment Load

As noted in section 5.3.1.4, there is evidence that suspended sediment suppresses turbulence and causes the value of von Kármán's constant, κ, to decrease below its clear-water value of $\kappa = 0.4$. Evidence analyzed by Einstein and Chien (1954) suggested values as low as $\kappa = 0.2$ at high sediment concentrations. Because the coefficient in equation 6.25 is κ, this suggests that resistance could be as little as 50% of its clear-water value in flows transporting sediment.

However, some researchers contend that κ remains constant and the observed resistance reduction in flows transporting sediment is due to an altered velocity distribution such that, in sediment-laden flows, velocities near the bed are reduced and those near the surface increased compared with the values given by the P-vK law (Coleman 1981; Lau 1983). Other studies have even suggested that resistance is generally increased sediment-laden flows compared with clear-water flows under identical conditions (Lyn 1991). Clearly this is a question that requires further research.

6.6.4.2 Effects of Bedforms

Observations of rivers and experiments in flumes (e.g., Simons and Richardson 1966) have revealed that in flows over sand beds, there is a typical sequence of bedforms that occurs as discharge changes. These forms are intimately related to processes of erosion that begin when the critical value of boundary shear stress, τ_0, is reached,[7] and in turn they strongly affect the velocity because of their effects on flow resistance.

The bedforms are described and illustrated in table 6.2 and figures 6.17–6.19, and figure 6.20 shows qualitatively how resistance changes through the sequence. In general, resistance increases directly with bedform height (amplitude) and inversely with bedform wavelength.

Bathurst (1993) developed an approach to accounting for these effects that involves computing the effective roughness height of the bedforms, y_{bf}, as a function of grain size, d_{84}; bedform amplitude, $\mathcal{A}_{bf}$; and bedform wavelength, λ_{bf}:

$$y_{bf} = 3 \cdot d_{84} + 1.1 \cdot \mathcal{A}_{bf} \cdot [1 - exp(-25 \cdot \mathcal{A}_{bf}/\lambda_{bf})] \tag{6.31}$$

Table 6.2 Bedforms in sand-bed streams (see figures 6.17–6.20).

	Bedform	Description	Amplitude	Wavelength	Migration velocity (mm/s)	Ω_{bf}
Lower flow regime, $Fr < 1$	Plane bed	Generally flat bed, often with irregularities due to deposition; occurs in absence of erosion.				0.05–0.06
	Ripples	Small wavelike bedforms; may be triangular to sinusoidal in longitudinal cross section. Crests are transverse to flow and may be short and irregular to long, parallel, regular ridges; typically migrate downstream at velocities much lower than stream velocity; may occur on upslope portions of dunes.	< 40 mm; mostly 10–20 mm	< 60 mm	0.1–1	0.07–0.1
	Dunes	Larger wavelike forms with crests transverse to flow, out of phase with surface waves; generally triangular in longitudinal cross section with gentle upstream slopes and steep downstream slopes. Crest lengths are approximately same magnitude as wavelength; migrate downstream at velocities much lower than stream velocity.	0.1–10 m; usually $\approx 0.1 \times Y$ to $0.3 \times Y$	0.1–100 m, usually $\approx 2 \times Y$ to $10 \times Y$	0.1–1	0.07–0.14
Upper flow regime, $Fr > 1$	Plane bed	Often occurs with heterogeneous, irregular forms; a mixture of flat areas and low-amplitude ripples and/or dunes.	< 3 mm	Irregular	10	0.05–0.06
	Antidunes	Large wavelike forms with triangular to sinusoidal longitudinal cross sections that are in phase with water-surface waves. Crest lengths approximately equal wavelength; may migrate upstream or downstream or remain stationary.	30–100 mm	$2 \cdot \pi \cdot Y$	Variable	0.05–0.06
	Chutes and pools	Large mounds of sediment that form steep chutes in which flow is supercritical, separated by pools in which flow may be subcritical or supercritical. Hydraulic jumps (see chapter 10) form at supercritical-to-subcritical transitions; migrate slowly upstream.			1–50	

After Task Force on Bed Forms in Alluvial Channels (1966) and Bridge (2003).

(a)

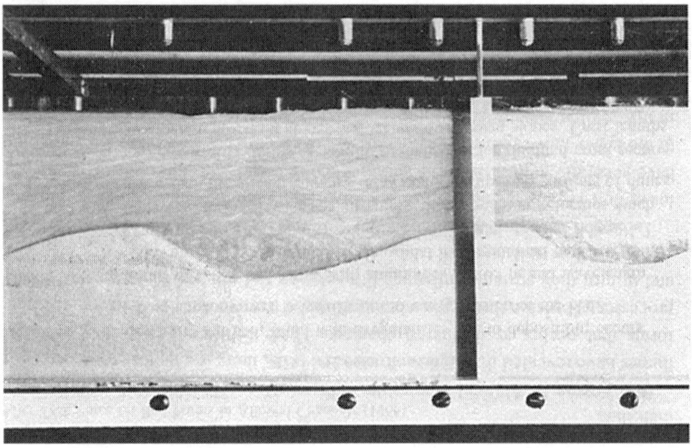

(b)

Figure 6.17 Ripples. (a) Side view of ripples in a laboratory flume. The flow is from left to right at a mean depth of 0.064 m and a mean velocity of 0.43 m/s ($Fr = 0.54$). Aluminum powder was added to the water to make the flow paths visible. Note that the water surface is unaffected by the ripples. Photograph courtesy of A. V. Jopling, University of Toronto. (b) Ripples on the bed of the Delta River in central Alaska. Flow was from left to right.

Resistance is then computed as

$$\Omega = 0.400 \cdot \left[-\ln\left(\frac{y_{\text{bf}}}{12.1 \cdot R} \right) \right]^{-1}, \tag{6.32}$$

where R is hydraulic radius ($\approx Y$ for wide channels).

In another approach, the resistance is separated into 1) that due to the bed material (the plane-bed resistance Ω_* given by equation 6.25) and 2) that due to the bedforms, Ω_{bf}:

$$\Omega = \Omega_* + \Omega_{\text{bf}}. \tag{6.33}$$

(a)

(b)

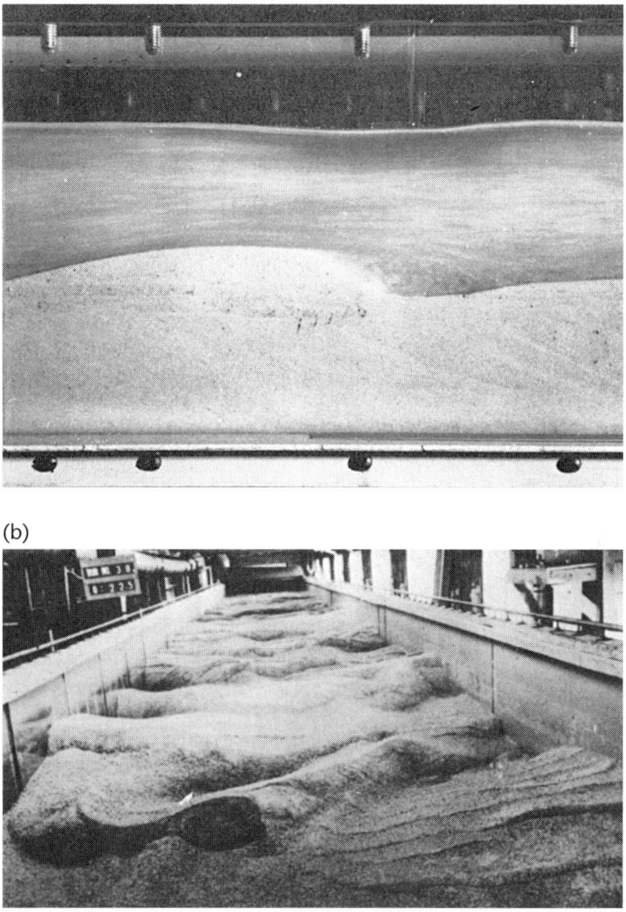

Figure 6.18 Dunes. (a) Side view of dunes in a laboratory flume. The flow is from left to right at a mean depth of 0.064 m and a mean velocity of 0.67 m/s ($Fr = 0.85$). Aluminum powder was added to the water to make the flow paths visible. Note that the water surface is out of phase with the bedforms. Photograph courtesy of A.V. Jopling, University of Toronto. (b) Dunes in a laboratory flume. Flow was toward the observer at a mean depth of 0.31 m and a mean velocity of 0.85 m/s ($Fr = 0.49$). Note ripples superimposed on some dunes. Photograph courtesy of D.B. Simons, Colorado State University.

Yen (2002) reviews several approaches to estimating Ω_{bf}; some typical values are indicated in table 6.2.

6.6.5 Effects of Ice

As noted in section 3.2.2.3, the presence of an ice cover or frazil ice can significantly increase resistance. For a uniform flow in a rectangular channel (figure 6.7), the effect

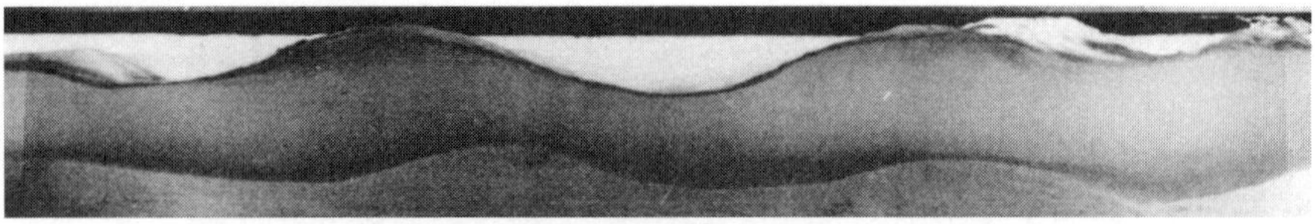

Figure 6.19 Side view of antidunes in a laboratory flume The flow is from left to right at a mean depth of 0.11 m and a mean velocity of 0.79 m/s ($Fr = 0.76$). Note that the surface waves are approximately in phase with the bedforms, which are also migrating to the right. Photograph courtesy of J. F. Kennedy, University of Iowa.

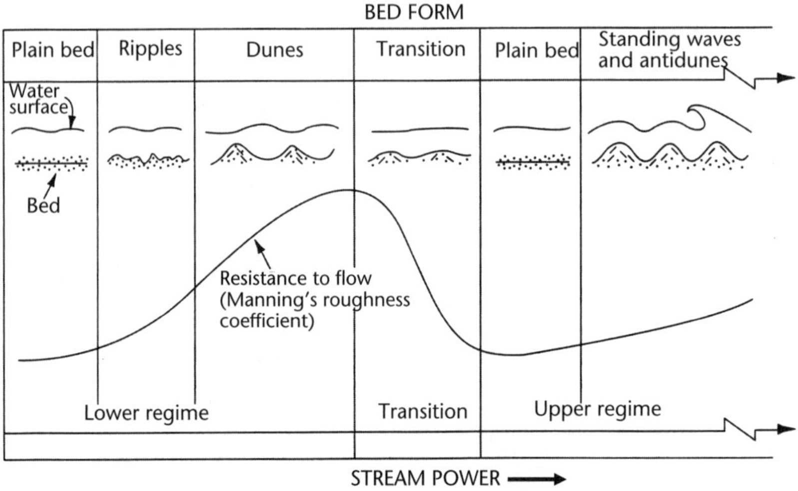

Figure 6.20 Sequence of bedforms and flow resistance in sand-bed streams. From Arcement and Schneider (1989). See table 6.2 for typical Ω values.

of an ice cover can be included in formulating the expression for the resisting forces, so that equation 6.10 becomes

$$F_R = \tau_B \cdot (2 \cdot Y + W) \cdot X + \tau_I \cdot W \cdot X, \tag{6.34}$$

where τ_B is the shear stress on the bed and τ_I is the shear stress on the ice cover. If this force balances the downstream-directed force (equation 6.7) and we assume a wide channel (i.e., $P_w = W$), the modified Chézy equation becomes

$$U = (\Omega_B^2 + \Omega_I^2)^{-1/2} \cdot u_*, \tag{6.35}$$

where Ω_B and Ω_I are the resistances due to the bed and the ice cover, respectively.

One would expect Ω_I to vary widely in natural streams due to 1) variations in the degree of ice cover, 2) development of ripplelike and dunelike bedforms on the underside of the ice cover (Ashton and Kennedy 1972), 3) development of partial or complete ice jamming, and 4) the concentration of frazil ice in the flow. An analysis of ice resistance on the St. Lawrence River by Tsang (1982) indicates that Ω_I is on the order of $0.7-1.5$ times Ω_B, and data presented by Chow (1959) suggest values in the range from $\Omega_I = 0.03$ for smooth ice without ice blocks to $\Omega_I = 0.085$ for rough ice with ice blocks. White (1999) and Brunner (2001b) summarized resistance due to ice given by several studies; these cover a very wide range of values.

6.7 Field Computation of Reach Resistance

Validation of methods of determining reach resistance requires comparison with actual resistance values. The method developed here to compute resistance in natural, nonprismatic channels is based closely on the concepts used to derive the Chézy equation for uniform flow in prismatic channels in section 6.3.

Designating X as the distance measured along the stream course, the cross-sectional area, A, wetted perimeter, P_w, hydraulic radius, R, and water-surface slope, S_S, vary through a natural-channel reach (figure 6.21) and so are written as functions of X: $A(X)$, $P_w(X)$, $R(X)$, and $S_S(X)$ respectively. With this notation, the downstream-directed force, F_D, is

$$F_D = \gamma \cdot \int_{X_0}^{X_N} A(X) \cdot S_S(X) \cdot \mathrm{d}X, \tag{6.36}$$

where X_0 and X_N are the locations of the upstream and downstream boundaries of the reach, respectively. Note that this expression is analogous to equation 6.7, but for nonprismatic rather than prismatic channels.

Similarly, the upstream-directed resistance force, F_R in a nonprismatic channel is

$$F_R = K_T \cdot \rho \cdot U^2 \cdot \int_{X_0}^{X_N} P_w(X) \cdot \mathrm{d}X, \tag{6.37}$$

where U is the reach-average velocity. This expression is analogous to equation 6.10.

For a given discharge, Q, the reach-average velocity is

$$U = \frac{Q}{\left(\frac{1}{\Delta X}\right) \cdot \int_{X_0}^{X_N} A(X) \cdot \mathrm{d}X}. \tag{6.38}$$

where $\Delta X \equiv X_N - X_0$.

Equating F_D and F_R as in equation 6.6, substituting equations 6.36–6.38, and solving for K_T gives

$$K_T = \frac{g \cdot \int_{X_0}^{X_N} A(X) \cdot S_S(X) \cdot \mathrm{d}X \cdot \left[\int_{X_0}^{X_N} A(X) \cdot \mathrm{d}X\right]^2}{Q^2 \cdot \Delta X^2 \cdot \int_{X_0}^{X_N} P_w(X) \cdot \mathrm{d}X} = \Omega^2; \tag{6.39a}$$

$$\Omega = \frac{g^{1/2} \cdot \left[\int_{X_0}^{X_N} A(X) \cdot S_S(X) \cdot \mathrm{d}X\right]^{1/2} \cdot \int_{X_0}^{X_N} A(X) \cdot \mathrm{d}X}{Q \cdot \Delta X \cdot \left[\int_{X0}^{X_N} P_w(X) \cdot \mathrm{d}X\right]^{1/2}}. \tag{6.39b}$$

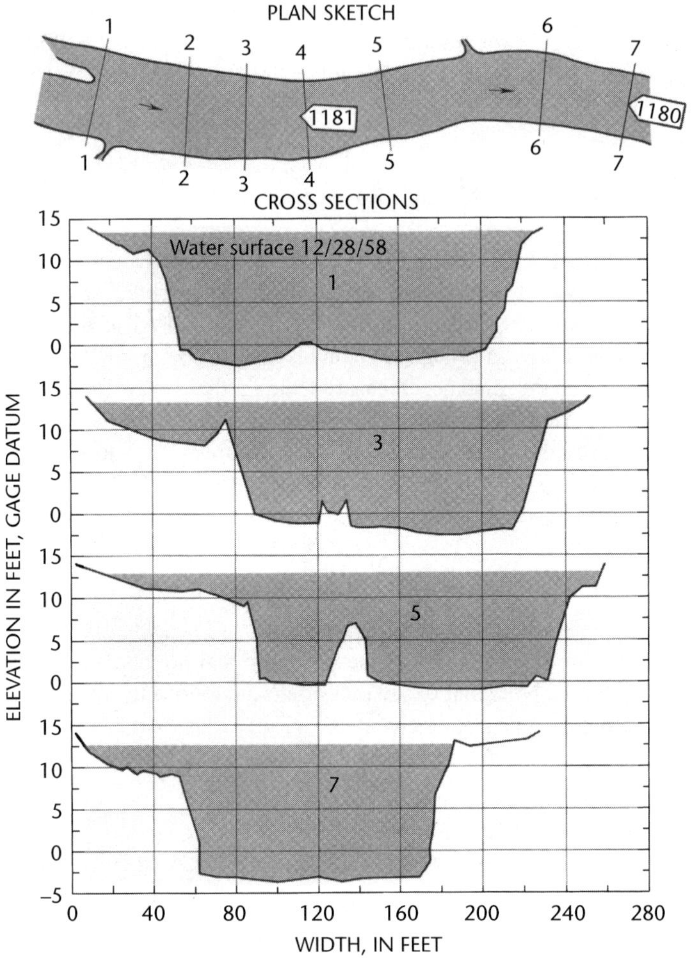

Figure 6.21 Plan view and cross sections of the Deep River at Ramseur, North Carolina, showing typical cross-section variability. From Barnes (1967).

In practice, the geometric functions $A(X)$, $S_S(X)$, and so on, can be approximated only by measurements at specific cross sections within the reach. Thus. for practical application, equation 6.39b becomes

$$\Omega = \frac{g^{1/2} \cdot \left[\sum_{i=1}^{N} A_i \cdot S_{Si} \cdot \Delta X_i \right]^{1/2} \cdot \sum_{i=1}^{N} A_i \cdot \Delta X_i}{Q \cdot \Delta X \cdot \left[\sum_{i=1}^{N} P_{wi} \cdot \Delta X_i \right]^{1/2}}, \tag{6.39c}$$

where the subscripts indicate the measured value of the variable at cross section i, $i = 1, 2, \ldots, N$, and ΔX_i is the downstream distance between successive cross sections.

Box 6.4 shows how field computations are used to compute resistance. It is important to be aware that careful field measurements are essential for accurate hydraulic computations. The manual by Harrelson et al. (1994) is an excellent illustrated guide to field technique.

6.8 The Manning Equation

6.8.1 Origin

In the century following the publication of the Chézy equation in 1769, European hydraulic engineers did considerable field and laboratory research to develop practical ways to estimate open-channel flow resistance (Rouse and Ince 1963; Dooge 1992). In 1889, Robert Manning (1816–1897), an Irish engineer, published an extensive review of that research (Manning 1889). He concluded that the simple equation that best fit the experimental results was

$$U = K_M \cdot R^{2/3} \cdot S_S^{1/2}, \tag{6.40a}$$

where K_M is a proportionality constant representing reach conductance. For historical reasons (see Dooge 1992), subsequent researchers replaced K_M by its inverse, $1/n_M$, and wrote the equation as

$$U = \left(\frac{1}{n_M}\right) \cdot R^{2/3} \cdot S_S^{1/2}, \tag{6.40b}$$

called **Manning's equation**, where the resistance factor n_M is called **Manning's *n***.

Manning's equation has come to be accepted as "the" resistance equation for open-channel flow, largely replacing the Chézy equation in practical applications. The essential difference between the two is that the hydraulic-radius exponent is 2/3 rather than 1/2. This difference is important because it makes the Manning equation dimensionally inhomogeneous.[8] As with Chézy's C (see box 6.1), values of n_M are treated as constants for all unit systems, and in order to give correct results, the Manning equation must be written as

$$U = u_M \cdot \left(\frac{1}{n_M}\right) \cdot R^{2/3} \cdot S^{1/2}, \tag{6.40c}$$

where u_M is a unit-adjustment factor that takes the following values:

Unit system	u_M
Système Internationale	1.00
British	1.49
Centimeter-gram-second	4.64

BOX 6.4 Calculation of Resistance, Deep River at Ramseur, North Carolina

The channel-geometry values in the table below were measured by Barnes (1967) at seven cross sections on the Deep River at Ramseur, North Carolina, on 28 December 1958, when the flow was $Q = 235\ m^3/s$ (figure 6.21). Note that $i = 0$ for the upstreammost cross section, so $N+1$ sections are measured, defining N subreaches (table 6B4.1).

Table 6B4.1

Section, i	$A_i\ (m^2)$	$R_i\ (m)$	$P_{wi}\ (m)$	$\Delta X_i (m)$	$\lvert\Delta Z_i\rvert\ (m)$	$S_{Si} = \lvert\Delta Z_i\rvert/\Delta X_i$
0	230.0	3.29	69.8			
1	198.4	3.17	62.6	66.8	0.052	0.000776
2	198.6	2.85	69.8	66.5	0.015	0.000229
3	223.4	2.66	83.9	55.5	0.037	0.000659
4	191.6	2.42	79.1	56.4	0.061	0.001081
5	210.5	3.29	63.9	102.7	0.091	0.000890
6	188.3	3.17	59.4	80.8	0.073	0.000906

(The quantity $\lvert\Delta Z_i\rvert$ is the decrease in water-surface elevation between successive sections.)

To compute the resistance via equation 6.39c, we calculate the quantities in table 6B4.2 from the above data.

Table 6B4.2

Section, i	$A_i{\cdot}S_{Si}{\cdot}\Delta X_i (m^3)$	$A_i{\cdot}\Delta X_i\ (m^3)$	$P_{wi}{\cdot}\Delta X_i (m^3)$
1	10.286	13,250	4178.9
2	3.029	13,202	4636.2
3	8.172	12,394	4656.5
4	11.681	10,805	4463.6
5	19.256	21,631	6569.4
6	13.779	15,215	4798.5
Sum	66.202	86,497	29,303.1

From the previous table, $\Delta X = \Sigma \Delta X_i = 428.7$ m. Substituting the appropriate values into 6.39c gives

$$\Omega = \frac{9.81^{1/2}{\cdot}[66.202]^{1/2}{\cdot}86497}{235{\cdot}428.7{\cdot}[29303.1]^{1/2}} = 0.128.$$

The Reynolds number for this flow, assuming kinematic viscosity $\nu = 1.5 \times 10^{-6}\ m^2/s$, is

$$Re = \frac{U{\cdot}R}{\nu} = \frac{1.15\ m/s \times 2.98\ m}{1.5\ \times\ 10^{-6} m^2/s} = 2.28 \times 10^6.$$

Referring to figure 6.8, we see that this flow was well into the "fully rough" range and that the actual resistance $\Omega = 0.128$ was well above the baseline value $\Omega_* \approx 0.04$ given by equation 6.25.

From equations 6.12, 6.19, 6B1.3, and 6.40c, we see that

$$n_M = \frac{u_M \cdot R^{1/6} \cdot K_T^{1/2}}{g^{1/2}} = \frac{u_M \cdot R^{1/6}}{u_C \cdot C} = \frac{u_M \cdot R^{1/6} \cdot \Omega}{g^{1/2}}. \tag{6.41}$$

A major justification for using the Manning equation instead of the Chézy equation has been that, because n_M depends on the hydraulic radius, it accounts for relative submergence effects and tends to be more constant for a given reach (i.e., changes less as discharge changes) than is C. However, this reasoning may not be compelling, because we have seen that we can write the Chézy equation using Ω^{-1} instead of $u_C \cdot C$ (equation 6.19) and that Ω, in fact, depends in large measure on relative submergence (equation 6.24). Another reason for the popularity of the Manning equation is that a number of methods have been developed that provide expedient (i.e., "quick-and-dirty") estimates of the resistance coefficient n_M. These methods are discussed in the following section.

6.8.2 Determination of Manning's n_M

In order to apply the Manning equation in practical problems, one must be able to determine a priori values of n_M. An overview of approaches to doing this are listed in table 6.3 and briefly described in the following subsections.

6.8.2.1 Visual Comparison with Photographs

Table 6.4 summarizes publications that provide guidance for field determination of n_M by means of photographs of reaches in which n_M values have been determined by measurement for one or more discharges. The books by Barnes (1967) and Hicks and Mason (1991) are specifically designed to provide visual guidance for the field determination of n_M for in-bank flows in natural rivers. Examples from Barnes (1967) are shown in figure 6.22.

6.8.2.2 Tables of Typical n_M Values

Chow (1959) provides tables that give a range of appropriate n_M values for various types of human-made canals and natural channels; the portions of those tables covering natural channels are reproduced here in table 6.5.

6.8.2.3 Formulas That Account for Components of Reach Resistance

Cowan (1956) introduced a formula that allowed for explicit consideration of many of the factors that determine resistance (see section 6.6) in determining an appropriate n_M value:

$$n_M = (n_0 + n_1 + n_2 + n_3 + n_4) \cdot m_\varsigma, \tag{6.42}$$

where n_0 is the base value for straight, uniform, smooth channel in natural material; n_1 is the factor for bed and bank roughness; n_2 is the factor for effect of

Table 6.3 General approaches to a priori estimation of Manning's n_M.

Approach	Comments	References
1. Visual comparison with photographs of channels for which n_M has been measured (see table 6.4)	Expedient method; subjective, dependent on operator experience; subject to considerable uncertainty	Faskin (1963), Barnes (1967), Arcement and Schneider (1989), Hicks and Mason (1991)
2. Tables of typical n_M values for reaches of various materials and types (see table 6.5)	Expedient method; subjective, dependent on operator experience; subject to considerable uncertainty	Chow (1959), French (1985)
3. Formulas that account for components of reach resistance (see table 6.6)	Expedient method; more objective than approaches 1 and 2 but lacks theoretical basis	Cowan (1956), Faskin (1963), Arcement and Schneider (1989)
4. Formulas that relate n_M to bed-sediment grain size d_p (see table 6.7)	Require measurement of bed sediment; reliable only for straight quasi-prismatic channels where bed roughness is the dominant factor contributing to resistance	Chang (1988), Marcus et al. (1992)
5. Formulas that relate n_M to hydraulic radius and relative smoothness	Require measurement of bed sediment, depth, and slope; forms are based on theory; coefficients are based on field measurement; can give good results in conditions similar to those for which established	Limerinos (1970), Bathurst (1985)
6. Statistical formulas that relate n_M to measurable flow parameters (see table 6.8)	Can provide good estimates, especially useful when bed-material information is lacking, as in remote sensing, but subject to considerable uncertainty	Riggs (1976), Jarrett (1984), Dingman and Sharma (1997), Bjerklie et al. (2003)

Table 6.4 Summary of reports presenting photographs of reaches for which Manning's n_M has been measured.

Types of reach	No. of reaches	No. of flows	Minimum n_M	Maximum n_M	Reference
Canals and dredged channels (USA)	48	326	0.014	0.162	Faskin (1963)
Natural rivers (USA)	51	62	0.024	0.075	Barnes (1967)
Flood plains (USA)	16	16	—[a]	—[a]	Arcement and Schneider (1989)
Natural rivers (New Zealand)	78	559	0.016	0.270	Hicks and Mason (1991)

[a] See reference for methodology for computing composite (channel plus flood plain) n_M values.

Figure 6.22 Photographs of U.S. river reaches covering a range of values of Manning's n_M, computed from measurements. (a) Columbia River at Vernita, Washington: $n_M = 0.024$; (b) West Fork Bitterroot River near Conner, Montana: $n_M = 0.036$; c) Moyie River at Eastport, Idaho: $n_M = 0.038$; (d) Tobesofkee Creek near Macon, Georgia: $n_M = 0.041$; (e) Grande Ronde River at La Grande, Oregon: $n_M = 0.043$; (f) Clear Creek near Golden, Colorado: $n_M = 0.050$; (g) Haw River near Benaja, North Carolina: $n_M = 0.059$; (h) Boundary Creek near Porthill, Idaho: $n_M = 0.073$. From Barnes (1967); photographs courtesy U.S. Geological Survey.

Table 6.5 Values of Manning's n_M for natural streams.

Channel description	Minimum	Normal	Maximum
Minor streams (bankfull width < 100 ft)			
Streams on plain			
1. Clean, straight, full stage, no riffles or deep pools	0.025	0.030	0.033
2. Same as above, but more stones and weeds	0.030	0.035	0.040
3. Clean, winding, some pools and shoals	0.033	0.040	0.045
4. Same as above, but some weeds and stones	0.035	0.045	0.050
5. Same as above, but lower stages, more ineffective slopes and sections	0.040	0.048	0.055
6. Same as item 4, but more stones	0.045	0.050	0.060
7. Sluggish reaches, weedy, deep pools	0.050	0.070	0.080
8. Very weedy reaches, deep pools, or floodways with heavy stand of timber and underbrush	0.075	0.100	0.150
Mountain Streams			
No vegetation in channel, banks usually steep, trees and brush along banks submerged at high stages			
1. Bottom: gravels, cobbles, and few boulders	0.030	0.040	0.050
2. Bottom: cobbles with large boulders	0.040	0.050	0.070
Major Streams (bankfull width > 100 ft)			
1. Regular section with no boulders or brush	0.025	—	0.060
2. Irregular and rough section	0.035	—	0.100
Floodplains			
1. Short grass, no brush	0.025	0.030	0.035
2. High grass, no brush	0.030	0.035	0.050
3. Cultivated area, no crop	0.020	0.030	0.040
4. Mature row crops	0.025	0.035	0.045
5. Mature field crops	0.030	0.040	0.050
6. Scattered brush, heavy weeds	0.035	0.050	0.070
7. Light brush and trees, in winter	0.035	0.050	0.060
8. Light brush and trees, in summer	0.040	0.060	0.080
9. Medium to dense brush, in winter	0.045	0.070	0.110
10. Medium to dense brush, in summer	0.070	0.100	0.160
11. Dense willows, summer, straight	0.110	0.150	0.200
12. Cleared land with tree stumps, no sprouts	0.030	0.040	0.050
13. Same as above, but with heavy growth of sprouts	0.050	0.060	0.080
14. Heavy stand of timber, a few down trees, little undergrowth, flood stage below branches	0.080	0.100	0.120
15. Same as above, but with flood stage reaching branches	0.100	0.120	0.160

From Chow (1959, table 5.6). Reproduced with permission of McGraw-Hill.

cross-section irregularity; n_3 is the factor for the effect of obstructions; n_4 is the factor for vegetation and flow conditions; and m_ς is the factor for sinuosity. Table 6.6 summarizes the determination of values for these factors.

Although equation 6.42 may provide a somewhat more objective method for considering the various factors that affect resistance than simply referring to tables or figures, note that there is no theoretical basis for assuming that n_M values are simply additive.

Table 6.6 Values of factors for estimating n_M via Cowan's (1956) formula (equation 6.42).

Material	n_0
Concrete	0.011–0.018
Rock cut	0.025
Firm soil	0.020–0.032
Sand ($d = 0.2$ mm)	0.012
Sand ($d = 0.5$ mm)	0.022
Sand ($d = 1.0$ mm)	0.026
Sand ($1.0 \leq d \leq 2.0$ mm)	0.026–0.035
Gravel	0.024–0.035
Cobbles	0.030–0.050
Boulders	0.040–0.070
Degree of Irregularity	n_1
Smooth	0.000
Minor	0.001–0.005
Moderate	0.006–0.010
Severe	0.011–0.020
Cross-Section Irregularity	n_2
Gradual	0.000
Alternating occasionally	0.001–0.005
Alternating frequently	0.010–0.015
Obstructions	n_3
Negligible	0.000–0.004
Minor	0.005–0.015
Appreciable	0.020–0.030
Severe	0.040–0.050
Amount of Vegetation	n_4
Small	0.002–0.010
Medium	0.010–0.025
Large	0.025–0.050
Very large	0.050–0.100
Sinuosity, ζ	m_ζ
$1.0 \leq \zeta \leq 1.2$	1.00
$1.2 \leq \zeta \leq 1.5$	1.15
$1.5 \leq \zeta$	1.30

6.8.2.4 Formulas That Relate n_M to Bed-Sediment Size and Relative Smoothness

From a study of flows over uniform sands and gravels, Strickler (1923) proposed that n_M is related to bed-sediment size as

$$n_M = 0.0150 \cdot d_{50}(mm)^{1/6}, \tag{6.43a}$$

where d_{50} is median grain diameter in mm, or

$$n_M = 0.0474 \cdot d_{50}(m)^{1/6}, \tag{6.43b}$$

where d_{50} is median grain diameter in m. Formulas of this form are called **Strickler formulas**, and several versions have been proffered by various researchers (see table 6.7). Although Strickler-type formulas are often invoked, experience shows that n_M values computed for natural channels from bed sediment alone are usually smaller than actual values.

It is interesting to note that, using equation 6.43b, the Manning equation (equation 6.40c) can be written as

$$U = 6.74 \cdot \left(\frac{R}{d_{50}}\right)^{1/6} \cdot u_* = 6.74 \cdot \left(\frac{R}{d_{50}}\right)^{0.167} \cdot (g \cdot Y \cdot S_S)^{1/2}; \tag{6.44}$$

which can be interpreted as an integrated 1/6-power-law velocity profile (see equation 5.46 with $m_{\mathrm{PL}} = 1/6$). This equation is of the same form as equation 4.74, which was developed from dimensional analysis and measured values, but has a considerably different coefficient (1.84) and exponent (0.704).

We have seen several formulas (equations 6.25, 6.27, 6.28, 6.30, and 6.32) that relate resistance in fully rough flows to relative roughness in the form

$$\Omega = \kappa \cdot \left[-\ln\left(\frac{y_r}{K_r \cdot R}\right)\right]^{-1}, \tag{6.45}$$

Table 6.7 Formulas relating Manning's n_M to bed-sediment size and relative smoothness (grain diameters d_p, in mm; hydraulic radius, R, in m).

Formula	Remarks	Source
n_M or $n_0 = 0.015 \cdot d^{1/6}$	Original "Strickler formula" for uniform sand	Strickler (1923) as reported by Chang (1988)
n_M or $n_0 = 0.0079 \cdot d_{90}^{1/6}$		Keulegan (1938) as reported by Marcus et al. (1992)
n_M or $n_0 = 0.0122 \cdot d_{90}^{1/6}$	Sand mixtures	Meyer-Peter and Muller (1948)
n_M or $n_0 = 0.015 \cdot d_{75}^{1/6}$	Gravel lined canals	Lane and Carlson (1938) as reported by Chang (1988)
n_M or $n_0 = \dfrac{R^{1/6}}{[7.69 \cdot \ln(R/d_{84}) + 63.4]}$		Limerinos (1970)
n_M or $n_0 = \dfrac{R^{1/6}}{[7.64 \cdot \ln(R/d_{84}) + 65.3]}$	Gravel streams with slope > 0.004	Bathurst (1985)
n_M or $n_0 = \dfrac{R^{1/6}}{[7.83 \cdot \ln(R/d_{84}) + 72.9]}$	Derived from P-vK law for wide channels	Dingman (1984)

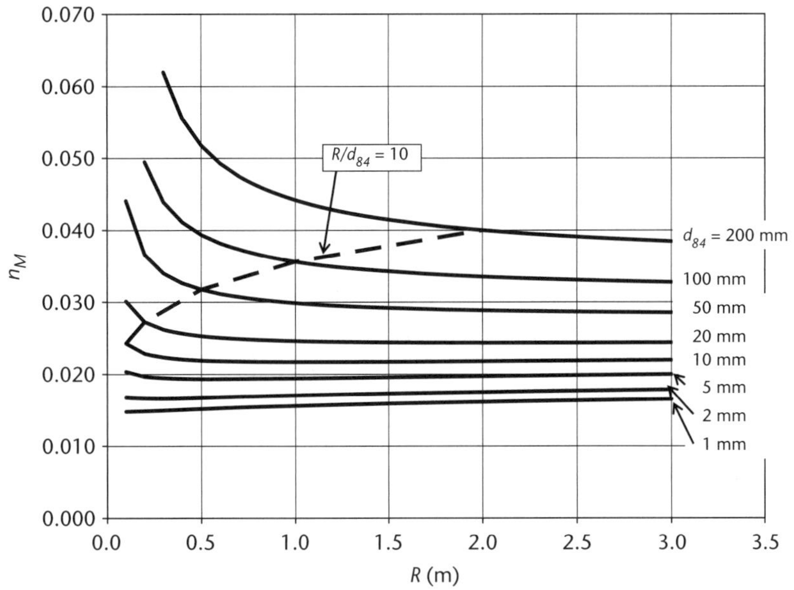

Figure 6.23 Variation of Manning's n_M (or n_0 in equation 6.42) with hydraulic radius, R, and bed grain diameter d_{84} as predicted by the Dingman (1984) version of equation 6.46 (see table 6.7). Manning's n_M is effectively independent of depth for $R/d_{84} > 10$.

where the values of κ, K_r, and y_r take different values in different contexts. If equation 6.45 is substituted into equation 6.41, we find that

$$n_M = \frac{u_M \cdot \kappa \cdot R^{1/6}}{g^{1/2} \cdot \ln\left(\frac{K_r \cdot R}{y_r}\right)}. \tag{6.46}$$

Thus, equation 6.46 can be used to provide estimates of n_M (or n_0 in equation 6.42) in those contexts. Table 6.7 lists versions of equation 6.46 derived by various authors, and figure 6.23 shows the relation of n_M to relative smoothness for various bed-sediment sizes in gravel-bed streams as given by the Dingman (1984) version of that equation. Note that the formula predicts little dependence of n_M on R/d_{84} when $R/d_{84} > 10$.

6.8.2.5 Statistically Derived Formulas That Relate n_M to Hydraulic Variables

A number of researchers have used statistical analysis (regression analysis, as described in section 4.8.3.1) to develop equations to predict n_M based on measurable flow variables. Three of these equations are listed in table 6.8. There is considerable uncertainty associated with estimates from such equations: The equation of Dingman and Sharma (1997), which is based on the most extensive data set, was found to give

Table 6.8 Statistically derived formulas for estimating Manning's n_M [A = cross-sectional area (m^2); R = hydraulic radius (m); S =slope].

Formula	Remarks	Source
$n_M = 0.210 \cdot A^{-0.33} \cdot R^{0.667} \cdot S^{0.095}$	Based on 62 flows in Barnes (1967); $0.024 \leq n_M \leq 0.075$	Riggs (1976)
$n_M = 0.32 \cdot R^{-0.16} \cdot S^{0.38}$	Mountain streams with $0.17 \text{ m} \leq R \leq 2.13 \text{ m}$ and $0.002 \leq S \leq 0.052$	Jarrett (1984)
$n_M = 0.217 \cdot A^{-0.173} \cdot R^{0.267} \cdot S^{0.156}$	Based on 520 flows from Hicks and Mason (1991); $0.015 \leq n_M \leq 0.290$	Dingman and Sharma (1997)

discharge estimates within ±50% of the true value 77% of the time. This topic is addressed further in section 6.9.

6.8.2.6 Field Measurement of Discharge and Hydraulic Variables

The only way that the value of Manning's n_M can be established with certainty is by measuring the discharge and hydraulic variables at a given time in a given reach, determining the prevailing reach-average velocity, and solving the Manning equation for n_M. Ideally, one would repeat the calculations over a range of discharges in a particular reach and use the n_M values so determined in future a priori estimates of velocity or discharge for that reach.

Barnes (1967) and Hicks and Mason (1991) give equations for direct computation of n_M from measured values of discharge and surveyed values of cross-sectional area, hydraulic radius, reach length, and water-surface slope at several cross sections within a reach. However, their methodology is based on energy considerations (sections 4.5 and 8.1), whereas the Manning equation is a modification of the Chézy equation, which was derived from momentum considerations (sections 4.4 and 8.2).[9] Thus, it is preferable to compute resistance via the method described in section 6.7 for computing Ω (equation 6.39c); if desired, the corresponding n_M value can then be determined via equation 6.41. In most cases, the two methods give very similar n_M values (within ±0.002).

6.8.3 Summary

As noted above, the Manning equation has been the most commonly used resistance relation for most engineering and many scientific purposes. It is common to use the expedient methods described in approaches 1–3 of table 6.3 to estimate n_M in these applications. However, it has been shown that even engineers with extensive field experience generate a wide range of n_M estimates for a given reach using these methods (Hydrologic Engineering Center 1986). Approach 4 is not usually appropriate for natural rivers because, as we have seen, resistance depends on many factors in addition to bed material. The various equations developed for approach 5 can be used for conditions similar to those for which the particular equation was established. Approach 6 can be useful, especially when trying to estimate discharge via remote sensing (Bjerklie et al. 2003), but may produce errors of ±50% or more (see section 6.9). As noted above, the only way to determine resistance (Ω or n_M) with certainty for a given reach is to measure discharge and reach-average values of hydraulic variables at a given discharge and use equation 6.39c and, if desired, equation 6.41.

The questionable theoretical basis for the Manning equation—reflected in its dimensional inhomogeneity—and the common reliance on expedient methods for estimating n_M significantly limit the confidence one can have in many applications of the Manning equation. As explained in section 6.3, the Chézy equation has a theoretical basis and, coupled with 1) the theoretical and empirical studies of resistance summarized in the Moody diagram (figure 6.8) and 2) the various studies described in sections 6.5 and 6.6, provides a sound and useful framework for understanding and estimating reach resistance. Thus, there seems to be no well-founded theoretical or empirical basis for preferring the Manning equation to the Chézy equation. However, as we will see in the following section, the theoretical basis for the Chézy equation may itself need reexamination.

6.9 Statistically Derived Resistance Equations

Because of the theoretical uncertainty associated with the Manning equation and the difficulty of formulating physically based approaches for characterizing resistance, some researchers have applied statistical techniques (regression analysis, section 4.8.3.1) to identify relations between discharge or velocity and other measurable hydraulic variables (Golubtsev 1969; Riggs 1976; Jarrett 1984; Dingman and Sharma 1997).

Box 6.5 describes a study that compares the performance of five statistically established resistance/conductance models for a large set of flow data. Overall, the study found that the best predictor was the "modified Manning" model:

$$Q = 7.14 \cdot W \cdot Y^{5/3} \cdot S_0^{1/3}, \tag{6.47}$$

where Q is discharge (m^3/s), W is width (m), Y is average depth (m), and S_0 is channel slope.

Interestingly, that study found that resistance models incorporating a slope exponent $q = 1/3$ (the "modified Manning" and "modified Chézy," as well as the pure regression relation) had greater predictive accuracy than those using the generally accepted theoretical value $q = 1/2$. A possible interpretation of this result is that the assumption that resistance (shear stress) is proportional to the square of velocity (equation 6.8), which is the basis of the derivation of the Chézy resistance relation, is not completely valid.

Measurements of resistance/conductance (e.g., Barnes 1967; Hicks and Mason 1991) clearly demonstrate that resistance varies strongly from reach to reach and with varying discharge in a given reach. The Bjerklie et al. (2005b) study in fact found that values of K_2 (equation 6B5.2a) for individual flows varied from about 1.0 to as high as 18, with about two-thirds of the values Between 4.6 and 9.6. Thus, the use of a universal conductance coefficient as in 6.47 is not correct.

BOX 6.5 Statistically Determined Resistance/Conductance Equations

Bjerklie et al. (2005b) used data for 1037 flows at 103 reaches to compare four resistance/conductance models incorporating various combinations of depth exponents and slope exponents.

Manning model:

$$Q = K_1 \cdot W \cdot Y^{5/3} \cdot S_0^{1/2} \tag{6B5.1a}$$

Modified Manning model:

$$Q = K_2 \cdot W \cdot Y^{5/3} \cdot S_0^{1/3} \tag{6B5.2a}$$

Chézy model:

$$Q = K_3 \cdot W \cdot Y^{3/2} \cdot S_0^{1/2} \tag{6B5.3a}$$

Modified Chézy model

$$Q = K_4 \cdot W \cdot Y^{3/2} \cdot S_0^{1/3} \tag{6B5.4a}$$

In these models, Q is discharge, $K_1 - K_4$ are conductance coefficients, W is width, Y is average depth, and S_0 is channel slope. These models can also be written as velocity predictors by dividing both sides by $W \cdot Y$.

The best-fit values of $K_1 - K_4$ were determined by statistical analysis of 680 of the flows.

Manning model:

$$Q = 23.3 \cdot W \cdot Y^{2/3} \cdot S_0^{1/2} \tag{6B5.1b}$$

Modified Manning model:

$$Q = 7.14 \cdot W \cdot Y^{5/3} \cdot S_0^{1/3} \tag{6B5.2b}$$

Chézy model:

$$Q = 25.2 \cdot W \cdot Y^{3/2} \cdot S_0^{1/2} \tag{6B5.3b}$$

Modified Chézy model:

$$Q = 7.73 \cdot W \cdot Y^{3/2} \cdot S_0^{1/3} \tag{6B5.4b}$$

SI units were used for all quantities. A fifth resistance model was determined by log-regression analysis (section 4.8.3.1) of the 680 flows.

Regression model:

$$Q = 4.84 \cdot W^{1.10} \cdot Y^{1.63} \cdot S_0^{0.330} \tag{6B5.5}$$

Note that the statistically determined exponent values in equation 6B5.5 are close to those of the "modified Manning" model (equation 6B5.2).

The predictive ability of these five equations was then compared for the 357 flows not used to establish the numerical values of $K_1 - K_4$ and equation 6B5.5 using several criteria. Overall, the "modified Manning" relation performed best, and the study found that resistance models incorporating a slope exponent $q = 1/3$ (the modified Manning and modified Chézy, as well as the pure regression relation) had greater predictive accuracy than those using the generally accepted theoretical value $q = 1/2$. For all models, there was a strong relation between prediction error and Froude number, *Fr*: The models tended to overestimate discharge for $Fr <\sim 0.15$, and underestimate for $Fr > 0.4$. Unfortunately, this information cannot be used to improve the predictions, because one needs to know velocity to compute *Fr*.

However, given the theoretical difficulties in characterizing resistance/conductance and the need to estimate discharge for cases where there is little or no reach-specific information available, "universal" equations such as 6.47 may be useful. This is particularly true attempting to estimate discharge from satellite or airborne remote-sensing information (Bjerklie et al. 2003). The statistical results (i.e., the suggestion that $q = 1/3$ rather than 1/2) may also point to a reexamination of some of the theoretical assumptions underlying the phenomenon of reach resistance—or to the fact that many natural flows are far from uniform.

6.10 Applications of Resistance Equations

As stated at the beginning of this chapter, the central problem of open-channel-flow hydraulics can be stated as that of determining the average velocity (or depth) associated with a specified discharge in a reach with a specified geometry and bed material. Two practical versions of that problem that commonly arise are:

1. Given a range of discharges due to hydrological processes upstream of the reach, what average velocity and depth will be associated with each discharge? Answers to this question provide information about the elevation and areal extent of flooding to be expected at future high discharges, the ability of the river to assimilate wastes, the amount of erosion to be expected at various discharges, and the suitability of riverine habitats at various discharges. These answers are in the form of reach-specific functions $U = f_U(Q)$ and/or $Y = f_Y(Q)$, where Q is discharge.
2. Given evidence of the water-surface elevation for a recent flood, what was the flood discharge? Answers to this question are important in determining regional flood magnitude–frequency relations. The answers may be expressed functionally as $Q = f_Q(Y)$.

This section shows how these problems are approached for a reach in which concurrent measurements of discharge and hydraulic parameters are not available, but where it is possible to obtain measurements of channel geometry, channel slope, and bed material.

Although both types of problems commonly arise in situations involving overbank flow on floodplains, the discussion here applies when flow is contained within the channel banks. When flow extends onto the floodplain, the channel and the floodplain usually have very different resistances, and the cross section is **compound**. Methods for treating flows in reaches with compound sections are discussed in Chow (1959), French (1985), and Yen (2002).

6.10.1 Determining the Velocity–Discharge and Depth–Discharge Relations

Box 6.6 summarizes the steps involved in determining velocity–discharge and depth–discharge relations for an ungaged reach. The process begins with a survey of channel geometry (boxes 2.1 and 2.2); this is demonstrated in box 6.7 for the Hutt River

BOX 6.6 Steps for Estimating Velocity–Discharge and Depth–Discharge Relations for an Ungaged Reach

1. Using the techniques of box 2.1, identify the bankfull elevation through the reach.
2. Using the techniques of box 2.2 [1. Channel (Bankfull) Geometry], survey a typical cross section to determine the channel geometry.
3. Determine the size distribution of bed sediment, d_p. [See section 2.3.2.1. Refer to Bunte and Abt (2001) for detailed field procedures.]
4. Survey water-surface elevation through the reach to determine water-surface slope, S_S. [Refer to Harrelson et al. (1994) for detailed survey procedures.]
5. Select a range of elevations up to bankfull.
6. Using the techniques of box 2.2 (2. Geometry at a Subbankfull Flow), determine water-surface width W, cross-sectional area A, and average depth $Y \equiv A/W$ associated with each selected elevation.
7. Estimate reach resistance: (a) If using the Chézy equation, use results of steps 3–6 to estimate Ω_* via equation 6.25 for each selected elevation and adjust to give Ω based on considerations of section 6.6. (b) If using the Manning equation, use one of the methods of section 6.8.2 to estimate Manning's n_M.
8. Assume hydraulic radius $R = Y$ and estimate average velocity U for each selected elevation via either the Chézy equation (equation 6.15a) or the Manning equation (equation 6.40).
9. Estimate discharge as $Q = U \cdot A$ for each selected elevation.
10. Use results to generate plots of U versus Q and Y versus Q.

BOX 6.7 **Example Computation of Channel Geometry: Hutt River at Kaitoke, New Zealand**

The line of a cross section is oriented at right angles to the general flow direction. An arbitrary zero point is established at one end of the line; by convention, this is usually on the left bank (facing downstream), but it can be on either bank. Points are selected along the line to define the cross-section shape; these are typically "slope breaks"—points where the ground-surface slope changes. An arbitrary elevation datum is established, and the elevations of these points above this datum are determined by surveying (see Harrelson et al. 1994). To illustrate the computations, we use data for a cross section of the Hutt River in New Zealand (figure 6.24). Section survey results are recorded as elevations, z_i, at distances along the section line, w_i. At each point, the local bankfull depth Y_{BFi} can be calculated as

$$Y_{BFi} = \Psi_{BF} - z_i, \tag{6B7.1}$$

where Ψ_{BF} is the bankfull maximum depth. The data for the Hutt River section are given in table 6B7.1 and are plotted in figure 6.25.

Table 6B7.1

$\mathbf{w}_i$ (m)	0.0	1.0	5.5	7.5	9.0	10.0	11.2	13.3	13.4	14.5
$\mathbf{z}_i$ (m)	3.78	3.71	2.72	2.18	1.92	1.50	0.96	0.86	0.85	0.54
Y_{BFi} (m)	0.00	0.07	1.06	1.60	1.86	2.28	2.82	2.92	3.13	3.24
$\mathbf{w}_i$ (m)	17.5	19.8	19.9	20.6	21.3	24.0	25.8	27.7	28.8	30.0
$\mathbf{z}_i$ (m)	0.53	0.58	0.32	0.28	0.41	0.30	0.44	0.12	0.00	0.24
Y_{BFi} (m)	3.25	3.20	3.46	3.50	3.37	3.49	3.34	3.66	3.78	3.54
$\mathbf{w}_i$ (m)	32.3	34.3	35.1	38.4	39.9	41.2	42.5	43.5	44.8	45.0
$\mathbf{z}_i$ (m)	0.23	0.29	0.50	0.64	0.80	1.84	2.41	2.90	3.71	3.78
Y_{BFi} (m)	3.55	3.49	3.28	3.14	2.98	1.94	1.37	0.88	0.07	0.00

Once the section is plotted, several arbitrary elevations are identified to represent water-surface elevations (the horizontal lines in figure 6.25). For each level, the horizontal positions of the left- and right-bank intersections of the level line with the channel bottom are determined and identified as w_L and w_R, respectively. For each selected elevation, the water-surface width W is

$$W = |w_R - w_L|. \tag{6B7.2}$$

Selecting the level $\Psi = 2$ m in the Hutt River cross section for example calculations, we see from figure 6.25 that

$$W = |41.5 - 8.5| = 33.0 \text{ m}.$$

(*Continued*)

BOX 6.7 *Continued*

The cross-sectional area A associated with a given level is found as

$$A = \sum_{i=1}^{N} A_i = \sum_{i=1}^{N} W_i \cdot Y_i, \qquad (6B7.3)$$

where W_i is the incremental width associated with each surveyed depth Y_i, N is the number of points for which we have observations, and $i = 1, 2, \ldots, N$. If we start from the left bank, $W_1 = w_L$, $W_N = w_R$, and $Y_1 = 0$, $Y_N = 0$ in all cases. The values of the incremental widths are determined as

$$W_1 = \frac{|w_2 - w_1|}{2}; \qquad (6B7.4a)$$

$$W_i = \frac{|w_{i+1} - w_{i-1}|}{2}, i = 2, 3, \ldots, N-1; \qquad (6B7.4b)$$

$$W_N = \frac{|w_N - w_{N-1}|}{2}. \qquad (6B7.6c)$$

Note that $\Sigma W_i = W$.

Table 6B7.2 gives the data for the $\Psi = 2$ m elevation in the Hutt River cross section.

Table 6B7.2

i	w_i(m)	Y_i(m)	W_i(m)	A_i(m^2)
1	8.5	0.00	0.25	0.000
2	9.0	0.08	0.75	0.063
3	10.0	0.50	1.10	0.553
4	11.2	1.04	1.65	1.721
5	13.3	1.14	1.10	1.253
6	13.4	1.35	0.60	0.809
7	14.5	1.46	2.05	2.999
8	17.5	1.47	2.65	3.903
9	19.8	1.42	1.20	1.708
10	19.9	1.68	0.40	0.671
11	20.6	1.72	0.70	1.206
12	21.3	1.59	1.70	2.701
13	24.0	1.71	2.25	3.836
14	25.8	1.56	1.85	2.882
15	27.7	1.88	1.50	2.817
16	28.8	2.00	1.15	2.300
17	30.0	1.76	1.75	3.080
18	32.3	1.77	2.15	3.812
19	34.3	1.71	1.40	2.395
20	35.1	1.50	2.05	3.073
21	38.4	1.36	2.40	3.257
22	39.9	1.20	1.40	1.680
23	41.2	0.16	0.80	0.130
24	41.5	0.00	0.15	0.000
Sum			$33.00 = W$	$46.851 = A$

The average depth, Y, associated with this elevation is

$$Y \equiv \frac{A}{W}, \qquad (6B7.5)$$

so for the example calculation,

$$Y = \frac{46.851}{33.0} = 1.42 \text{ m}.$$

These computations are repeated for each of the selected elevations.

in New Zealand (figures 6.24 and 6.25). The construction of the velocity–discharge and depth–discharge relations is demonstrated for the Hutt River in box 6.8; the results are shown in figure 6.26.

6.10.2 Determining Past Flood Discharge (Slope-Area Measurements)

As noted above, knowledge of past flood discharges in reaches where discharge is not measured is helpful in understanding regional flood-frequency relations. A flood wave passing through a reach typically leaves evidence of the maximum water level in the form of scour marks, removal of leaves and other vegetative material, and/or deposition of silt. Where such evidence is present one can survey the flow cross sections at locations through the reach and estimate the peak flood discharge by inverting equation 6.39c:

$$Q = \frac{g^{1/2} \cdot \left[\sum_{i=1}^{N} A_i \cdot S_i \cdot \Delta X_i \right]^{1/2} \cdot \sum_{i=1}^{N} A_i \cdot \Delta X_i}{\Omega \cdot \Delta X \cdot \left[\sum_{i=1}^{N} P_{wi} \cdot \Delta X_i \right]^{1/2}} \qquad (6.48)$$

This a posteriori application of the resistance relation is called a **slope-area computation**.

The critical practical issue in slope-area computations is in determining the appropriate value of Ω. The standard approach is to use the Manning equation after determining n_M via one of the methods described in section 6.8.2; one can then compute Ω via equation 6.41 or compute Q directly via the Manning equation.

Box 6.9 illustrates the application of equation 6.48 in a slope-area computation, first using a resistance estimated using one of the formulas based on grain size and relative smoothness, and then using a resistance measured in the reach at a lower flow. In this case, the discharge using the estimated resistance was several times too

Figure 6.24 The Hutt River at Kaitoke, New Zealand. (a) View downstream at middle of reach. (b) View upstream at middle of reach. From Hicks and Mason (1991); reproduced with permission of New Zealand National Institute of Water and Atmospheric Research Ltd.

large (i.e., resistance was severely underestimated), while the discharge using the measured resistance was within 2% of the actual value. However, such good results may not always be obtained even with resistance values measured in the reach of interest, because one or more of the factors discussed in section 6.6 may have been significantly different at the time of the peak flow than at the time of measurement (Kirby 1987):

Cross-section geometry: The peak flow may have scoured the channel bed and subsequent lower flows deposited bed sediment. If this happened, the cross-sectional area that existed at the time of the peak flow was larger than the surveyed values and the peak discharge will be underestimated.

Plan-view irregularity: In meandering streams, high flows may "short-circuit" the bends, leading to lower resistance at the high flow than when measured at lower flows.

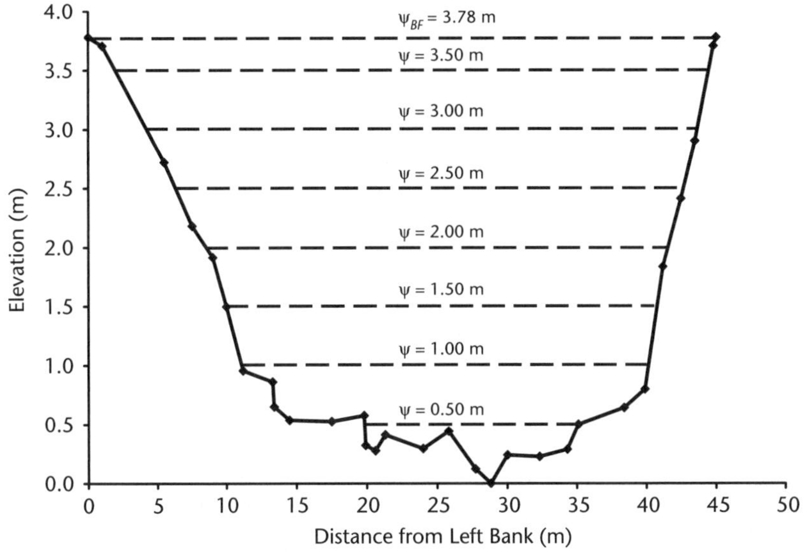

Figure 6.25 Surveyed cross section in the center of the Hutt River reach shown in figure 6.24. Elevations are relative to the lowest elevation in the cross section. The dashed lines are the water levels at the maximum depths (Ψ) indicated; Ψ_{BF} is the bankfull maximum depth. Note approximately 10-fold vertical exaggeration.

BOX 6.8 Example Computation of Velocity–Discharge and Depth–Discharge Relations for an Ungaged Reach: Hutt River at Kaitoke, New Zealand

Using the procedure described in boxes 6.6 and 6.7, the following values of average depth Y have been computed for selected maximum-depth levels Ψ for the cross section of the Hutt River at Kaitoke, New Zealand, shown in figure 6.25:

Ψ (m)	0.50	1.00	1.50	2.00	2.50	3.00	3.50	3.78
Y (m)	0.22	0.55	1.01	1.42	1.77	2.11	2.44	2.57

The bed-sediment material consists of gravel, cobbles, and boulders; $d_{84} =$ 212 mm. The average channel slope through the reach is $S = 0.00539$. We estimate the velocity–discharge and depth–discharge relations for this cross section via 1) the Chézy equation and 2) the Manning equation.

Chézy Equation

There is a range of bed-material sizes; we select the resistance relation for gravel-bed streams suggested by Bathurst (1993) (equation 6.27).

(Continued)

BOX 6.8 *Continued*

We assume $R = Y$ and estimate Ω as

$$\Omega = 0.400 \cdot \left[-\ln\left(\frac{0.212}{3.60 \cdot R} \right) \right]^{-1}.$$

Values of u_* are determined via equation 6.16:

$$u_* = (9.81 \cdot R \cdot 0.00539)^{1/2}$$

Average velocity U is then computed via equation 6.19 and discharge Q via equation 6.3. The results are tabulated in table 6B8.1.

Table 6B8.1

Ψ (m)	0.50	1.00	1.50	2.00	2.50	3.00	3.50	3.78
R (m)	0.22	0.55	1.01	1.42	1.77	2.11	2.44	2.57
Ω	0.301	0.179	0.141	0.126	0.118	0.112	0.107	0.106
U (m/s)	0.359	0.956	1.642	2.180	2.603	2.988	3.344	3.480
Q (m^3/s)	1.19	15.3	50.9	102	167	249	348	404

Manning Equation

In practice, one would use one of the approaches listed in table 6.3 and discussed in section 6.8.2 to estimate the appropriate n_M for this reach. In this example, we will use the value determined for the reach by measurement and reported in Hicks and Mason (1991): $n_M = 0.037$. Using this value and the measured slope in the Manning equation (equation 6.40c), we compute the values in table 6B8.2.

Table 6B8.2

Ψ (m)	0.50	1.00	1.50	2.00	2.50	3.00	3.50	3.78
R (m)	0.22	0.55	1.01	1.42	1.77	2.11	2.44	2.57
U (m/s)	0.727	1.335	2.000	2.507	2.903	3.264	3.596	3.723
Q (m^3/s)	2.42	21.4	61.9	118	187	272	374	432

Comparison of Estimates with Measured Values

Hicks and Mason (1991) provided measured values of R, U, and Q for this reach, so we can compare the two estimates with actual values, as shown in figure 6.26. The Chézy estimate, which uses only measured quantities (R, S, d_{84}) fits the measured values very closely except at the highest flow, while the Manning estimate of velocity is slightly too high (and depth too low) over most of the range. Recall though that the Manning estimate is based on a value of n_M determined by measurement in the reach; in many actual applications, such measurements would not be available, and we would be forced to estimate n_M by other means (section 6.8.2), probably leading to greater error.

In this example, the Chézy relation appears to give better results than the Manning relation.

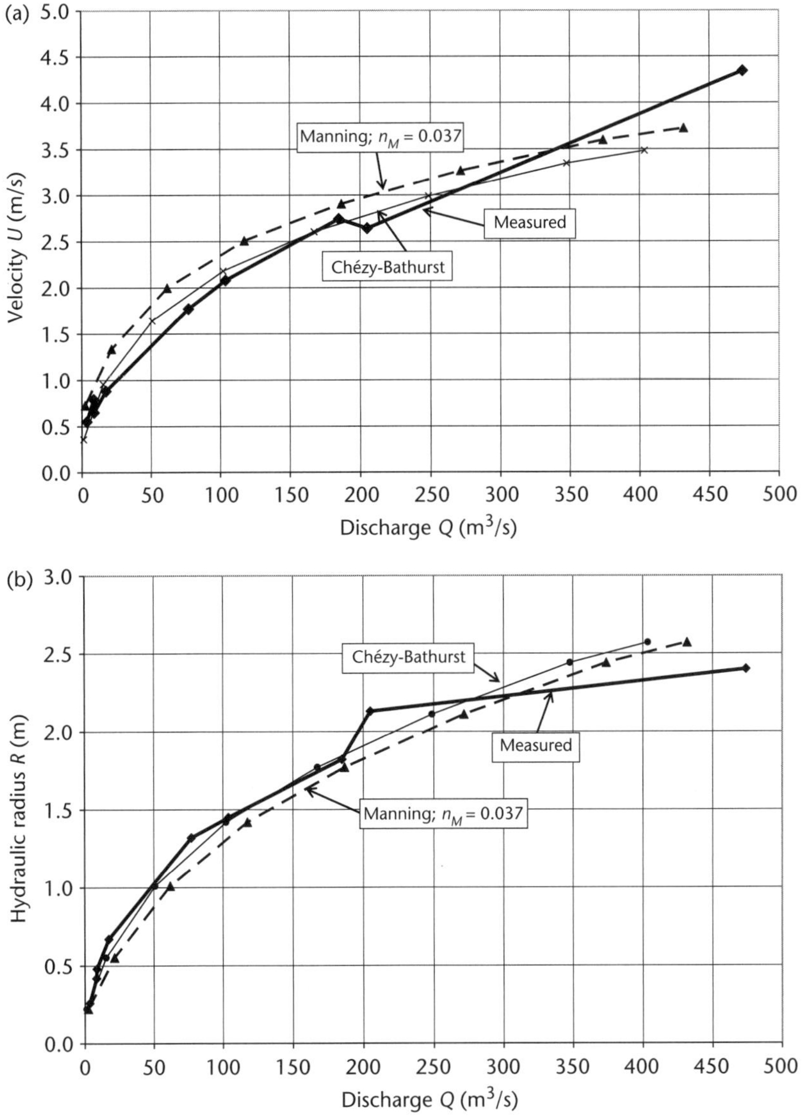

Figure 6.26 Comparison of estimated and actual hydraulic relations for the Hutt River cross section shown in figures 6.24 and 6.25. (a) Velocity–discharge relation. (b) Hydraulic radius (depth)–discharge relation. Heavy lines are measured; lighter solid line is calculated via Chézy equation with Bathurst (1993) resistance relation for gravel-bed streams (equation 6.27); dashed line is calculated via Manning equation using measured value of $n_M = 0.037$.

BOX 6.9 Slope-Area Computations, South Beaverdam Creek Near Dewy Rose, Georgia

A peak flood on 26 November 1957 left high water marks in a reach of South Beaverdam Creek near Dewy Rose, Georgia. The peak flood discharge was measured at $Q = 23.2$ m^3/s. The cross-sectional area, width, average depth, hydraulic radius, wetted perimeter, and water-surface slope defined by these high-water marks were surveyed by Barnes (1967) at five cross sections and are summarized in table 6B9.1.

Table 6B9.1

Section, i	A_i (m^2)	W_i (m)	Y_i (m)	R_i (m)	P_{wi} (m)	ΔX_i (m)	ΔZ_i (m)	$S_{Si} = \lvert\Delta Z_i\rvert/\Delta X_i$
0	24.9	21.6	1.16	1.10	22.6			
1	26.8	17.1	1.55	1.52	17.7	21.6	0.043	0.00197
2	25.8	18.0	1.43	1.32	19.5	20.1	0.037	0.00182
3	26.1	18.0	1.46	1.34	19.4	24.7	0.040	0.00161
4	24.2	17.7	1.37	1.26	19.2	19.5	0.018	0.00094
Average or sum	$A =$ 25.6	$W =$ 18.5	$Y =$ 1.40	$R =$ 1.31	$P_W =$ 19.7	$\Delta X =$ 85.9	$\Delta Z =$ 0.137	$S_S = 0.00160$

To illustrate slope-area computations, we assume the discharge is unknown and apply three approaches that could be used to estimate a past flood discharge from high-water marks.

Standard Approach

This is the method described in section 6.8.2. We first assume we do not have a resistance determined by measurement in the reach. Table 6B9.2 gives the values of the quantities that are summed in equation 6.39c.

Table 6B9.2

Section, i	$A_i \cdot S_{Si} \cdot \Delta X_i$ (m^3)	$A_i \cdot \Delta X_i$ (m^3)	$P_{wi} \cdot \Delta X_i$ (m^3)
1	1.143	579	382
2	0.945	520	393
3	1.035	645	480
4	0.442	472	375
Sum	3.465	2216	1630

The channel bed "consists of sand about 1 ft deep over clay and rock. Banks are irregular with trees and bushes growing down to the low water line" Barnes (1967, p. 142). Because this is a sand-bed reach, we estimate Ω via equation 6.25 assuming $Y = R$ and $y_r = d_{84} = 0.002$ m (the upper limit for sand), and compute

$$\Omega = 0.400 \cdot \left[-\ln\left(\frac{0.002}{11 \cdot 1.31}\right)\right]^{-1} = 0.045.$$

Substituting the appropriate values into equation 6.48 gives

$$Q = \frac{9.81^{1/2} \cdot [3.465]^{1/2} \cdot 2216}{0.045 \cdot 85.9 \cdot [1630]^{1/2}} = 82.8 \text{ m}^3/\text{s}$$

as our estimate of peak discharge.

This estimate is several times too high. Thus, it appears that we severely underestimated the resistance using equation 6.25. Some of the "excess" resistance probably comes from the bank vegetation that extended into the flow, and some may be due to the development of ripples or dunes on the sand bed. Perhaps we could have come up with a better estimate using another of the approaches of section 6.8.2, or had accounted for effects of bedforms on the resistance (see section 6.6.4.2).

A better approach would be to determine the reach resistance via measurement before applying equation 6.48. On the day after the 26 November flood, when the flow was $Q = 6.26$ m^3/s, Barnes (1967) surveyed the same cross sections and obtained the values in table 6B9.3.

Table 6B9.3

Section, i	A_i (m^2)	R_i (m)	P_{wi} (m)	ΔX_i(m)	$\lvert\Delta Z_i\rvert$(m)	$S_{Si} = \lvert\Delta Z_i\rvert/\Delta X_i$
0	8.5	0.62	13.7			
1	11.9	0.82	14.5	21.6	0.034	0.00155
2	10.0	0.61	16.5	20.1	0.030	0.00152
3	10.0	0.60	16.6	24.7	0.024	0.00099
4	9.4	0.62	15.1	19.5	0.043	0.00219
Average or sum	$A = 9.96$	$R = 0.65$	$P_w = 15.3$	$\Delta X = 85.9$	$\lvert\Delta Z\rvert = 0.131$	$S_S = 0.00153$

We want to determine the value of Ω for this flow and use that value to estimate the flood peak on 26 November 1957. Table 6B9.4 gives the values of the quantities that are summed in equation 6.39c.

Table 6B9.4

Section, i	$A_i \cdot S_{Si} \cdot \Delta X_i$ (m^3)	$A_i \cdot \Delta X_i$ (m^3)	$P_{wi} \cdot \Delta X_i$ (m^3)
1	0.399	258	314
2	0.306	202	331
3	0.245	248	411
4	0.401	183	295
Sum	1.351	891	1351

Substituting the appropriate values into equation 6.39c yields

$$\Omega = \frac{9.81^{1/2} \cdot [1.351]^{1/2} \cdot 891}{6.26 \cdot 85.9 \cdot [1351]^{1/2}} = 0.164.$$

(Continued)

BOX 6.9 *Continued*

Thus, the measured reach resistance is several times higher than that based on equation 6.25. Finally, we use this measured value of Ω to estimate the peak discharge of 26 November 1957 via equation 6.48:

$$Q = \frac{9.81^{1/2} \cdot [3.465]^{1/2} \cdot 2216}{0.164 \cdot 85.9 \cdot [1630]^{1/2}} = 22.7 \text{ m}^3/\text{s}$$

The value of Q estimated using the Ω value measured in the reach is within 2% of the actual value.

Application of General Statistically Derived Relation

It is of interest to see how well the statistically developed "modified Manning" equation (equation 6.47) does in estimating the peak flood discharge from the high-water marks. Using values from table 6B9.1, that equation gives

$$Q = 7.14 \cdot 18.5 \cdot 1.40^{5/3} \cdot 0.00160^{1/3} = 27.1 \text{ m}^3/\text{s}.$$

The estimate for this case is quite good, about 17% higher than actual. The Froude number for this flow can be calculated from data in table 6B9.1:

$$Fr = \frac{U}{(g \cdot Y)^{1/2}} = \frac{Q/A}{(g \cdot Y)^{1/2}} = \frac{23.2/25.6}{(9.81 \cdot 1.40)^{1/2}} = 0.24$$

This value is in the range where equation 6.47 was found to give generally good predictions.

Application of Relation Developed from Dimensional Analysis

It is also of interest to see how well equation 4.74, developed by dimensional analysis and measurement data from New Zealand rivers, does in predicting the flood-peak discharge. Recall that that relation, written in terms of discharge, is

$$Q = 1.84 \cdot \left(\frac{Y}{y_r}\right)^{0.704} \cdot g^{1/2} \cdot W \cdot Y^{3/2} \cdot S_0^{1/2}, \; Y/y_r \leq 10; \qquad (6B9.1a)$$

$$Q = 9.51 \cdot g^{1/2} \cdot W \cdot Y^{3/2} \cdot S_0^{1/2}, \; Y/y_r > 10. \qquad (6B9.1b)$$

Since $y_r = 0.002$ m, $Y/y_r > 10$, and we use equation 6B9.1b with data from table 6B9.1:

$$Q = 9.51 \cdot 9.81^{1/2} \cdot 18.5 \cdot 1.40^{3/2} \cdot 0.00160^{1/2} = 36.5 \text{ m}^3/\text{s}$$

This estimate is 57% greater than actual, suggesting that equation 4.74 is not sufficiently precise to use for prediction (note the scatter in figure 4.14).

Longitudinal-profile irregularity: At high flows, the pool/riffle alteration tends to become submerged, tending to decrease resistance at higher flows (figure 6.11c).

Vegetation: Resistance may decrease at higher flows because flexible vegetation is bent further or because low vegetation becomes more submerged, or increase because more of the flow encounters bank and floodplain vegetation.

Surface stability: Resistance may increase at higher flows due to surface irregularities, particularly at bends or abrupt obstructions.

Sediment: In sand-bed streams, bedforms may be different at high flows than when flow is measured, leading to higher or lower resistance (figure 6.20).

Ice: During breakup of an ice cover, there may be large and unknown differences in resistance between the time of a high flow and when reach resistance is measured.

6.11 Summary

The standard approach to open-channel flow resistance is usually presented in terms of the Manning equation, with focus on determining appropriate values of Manning's n_M in various applications. However, the Manning equation was not derived from first principles, nor was it established by rigorous statistical analysis. Thus, this chapter has explored the fundamentals and practical aspects of resistance via the Chézy equation, which is derived from straightforward macroscopic force-balance considerations. This approach is consistent with fundamental fluid-mechanics principles:

- The Chézy derivation incorporates assumptions consistent with the models of turbulence presented in section 3.3.4 .
- Formulating the resistance as the dimensionless quantity Ω allows us to consider the subject in a way that is consistent with theoretical and observational approaches that are applicable in a wide range of fluid-mechanics contexts (summarized by the Moody diagram, figure 6.8).
- At least for the simplest flow situations, resistance can be related to measurable variables via physically based expressions for the velocity profile discussed in chapter 5 (equation 6.25).

As noted at the beginning of this chapter, our goal has been to develop relations for computing the average velocity U in a channel reach given the reach geometry, material, and slope and the depth or discharge. We expressed this relation as

$$U = \Omega^{-1} \cdot u_* = \Omega^{-1} \cdot (g \cdot R \cdot S)^{1/2} \approx \Omega^{-1} \cdot (g \cdot Y \cdot S)^{1/2} \tag{6.49}$$

and explored the factors that control Ω. Following Rouse (1965) and Yen (2002), we can summarize these factors for quasi-uniform flows in natural channels:

$$\Omega = f_\Omega(Y/y_r, Re, Y/W, \boldsymbol{\Psi}, \boldsymbol{\zeta}, \boldsymbol{\Theta}, \boldsymbol{V}, Fr, \boldsymbol{\Sigma}, \boldsymbol{I}), \tag{6.50a}$$

where $\boldsymbol{\Psi}$ represents the effects of cross-section irregularities, $\boldsymbol{\zeta}$ the effects of planform irregularities, $\boldsymbol{\Theta}$ the effects of longitudinal-profile irregularities, $\boldsymbol{V}$ the effects of vegetation, $\boldsymbol{\Sigma}$ the effects of sediment transport, and $\boldsymbol{I}$ the effects of ice.

Considering only ice-free channels and noting that the effects of Y/W are generally minor in natural channels (figure 6.10), we can write

$$\boldsymbol{\Omega} \approx f_{\Omega}(Y/y_r, Re, \boldsymbol{\Psi}, \boldsymbol{\zeta}, \boldsymbol{\Theta}, V, Fr, \boldsymbol{\Sigma}). \tag{6.50b}$$

Further simplification may be possible if we recall that the effects of cross-sectional variability $\boldsymbol{\Psi}$ and longitudinal variability $\boldsymbol{\Theta}$ are at least in part captured by the relative submergence Y/y_r, so that

$$\Omega \approx f_{\Omega}(Y/y_r, Re, \boldsymbol{\zeta}, V, Fr, \boldsymbol{\Sigma}). \tag{6.50c}$$

One barrier to using 6.50c to determine velocity via 6.49 is that Re, $\boldsymbol{\zeta}$, Fr, $\boldsymbol{\Sigma}$, and to some extent V all depend on velocity—so we are faced with a logical circularity. However, if we confine ourselves to fully rough flows in wide, reasonably straight channels at low to moderate Froude numbers and insignificant sediment transport, the problem becomes more tractable:

$$\Omega \approx f_{\Omega}(Y/y_r). \tag{6.50d}$$

Based on the P-vK law and the analyses in section 6.6, we can be reasonably confident that the form of this relation is given by

$$\Omega \approx \kappa \cdot \left[\ln\left(\frac{K_r \cdot Y}{y_r}\right)\right]^{-1}. \tag{6.51}$$

The standard form of this relation is the C-K equation, in which $\kappa = 0.400$ and $K_r = 11$. However, as we have seen in equations 6.27, 6.28, 6.30, and 6.32, the values of κ and K_r may vary from reach to reach—and maybe even for different flows in the same reach.

We saw in box 6.7 that the Chézy approach incorporating an appropriate resistance relation can provide good estimates of velocity-discharge and depth-discharge relations that can be used to solve practical problems.

Approaching resistance via the Chézy equation also provides a straightforward formula for computing reach resistance from field data (equation 6.39). This formula can be inverted to give a relation for estimating past flood discharges in slope-area computations (equation 6.48). However, we saw in box 6.9 that such estimates can be erroneous in the absence of appropriate resistance estimates.

Clearly, although we have learned much about the factors that determine reach resistance, there are still many uncertainties to be faced in obtaining reliable a priori and a posteriori resistance estimates for practical use and much need for additional research in this area.

7

Forces and Flow Classification

7.0 Introduction and Overview

The forces involved in open-channel flow are introduced in section 4.2.2.1. The goals of this chapter are 1) to develop expressions to evaluate the magnitudes of those forces at the macroscopic scale, 2) to examine the relative magnitudes of the various forces in natural channels and show how they change with the flow scale, and 3) to show that the Reynolds number (introduced in section 3.4.2) and the Froude number (introduced in section 6.2.2.2) can be interpreted in terms of force ratios. Understanding the relative magnitudes of forces provides a helpful perspective for developing quantitative solutions to practical problems.

Open-channel flows are induced by gradients of potential energy proportional to the sine of the water-surface slope (section 4.7). This chapter shows that the water-surface slope reflects the magnitude of the driving forces due to gravity and pressure. Once motion begins, frictional forces resisting the flow arise due to molecular viscosity and, usually, turbulence; these forces are increasing functions of velocity. In steady uniform flow, which was assumed in the developments of chapters 5 and 6, the gravitational driving force is balanced by the frictional forces, so there is no acceleration and no other forces are involved. However, in general, the forces affecting open-channel flows are not in balance, so the flow experiences convective acceleration (spatial change in velocity) and/or local acceleration (temporal change in velocity)—concepts introduced in section 4.2.1.2 at the "microscopic" scale (fluid elements).

In this chapter, as in chapters 5 and 6, we continue to analyze the flow on a macroscopic scale; that is, the physical relations are developed for the entire flow in a reach in an idealized channel rather than for a fluid element. We consider changes only in time and in one spatial dimension (the downstream direction), so the resulting equations are characterized as "one-dimensional."

The chapter begins by reviewing the forces that induce and oppose fluid motion in open channels and presenting the basic force-balance equations for various flow categories. Next we lay out the basic geometry of an idealized reach and then formulate quantitative expressions for the magnitudes of the various forces as functions of fluid properties and flow parameters. We also develop expressions for the convective and local accelerations so that we can ultimately formulate the complete macroscopic force-balance equation for one-dimensional open-channel flow.

Using data for a range of flows, we examine the typical values of each of the forces in natural streams and compare their magnitudes. We also compare the relative magnitudes of the forces as a function of scale, from small laboratory flumes to the Gulf Stream. This comparison provides guidance for identifying conditions under which the force balance may be simplified by omitting particular forces due to their relative insignificance. The chapter concludes by showing how the Reynolds and Froude numbers can be interpreted in terms of force ratios.

7.1 Force Classification and the Overall Force Balance

In this section we formulate the overall force-balance relations for flows of various categories. To simplify the development, these relations are formulated for the simple open-channel flow shown in figure 7.1: a wide rectangular channel ($Y = R$) with constant width ($W_1 = W_2 = W$) but spatially varying depth. At any instant, the reach contains a spatially constant discharge Q, so

$$Q = W \cdot Y_1 \cdot U_1 = W \cdot Y_2 \cdot U_2, \tag{7.1}$$

where Y_i is the average depth and U_i is the average velocity at section i.

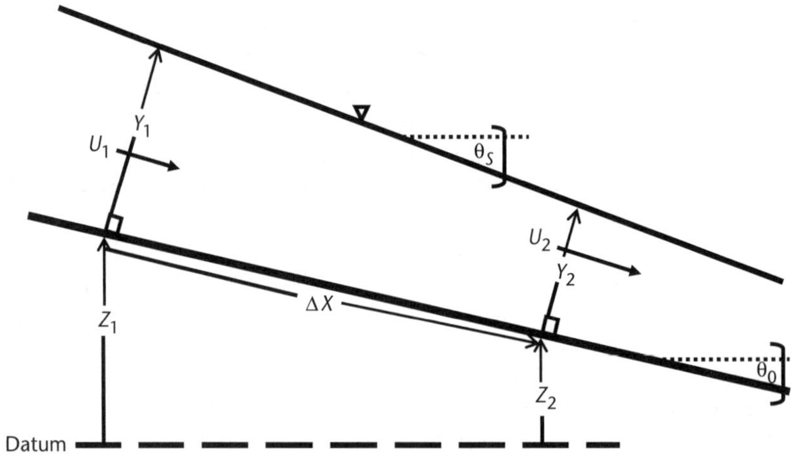

Figure 7.1 Definition diagram for deriving expressions to calculate force magnitudes for a nonuniform flow in a prismatic channel. Width and discharge are assumed constant.

To generalize the force expressions, individual forces are expressed as the force magnitude divided by the mass of water in the reach between cross sections 1 and 2 in figure 7.1. Since force/mass = acceleration, we use the symbol a for these quantities, with a subscript identifying each force.

7.1.1 Classification of Forces

The forces that affect open-channel flows are listed and characterized in table 7.1. Forces and accelerations are vector quantities and so are completely specified by their magnitude and direction. Here we use a simplified specification of direction, classed as downstream (**driving forces**), upstream (**resisting forces**), or at right angles to the flow (**perpendicular forces**). As explained further below, the perpendicular forces are "pseudoforces."

As we saw in section 4.2.2.1, forces are also classified as to whether they act on all the matter in a fluid element (**body forces**) or on the element surface (**surface forces**), and this aspect of each of the forces is identified in table 7.1. The expressions for computing the force magnitudes are derived and discussed in section 7.3.

The coordinate system that we use to describe fluid motion is usually fixed relative to the earth's surface. However, the earth is rotating, so the coordinate system is rotating, and this rotation gives rise to an apparent force, or **pseudoforce**, perpendicular to the original direction of motion. This force, called the **Coriolis force** or **Coriolis acceleration**,[1] is directed to the right in the northern hemisphere

Table 7.1 Summary of expressions for forces per unit mass (accelerations) for figure 7.1 (symbols are defined in the text and in figure 7.1).

Acceleration	Direction	Body/surface	Expression	Comments
Gravitational, a_G	Downstream	Body	$g \cdot S_0$	Acts upstream if $S_0 < 0^a$
Pressure, a_P	Downstream	Surface	$-g \cdot \left(\cos \theta_0 \cdot \frac{\Delta Y}{\Delta X}\right)$	Acts upstream if $\frac{\Delta Y}{\Delta X} > 0^a$
Viscous, a_V	Upstream	Surface	$3 \cdot \nu \cdot \frac{U}{Y^2}$	Always present
Turbulent, a_T	Upstream	Surface	$\Omega^2 \cdot \frac{U^2}{Y}$	Absent in purely viscous (laminar) flow
Coriolis, a_{CO}	Perpendicular	Body	$2 \cdot \omega \cdot U \cdot \sin \Lambda$	Always present
Centrifugal, a_C	Perpendicular	Body	$\left[1 + (12 \cdot \Omega + 30 \cdot \Omega^2) \cdot \left(\frac{Y}{r_c}\right)\right] \cdot \left(\frac{U^2}{r_c}\right)$	Absent in straight reaches ($r_c \to \infty$)
Convective, a_X	Upstream	Body	$U \cdot \frac{\Delta U}{\Delta X}$	Acts downstream if $\Delta U < 0$
Local, a_t	Upstream	Body	$\frac{\Delta U}{\Delta t}$	Acts downstream if $\Delta U < 0$

[a] The sum $a_G + a_P$ must be > 0.

and to the left in the southern hemisphere and, as we will see, depends on the latitude and the velocity. We will show in section 7.3 that this acceleration is of negligible relative magnitude in all but the largest water motions on the earth's surface—the very largest rivers and ocean currents. Thus, the force-balance equations formulated in this section do not include the Coriolis acceleration.

As we saw in section 6.6.1.2, flow in a curved channel gives rise to an apparent force perpendicular to the streamwise direction, the **centrifugal force** or **centrifugal acceleration**. This pseudoforce represents a deviation from straight-line motion and hence contributes to the resisting forces opposing downstream flow. This force varies with the radius of curvature of a channel bend as well as the velocity (equation 6.29). We will compare the magnitudes of centrifugal accelerations in typical channel bends with other accelerations in section 7.4, but the force-balance equations formulated in this section are for straight-channel reaches and do not include the centrifugal acceleration.

The major categories used to classify open-channel flows are reviewed in sections 7.1.2–7.1.4.

7.1.2 Steady Uniform Flow

As noted above, the only driving force involved in the steady uniform flows discussed in chapters 5 and 6 is gravity, a_G. In a straight channel, the only resisting forces are those due to boundary friction transmitted into the flow by molecular viscosity, a_V, and, in most flows, by turbulence, a_T [$a_T = 0$ in purely viscous (laminar) flows]. Thus, the force balance for a steady uniform flow is

$$a_G - (a_V + a_T) = 0. \tag{7.2}$$

7.1.3 Steady Nonuniform Flow

With a constant discharge and width (equation 7.1), a spatial change in depth implies a spatial change in velocity such that

$$\frac{Q}{W} = U_1 \cdot Y_1 = U_2 \cdot Y_2, \tag{7.3}$$

and the flow is nonuniform. The pressure force, a_P, that arises due to the spatial change in depth then also contributes to the driving force, either increasing or decreasing it. Therefore, a steady nonuniform flow also involves a convective acceleration, a_X, and the force balance for a steady nonuniform flow is

$$(a_G + a_P) - (a_V + a_T) = a_X, \tag{7.4}$$

where a_P is positive if depth decreases downstream and negative if depth increases downstream.

7.1.4 Unsteady Nonuniform Flow

In unsteady flows, discharge, depth, and velocity change with time, so there is local acceleration, a_t, as well as convective acceleration (unsteady uniform flow is virtually impossible). Thus, the force balance for an unsteady nonuniform flow is

$$(a_G + a_P) - (a_V + a_T) = a_X + a_t. \tag{7.5}$$

7.2 Basic Geometric Relations

Here we develop the basic geometric relations that are used to formulate the quantitative expressions for the various forces in section 7.3.

In figure 7.1, the volume V of water between cross sections 1 and 2 separated by the streamwise distance ΔX is

$$V = W \cdot Y \cdot \Delta X, \tag{7.6}$$

where Y is the reach-average depth, given by

$$Y \equiv \left(\frac{Y_1 + Y_2}{2}\right). \tag{7.7}$$

The reach-average velocity, U, is similarly

$$U \equiv \left(\frac{U_1 + U_2}{2}\right). \tag{7.8}$$

The mass, M, and weight, Wt, of water between the two sections are given by

$$M = \rho \cdot W \cdot \Delta X \cdot Y \tag{7.9}$$

$$Wt = \gamma \cdot W \cdot \Delta X \cdot Y, \tag{7.10}$$

where ρ is the mass density and γ the weight density of water.

The channel slope, S_0, is defined as positive downstream, so

$$-\frac{\Delta Z}{\Delta X} = \sin\theta_0 \equiv S_0, \tag{7.11}$$

where $\Delta Z \equiv Z_2 - Z_1$. Of course, river channels almost always slope downstream ($\Delta Z < 0$) when measured over distances equal to several widths, but locally the bottom can be horizontal ($\Delta Z = 0$) or slope upward ($\Delta Z > 0$). When the local bottom slopes upstream the slope is said to be **adverse**; then, $\sin\theta_0 \equiv S_0 < 0$. However, the value of $\cos\theta_0$ is > 0 for adverse as well as downstream slopes.

The water-surface slope, S_S, is given by

$$\begin{aligned} S_S \equiv \sin\theta_S &= -\frac{(Z_1 + Y_1 \cdot \cos\theta_0) - (Z_2 + Y_2 \cdot \cos\theta_0)}{\Delta X} \\ &= \left(-\frac{\Delta Z}{\Delta X} - \cos\theta_0 \cdot \frac{\Delta Y}{\Delta X}\right) = \left(\sin\theta_0 - \cos\theta_0 \cdot \frac{\Delta Y}{\Delta X}\right) \\ &= \left(S_0 - \cos\theta_0 \cdot \frac{\Delta Y}{\Delta X}\right), \end{aligned} \tag{7.12}$$

where $\Delta Y \equiv Y_2 - Y_1$. ΔZ or ΔY can be either positive or negative, but the sum of the terms in parentheses in equation 7.12 must be positive. In other words, the water-surface elevation must decrease in the downstream direction if flow is occurring.

7.3 Magnitudes of Forces per Unit Mass

7.3.1 Driving Forces

7.3.1.1 Gravitational Force

The gravitational driving force F_G is the downslope component of the weight:

$$F_G = Wt \cdot \sin\theta_0 = \gamma \cdot W \cdot \Delta X \cdot Y \cdot \sin\theta_0 \tag{7.13}$$

The gravitational force per unit mass, or acceleration due to gravity, a_G, is found from 7.13 and 7.9 as

$$a_G = \frac{F_G}{M} = -g \cdot \frac{\Delta Z}{\Delta X} = g \cdot \sin\theta_0 \equiv g \cdot S_0. \tag{7.14}$$

If the local bottom slopes downstream, the gravitational force acts to accelerate flow; if the slope is adverse, it acts in the upstream direction, opposing flow.

7.3.1.2 Pressure Force

We assume that the pressure distribution is hydrostatic (see figures 4.4 and 4.5); that is, at any distance y above the bottom, the pressure $P_i(y)$ at section i is given by

$$P_i(y) = \gamma \cdot (Y_i - y) \cdot \cos\theta_0, \tag{7.15}$$

where Y_i is the total depth at section i. Thus, the average pressure at a given cross section, P_i, is

$$P_i = \gamma \cdot \frac{Y_i}{2} \cdot \cos\theta_0. \tag{7.16}$$

The pressure force on face i, F_{Pi}, is given by

$$F_{Pi} = P_i \cdot Y_i \cdot W = \gamma \cdot \frac{Y_i}{2} \cdot \cos\theta_0 \cdot Y_i \cdot W. \tag{7.17}$$

The net downstream-directed pressure force operating on the water between the upstream and downstream sections, F_P, is thus

$$F_P = F_{P1} - F_{P2} = \frac{\gamma \cdot \cos\theta_0 \cdot W}{2} \cdot (Y_1^2 - Y_2^2). \tag{7.18}$$

Defining $\Delta Y \equiv (Y_2 - Y_1)$ and noting that $(Y_1^2 - Y_2^2) = -(Y_1 + Y_2) \cdot (Y_2 - Y_1) = -2 \cdot Y \cdot \Delta Y$, we can rewrite 7.12 as

$$F_P = -\gamma \cdot \cos\theta_0 \cdot W \cdot Y \cdot \Delta Y. \tag{7.19}$$

Dividing equation 7.19 by equation 7.9 then gives the acceleration due to pressure, a_P:

$$a_P = -g \cdot \cos\theta_0 \cdot \frac{\Delta Y}{\Delta X} \tag{7.20}$$

When depth decreases downstream (i.e., the water surface slopes downstream more steeply than the bottom slope), $\Delta Y < 0$ and $a_P > 0$, and the pressure force acts to accelerate flow. When the depth increases downstream, $\Delta Y > 0$ and $a_P < 0$, and the pressure force acts to oppose flow. Note that for most rivers, $S_0 < 0.1$ so that $\cos\theta_0 \approx 1$.

7.3.1.3 *Total Driving Force*

Now we can write the total driving force per unit mass, a_D, as the sum of the downstream-directed gravitational and pressure accelerations:

$$a_D = a_G + a_P = g \cdot \left(-\frac{\Delta Z}{\Delta X} - \cos\theta_0 \cdot \frac{\Delta Y}{\Delta X} \right) = g \cdot \left(S_0 - \cos\theta_0 \cdot \frac{\Delta Y}{\Delta X} \right). \quad (7.21)$$

As noted for equation 7.12, the term in parentheses equals the water-surface slope and must always be positive.

7.3.2 Frictional Resisting Forces

As we saw in chapters 5 and 6, the frictional resisting forces are due to the retarding effect of the boundary (the no-slip condition) that is transmitted into the flow by molecular viscosity and, if the flow is deep enough and fast enough, by turbulence (eddy viscosity). The frictional forces are always directed upstream and, as shown below, are increasing functions of the flow velocity.

7.3.2.1 *Viscous Force*

Equation 5.8 gives the relation between the frictional force per unit boundary area due to molecular viscosity, τ_V, and the local velocity gradient normal to the boundary, du/dy, as

$$\tau_V = \mu \cdot \frac{du(y)}{dy}. \quad (7.22)$$

Because we are treating the flow macroscopically, we replace the local derivative with an "average" gradient U/Y and write

$$\tau_V = k_V \cdot \mu \cdot \frac{U}{Y}, \quad (7.23a)$$

where k_V is a proportionality constant to account for the change from du/dy to U/Y. Box 7.1 shows that $k_V = 3$, so

$$\tau_V = 3 \cdot \mu \cdot \frac{U}{Y}, \quad (7.23b)$$

Thus, for the flow of figure 7.1, the total viscous resisting force F_V equals the force per unit area τ_V times the area of the boundary:[2]

$$F_V = 3 \cdot \mu \cdot \frac{U}{Y} \cdot W \cdot \Delta X \quad (7.24)$$

Dividing equation 7.24 by equation 7.9 gives the viscous force per unit mass acting to resist the flow, a_V:

$$a_V = 3 \cdot \nu \cdot \frac{U}{Y^2}, \quad (7.25)$$

where $\nu \equiv \mu/\rho$ (kinematic viscosity). Thus, we see that the frictional force due to molecular viscosity is proportional to the first power of the velocity.

BOX 7.1 Evaluation of k_V in Equation 7.23a

At the boundary, equation 7.22 becomes

$$\tau_0 = \mu \cdot \left.\frac{du(y)}{dy}\right|_{y=0}, \tag{7B1.1}$$

where τ_0 is the boundary shear stress. From equation 5.7,

$$\tau_0 = \gamma \cdot Y \cdot S. \tag{7B1.2}$$

Because τ_0 should be the same value when we use the macroscopic formulation of equation 7.23a,

$$\gamma \cdot Y \cdot S = k_V \cdot \mu \cdot \frac{U}{Y}, \tag{7B1.3}$$

and

$$k_V = \frac{\gamma \cdot Y^2 \cdot S}{\mu \cdot U}. \tag{7B1.4}$$

Equation 5.14 shows that

$$U = \frac{\gamma \cdot Y^2 \cdot S}{3 \cdot \mu}, \tag{7B1.5}$$

and substituting equation 7B1.5 into equation 7B1.4 gives $k_V = 3$.

As we saw in table 3.4, viscosity is a strong function of temperature, so note that the frictional force due to molecular viscosity depends on the temperature.

7.3.2.2 Turbulent Force

As indicated in equation 6.10, the shear stress due to turbulence, τ_T, is

$$\tau_T = K_T \cdot \rho \cdot U^2, \tag{7.26a}$$

where K_T is a constant of proportionality that depends on boundary conditions. Using the definition of resistance Ω and equation 6.18,

$$\tau_T = \Omega^2 \cdot \rho \cdot U^2. \tag{7.26b}$$

Thus, the turbulent resisting force F_T is

$$F_T = \Omega^2 \cdot \rho \cdot U^2 \cdot W \cdot \Delta X, \tag{7.27}$$

and the turbulent resisting force per unit mass a_T is

$$a_T = \Omega^2 \cdot \frac{U^2}{Y} = \frac{u_*^2}{Y} = g \cdot S_s. \tag{7.28}$$

Thus, we see that the frictional force due to turbulence is proportional to the square of the velocity and to the square of the resistance. From the discussions in section 6.6, recall that resistance depends on Reynolds number, relative roughness, the nature of the channel boundary, and other factors.

7.3.2.3 Total Frictional Resisting Force

The total frictional resisting force per unit mass, a_R, is the sum of the viscous and turbulent forces:

$$a_R = a_V + a_T = 3 \cdot \nu \cdot \frac{U}{Y^2} + \Omega^2 \cdot \frac{U^2}{Y} \tag{7.29}$$

As noted above, this force is always directed upstream.

7.3.3 Perpendicular Forces

7.3.3.1 Coriolis Force

As explained in section 4.1.3, motion is measured by reference to a coordinate system that is fixed relative to the earth's surface. In such a system, a mass moving on the surface is subject to an apparent deflecting force called the Coriolis force due to the earth's rotation. The magnitude of this force per unit mass, a_{CO}, is given by

$$a_{CO} = 2 \cdot \omega \cdot \sin \Lambda \cdot U, \tag{7.30}$$

where ω is the angular velocity of the earth's rotation (7.27×10^{-5} s^{-1}), and Λ is latitude.

The Coriolis force is always present and acts perpendicularly to the velocity, to the right (left) in the Northern (Southern) Hemisphere. The vector addition of the Coriolis force to the downstream force results in a deflection that affects the magnitude as well as the direction of flow (figure 7.2); this apparent force tends to make the flow follow a curved path and hence adds to the flow resistance.

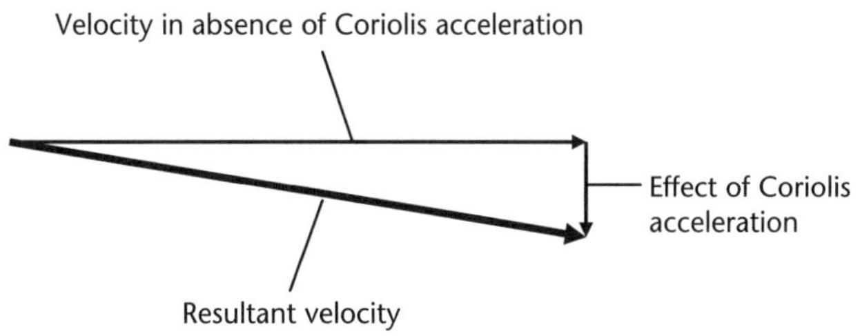

Figure 7.2 Vector diagram showing effect of Coriolis force on velocity direction and magnitude in the northern hemisphere. The magnitude of the force depends on the latitude and the velocity (equation 7.30). The Coriolis force acts to the left in the Southern Hemisphere.

7.3.3.2 Centrifugal Force

As discussed in section 6.6.1.2 (equation 6.29), a mass of water traveling in a curved channel is subject to a centrifugal acceleration a_c,

$$a_c = \frac{U^2}{r_c}, \tag{7.31}$$

where r_c is the radius of curvature of the channel (see figure 6.11b). Since this acceleration tends to cause a deviation from flow in a straight-line path, the water is subject to an oppositely directed centrifugal force that is an addition to the resisting forces.

Equation 7.31 accounts for the resistance that arises because the entire mass of water is flowing in a curved path. In a stream bend, additional resistance arises due to the velocity distribution: Faster-flowing water near the surface is subject to a higher centrifugal acceleration than is slower water near the bottom, and this sets up a secondary circulation as described in section 5.4.2.2 (see figure 5.21). Some of the driving force must be used to sustain this circulation, and thus it contributes to the flow resistance.

Chang (1988) presented a formula derived by Rozovskii (1957) for computing the force per unit mass diverted to maintaining the circulation, a_{CC}:

$$a_{CC} = \left(12 \cdot \Omega + 30 \cdot \Omega^2\right) \cdot \left(\frac{Y}{r_c}\right) \cdot \left(\frac{U^2}{r_c}\right). \tag{7.32}$$

Incorporating this relation, the total force per unit mass involved in flow in a curved reach, a_C, is

$$a_C = a_c + a_{CC} = \left[1 + (12 \cdot \Omega + 30 \cdot \Omega^2) \cdot \left(\frac{Y}{r_c}\right)\right] \cdot \left(\frac{U^2}{r_c}\right). \tag{7.33}$$

7.3.4 Accelerations

Here we formulate the expressions for the convective and local accelerations in the overall force balance of equation 7.5. Following the developments in section 4.2.1.2 for fluid elements, note that velocity is a function of the spatial dimension X and time, t, so

$$U = f(X, t). \tag{7.34}$$

From the rules of differentials, equation 7.34 implies that

$$\mathrm{d}U = \frac{\partial U}{\partial X} \cdot \mathrm{d}X + \frac{\partial U}{\partial t} \cdot \mathrm{d}t. \tag{7.35}$$

Acceleration, a, is defined as $\mathrm{d}U/\mathrm{d}t$, and if we divide equation 7.35 by $\mathrm{d}t$ and note that $U \equiv \mathrm{d}X/\mathrm{d}t$, we have

$$a \equiv \frac{\mathrm{d}U}{\mathrm{d}t} = \frac{\partial U}{\partial X} \cdot \frac{\mathrm{d}X}{\mathrm{d}t} + \frac{\partial U}{\partial t} \cdot \frac{\mathrm{d}t}{\mathrm{d}t} = U \cdot \frac{\partial U}{\partial X} + \frac{\partial U}{\partial t}. \tag{7.36}$$

7.3.4.1 Convective Acceleration

The first term on the right-hand side of equation 7.36 is the convective acceleration, a_X:

$$a_X \equiv U \cdot \frac{\partial U}{\partial X}, \tag{7.37a}$$

which in the macroscopic context is

$$a_X \equiv U \cdot \frac{\Delta U}{\Delta X}, \tag{7.37b}$$

where $\Delta U \equiv U_2 - U_1$.

7.3.4.2 Local Acceleration

The second term on the right-hand side of equation 7.36 is the local acceleration, a_t:

$$a_t \equiv \frac{\partial U}{\partial t}, \tag{7.38a}$$

which we can write for a finite time interval Δt as

$$a_t \equiv \frac{\Delta U}{\Delta t}. \tag{7.38b}$$

7.4 Overall Force Balance and Velocity Relations

Note that the resisting and perpendicular forces are functions either of U or U^2, so we can rewrite the overall force balance of equation 7.5 as

$$(a_G + a_P) - [a_V(U) + a_T(U^2) + a_{CO}(U) + a_C(U^2)] = a_X(U) + a_t. \tag{7.39}$$

Thus, for steady flows ($a_t = 0$), the force-balance relation can be written as a quadratic equation in U:

$$(a_T' + a_C') \cdot U^2 + (a_{CO}' + a_V' + a_X') \cdot U - (a_G + a_P) = 0, \tag{7.40}$$

where the primes indicate the respective accelerations divided by U^2 or U (e.g., $a_T' \equiv a_T/U^2$; $a_V' \equiv a_V/U$). The solution to equation 7.40 can be found via the quadratic formula:

$$U = \frac{-(a_{CO}' + a_V' + a_X') \pm \left[(a_{CO}' + a_V' + a_X')^2 + 4 \cdot (a_T' + a_C') \cdot (a_G + a_P)\right]^{1/2}}{2 \cdot (a_T' + a_C')} \tag{7.41}$$

Equation 7.41 states that average velocity is a somewhat cumbersome expression involving the terms listed in table 7.1. However, we can show that the solutions to equation 7.40 are consistent with the expressions for the mean velocities of uniform ($a_P = 0$; $a_X' = 0$) laminar and turbulent flows developed in chapters 5 and 6 if we restrict our attention to straight flows ($a_C' = 0$) and ignore the Coriolis acceleration ($a_{CO}' = 0$).[3] Then, equation 7.40 can be simplified to

$$a_T' \cdot U^2 + a_V' \cdot U - a_G = 0. \tag{7.42}$$

For laminar flows, $a_T{}' = 0$, and substituting equations 7.14 and 7.25 into equation 7.42 yields

$$3 \cdot \nu \cdot \frac{U}{Y^2} - g \cdot S_0 = 0, \tag{7.43a}$$

$$U = \left(\frac{g}{3 \cdot \nu}\right) \cdot Y^2 \cdot S_0, \tag{7.43b}$$

which is identical to equation 5.14. For turbulent flows, $a_V{}' \ll a_T{}'$, and substituting equations 7.14 and 7.28 into equation 7.42 yields

$$\Omega^2 \cdot \frac{U^2}{Y} - g \cdot S_0 = 0, \tag{7.44a}$$

$$U = \Omega^{-1} \cdot (g \cdot Y \cdot S_0)^{1/2}, \tag{7.44b}$$

which is identical to the Chézy equation (equation 6.19).

7.5 Magnitudes of Forces in Natural Streams

7.5.1 Database

In this section we use measurements made on a sample of natural stream reaches to explore typical values of the force-magnitude terms derived in section 7.4. The data are from Barnes (1967), who presented measurements of channel geometry and velocity for 61 flows in 51 natural river reaches in the United States. A total of 242 cross sections were surveyed; these data can be used to compute the magnitudes of the forces for 181 subreaches. Table 7.2 summarizes the range of channel sizes included, and table 7.3 gives an example of the data presentation. Although these

Table 7.2 Summary of range of flow parameters in the 181 subreaches measured by Barnes (1967).

	Discharge (m^3/s)	Width (m)	Depth (m)	Velocity (m/s)	Surface slope	Channel slope
Maximum	28,300	529	16.4	3.33	4.05×10^{-2}	9.38×10^{-2}
Median	69.7	34.7	1.84	1.78	2.34×10^{-3}	3.12×10^{-2}
Minimum	1.84	6.75	0.27	0.16	1.58×10^{-4}	-4.74×10^{-2}

Table 7.3 Example of stream reach data from Barnes (1967).[a]

Section	Area (m^2)	Top width (m)	Mean depth (m)	Hydraulic radius (m)	Mean velocity (m/s)	Distance between sections (m)	Fall[b] between sections (m)
1	230.5	68.3	3.38	3.31	2.79		
2	229.6	69.5	3.29	3.23	2.80	94.8	0.229
3	226.8	72.3	3.14	3.06	2.84	99.1	0.229

[a]U.S. Geological Survey station 12-4570, Wenatchee River at Plain, Washington. Flood of 12 May 1948; discharge $Q = 643$ m^3/s. Bed is boulders; $d_{50} = 162$ mm, $d_{84} = 320$ mm; banks are lined with trees and bushes.
[b]Decrease in water-surface elevation.

data certainly do not cover the full range of stream types and sizes, they do provide some quantitative feeling for the absolute and relative magnitudes of forces likely to be encountered in natural streams. (These data are accessible via the Internet, as described in appendix B.)

7.5.2 Driving Forces

Figure 7.3 shows the distribution of gravitational force per unit mass, a_G, values for the Barnes (1967) data. Note that about 25% of the subreaches have a negative value, indicating an adverse slope. More than 80% of the values are less than 0.1 m/s^2, corresponding to channel slopes less than 0.01. The maximum value was 0.92 m/s^2, corresponding to a channel slope of 0.094.

The distribution of pressure force per unit mass for the Barnes data is shown in figure 7.4. Note that there are equal values of positive (depth decreases downstream) and negative (depth increases downstream) values. The pressure force is typically in the range from 0.01 to 0.1 m/s^2, and all but a handful of values are less than 0.2 m/s^2.

The magnitudes of the two driving forces are compared in figure 7.5. In about 73% of the subreaches, gravitational-force magnitude exceeded pressure-force magnitude. However, the ratio $|a_P/a_G|$ ranged from less than 0.1 to more than 10. Clearly, pressure forces are generally significant in natural channels; or, stated another way, natural-channel reaches are generally significantly nonuniform.

7.5.3 Resisting Forces

For the 181 subreaches in the Barnes (1967) data, the value of the viscous force per unit mass a_V ranges from 5.31×10^{-7} m/s^2 to 1.49×10^{-5} m/s^2, with a median value of 2.91×10^{-6} m/s^2. These values are several orders of magnitude smaller than a_G and a_P shown in figures 7.3 and 7.4.

Figure 7.6 shows the distribution of turbulent resisting forces in the Barnes (1967) sample: The values of a_T extend over several orders of magnitude, ranging from 3.29×10^{-3} m/s^2 to 4.81 m/s^2. The distribution of the ratio of turbulent to viscous resisting forces is shown in figure 7.7, confirming that viscosity plays a negligible role in the force balance of natural open channels.

7.5.4 Perpendicular Forces

7.5.4.1 Coriolis Force

The reaches measured by Barnes (1967) were at latitudes ranging from about 33° N to 47° N. We can get a sense of the magnitude of the Coriolis acceleration by assuming a latitude of 40° for all sites and using the measured velocities to compute a_{CO} via equation 7.30. We find that a_{CO} ranges from 4.84×10^{-5} m/s^2 to 3.7×10^{-4} m/s^2, orders of magnitude smaller than the principal driving and resisting forces. If we had assumed the maximum possible latitude of 90°, the maximum value would have risen to only 5.8×10^{-4} m/s^2; thus, we can conclude that the Coriolis force can be neglected in the force balance of natural open channels.

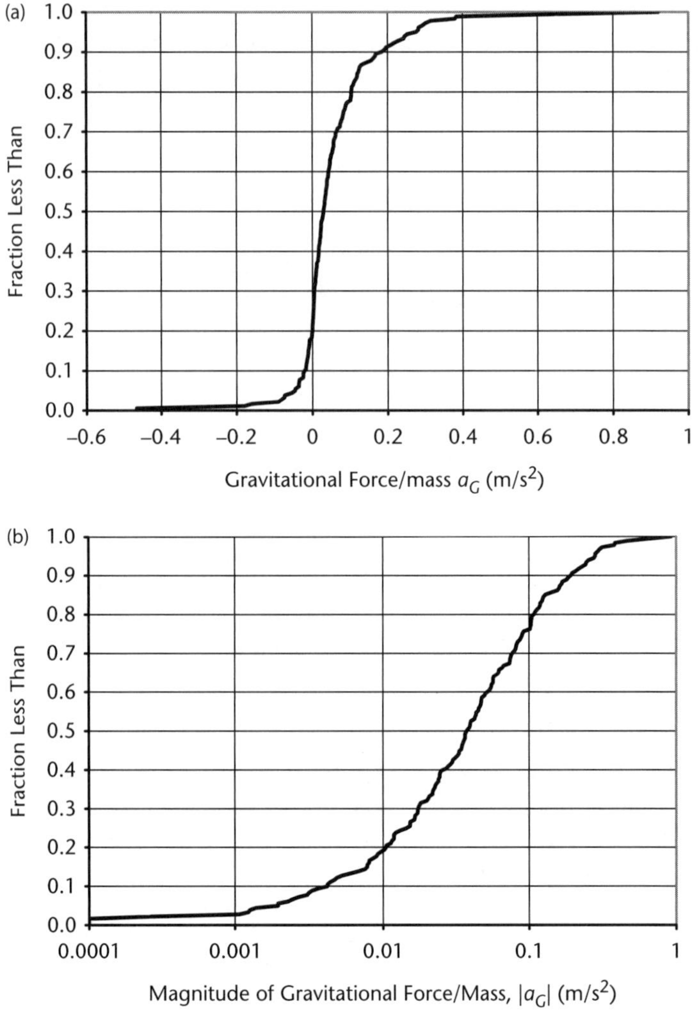

Figure 7.3 (a) Cumulative distribution of gravitational force per unit mass, a_G, and (b) cumulative distribution of absolute value of gravitational force per unit mass, $|a_G|$ (note logarithmic scale), for 181 natural-stream reaches measured by Barnes (1967).

7.5.4.2 Centrifugal Force

The reaches measured by Barnes (1967) were fairly straight. However, we can get a feel for the potential magnitude of centrifugal force likely to be encountered in natural channels by assuming that the channels were curved and using equation 7.33. As noted in section 2.2.3, meander radii of curvature r_c are typically about 2.3 times the channel width, so we use that value in calculating a_C. The distribution of values is shown in figure 7.8; almost all the values are between 0.01 m/s^2 and 1 m/s^2. Figure 7.9 shows the ratio of centrifugal to turbulent forces; a_C values tend to be somewhat smaller

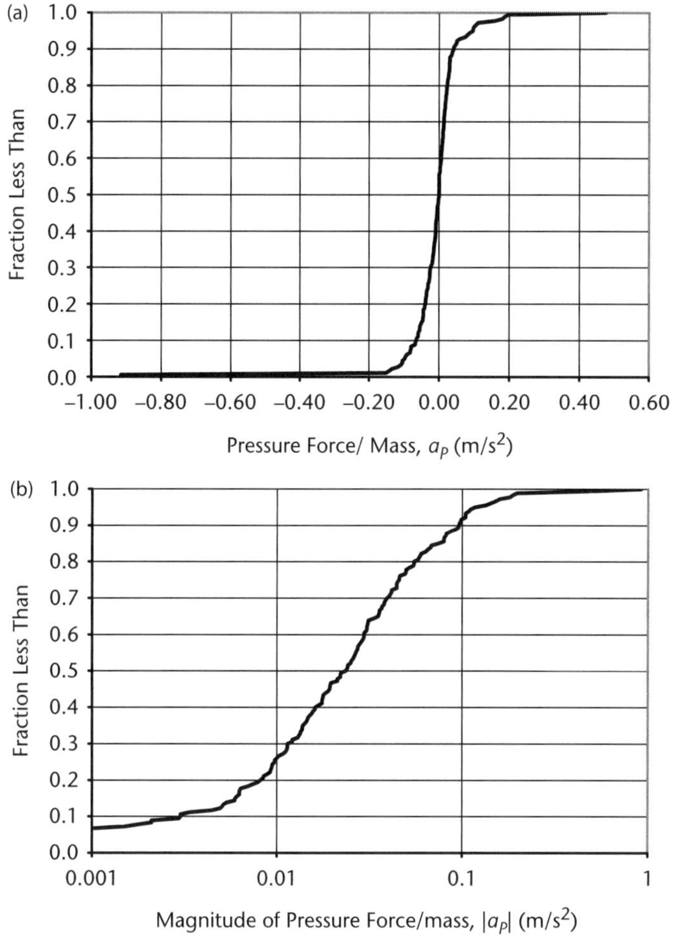

Figure 7.4 (a) Cumulative distribution of pressure force per unit mass, a_P, and (b) cumulative distribution of absolute value of pressure force per unit mass, $|a_P|$ (note logarithmic scale), for 181 natural-stream reaches measured by Barnes (1967).

than a_T values but are generally of similar magnitude. Hence, we conclude that centrifugal forces are generally a significant addition to resistance in typical curved (meandering) channels. This was also the conclusion of the laboratory experiments described in section 6.6.1.2 (see box 6.3).

7.5.5 Accelerations

7.5.5.1 Convective Acceleration

The data of Barnes (1967) can be used to compute the convective acceleration through each subreach via equation 7.37. The distribution of these values is shown

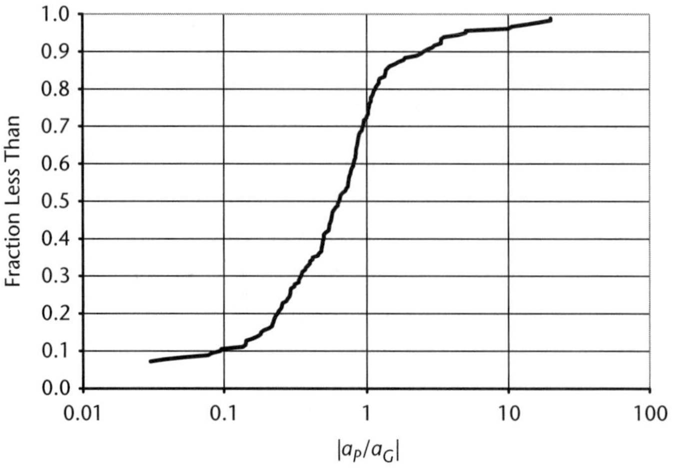

Figure 7.5 Cumulative distribution of the absolute value of the ratio of pressure force to gravitational force, $|a_P/a_G|$, for 181 natural-stream reaches measured by Barnes (1967). Note the logarithmic scale.

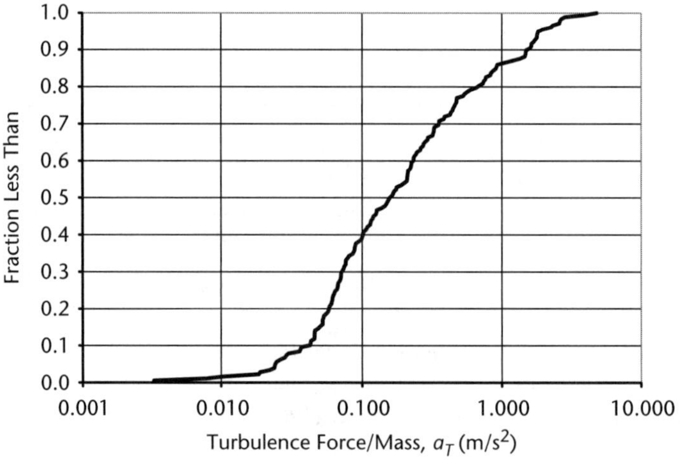

Figure 7.6 Cumulative distribution of turbulence force per unit mass, a_T, for 181 natural-stream reaches measured by Barnes (1967). Note the logarithmic scale.

in figure 7.10. The absolute value of the convective acceleration $|a_X|$ in natural rivers is typically in the range from 0.0001 m/s^2 to 0.01 m/s^2, with a median value near 0.001 m/s^2. Figure 7.11 shows that the ratio of convective to gravitational acceleration $|a_X/a_G|$ is usually in the range from 0.005 to 0.5, with a median value of about 0.05. Thus, although we concluded that most natural reaches are significantly nonuniform, it appears that convective acceleration can often—but certainly not always—be neglected in the force balance of natural river reaches.

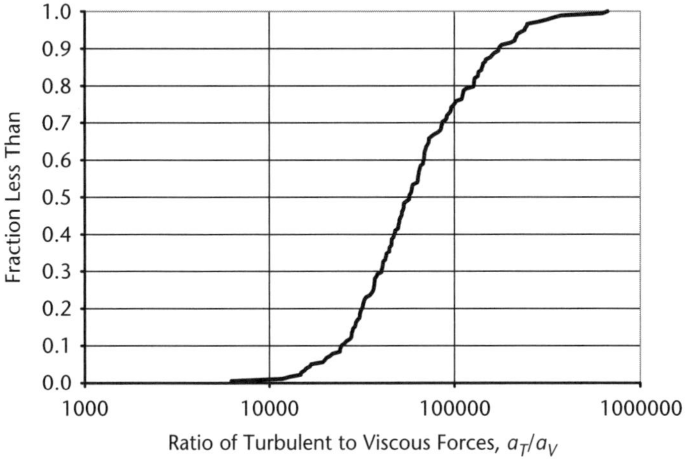

Figure 7.7 Cumulative distribution of the ratio of turbulent to viscous forces, a_T/a_V, for the 181 subreaches measured by Barnes (1967). As shown in section 7.6.1, this ratio is equal to the Reynolds number, *Re*. Note the logarithmic scale.

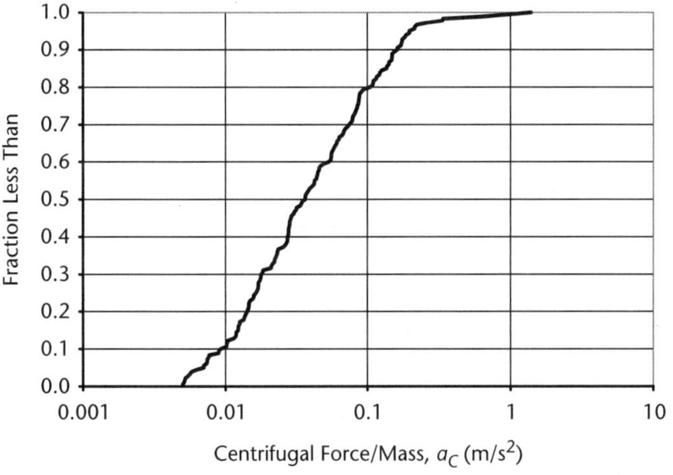

Figure 7.8 Cumulative distribution of typical centrifugal force per unit mass, a_C, (calculated by assuming that radius of curvature is 2.3 times width) for 181 natural-stream reaches measured by Barnes (1967). Note the logarithmic scale.

7.5.5.2 Local Acceleration

The value of local acceleration a_t depends on the local rapidity of response to streamflow-generating events in the drainage basin—rain and snowmelt events or the breaching of natural or artificial dams—and thus is difficult to generalize. The Barnes (1967) data cannot be used to calculate changes with time, so to get a feeling for

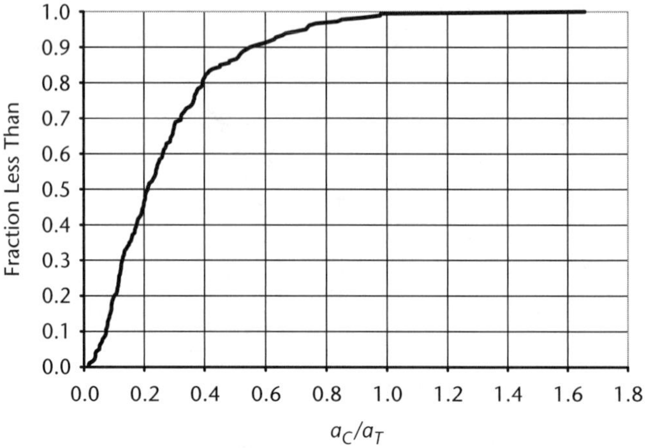

Figure 7.9 Cumulative distribution of the ratio of typical centrifugal force (calculated by assuming that radius of curvature is 2.3 times width) to turbulent force, a_C/a_T, for 181 natural-stream reaches measured by Barnes (1967).

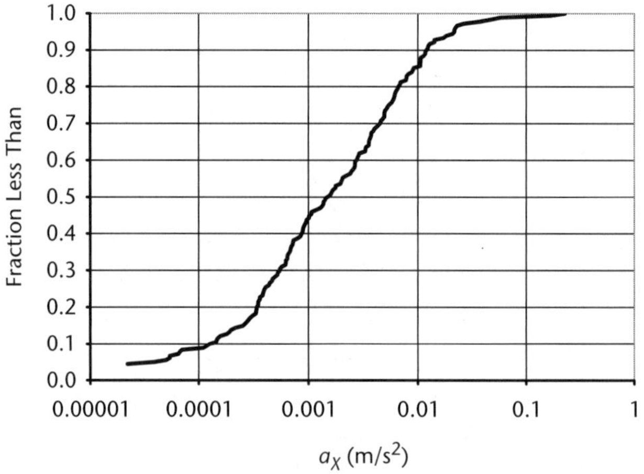

Figure 7.10 Cumulative distribution of the magnitude of convective acceleration, $|a_X|$, for 181 natural-stream reaches measured by Barnes (1967). Note the logarithmic scale.

the magnitude of a_t, we examine the response of the Diamond River near Wentworth Location, New Hampshire, to a large rainstorm (figure 7.12).

At the gaging station, the Diamond River drains an area of 153 mi^2 (395 km^2). On 23 July 2004, the discharge increased rapidly from 82 ft^3/s (2.3 m^3/s) to 910 ft^3/s (25.8 m^3/s) in a period of 7.3 h (26,280 s). We can evaluate the change in velocity accompanying this response from the relation between average velocity and discharge established as part of the at-a-station hydraulic geometry relations (section 2.6.3) for this location; this relation is shown in figure 7.13. In response to the increase in

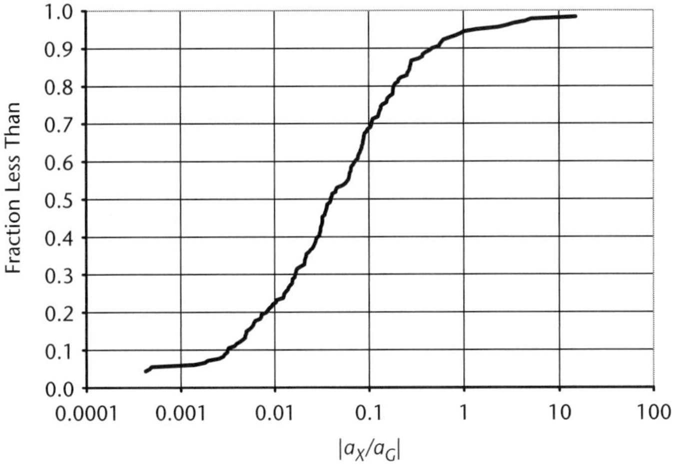

Figure 7.11 Cumulative distribution of the absolute value of the ratio of convective acceleration to gravitational force, $|a_X/a_G|$, for 181 natural-stream reaches measured by Barnes (1967). Note the logarithmic scale.

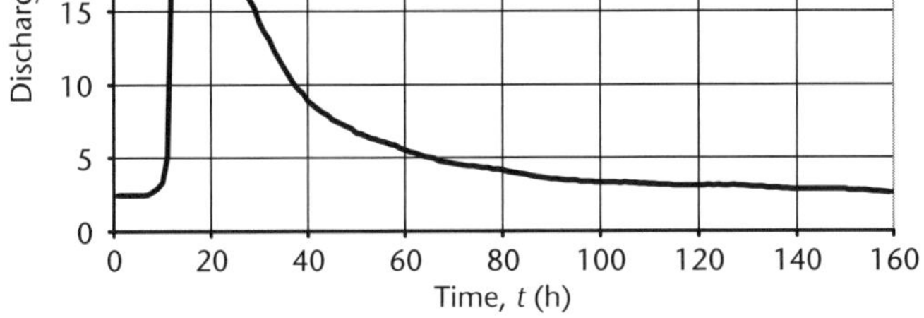

Figure 7.12 Discharge hydrograph of the Diamond River near Wentworth Location, New Hampshire, from 08:00 23 July to 24:00 31 July 2004 showing very rapid increase in discharge in response to a rainstorm.

discharge, velocity increased from 0.26 m/s to 0.82 m/s, so the local acceleration was $a_t = (0.82 - 0.26)/26{,}280 = 2.1 \times 10^{-5}$ m/s^2.

Although this is only one case, the increase in discharge was quite rapid, yet the local acceleration was several orders of magnitude smaller than the typical values of gravitational, pressure, and turbulent forces as calculated for the Barnes (1967) database. Thus, we conclude that local acceleration is typically several orders of

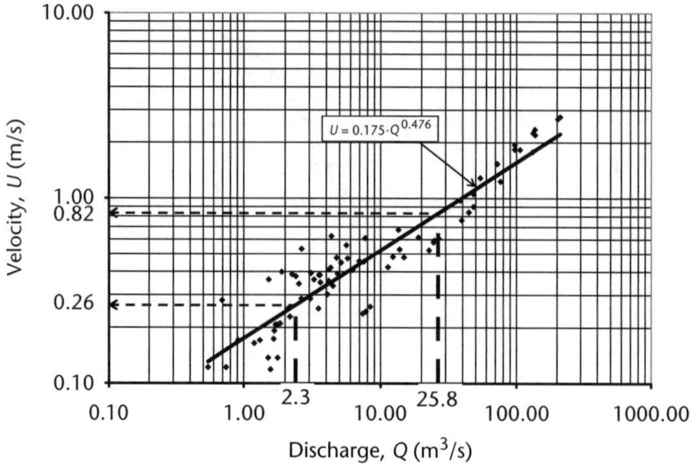

Figure 7.13 At-a-station hydraulic geometry relation between average velocity, U, and discharge, Q, for the Diamond River near Wentworth Location, New Hampshire: $U = 0.175 \cdot Q^{0.476}$. The change in discharge from 2.3 m^3/s to 25.8 m^3/s on 23 July 2004 was accompanied by a change in velocity from 0.26 m/s to 0.82 m/s.

magnitude less than other forces and can often be neglected. Or, stated another way, natural stream flows can often be considered approximately steady.

However, it is important to include the local acceleration when characterizing the movement of steep flood waves through channels—especially those generated by dam breaks, which can involve very rapid velocity changes. We examine the modeling of unsteady flows in chapter 11.

7.5.6 Summary of Force Magnitudes

The ranges of the magnitudes of the forces and convective acceleration computed for the Barnes (1967) database are summarized in figure 7.14. The probable range of values of local accelerations is also shown.

7.5.7 Forces as a Function of Scale

A principal motivation for developing expressions for the magnitudes of various forces is to explore how the relative importance of the forces changes with the spatial scale of the flow. To do this, we tabulate some "typical" values of width, depth, velocity, slope, Reynolds number, and resistance for flows ranging from laboratory flumes to the Gulf Stream (table 7.4, figure 7.15) Depth increases by several orders of magnitude along with width, and this produces a strong increasing trend in Reynolds number. Resistance is calculated by assuming a smooth flow and using equation 6.23 for the flume and the Gulf Stream, and by assuming a rough flow and using equation 6.24 with $y_r = 2$ mm for the stream flows; its decreasing trend is due to the increasing Reynolds numbers and relative smoothness as depth increases

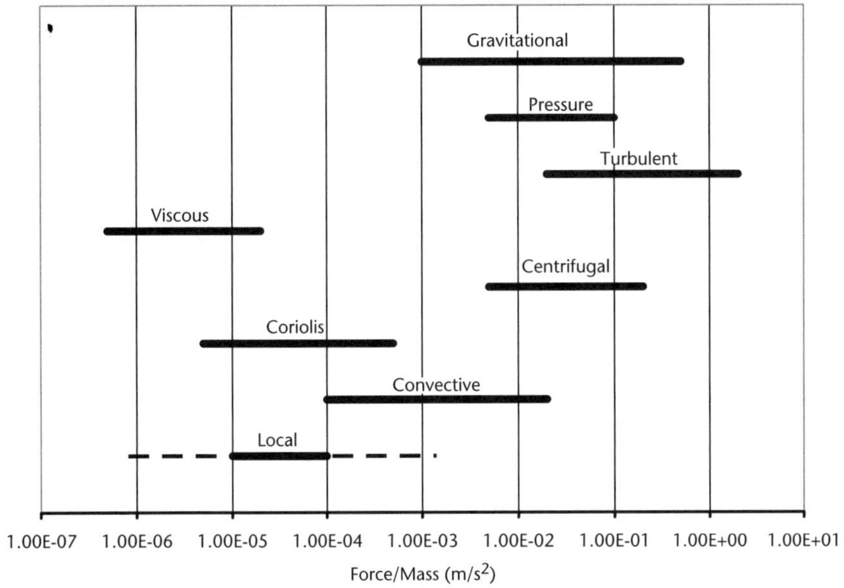

Figure 7.14 Range of values of forces per unit mass (accelerations) typical of natural channels as calculated for the Barnes (1967) data. The probable range of local accelerations is also shown.

Table 7.4 Typical values of flow parameters used to calculate forces over a range of spatial scales.[a]

Flow	Width, W (m)	Depth, Y (m)	Velocity, U (m/s)	Slope, S_0	Reynolds number, Re	Resistance, Ω
Small flume	0.22	0.03	0.28	6.1×10^{-3}	5.8×10^{3}	0.057
Large flume	0.76	0.07	0.41	4.3×10^{-2}	2.3×10^{4}	0.048
Small stream	2	0.1	0.5	1.0×10^{-2}	3.8×10^{4}	0.064
Medium river	10	0.5	1	3.8×10^{-3}	3.8×10^{5}	0.051
Large river	100	5	1.5	8.8×10^{-4}	5.7×10^{6}	0.039
Larger river	500	25	2	3.2×10^{-4}	3.8×10^{7}	0.034
Gulf Stream	50,000	700	2	1.4×10^{-5}	1.1×10^{9}	0.012

[a]See figure 7.15 for plot of values; see section 7.5.7 for details.

(see figure 6.8). Typically, river slopes decrease with width, while velocity increases slightly.

The values in table 7.4 are used in the equations of table 7.1 to calculate the various forces per unit mass. The results are summarized in table 7.5 and figure 7.16, but before examining them, we should note the following:

1. Pressure force and acceleration is not shown. This is discussed further in the following sections.

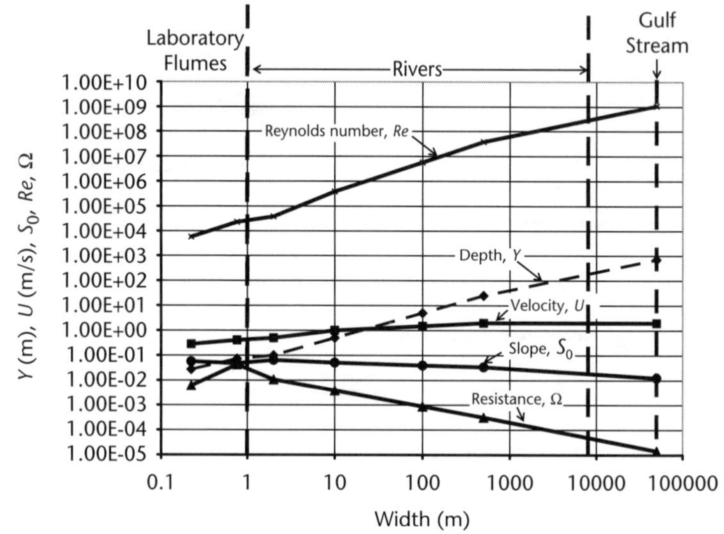

Figure 7.15 Trends in depth, velocity, slope, Reynolds number, and resistance over the spatial scale (width) of flumes and natural open-channel flows. For data, see table 7.4.

Table 7.5 Forces per unit mass (m/s^2) in flows of various scales calculated from values in table 7.4.[a]

Flow	a_G	a_V	a_T	a_{CO}	a_C
Small flume	6.0×10^{-2}	1.5×10^{-3}	9.4×10^{-3}	3.5×10^{-5}	1.5×10^{-1}
Large flume	4.2×10^{-1}	3.0×10^{-4}	5.2×10^{-3}	5.1×10^{-5}	9.4×10^{-2}
Small stream	1.0×10^{-1}	2.0×10^{-4}	1.0×10^{-2}	6.2×10^{-5}	5.3×10^{-2}
Medium river	3.5×10^{-2}	1.6×10^{-5}	5.2×10^{-3}	1.2×10^{-4}	4.2×10^{-2}
Large river	8.6×10^{-3}	2.4×10^{-7}	7.0×10^{-4}	1.9×10^{-4}	9.5×10^{-3}
Larger river	3.1×10^{-3}	1.3×10^{-8}	1.8×10^{-4}	2.5×10^{-4}	3.4×10^{-3}
Gulf Stream	1.4×10^{-4}	1.6×10^{-11}	8.8×10^{-7}	2.5×10^{-4}	3.3×10^{-5}

[a]Coriolis forces are calculated for latitude 45°. Centrifugal forces are calculated by assuming that the radius of curvature equals 2.3 times the width. Flows in flumes and Gulf Stream assumed hydraulically smooth; flows in streams and rivers assumed hydraulically rough with $y_r = 2$ mm. See section 7.5.7 for other details.

2. The viscous force is calculated via equation 7.25 assuming the kinematic viscosity at 10°C. Note from table 3.4 that this value could be considerably larger or smaller depending on temperature.
3. The turbulent force is calculated via equation 7.28 using the value of resistance shown in table 7.4. This value can vary by an order of magnitude due to variations in resistance.
4. The Coriolis force is calculated via equation 7.30 for latitude 45°; this force varies from zero at the equator to $2 \cdot \omega \cdot U = 1.5 \times 10^{-4} \cdot U$ m s^{-2} at the poles.

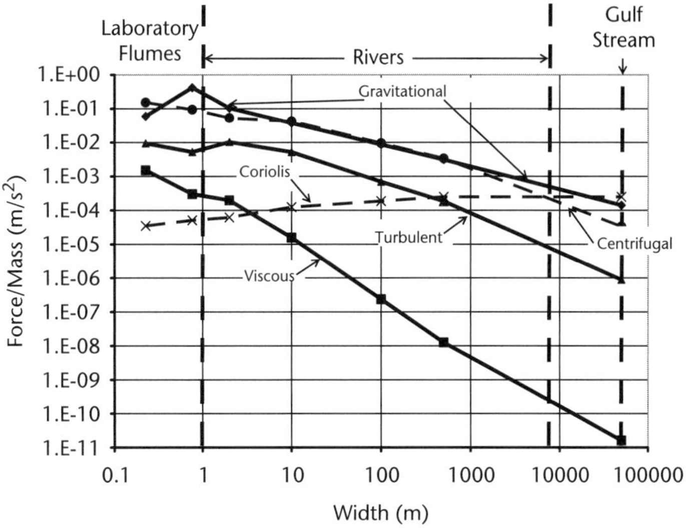

Figure 7.16 Magnitudes of gravitational, viscous, turbulent, Coriolis, and centrifugal forces per unit mass as a function of flow scale (width) computed using expressions in table 7.1 and representative values in table 7.4. See text for discussion.

5. The centrifugal force is calculated via equation 7.33, assuming that the radius of curvature equals 2.3 times the width (a typical value for river meanders, as discussed in section 6.6.1.2). This force of course equals zero in straight channels and could be somewhat higher than the value in table 7.5 in highly sinuous reaches.

Because of the above considerations, the values in table 7.5 and figure 7.16 should be taken only as very general indications of the relative force values for flows of different scales. However, these values are instructive; note the following important generalities:

1. Gravitational force is usually the largest force in all flows. However, it can be exceeded by the pressure force, as shown in section 7.5.2.
2. Centrifugal force can be of the same order of magnitude as gravitational force.
3. Turbulent resisting force is orders of magnitude larger than viscous resisting force, and the difference between the two increases with flow scale. Turbulence is usually the main resisting force and viscous force can be neglected in most (but not all, as discussed in section 5.1) natural open-channel flows.
4. Coriolis force is orders of magnitude less than gravitational and turbulent force and therefore has no influence on river flows, except perhaps in the very largest rivers. It is of the same order as the gravitational force for the Gulf Stream and other ocean currents, and hence causes the paths of these flows to curve to the right (left) in the Northern (Southern) Hemisphere.

7.6 Force Ratios and the Reynolds and Froude Numbers

7.6.1 The Reynolds Number

The Reynolds number, *Re*, where

$$Re \equiv \frac{U \cdot Y}{\nu}, \tag{7.45}$$

was introduced in section 3.4, where this quantity was shown to be proportional to the ratio of eddy viscosity to molecular viscosity. We can show that it is also proportional to the ratio of turbulent force to viscous force, a_T/a_V, by referring to equations 7.25 and 7.28 and writing

$$\frac{a_T}{a_V} = \frac{\frac{\Omega^2 \cdot \rho \cdot U^2}{Y}}{\frac{3 \cdot \mu \cdot U}{Y^2}} = \frac{\Omega^2 \cdot \rho \cdot U \cdot Y}{3 \cdot \mu} = \frac{\Omega^2 \cdot U \cdot Y}{3 \cdot \nu} = \left(\frac{\Omega^2}{3}\right) \cdot Re. \tag{7.46}$$

Thus, we see that the Reynolds number is proportional to the ratio of turbulent resisting force to viscous resisting force as well as to the ratio of eddy viscosity to molecular viscosity.

Recall that the transition from laminar to turbulent flow takes place when $Re \approx 500$. One might reason on physical grounds that this transition should occur when $a_T/a_V \approx 1$. To see if this is true, we substitute $Re = 500$ and a typical value of $\Omega = 0.07$ (see figure 6.8) into equation 7.46. Solving this gives $a_T/a_V = 0.82$, which is close to 1. This confirms our reasoning and we conclude that a Reynolds number of 500 represents a near equality of turbulent and viscous resisting forces and a near equality of eddy viscosity and molecular viscosity.

7.6.2 The Froude Number

The Froude number, *Fr*, where

$$Fr \equiv \frac{U}{(g \cdot Y)^{1/2}}, \tag{7.47}$$

was introduced in section 6.2.2.2 (equation 6.5) as the ratio of flow velocity to the celerity of a surface wave in shallow water. The Froude number can also be related to the ratio of turbulent to total driving force:

$$\frac{a_T}{a_D} = \frac{\frac{\Omega^2 \cdot U^2}{Y}}{g \cdot S_S} = \frac{\Omega^2 \cdot U^2}{S_S \cdot g \cdot Y} = \left(\frac{\Omega^2}{S_S}\right) \cdot Fr^2 \tag{7.48}$$

In a uniform turbulent flow, $a_T \approx a_D$, $a_T/a_D \approx 1$, so

$$Fr \approx \frac{S_S^{1/2}}{\Omega}. \tag{7.49}$$

As noted above, a typical value of Ω is 0.07, and a typical value of S_S is 0.0023 (table 7.2). Substituting these values into equation 7.49 and solving yields $Fr = 0.68$. Figure 7.17 shows the distribution of Froude numbers in the 181 subreaches in the Barnes (1967) database; it shows that *Fr* values in natural rivers are in this general range, though usually somewhat less than 0.68 and almost always less than 1.

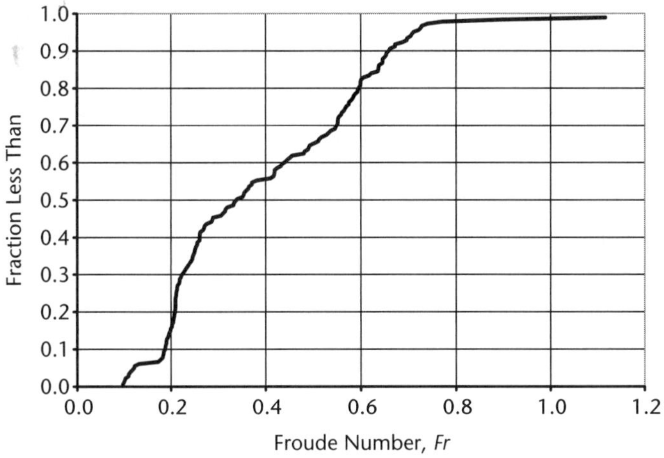

Figure 7.17 Cumulative distribution of Froude numbers for the 181 subreaches measured by Barnes (1967).

7.7 Summary

We have identified six forces that act on water and thus determine its acceleration. We have derived expressions that can be used to calculate the magnitudes of each of these forces per unit mass in a macroscopic, one-dimensional formulation and have shown the typical ranges of these forces, their relative magnitudes, and how their relative magnitudes tend to change as a function of flow size (scale).

The total motion-inducing (driving) force is the sum of the gravitational and pressure forces. The **gravitational force** is proportional to the sine of the bottom slope, the **pressure force** is proportional to the spatial rate of change of depth, and the **total driving force** is proportional to the water-surface slope. In natural channels, the pressure force is typically of the same order of magnitude as the gravitational force.

Once motion begins, forces that are functions of the velocity arise to resist the motion. Two of these resisting forces arise from boundary friction: the viscous and turbulent force. The **viscous force** (proportional to the molecular viscosity and the first power of the velocity) is present in all flows but is overwhelmed by the **turbulent force** (proportional to the channel resistance and the second power of the velocity) in almost all natural rivers.

Flows are described in a nonrotating coordinate system, but because the earth rotates, all flows are affected by the **Coriolis pseudoforce** (proportional to the velocity and the sine of the latitude). This deflecting force adds to the forces resisting the flow; however, it is very small relative to the driving and frictional resisting forces and can be neglected in all but the very largest rivers. In curved channels another "pseudoforce," the **centrifugal force** (proportional to the second power of the velocity and inversely proportional to the radius of curvature), adds to the resisting forces because the flow paths are not straight lines: The mass of the flow follows the curved path of the channel, and the water within the flow follows a spiral path.

The difference between the driving and resisting forces is acceleration. **Convective acceleration** (spatial change in velocity) occurs in most natural reaches due to changes in channel geometry, but is often of negligible magnitude. Processes in a river's watershed may cause a temporal change in discharge and hence velocity (unsteady flow); this is **local acceleration**. Local acceleration itself is usually of negligible magnitude, but the propagation of temporal changes through a river channel produces spatial changes in discharge and velocity (and other parameters) and thus is always accompanied by convective acceleration.

We saw that the force-balance relations derived here reduce to the velocity relations derived for steady uniform laminar and turbulent flows described in preceding chapters. We also saw that the Reynolds number can be interpreted as the ratio of turbulent to viscous resisting forces, and that the Froude number is related to the ratio of turbulent to driving forces.

8

Energy and Momentum Principles

8.0 Introduction and Overview

The momentum and energy principles for a fluid element were introduced in sections 4.4 and 4.5, respectively. Here, we integrate those principles across a channel reach to apply to macroscopic one-dimensional steady flows. We conclude the chapter by comparing the theoretical and practical differences between the energy and momentum principles. We show in subsequent chapters how these energy and momentum relations can be applied to solve practical problems.

8.1 The Energy Principle in One-Dimensional Flows

Section 4.5 established the laws of mechanical energy for a fluid element. We saw (equation 4.41) that the total energy $\hbar$ of an element is the sum of its potential energy $\hbar_{PE}$ and its kinetic energy $\hbar_{KE}$:

$$\hbar = \hbar_{PE} + \hbar_{KE} \tag{8.1}$$

We also saw that its potential energy consists of gravitational potential energy $\hbar_G$ and pressure potential energy, $\hbar_P$:

$$\hbar_{PE} = \hbar_G + \hbar_P. \tag{8.2}$$

In equations 8.1 and 8.2, the energy quantities are expressed as energy [F L] divided by weight [F], which is called **head** [L].

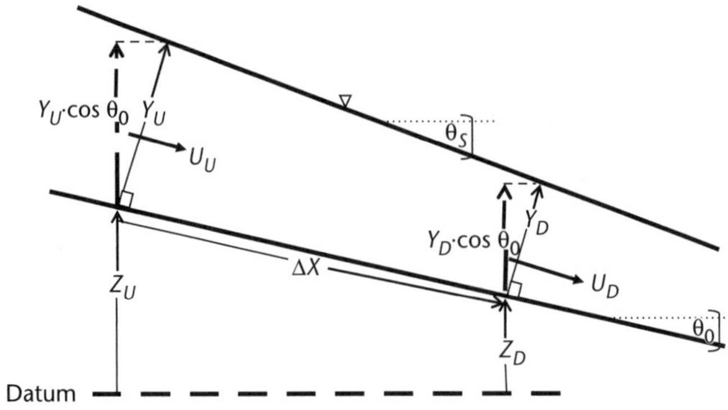

Figure 8.1 Definition diagram for derivation of the macroscopic one-dimensional energy equation.

8.1.1 The Energy Equation

Figure 8.1 defines the geometry of the wide rectangular channel reach with constant width W that we will use to formulate the macroscopic energy relations. The flow through the reach is steady with constant discharge Q. Although the geometry of figure 8.1 is simple, the relations we derive are general; that is, they apply to steady flows in nonprismatic channels also.

8.1.1.1 Total Mechanical Energy at a Cross Section

We saw in section 4.5.1 that the gravitational potential head for a fluid element equals its elevation above a datum (equation 4.33) and that, assuming a hydrostatic pressure distribution, the pressure potential head equals its distance below the water surface (equation 4.34). Thus, the total potential head has the same value at all elements and at all points in a cross section. For convenience, we choose the channel bottom as our reference point so that for cross section i we can write the integrated **gravitational head** (also called **elevation head**), $\mathcal{H}_{Gi}$, as

$$\mathcal{H}_{Gi} = Z_i, \tag{8.3}$$

where Z_i is the elevation of the channel bottom, and the integrated **pressure head**, $\mathcal{H}_{Pi}$, as

$$\mathcal{H}_{Pi} = Y_i \cdot \cos \theta_0, \tag{8.4}$$

where Y_i is the flow depth and θ_0 is the channel slope (see equation 4.13b).

We saw in equation 4.40 that the kinetic energy head $\hbar_{KE}$ for a fluid element with velocity u is given by

$$\hbar_{KE} = \frac{u^2}{2 \cdot g}, \tag{8.5}$$

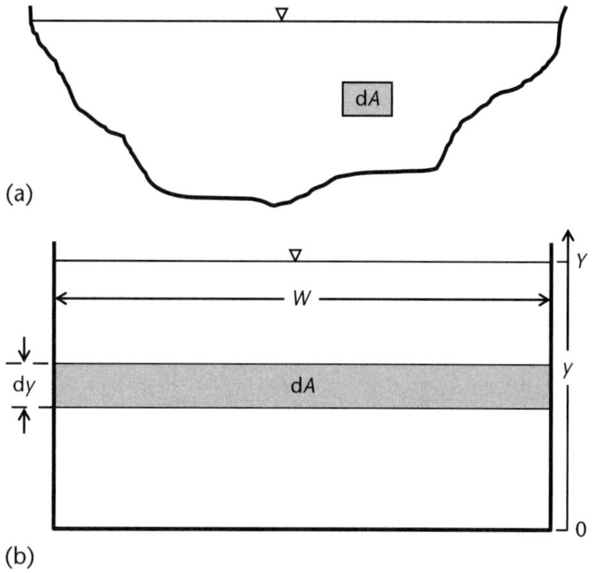

Figure 8.2 Definition diagrams for deriving expressions for deriving and evaluating the energy coefficient α and the momentum coefficient β. (a) An elemental area d*A* in a cross section of arbitrary shape (see box 8.1). (b) An elemental area d*A* extending across the entire width of a rectangular channel (see box 8.2).

where *g* is gravitational acceleration. In general, of course, velocity varies from point to point in a cross section, and as explained in box 8.1, we must account for this variation by computing the kinetic energy for cross-section *i*, $\mathcal{H}_{KEi}$, as

$$\mathcal{H}_{KEi} = \frac{\alpha_i \cdot U_i^2}{2 \cdot g}, \tag{8.6}$$

BOX 8.1 Velocity Coefficients for Energy and Momentum: Definitions

In macroscopic one-dimensional formulations of the energy (section 8.1) and momentum (section 8.2) relations, the velocity U_i is the velocity averaged over cross section *i*. This is the velocity that we use to compute the kinetic energy flux and the momentum flux through each section. However, because in general velocity *u* varies from point to point in each section, coefficients are required to compute the true kinetic energy and momentum fluxes using the average velocity. Here, we derive the general expressions for these coefficients for cross sections of arbitrary shape and

(Continued)

BOX 8.1 *Continued*

velocity distribution. Box 8.2 describes approaches to estimating α and β and computes their values for the case of a wide rectangular channel and the Prandtl-von Kármán (P-vK) velocity distribution.

Discharge

Referring to figure 8.2a, the elemental discharge, dQ, through an elemental area, dA, is

$$\mathrm{d}Q = u \cdot \mathrm{d}A, \tag{8B1.1}$$

where u is the elemental velocity. The total discharge, Q, is

$$Q = \int_A \mathrm{d}Q = \int_A u \cdot \mathrm{d}A = U \cdot A, \tag{8B1.2}$$

where A is the flow cross-sectional area.

Energy Coefficient, α

Referring to figure 8.2a, the weight of water passing through dA per unit time with velocity u is $\gamma \cdot u \cdot$ dA, where γ is weight density. From equation 4.39, the kinetic energy passing through the element per unit time equals

$$\frac{u^2}{2 \cdot g} \cdot \gamma \cdot u \cdot \mathrm{d}A = \frac{\gamma}{2 \cdot g} \cdot u^3 \cdot \mathrm{d}A. \tag{8B1.3}$$

The total flow rate of kinetic energy through the cross section is found by integrating (8B1.3):

$$\int_A \frac{\gamma}{2 \cdot g} \cdot u^3 \cdot \mathrm{d}A = \left(\frac{\gamma}{2 \cdot g}\right) \cdot \int_A u^3 \cdot \mathrm{d}A. \tag{8B1.4}$$

If we simply use the average velocity U to compute the flow rate of kinetic energy through a section, we get

$$\gamma \cdot Q \cdot \left(\frac{U^2}{2 \cdot g}\right) = \gamma \cdot A \cdot \left(\frac{U^3}{2 \cdot g}\right). \tag{8B1.5}$$

The **energy coefficient**, α, is defined as the ratio of the true kinetic-energy flow rate (equation 8B1.4) to the flow rate computed using the average velocity (equation 8B1.5):

$$\alpha \equiv \frac{\left(\frac{\gamma}{2 \cdot g}\right) \cdot \int_A u^3 \cdot \mathrm{d}A}{\left(\frac{\gamma}{2 \cdot g}\right) \cdot U^3 \cdot A} = \frac{\left(\frac{1}{A}\right) \cdot \int_A u^3 \cdot \mathrm{d}A}{U^3}. \tag{8B1.6a}$$

That is, it is the ratio

$$\alpha \equiv \frac{\text{average of cubed velocities}}{\text{cube of average velocity}}. \tag{8B1.6b}$$

If the velocity u is identical for all elements, then $\alpha = 1$; otherwise, $\alpha > 1$. Thus, if we use the average velocity U in computing the kinetic energy at a cross section, it must generally be multiplied by $\alpha \geq 1$ to give the true value.

Gaspard de Coriolis, for whom the Coriolis force (section 7.3.3.1) is named, first proposed the use of the energy coefficient, and α is sometimes called the **Coriolis coefficient**.

Momentum Coefficient, β

The expression for the **momentum coefficient**, β, is developed using reasoning analogous to that used for the energy coefficient. Again referring to figure 8.2a, the rate at which momentum passes through dA per unit time with velocity u is

$$\rho \cdot u^2 \cdot dA, \qquad (8B1.7)$$

where ρ is mass density. Integrating equation 8B1.7 gives the rate at which momentum passes through the cross section:

$$\int_A \rho \cdot u^2 \cdot dA = \rho \cdot \int_A u^2 \cdot dA \qquad (8B1.8)$$

If we simply use the average velocity U to compute the rate of flow of momentum through a section we get

$$\rho \cdot Q \cdot U = \rho \cdot A \cdot U^2. \qquad (8B1.9)$$

The momentum coefficient, β, is defined as the ratio of the true momentum flow rate to the flow rate computed using the average velocity:

$$\beta \equiv \frac{\rho \cdot \int_A u^2 \cdot dA}{\rho \cdot U^2 \cdot A} = \frac{\left(\frac{1}{A}\right) \cdot \int_A u^2 \cdot dA}{U^2}. \qquad (8.10a)$$

That is, it is the ratio

$$\beta \equiv \frac{\text{average of squared velocities}}{\text{square of average velocity}}. \qquad (8.10b)$$

If the velocity u is identical for all elements, then $\beta = 1$; otherwise, $\beta > 1$. Thus, if we use the average velocity U in computing the momentum at a cross section, it must generally be multiplied by $\beta \geq 1$ to give the true value.

Joseph Boussinesq (1842–1929), a French hydraulic engineer, first proposed the use of the momentum coefficient, and β is sometimes called the **Boussinesq coefficient**.

where α_i is the **energy coefficient** for the section. $\mathcal{H}_{KEi}$ is usually called the **velocity head**. Box 8.2 gives an idea of the numerical magnitude of the energy coefficient in natural channels.

The total mechanical energy-per-weight, or **total head**, at cross-section i, $\mathcal{H}_i$, is the sum of the gravitational, pressure, and velocity heads:

$$\mathcal{H}_i = \mathcal{H}_{Gi} + \mathcal{H}_{Pi} + \mathcal{H}_{KEi} = Z_i + Y_i \cdot \cos\theta_0 + \frac{\alpha_i \cdot U_i^2}{2 \cdot g} \tag{8.7}$$

BOX 8.2 Velocity Coefficients for Energy and Momentum: Evaluation

Here we describe approaches to evaluating the energy and momentum coefficients. Hulsing et al. (1966) reported values of α determined from 371 discharge measurements on natural streams; the range observed was $1.03 \leq \alpha \leq 4.70$.

Conventional Empirical Approach

This is the approach used by Hulsing et al. (1966). The general resistance relation can be written as

$$Q = \mathcal{K} \cdot S_f^{1/2}, \tag{8B2.1}$$

where Q is discharge, S_f is the friction slope, and $\mathcal{K}$ is called the **conveyance**, defined as

$$\mathcal{K} \equiv \frac{Q}{S_f^{1/2}}. \tag{8B2.2}$$

Thus, if the Chézy equation is used,

$$\mathcal{K} = \Omega^{-1} \cdot g^{1/2} \cdot A \cdot Y^{1/2} \tag{8B2.3C}$$

where Ω is resistance, g is gravitational acceleration, A is cross-sectional area, and Y is average depth. If the Manning equation is used,

$$\mathcal{K} = u_M \cdot n_M^{-1} \cdot A \cdot Y^{2/3}, \tag{8B2.3M}$$

where u_M is a unit-conversion factor, and n_M is the resistance factor. Noting that $U = Q/A$, invoking equation 8B2.1 and the definitions of α and β in box 8.1, if a given cross section is divided into I subsections, then α and β can be estimated as

$$\alpha = \frac{\sum_{i=1}^{I} (\mathcal{K}_i^3 / A_i^2)}{\mathcal{K}^3 / A^2} \tag{8B2.4}$$

and

$$\beta = \frac{\sum_{i=1}^{I} (\mathcal{K}_i^2/A_i)}{\mathcal{K}^2/A}, \tag{8B2.5}$$

where the $\mathcal{K}$ and A denote the values for the entire cross section. Note that the values calculated by equations 8B2.4 and 8B2.5 for a given section will generally increase as the number of subsections (I) increases.

In using equation 8B2.4 or 8B2.5, A and Y are measured, and the resistance is either 1) computed using the appropriate relation from section 6.6 (Chézy) or 2) estimated using one of the techniques described in table 6.3 (Manning).

Relation to Ratio of Maximum to Average Velocity

Chow (1959) suggested evaluating α from knowledge of the maximum cross-sectional velocity u_m and the average velocity U. By defining

$$\xi \equiv \frac{u_m}{U} - 1 \tag{8B2.6}$$

and assuming that the P-vK law applies across a wide rectangular channel, it can be shown that

$$\alpha = 1 + 3 \cdot \xi^2 - \xi^3, \tag{8B2.7}$$

and

$$\beta = 1 + \xi^2, \tag{8B2.8}$$

as long as $Y >> y_0$. The relations between α and β and U/u_m given by equations 8B2.7 and 8B2.8 are plotted in figure 8.3a.

Dingman (1989, 2007b) found that velocities in natural-stream cross sections tend to follow a power-law frequency distribution, from which it can be shown that α and β are related to ξ as

$$\alpha = \frac{(1+\xi)^3}{1 + 3 \cdot \xi} \tag{8B2.9}$$

and

$$\beta = \frac{(1+\xi)^2}{1 + 2 \cdot \xi}; \tag{8B2.10}$$

these relations are also plotted in figure 8.3a.

Relation to Resistance

Using the definition of resistance, Ω, and the P-vK law, equations 8B2.7 and 8B2.8 can also be expressed as

$$\alpha = 1 + 15.75 \cdot \Omega^2 - 31.25 \cdot \Omega^3 \tag{8B2.11}$$

(*Continued*)

BOX 8.2 *Continued*

and

$$\beta = 1 + 5.25 \cdot \Omega^2. \qquad (8B2.12)$$

Hulsing et al. (1966) used regression analysis (section 4.8.3.1) of velocities measured during 371 discharge measurements to find an empirical relation between α and Manning's n_M:

$$\alpha = 0.884 + 14.8 \cdot n_M. \qquad (8B2.13)$$

However, there was a lot of scatter in the plot of α versus n_M, and equation 8B2.13 explained only about 25% of the variability of α.

Statistical Approach

Dingman (1989, 2007b) showed that, regardless of channel shape or velocity distribution, α is related to statistical quantities of the frequency distribution of velocity in a cross section:

$$\alpha = 1 + SK(u) \cdot CV^3(u) + 3 \cdot CV^2(u), \qquad (8B2.14)$$

and

$$\beta = 1 + CV^2(u), \qquad (8B2.15)$$

where $SK(u)$ and $CV(u)$ are the skewness and the coefficient of variation, respectively, of velocity in the cross section. One can estimate CV and SK by measuring velocities at a representative sampling of points in a cross section and using conventional statistical formulas (see, e.g., appendix C in Dingman 2002).

Velocity Coefficients for Flow in a Wide Rectangular Channel

For the case of a wide rectangular channel in which the velocity distribution follows the P-vK law (figure 8.2b),

$$dA = W \cdot dy, \qquad (8B2.16)$$

$$A = W \cdot Y, \qquad (8B2.17)$$

and

$$u = u(y) = 2.5 \cdot u_* \cdot \ln\left(\frac{y}{y_0}\right), \qquad (8B2.18)$$

where u_* is the shear velocity, and y_0 depends on bed roughness as described in section 5.3.1.6. From equations 5.39 and 5.41, the average cross-sectional velocity U is then

$$U = 2.5 \cdot u_* \cdot \ln\left(\frac{Y}{e \cdot y_0}\right), \qquad (8B2.19)$$

where $e = 2.718$.

Using equations 8B2.16–8B2.19, the numerator of equation 8B1.6a is

$$\left(\frac{1}{A}\right)\cdot\int_A u^3\cdot dA = 15.625\cdot u_*^3\cdot\left[\ln^3\left(\frac{Y}{y_0}\right) - 3\cdot\ln^2\left(\frac{Y}{y_0}\right)\right.$$

$$\left. + 6\cdot\ln\left(\frac{Y}{y_0}\right) - 6 + 6\cdot\left(\frac{y_0}{Y}\right)\right], \tag{8B2.20}$$

and the denominator is

$$U^3 = 15.625\cdot u_*^3\cdot\ln^3\left(\frac{Y}{e\cdot y_0}\right). \tag{8B2.21}$$

Substituting equations 8B2.20 and 8B2.21 into equation 8B1.6a, we can evaluate α as a function of (Y/y_0); the results are shown in the upper curve of figure 8.3b.

Using equations 8B2.16–8B2.19, the numerator of equation 8B1.10a is

$$\left(\frac{1}{A}\right)\cdot\int_A u^2\cdot dA = 6.25\cdot u_*^2\cdot\left[\ln^2\left(\frac{Y}{y_0}\right) - 2\cdot\ln\left(\frac{Y}{y_0}\right) + 2 - 2\cdot\left(\frac{y_0}{Y}\right)\right], \tag{8B2.22}$$

and the denominator is

$$U^2 = 6.25\cdot u_*^2\cdot\ln^2\left(\frac{Y}{e\cdot y_0}\right). \tag{8B2.23}$$

Substituting equations 8B2.22 and 8B2.23 into equation 8B1.10a, we can evaluate β as a function of (Y/y_0); the results are shown in the lower curve of figure 8.3b.

Figure 8.4 compares velocity heads and pressure heads for a database of measurements on 931 reaches.[1] A value of $\alpha = 1.3$ is assumed in calculating velocity head. Figure 8.4 reveals that, typically, velocity head is less than 10% of pressure head. Because velocity head is often relatively small, determining the exact value of α is not usually a critical concern.

8.1.1.2 The Energy Equation

Section 4.5.3 derived the equation for the change in mechanical energy for a fluid element moving from an upstream to a downstream location (equation 4.45). Following the reasoning developed there, and using equation 8.7, we can write an expression for the change in cross-sectional integrated energy from an upstream section ($i = U$) to a downstream section ($i = D$):

$$\mathcal{H}_{GU} + \mathcal{H}_{PU} + \mathcal{H}_{KEU} = \mathcal{H}_{GD} + \mathcal{H}_{PD} + \mathcal{H}_{KED} + \Delta\mathcal{H}; \tag{8.8a}$$

$$Z_U + Y_U\cdot\cos\theta + \frac{\alpha_U\cdot U_U^2}{2\cdot g} = Z_D + Y_D\cdot\cos\theta + \frac{\alpha_D\cdot U_D^2}{2\cdot g} + \Delta H, \tag{8.8b}$$

where $\Delta\mathcal{H}$ is the energy lost (converted to heat) per weight of fluid, or **head loss**.

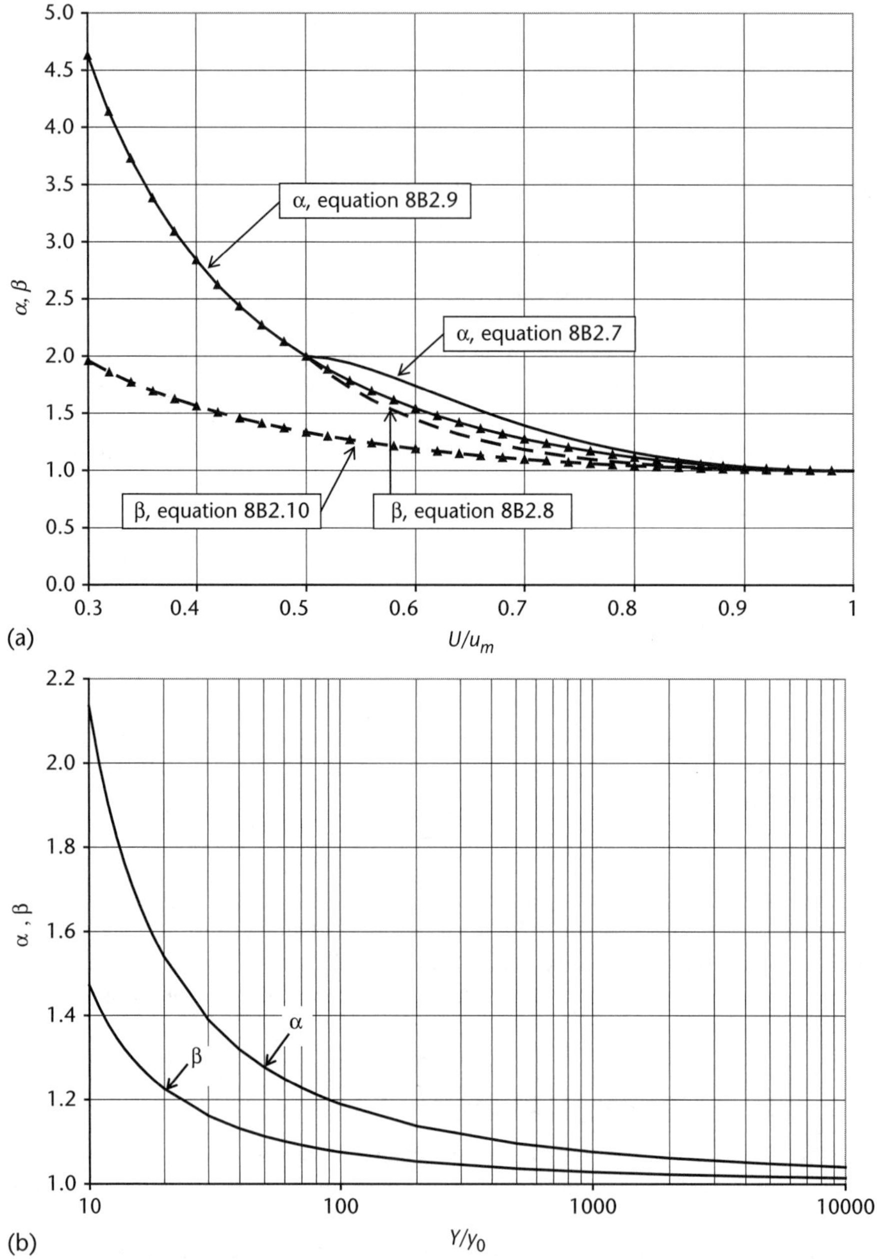

Figure 8.3 (a) α and β as functions of the ratio of average velocity U to maximum velocity u_m. Equations 8B2.7 (α) and 8B2.8 (β) are for the P-vK law in a wide rectangular channel; equations 8B2.9 (α) and 8B2.10 (β) assume a power-law distribution of velocity. (b) The energy coefficient α and the momentum coefficient β as functions of Y/y_0 for the P-vK velocity distribution in a wide rectangular channel (box 8.2).

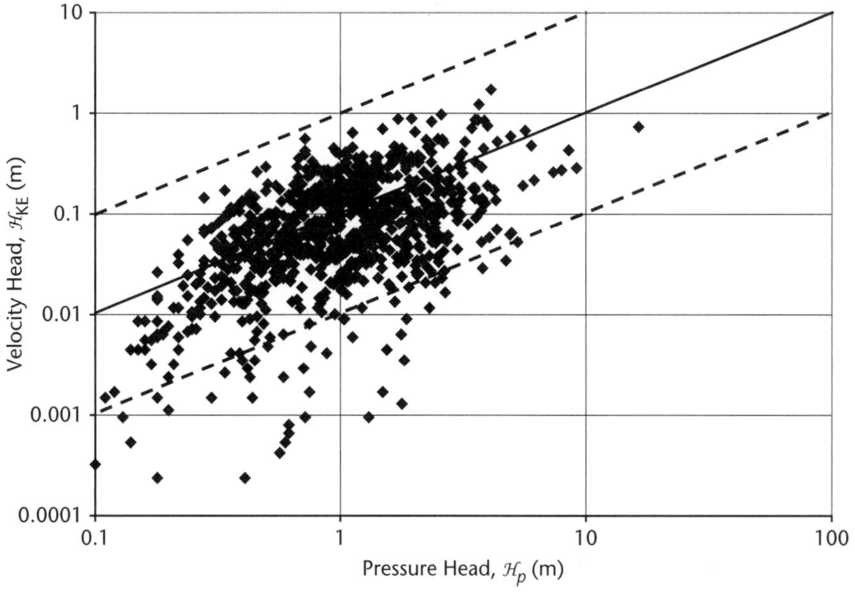

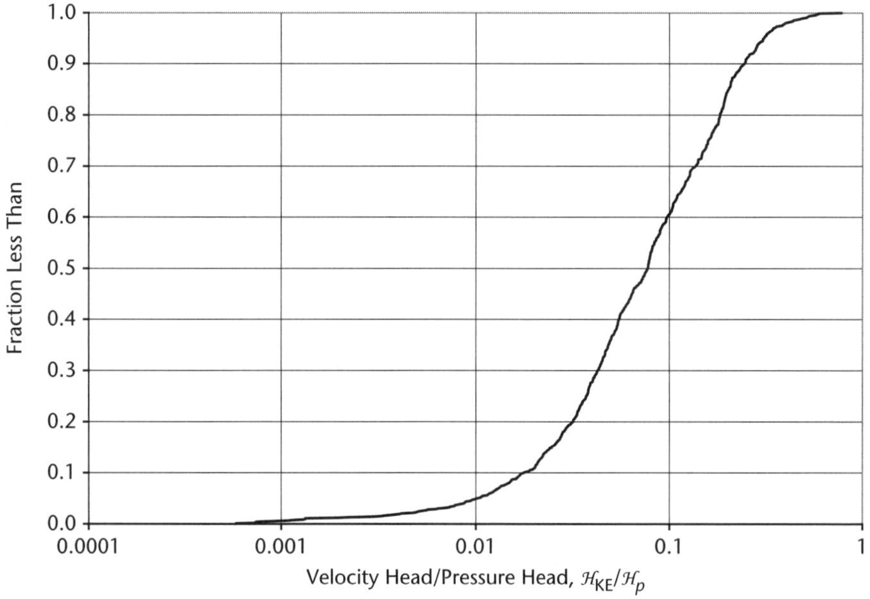

Figure 8.4 (a) Scatter plot of velocity head, $\mathcal{H}_{KE}$, versus pressure head, $\mathcal{H}_P$, for 931 flows in natural channels. The upper dashed line represents $\mathcal{H}_{KE} = \mathcal{H}_P$; the solid line, $\mathcal{H}_{KE} = 0.1 \cdot \mathcal{H}_P$; and the lower dashed line, $\mathcal{H}_{KE} = 0.01 \cdot \mathcal{H}_P$. (b) Cumulative-frequency diagram for the ratio $\mathcal{H}_{KE}/\mathcal{H}_p$ for the flows plotted in (a). These data show that $\mathcal{H}_{KE}$ is almost always less than $0.5 \cdot \mathcal{H}_P$ and is commonly less than $0.1 \cdot \mathcal{H}_P$.

Equation 8.8 is called the **energy equation**. As explained in section 4.5.3, it is an expression of the first and second laws of thermodynamics. Note that from the second law of thermodynamics, $\Delta\mathcal{H} > 0$ if flow is occurring. The energy equation applies to steady flows that are in the laminar, transitional, or turbulent flow states and in the subcritical, critical, or supercritical flow regimes.

The derivation assumed that the pressure distribution is hydrostatic (i.e., that the streamlines in the reach are not significantly curved); flows fitting this description are called **steady gradually varied flows**. We will see later, particularly in chapter 9, how this equation is used to solve important practical and scientific problems.

The terms of equation 8.8 are illustrated in figure 8.5. The line representing the total head from section to section is called the **energy grade line**. The change in total head from section U to section D defines the **energy slope**, S_e:

$$S_e \equiv -\left(\frac{\mathcal{H}_D - \mathcal{H}_U}{\Delta X}\right) = \frac{\Delta\mathcal{H}}{\Delta X}, \tag{8.9}$$

and because $\Delta\mathcal{H} > 0$, $S_e > 0$ if flow is occurring. For uniform flows, the depth and velocity are the same at all sections, so $S_e = S_S = S_0$.

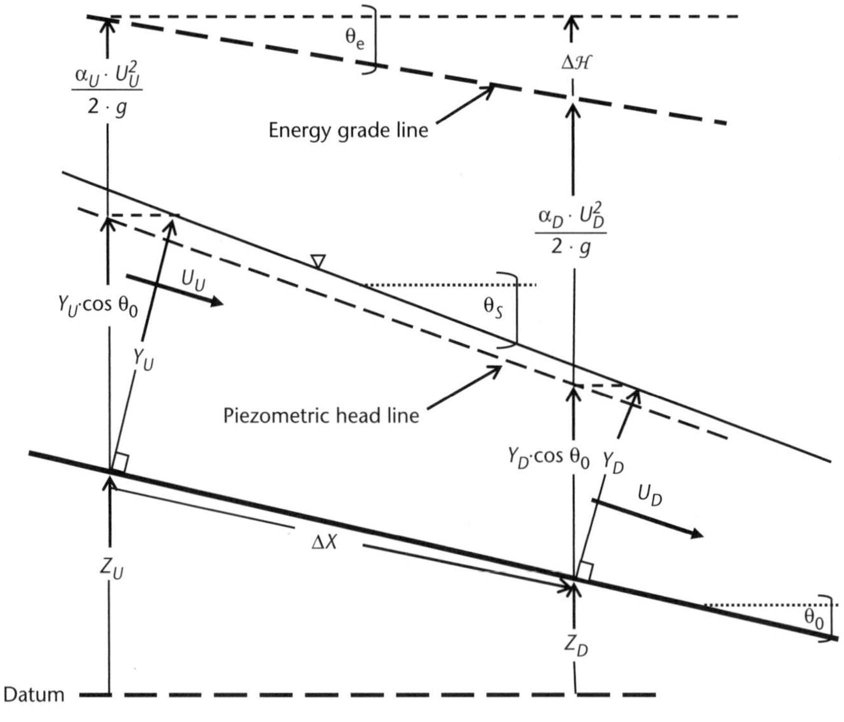

Figure 8.5 Definition diagram for the one-dimensional energy equation 8.8(b).

The line representing the total potential energy from section to section is called the **piezometric head line**. The slope of this line represents the gradient of potential energy that induces flow; therefore, the line must always slope downstream. Because $\cos\theta_0 < 1$, the piezometric head line lies some distance below the water surface. However, the slopes of most streams are almost always less than 0.1, so $\cos\theta_0 > 0.995$ and can be taken to be equal to 1; that is, the piezometric head line is essentially coincident with the water surface and has a slope equal to the surface slope S_S. Recall from section 7.3.1.3 that the surface slope represents the total driving force for the flow (i.e., the sum of the gravitational and pressure forces per unit mass).

8.1.2 Specific Energy

8.1.2.1 Definition

The **specific energy** at a cross section is the total mechanical energy measured with respect to the channel bottom rather than to a horizontal datum. Thus, the elevation-head term of equation 8.7 disappears, and the specific head for cross section i, $\mathcal{H}_{Si}$, is

$$\mathcal{H}_{Si} = \mathcal{H}_{Pi} + \mathcal{H}_{KEi} = Y_i \cdot \cos\theta_0 + \frac{\alpha_i \cdot U_i^2}{2 \cdot g}. \tag{8.10}$$

Note that, because of the elimination of one component of the total mechanical energy, equation 8.10 is no longer an expression of the conservation of energy. Thus, specific head may increase or decrease downstream, and the relative magnitudes of the two components of specific head can vary as we move downstream.

As we will see in chapter 10, the concept of specific energy is useful for understanding how water-surface profiles change through abrupt changes in channel depth and width. It also provides further insight into the distinction between subcritical, critical, and supercritical flow regimes, and we explore this aspect of the concept here.

If we consider flow of discharge Q in a channel of constant width W, we can use the fact that

$$Q = W \cdot Y_i \cdot U_i \tag{8.11}$$

to rewrite equation 8.10[2] as

$$\mathcal{H}_S = Y + \frac{\alpha \cdot Q^2}{2 \cdot g \cdot W^2 \cdot Y^2}. \tag{8.12}$$

With Q and W constant, equation 8.12 shows that specific head depends only on flow depth. However, since $\mathcal{H}_S$ is a function of both Y and Y^{-2}, it can be solved with two different positive values of Y. Thus a graph of equation 8.12 looks like figure 8.6: For all values of $\mathcal{H}_S$ greater than a minimum value, $\mathcal{H}_{Smin}$, the solutions define an upper limb asymptotic to the line $Y = \mathcal{H}_S$ and a lower limb asymptotic to the line $Y = 0$.

Figure 8.6 is a **specific-head diagram.** The curve represents all possible depths for a given discharge in a channel of specified width. As discharge changes in a given channel, the specific-head curve shifts, as shown in figure 8.7 (note that the axes are reversed in this figure).

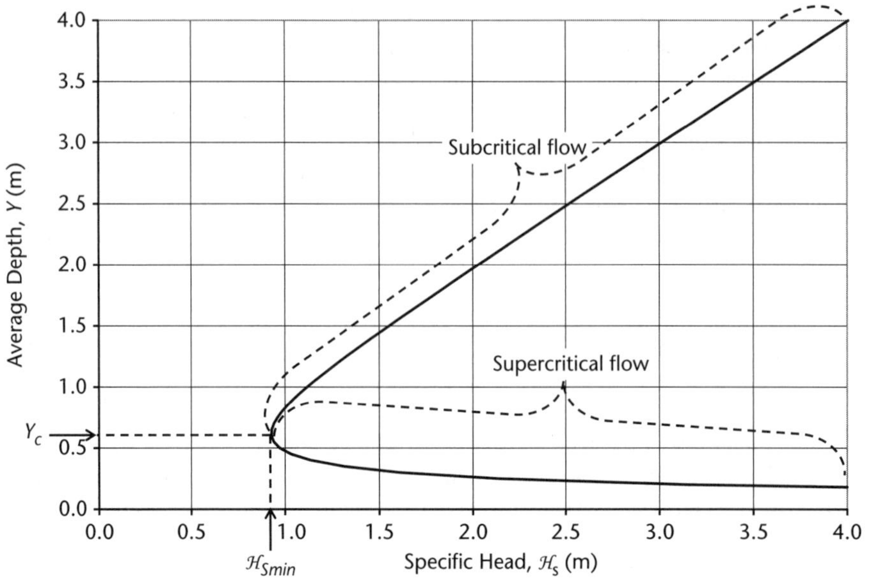

Figure 8.6 A specific-head diagram for a discharge $Q = 4\ \text{m}^3/\text{s}$ in a channel of width $W = 3$ m. The curve represents solutions to equation 8.12. The upper limb of the curve are depths for subcritical flows; the lower limb, for supercritical flows. Velocity head is computed assuming $\alpha = 1.3$. $\mathcal{H}_{Smin}$ is the minimum possible specific head for this discharge; the corresponding depth is the critical depth, $Y_c = 0.57$ m.

8.1.2.2 Alternate Depths, Critical Depth, and the Froude Number

The two solutions to equation 8.12 are called **alternate depths**. Their significance will become apparent after we determine the single value Y_c that gives the minimum specific head, $\mathcal{H}_{Smin}$. We do this by taking the derivative of 8.12, setting the result = 0, and solving for Y_c:

$$\frac{dH_S}{dY} = 1 - \frac{Q^2}{g \cdot W^2 \cdot Y_c^3} = 0; \tag{8.13}$$

$$Y_c = \left(\frac{Q^2}{g \cdot W^2} \right)^{1/3}. \tag{8.14}$$

Y_c is called the **critical depth**. Equation 8.14 shows that the critical depth is determined by the discharge and the width; thus, for a channel of a given width, the critical depth increases as the 2/3 power of the discharge.

From equation 8.11, $Q = W \cdot Y_c \cdot U_c$, where U_c is the velocity corresponding to the critical depth, and substituting this into equation 8.14 gives

$$Y_c = \frac{U_c^2}{g}; \tag{8.15a}$$

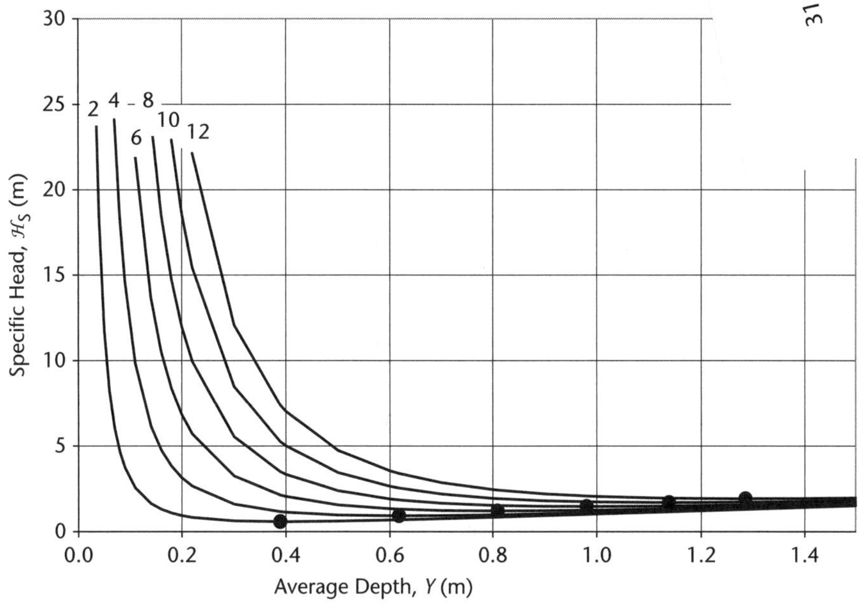

Figure 8.7 Specific-head relations for discharges of 2, 4, 6, 8, 10, and 12 m^3/s in a channel of width $W = 3$ m. Note that the curves represent specific-head diagrams with the axes reversed. The curves for a given channel move away from the origin, and the critical depth increases as discharge increases. The points show the critical depths for the flows, equal to the minimum values of $\mathcal{H}_S$: 0.39, 0.62, 0.81, 0.98, 1.14, and 1.28 m, respectively.

thus,

$$1 = \frac{U_c^2}{g \cdot Y_c}. \tag{8.15b}$$

Recall from equation 6.5 the definition of the Froude number, *Fr*:

$$Fr \equiv \frac{U}{(g \cdot Y)^{1/2}}. \tag{8.16}$$

Thus, the development of equations 8.13–8.15 tells us that the minimum value of $\mathcal{H}_S$ occurs when $Fr^2 = 1$ (and $Fr = 1$). As noted in section 6.2.2.2, the value $Fr = 1$ represents **critical flow**. When $Fr > 1$ the flow is **supercritical**, when $Fr < 1$ the flow is **subcritical**. Thus, for a given discharge in a given channel reach, critical flow represents the flow with minimum possible specific head.

Solutions of 8.12 that lie along the lower limb of the specific-head diagram represent supercritical flows, and solutions that lie along the upper limb are subcritical flows. For a given value of $\mathcal{H}_S > \mathcal{H}_{Smin}$, the upper alternate depth is the depth for subcritical flow, and the lower is the depth for supercritical flow.

Note also that the ratio of the velocity head to the pressure head is

$$\frac{U^2}{2 \cdot g \cdot Y} = \frac{Fr^2}{2}, \tag{8.17}$$

so the Froude number is also related to the ratio of velocity head to pressure head.

Thus, we have now identified four aspects of the significance of the Froude number:

- The Froude number is the ratio of the average flow velocity to the celerity of a gravity wave in shallow water (section 6.2.2.2).
- The Froude number is proportional to a measure of the ratio of driving force to resistance, $S_S^{1/2} / \Omega$ (section 7.6.2).
- The Froude number is a measure of the ratio of velocity head to pressure head (equation 8.17).
- When the Froude number $= 1$, the flow attains the minimum specific energy possible for a given discharge.

8.1.2.3 Which Alternate Depth?

Equation 8.12 shows that the specific head is determined by the channel width, the prevailing discharge, and the depth, but does not explain which value of depth is appropriate in a given situation. Here we address this question.

For uniform flow, depth is determined by the channel slope, S_0, and resistance Ω via the Chézy equation (equation 6.15a),

$$U = \Omega^{-1} \cdot (g \cdot Y \cdot S_0)^{1/2}. \tag{8.18}$$

For a given discharge in a channel of a given width, $U = Q/(W \cdot Y)$, so

$$\frac{Q}{W \cdot Y} = \Omega^{-1} \cdot (g \cdot Y \cdot S_0)^{1/2}, \tag{8.19}$$

and

$$Y = \left(\frac{Q \cdot \Omega}{g^{1/2} \cdot W \cdot S_0^{1/2}} \right)^{2/3}; \tag{8.20a}$$

$$Q = \frac{g^{1/2} \cdot W \cdot S_0^{1/2} \cdot Y^{3/2}}{\Omega}. \tag{8.20b}$$

Equation 8.20 indicates that, in a rectangular channel of width W, slope S_0, and constant resistance, the depth is proportional to the 2/3 power of the discharge, or conversely, the discharge is proportional to the 3/2 power of the depth. However, recall from equation 6.25 that for fully rough flow, Ω is not constant but is a function of relative roughness, which decreases as depth increases:

$$\Omega = \kappa \cdot \left[\ln \left(\frac{11 \cdot Y}{y_r} \right) \right]^{-1}, \tag{8.21}$$

where κ is von Kármán's constant (= 0.400), and y_r is the characteristic height of bed-roughness elements. Substituting equation 8.21 into equation 8.20 and rearranging yields

$$Q = 2.5 \cdot g^{1/2} \cdot W \cdot S_0^{1/2} \cdot Y^{3/2} \cdot \ln\left(\frac{11 \cdot Y}{y_r}\right). \tag{8.22}$$

This relation is somewhat more complicated than 8.20b and cannot be solved explicitly for Y as a function of Q.

The best way to explore the relation between Y and Q given by equation 8.22 is by means of a concrete example. Consider a rectangular channel of width $W = 50$ m, slope $S_0 = 0.001$, and bed-roughness height $y_r = 0.002$ m (2 mm). Substituting the appropriate values into equation 8.22, we generate the relation between Q and Y shown in figure 8.8a. (Note that this relation has the same shape as found for the natural-channel cross section of figure 6.26b.) If we replot the data using logarithmic axes, we have figure 8.8b, which reveals that the discharge-depth relation is essentially a straight line when plotted against logarithmic axes, and hence can be represented as a power law. The equation for this relation for this example is

$$Q = 106 \cdot Y^{1.62}, \tag{8.23a}$$

where Q is in m^3/s and Y is in m.[3] Thus, we see that equation 8.22 implies that the depth-discharge relation remains essentially a power law, but that the exponent on Y is somewhat greater than the value 1.5 given by equation 8.20b. The exact value is determined by the other parameters (W, S_0, y_r) and by the actual channel shape.

Thus, we see that even though we cannot solve 8.22 explicitly for Y as a function of Q and the other parameters, we can usefully approximate that relation by plotting the results of 8.22 in terms of Y versus Q. Since the Q versus Y relation is essentially a power law, Y versus Q is also a power law (figure 8.8c); it is given for this case by

$$Y = 0.056 \cdot Q^{0.619}, \tag{8.23b}$$

where Q is in cubic meters per second and Y is in meters. This relation is the at-a-station hydraulic geometry relation between depth and discharge, as described in section 2.6.3.1.

Continuing with this example, we can now show how the hydraulic geometry relation of equation 8.23 can be used to determine where a particular flow—say, $Q = 326$ m^3/s—plots on the specific-head curve. First, we plot the specific-head diagram for $Q = 326$ m^3/s via equation 8.12 (figure 8.9). Substituting $Q = 326$ into equation 8.23b yields $Y = 2.01$ m. This point is plotted on figure 8.9. As it plots on the upper limb, the flow is subcritical. (This can be checked by computing the Froude number for this flow.)

Thus, while the general specific-head curve for a channel of a given width is determined by discharge (equation 8.12), the particular point on the curve that applies to a specific flow is determined by the channel slope and boundary roughness.

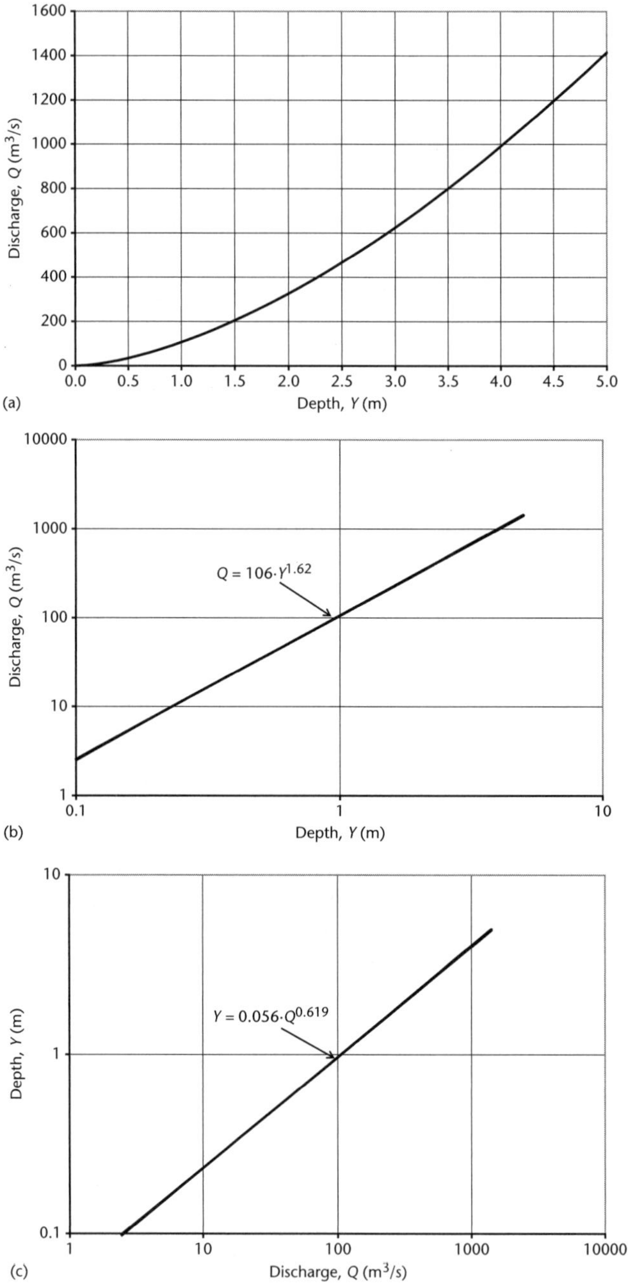

Figure 8.8 Relations between depth, Y, and discharge, Q, for a rectangular channel with width $W = 50$ m, slope $S_0 = 0.001$, and roughness height $y_r = 2$ mm as computed by equation 8.22. (a) Q versus Y plotted on arithmetic axes. (b) Q versus Y plotted on logarithmic axes. (c) Y versus Q plotted on logarithmic axes.

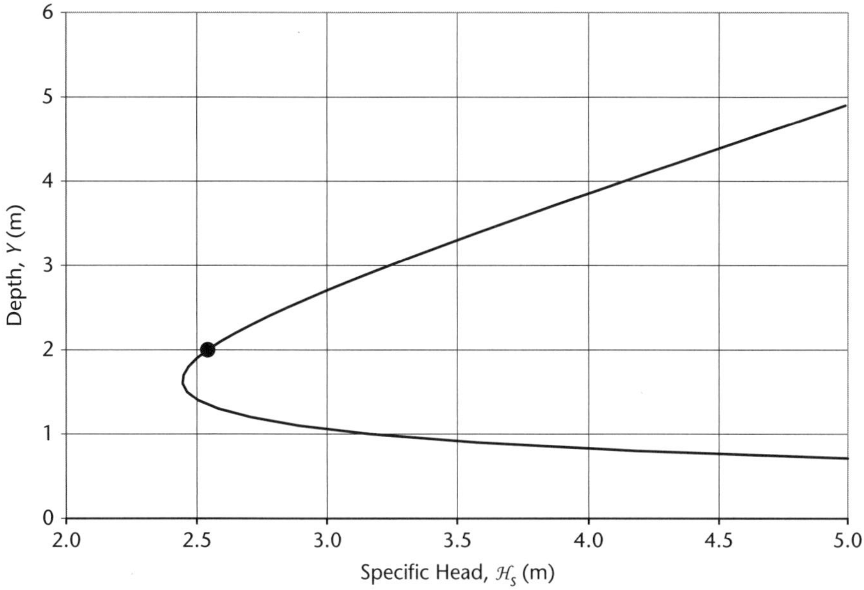

Figure 8.9 Specific-head diagram for the example discussed in section 8.1.2.3. The point gives the depth and specific head for a flow of $Q = 326\ \text{m}^3/\text{s}$.

This point may be on the upper or lower limb of the curve. From equation 8.20, we see that depth is positively related to resistance and inversely related to slope. If the resistance is small enough and/or the slope steep enough, the depth for a given discharge will be on the lower limb of the curve and the flow will be supercritical.

8.1.3 Stream Power

8.1.3.1 Definitions

Power is the time rate of energy expenditure or, equivalently, the time rate of doing work; its dimensions are [F L T^{-1}] or [M L^2 T^{-3}]. Here we derive expressions for stream power in the steady uniform flow shown in figure 8.10.

The channel slope $S_0 = -\Delta Z/\Delta X$ is the vertical distance that the water falls while traveling a unit distance. The time rate of fall, $|\Delta Z/\Delta t|$, is

$$\left|\frac{\Delta Z}{\Delta t}\right| = \left|\frac{\Delta Z}{\Delta X}\right| \cdot \frac{\Delta X}{\Delta t} = \left|\frac{\Delta Z}{\Delta X}\right| \cdot U = S_0 \cdot U. \tag{8.24}$$

The weight of water in length of channel X, Wt, is

$$Wt = \gamma \cdot W \cdot Y \cdot X, \tag{8.25}$$

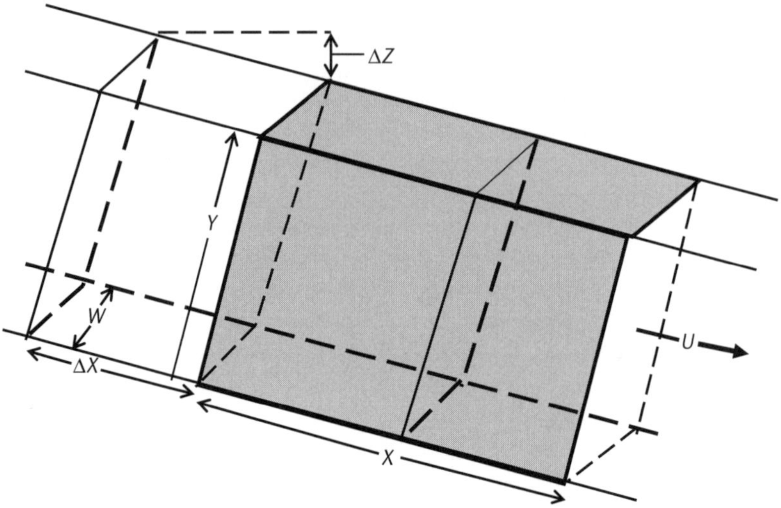

Figure 8.10 Definition diagram for deriving expressions for stream power (equations 8.24–8.28). The shaded block represents the position of a volume of water $W{\cdot}Y{\cdot}X$ after it has moved a distance ΔX.

where γ is the weight density of water. The fall of this water represents a loss in gravitational potential energy, and the time rate of this energy loss per unit channel length, Π, is

$$\Pi = \frac{Wt{\cdot}U{\cdot}S_0}{X} = \gamma{\cdot}(W{\cdot}Y{\cdot}U){\cdot}S_0 = \gamma{\cdot}Q{\cdot}S_0, \tag{8.26}$$

where Q is discharge. Π is called the **stream power per unit channel length**.

It has proved useful to define two additional expressions for stream power. The first of these is **stream power per unit bed area**, Π_A:

$$\Pi_A \equiv \frac{Wt{\cdot}U{\cdot}S_0}{W{\cdot}X} = \frac{\Pi}{W} = \gamma{\cdot}Y{\cdot}S_0{\cdot}U. \tag{8.27a}$$

But recall (equation 5.7) that the boundary shear stress $\tau_0 = \gamma{\cdot}Y{\cdot}S_0$, so this can also be written as

$$\Pi_A = \tau_0{\cdot}U. \tag{8.27b}$$

The third version of stream power is the stream power per weight of water flowing, or **unit stream power**, Π_B:

$$\Pi_B \equiv \frac{Wt{\cdot}U{\cdot}S_0}{Wt} = U{\cdot}S_0, \tag{8.28}$$

which is identical to equation 8.24.

8.1.3.2 Applications

Stream power has been invoked in theories that attempt to predict the cross-sectional shape and planform of rivers. Langbein and Leopold (1964) suggested that two basic tendencies underlie the behavior of streams and, along with the principles of conservation of mass and energy, determine channel shape: 1) the tendency toward equal rate of expenditure of energy on each unit area of the channel bed, which requires that Π_A be constant along a river; and 2) the tendency toward minimization of the total energy expenditure over the river's length, X_L, which requires that $\int_0^{X_L} \Pi \cdot dX$ achieve a minimum value. They pointed out, however, that these two conditions cannot be simultaneously satisfied because of physical constraints, and therefore, the shapes of longitudinal profiles and the downstream changes in channel geometry that are observed in nature are the result of "compromises" between the two opposing tendencies.

These concepts have been extended by others. For example, Song and Yang (1980, p. 1484) stated that

> a river may adjust its flow as well as its boundary such that the total energy loss (or, for a fixed bed the total stream power) in minimized. The principal means of adjusting the boundary is sediment transport. If there is no sediment transport, then the river can only adjust its velocity distribution. In achieving the condition of minimum stream power, the river is constrained by the law of conservation of mass and the sediment transport relations.

Chang (1980, p. 1445) proposed the following:

> For an alluvial channel, the necessary and sufficient condition of equilibrium occurs when the stream power per unit length of channel $\gamma \cdot Q \cdot S$ is a minimum subject to given constraints. Hence an alluvial channel with water discharge Q and sediment load L as independent variables, tends to establish its width, depth and slope such that $\gamma \cdot Q \cdot S$ is a minimum. Since Q is a given parameter, minimum $\gamma \cdot Q \cdot S$ also means minimum channel slope S.

Developing similar ideas, Huang et al. (2004) stated that there is a unique equilibrium channel shape (width/depth ratio) associated with the minimum slope at a given water discharge and sediment load. This minimum slope condition is equivalent to minimum stream power (Π).

Stream power per unit bed area, Π_A, has also been used as a predictor of which of the types of bedform described in table 6.2 and illustrated in figures 6.17–6.20 are present in sand-bed streams, and as a predictor of sediment-transport rates. We will explore those applications in chapter 12.

8.2 The Momentum Principle in One-Dimensional Flows

The momentum principle given in section 4.4 can also be stated as "the impulse (force times time) applied to a fluid element equals its change in momentum (mass times velocity)." For a steady flow, in which the force magnitudes do not change with time, we can write this as

$$\Sigma F \cdot \Delta t = \Delta \mathrm{M}, \tag{8.29}$$

where ΣF is the sum of forces acting over the time period Δt and $\Delta \mathrm{M}$ is the change in momentum. (This relation is identical to equation 4.21.) Here, we integrate this principle to apply to a steady one-dimensional macroscopic flow.

8.2.1 The Momentum Equation

Consider again the steady flow in the straight rectangular channel depicted in figure 8.1. Recall from chapter 7 that if the flow is fully turbulent and its scale not too large, the only forces acting on the mass of water between upstream and downstream cross sections are the driving forces of gravity (F_G) and pressure (F_P) opposed by the resisting force due to turbulence (F_T). The mass, M, of water between the two sections remains constant and equal to

$$M = \rho \cdot W \cdot Y \cdot \Delta X, \tag{8.30}$$

where ρ is mass density. Thus, we can write equation 8.29 for this situation as

$$(F_G + F_P - F_T) \cdot \Delta t = [\rho \cdot W \cdot Y \cdot \Delta X] \cdot (\beta_D \cdot U_D - \beta_U \cdot U_U), \tag{8.31}$$

where the momentum coefficient β is necessary in order to account for the use of the cross-section-averaged velocity, as explained in box 8.1. Dividing through by Δt and noting that $W \cdot Y \cdot \Delta X / \Delta t \equiv Q$, the constant discharge, we can write the **momentum equation** for a steady one-dimensional macroscopic flow as

$$F_G + F_P - F_T = \rho \cdot Q \cdot (\beta_D \cdot U_D - \beta_U \cdot U_U). \tag{8.32}$$

An alternative version of the momentum equation can be derived by the following steps:

1. Divide 8.31 by the mass of water in the reach, $\rho \cdot W \cdot Y \cdot \Delta X$, to give

$$(a_G + a_P - a_T) \cdot \Delta t = \beta_D \cdot U_D - \beta_U \cdot U_U, \tag{8.33a}$$

where the a-terms are the respective forces per unit mass.

2. Replace these terms with their equivalents from table 7.1 and assume $\cos \theta_0 = 1$:

$$\left(-g \cdot \frac{Z_D - Z_U}{\Delta X} - g \cdot \frac{Y_D - Y_U}{\Delta X} - \Omega^2 \cdot \frac{U^2}{Y} \right) \cdot \Delta t = \beta_D \cdot U_D - \beta_U \cdot U_U, \tag{8.33b}$$

where U is the average velocity given by $U = (U_D + U_U)/2$.

3. Divide through by g, multiply through by ΔX, divide through by Δt, and note that $\Delta X / \Delta t \equiv U$:

$$-Z_D + Z_U - Y_D + Y_U - \frac{\Omega^2 \cdot U^2 \cdot \Delta X}{g \cdot Y} = \frac{\beta_D \cdot U_D^2}{2 \cdot g} - \frac{\beta_U \cdot U_U^2}{2 \cdot g} \tag{8.33c}$$

4. Rearrange to give

$$Z_U + Y_U + \frac{\beta_U \cdot U_U^2}{2 \cdot g} = Z_D + Y_D + \frac{\beta_D \cdot U_D^2}{2 \cdot g} + \frac{\Omega^2 \cdot U^2 \cdot \Delta X}{g \cdot Y}. \tag{8.33d}$$

5. The last term on the right-hand side is the "friction-head loss", i.e. the momentum loss per unit mass due to boundary friction during the time the water moves from the upstream to the downstream section. Defining $\Delta\mathcal{M} \equiv \Omega^2 \cdot U^2 \cdot \Delta X/(g \cdot Y)$, we can write

$$Z_U + Y_U + \frac{\beta_U \cdot U_U^2}{2 \cdot g} = Z_D + Y_D + \frac{\beta_D \cdot U_D^2}{2 \cdot g} + \Delta\mathcal{M}. \tag{8.33e}$$

Equation 8.33e is very similar to the energy equation, equation 8.8. It differs in that 1) the velocity-head terms contain the momentum coefficient rather than the energy coefficient, and 2) $\Delta\mathcal{M}$ term represents the change in momentum per mass of flowing water rather than the change in energy per weight of flowing water, $\Delta\mathcal{H}$. We examine the similarities and differences between the energy and momentum equations further in section 8.3.

8.2.2 Specific Force

The concept of **specific force** is analogous to the concept of specific energy discussed in section 8.1.2. The concept is developed for a short reach (small ΔX) in a horizontal channel ($Z_D = Z_U$), so that the gravitational force F_G and the resisting force F_T in equation 8.32 are neglected and

$$F_P = \rho \cdot Q \cdot (\beta_D \cdot U_D - \beta_U \cdot U_U). \tag{8.34}$$

The pressure distribution is assumed hydrostatic so that

$$F_P = \gamma \cdot \frac{Y_U}{2} \cdot A_U - \gamma \cdot \frac{Y_D}{2} \cdot A_D, \tag{8.35}$$

where A_i is the cross-sectional area of section i. If we write the average velocity of section i as $U_i = Q/A_i$ and assume $\beta_U = \beta_D = 1$, 8.35 becomes

$$\gamma \cdot \frac{Y_U}{2} \cdot A_U - \gamma \cdot \frac{Y_D}{2} \cdot A_D = \rho \cdot Q \cdot \frac{Q}{A_D} - \rho \cdot Q \cdot \frac{Q}{A_U}, \tag{8.36a}$$

which can be rearranged to

$$\frac{Y_U}{2} \cdot A_U + \frac{Q^2}{g \cdot A_U} = \frac{Y_D}{2} \cdot A_D + \frac{Q^2}{g \cdot A_D}. \tag{8.36b}$$

Referring to equation 8.36b, we define the specific force F_S at a cross section as

$$F_S = \frac{Y}{2} \cdot A + \frac{Q^2}{g \cdot A}. \tag{8.37a}$$

Note that the dimensions of F_S are $[L^3]$.

For a rectangular section in which $A = W \cdot Y$,

$$F_S = \frac{Y^2 \cdot W}{2} + \frac{Q^2}{g \cdot W \cdot Y}. \tag{8.37b}$$

As with specific energy (equation 8.12), for a channel with a specified width and a given discharge, there are two values of Y that satisfy equation 8.37b, and we can

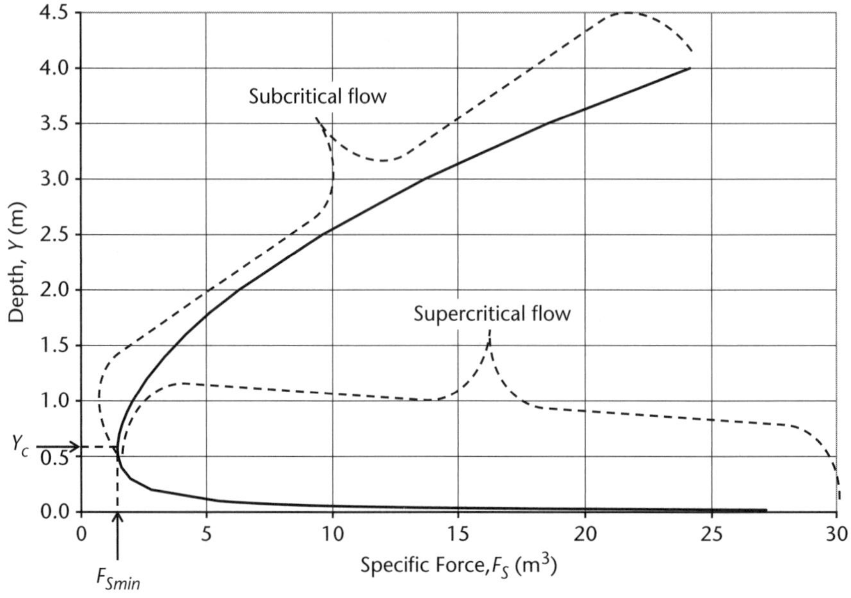

Figure 8.11 A specific-force diagram for a discharge $Q = 4$ m^3/s in a channel of width $W = 3$ m (as in figure 8.6). F_{Smin} is the minimum possible specific head for this discharge; the corresponding depth is the critical depth, $Y_c = 0.57$ m. The curve represents solutions to equation 8.37(b). The upper limb of the curve are depths for subcritical flows; the lower limb, for supercritical flows.

construct a **specific-force diagram** like that of figure 8.11. This curve has many similarities with the specific-head diagram:

1. As with specific head, there is a minimum value of specific force, F_{Smin}, that can be evaluated by differentiating equation 8.37b with respect to Y and setting the result equal to 0, and as with specific head, minimum specific force occurs at critical flow ($Fr = 1, Y = Y_c$).
2. The lower limb of the specific-force curve represents supercritical flows and is asymptotic to $Y = 0$.
3. The upper limb of the specific-force curve represents subcritical flows.

However, there are important differences between the two types of diagrams. Unlike the specific-head diagram,

1. The upper limb of the specific-force diagram has no asymptote, but curves indefinitely to the right. (Note that whereas specific energy depends on Y and Y^{-2}, specific force depends on Y^2 and Y^{-1}.)
2. For a given specific force, the two depths represent the depths before and after a transition from supercritical to subcritical flow, and are called **sequent depths**.

As we will see in chapter 10, the specific-force diagram is useful in determining how the water-surface profile changes through a transition from supercritical to subcritical flow.

Table 8.1 The energy and momentum equations 8.8 and 8.33e.[a]

Symbol	Definition	Dimensions
Energy:	$Z_U + Y_U + \alpha_U \cdot U_U^2/(2 \cdot g) = Z_D + Y_D + \alpha_D \cdot U_D^2/(2 \cdot g) + \Delta\mathcal{H}$	
Momentum:	$Z_U + Y_U + \beta_U \cdot U_U^2/(2 \cdot g) = Z_D + Y_D + \beta_D \cdot U_D^2/(2 \cdot g) + \Delta\mathcal{M}$	
g	Acceleration due to gravity	[L T^{-2}]
$\Delta\mathcal{H}$	Loss of energy per weight of flowing water (head loss) (total internal energy loss)	[L]
$\Delta\mathcal{M}$	Loss of momentum per mass of flowing water in travel time between sections due to boundary friction	[L]
U	Cross-sectional average velocity	[L T^{-1}]
Y	Cross-sectional average depth	[L]
Z	Elevation of channel bottom	[L]
α	Energy (Coriolis) coefficient to account for variation of velocity in cross section	[1]
β	Momentum (Boussinesq) coefficient to account for variation of velocity in cross section	[1]

[a] Subscripts in equations indicate upstream (U) and downstream (D) cross sections.

The following section explores more fully the differences and similarities between the energy and momentum principles.

8.3 Comparison of the Energy and Momentum Principles

To facilitate comparison, the energy and momentum equations are displayed together in table 8.1. A conceptually important difference between them is that energy is a scalar quantity and momentum is a vector quantity; however, this distinction has little practical import in describing one-dimensional macroscopic flows. Aside from this, the two equations are identical except for 1) the velocity-distribution coefficients and 2) the loss terms (last terms on the right-hand side). As indicated in figure 8.3, the values of α and β do not differ greatly, and as noted in section 8.1.1.1, the term involving velocity is usually relatively small, so this difference is usually numerically minor. The major theoretical and practical distinction between the energy and momentum principles is in the interpretation of the loss terms.

In the energy equation, $\Delta\mathcal{H}$ represents all the conversion of kinetic energy of the flow to heat between the two cross sections. This energy loss is the internal energy loss due to viscosity and turbulence. At least a portion of this energy loss originates as the external friction between the flowing water and the channel boundary, but turbulence can also be generated in rapid increases or decreases in depth or width. When the flow cross-sectional area increases significantly over a short distance, eddies form (figure 8.12). The circulation in these eddies represents a conversion of potential to kinetic energy and of kinetic energy to heat due to the internal velocity gradients. At rapid decreases in cross-sectional area the convergence of stream lines increases internal velocity gradients and thus adds to the energy loss. Energy losses due to expansion and contraction are collectively called **eddy losses**, and we will present methods for estimating them in chapter 9.

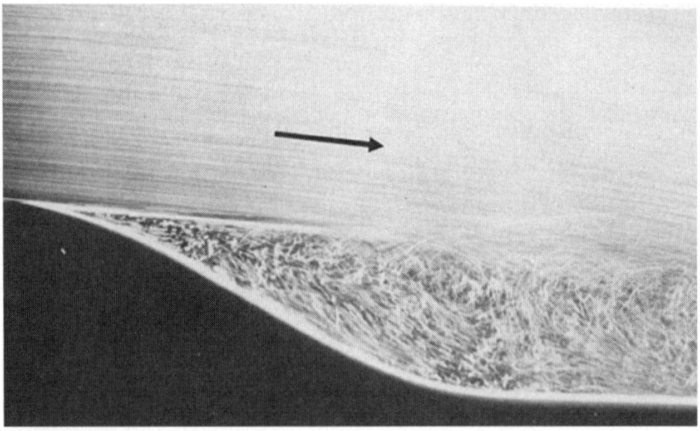

(a)

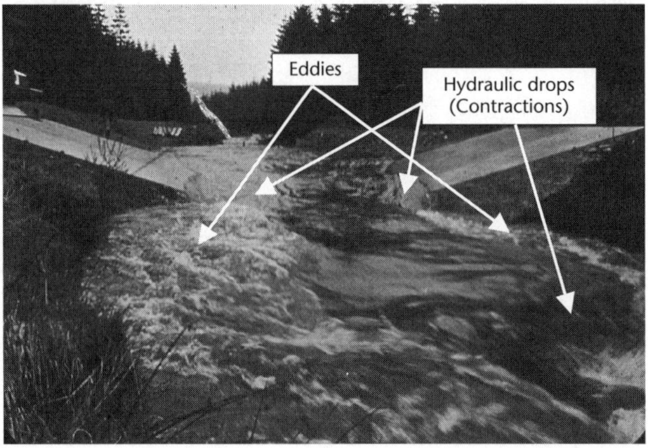

(b)

Figure 8.12 (a) Expansion eddies in laminar flow in a laboratory flume. From Van Dyke (1982). Original photo by Henri Werlé; reproduced with permission of ONERA, the French Aerospace Labatory. (b) Hydraulic drops and expansion eddies induced in flow downstream of a measurement structure on a stream in Wales.

In contrast to $\Delta\mathcal{H}$, the $\Delta\mathcal{M}$ term in the momentum equation represents only the loss of momentum induced by boundary friction, that is, external losses. Thus,

$$\Delta\mathcal{H} \geq \Delta\mathcal{M}. \tag{8.38}$$

For a given flow and channel reach, the difference $\Delta\mathcal{H} - \Delta\mathcal{M}$ is

$$\Delta\mathcal{H} - \Delta\mathcal{M} = \left(\frac{1}{2\cdot g}\right)[(\alpha_U - \beta_U)\cdot U_U^2 - (\alpha_D - \beta_D)\cdot U_D^2]. \tag{8.39}$$

If cross-sectional shape does not change drastically between the two cross sections, it may be reasonable to assume $\alpha_U = \alpha_D = \alpha$ and $\beta_U = \beta_D = \beta$, in which case

$$\Delta\mathcal{H} - \Delta\mathcal{M} = \left(\frac{1}{2\cdot g}\right)\cdot(\alpha - \beta)\cdot(U_U^2 - U_D^2). \tag{8.40}$$

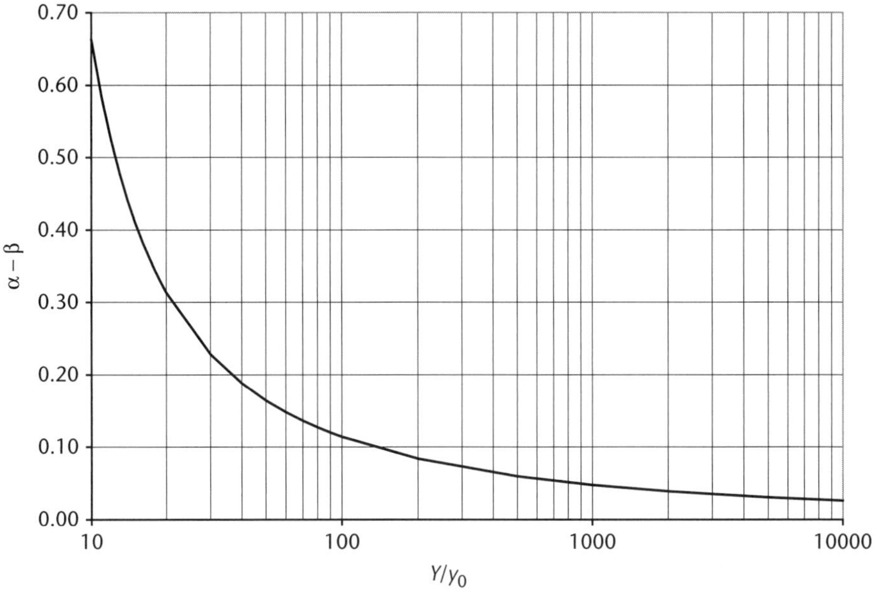

Figure 8.13 The difference between the energy coefficient and the momentum coefficient, $\alpha - \beta$, as a function of Y/y_0. α and β are computed assuming the Prandtl-von Kármán velocity distribution in a rectangular channel (box 8.2).

In uniform flow there is no change in cross-sectional area or velocity, so $U_U = U_D$ and $\Delta\mathcal{H} = \Delta\mathcal{M}$. Thus, in uniform flow, the energy and momentum equations, although representing scalar and vector quantities respectively, give identical numerical values. And, even if the flow is not strictly uniform, the value of $\alpha - \beta$ is usually a small number (figure 8.13), so in natural streams, the difference $\Delta\mathcal{H} - \Delta\mathcal{M}$ will often be smaller than the uncertainties in determining other quantities in the energy or momentum equation and thus of little practical import.

As Chow (1959, pp. 51–52) pointed out, "generally speaking, the energy principle offers a simpler and clearer explanation than does the momentum principle." However, the energy and momentum principles, used separately or together, can both be useful in solving practical problems. For example, in situations that involve high internal energy losses over short distances (e.g., the hydraulic jump, section 10.1), there is no practical way to quantify $\Delta\mathcal{H}$, and the energy equation cannot be applied. However, because the channel distance is short, it may be acceptable to assume that external (friction) losses are negligible and apply the momentum equation with $\Delta\mathcal{M} = 0$.

Henderson (1961, p. 11) also provided useful insight to this question:

> The general conclusion is that the energy and momentum equations play complementary parts in the analysis of a flow situation: Whatever information is not supplied by one is usually supplied by the other. One of the most common uses of the momentum equation is in situations where the energy equation breaks down because of the presence of an

unknown energy loss; the momentum equation can then supply results which can be fed back into the energy equation, enabling the energy loss to be calculated.

We will show how the energy and momentum equations are applied in analyzing situations of rapidly varied flow, where the cross-sectional area changes significantly between upstream and downstream sections in chapter 10.

9

Gradually Varied Flow and Water-Surface Profiles

9.0 Introduction and Overview

Gradually varied flow is flow in which 1) downstream changes in velocity and depth are gradual enough that the flow can be considered to be uniform, and 2) the temporal changes in velocity and depth are gradual enough that the flow can be considered to be steady. Under gradually varied flow conditions, we can assume that 1) the pressure distribution is hydrostatic, 2) the one-dimensional energy equation (equation 8.8b) applies, and 3) a uniform-flow resistance equation (i.e., Chézy equation 6.19 or Manning equation 6.40c) applies.

We have seen in section 7.5 that these conditions are commonly satisfied in natural stream reaches. In particular, recall from section 7.5.5.2 that the local acceleration (time rate of change of velocity) is typically much smaller than other accelerations. This is the justification for applying gradually varied flow computations in modeling water-surface profiles associated with flows that are not strictly steady.

Application of gradually varied flow concepts allows one to apply the hydraulic principles developed in preceding chapters in a linked manner over an extended portion of a stream profile, rather than at an isolated cross section or reach. This linkage provides a model of how the water-surface elevation and hence the depth and velocity change along a channel carrying a specified discharge.

Gradually varied flow computations play an essential role in the strategy for reducing future flood damages. According to the U.S. National Weather Service, floods are among the most frequent and costly natural disasters in terms of human hardship and economic loss. Between 1970 and 2003, annual flood damages in the

Figure 9.1 Computation of water-surface profiles by application of the concepts of gradually varied flow is an essential step in identifying flood-prone areas that should be restricted from development to prevent occurrences like the one in this photograph.

United States averaged $3.8 billion (1995 dollars) and took about 100 lives per year (University Corporation for Atmospheric Research 2003) (this was before hurricanes Katrina and Rita devastated the U.S. Gulf Coast in August 2005). It is widely accepted among water-resource planners that the most cost-effective way to reduce future flood damages is to prevent damageable development in flood-prone areas (figure 9.1). The process of identifying such areas involves the steps below; concepts of gradually varied flow are the basis for step 4 of this sequence.

1. Select the design flood. The design flood is usually specified in terms of the probability that it will be exceeded in any year. Federal regulations in the United States specify that the design flood will be the flood discharge with an annual exceedence probability of 0.01 (i.e., there is a 1 % chance that this discharge will be exceeded in any year; this is called **100-year flood**; see section 2.5.6.3).
2. Conduct hydrologic studies to determine the design-flood discharge along the significant streams in the study area.
3. Determine stream cross-section geometry at selected locations along streams in the study area via field surveys, airborne laser altimetry (LIDAR), aerial photographs, or topographic maps
4. Using the surveyed cross-section data, compute the elevation of the water surface associated with the design flood at each cross section via application of gradually varied flow concepts.
5. Use the design-flood-surface elevations in conjunction with topographic data to delineate areas lying below the elevation of the design flood.

Gradually varied flow methodology has several other important practical applications:

- It provides insight for identifying where sediment erosion and deposition may occur.
- It allows us to use known relations between depth and discharge at a particular section to develop predictions of those relations at other locations along the stream profile.
- It allows us to predict the effects of engineering structures (dams, bridges, etc.) on water-surface elevations and velocity and depth over significant distances.
- It provides physically correct initial conditions for modeling unsteady flows (chapter 11).
- Used in an inverse manner, it provides a tool for estimating the discharge of a past flood from high-water marks left by that flood.

We begin this chapter by recalling from preceding chapters the basic equations underlying gradually varied flow computations, and then use these equations to 1) develop a classification of water-surface profiles, 2) develop the basic mathematics of profile computations, and 3) present a standard method for practical computation of profiles.

9.1 The Basic Equations

Gradually varied flow computations are based on 1) the fundamental principles of conservation of mass and conservation of energy and 2) a resistance relation. As elaborated in the following subsections, these relations are formulated in finite-difference form for one-dimensional steady flows.

The computations require that we have the following information for an extended distance along the channel of interest:

1. The elevation of the channel bottom and the configuration of cross sections (usually including the floodplain adjacent to the channel proper) at selected locations
2. Information for determination of resistance at each cross section
3. A specified design discharge
4. The water-surface elevation associated with the design discharge at the downstream-most (for subcritical flow) or upstream-most (for supercritical flow) cross section

In the discussion here, we assume subcritical flow and number the cross sections in the upstream direction, beginning at section $i = 0$ where the water-surface elevation is known for the design discharge. To further simplify the developments, we assume that 1) the design discharge, Q, is constant through the reach; 2) the channel is rectangular; and 3) the channel slope S_0 is small enough that $\cos \theta_0 \approx 1$.

9.1.1 Continuity (Conservation-of-Mass) Equation

In gradually varied flow computations, the design discharge, Q, at and between successive cross sections is specified. This implies that there are no significant tributaries and no significant inflows or outflows of groundwater between successive sections. Thus, at cross section i, the continuity equation is

$$Q = W_i \cdot Y_i \cdot U_i, \tag{9.1}$$

where W_i is channel width, Y_i is cross-section-average depth, and U_i is cross-section-average velocity.

9.1.2 Energy Equation

The one-dimensional energy equation for steady flow between an upstream cross-section (subscript i) and a downstream cross-section (subscript $i-1$) is given by equation 8.8b:

$$Z_i + Y_i + \frac{\alpha_i \cdot U_i^2}{2 \cdot g} = Z_{i-1} + Y_{i-1} + \frac{\alpha_{i-1} \cdot U_{i-1}^2}{2 \cdot g} + \Delta\mathcal{H}_{i,i-1}, \tag{9.2}$$

where Z is the channel-bottom elevation, α is the velocity-head coefficient (see box 8.1), g is gravitational acceleration, and $\Delta\mathcal{H}_{i,i-1}$ is the head loss between section i and section $i-1$.

As discussed in section 8.3, $\Delta\mathcal{H}_{i,i-1}$ is the total energy loss between the two sections. At least a portion of this total loss is due to the friction of the channel boundary; this **friction loss**, $\Delta\mathcal{M}_{i,i-1}$, is the "external" energy loss given by the momentum equation (equation 8.33e). The difference, $\Delta\mathcal{H}_{i,i-1} - \Delta\mathcal{M}_{i,i-1}$, is due to internal energy losses that arise when the streamlines diverge (producing eddies) or converge (producing increased shear) (see figure 8.12); both types of loss are collectively called **eddy loss** (or **contraction/expansion loss**), $\Delta\mathcal{H}_{\text{eddy}:i,i-1}$. Thus,

$$\text{Total energy loss}\,(\Delta\mathcal{H}_{i,i-1}) = \text{friction loss}\,(\Delta\mathcal{M}_{i,i-1}) + \text{eddy loss}\,(\Delta\mathcal{H}_{\text{eddy}:i,i-1}), \tag{9.3a}$$

and

$$\Delta\mathcal{H}_{\text{eddy}:i,i-1} \equiv \Delta\mathcal{H}_{i,i-1} - \Delta\mathcal{M}_{i,i-1}. \tag{9.3b}$$

As explained in the following section, $\Delta\mathcal{M}_{i,i-1}$ is the resistance accounted for in the uniform-flow (Chézy and Manning) equations. Thus, we use equation 9.3 to write equation 9.2 as

$$Z_i + Y_i + \frac{\alpha_i \cdot U_i^2}{2 \cdot g} = Z_{i-1} + Y_{i-1} + \frac{\alpha_{i-1} \cdot U_{i-1}^2}{2 \cdot g} + \Delta\mathcal{M}_{i,i-1} + \Delta\mathcal{H}_{\text{eddy}:i,i-1}. \tag{9.4}$$

The eddy loss is always positive and, from equation 8.39, can be approximated as

$$\Delta\mathcal{H}_{\text{eddy}:i,i-1} = (\alpha - \beta) \cdot \left| \left(\frac{U_i^2}{2 \cdot g} - \frac{U_{i-1}^2}{2 \cdot g} \right) \right|, \tag{9.5a}$$

where α and β are the energy and momentum coefficients, respectively. Since little information is typically available for evaluating α and β, conventional practice is to estimate eddy losses as

$$\Delta\mathcal{H}_{\text{eddy}:i,i-1} = k_{\text{eddy}} \cdot \left| \left(\frac{U_i^2}{2 \cdot g} - \frac{U_{i-1}^2}{2 \cdot g} \right) \right|, \tag{9.5b}$$

where k_{eddy} values are estimated as described in section 9.4.2.1.

9.1.3 Resistance Relations

As developed in chapter 6, a uniform flow is one in which the driving force due to gravity is balanced by resisting forces originating as boundary friction. In natural rivers, the resisting forces can be considered to be those due to turbulence only. We formulated the Chézy equation (equation 6.19) as the preferred uniform-flow equation:

$$U = \Omega^{-1} \cdot u_* = \Omega^{-1} \cdot g^{1/2} \cdot Y^{1/2} \cdot S_0{}^{1/2}, \tag{9.6}$$

where S_0 is local channel slope, and Ω is local flow resistance. For fully rough flow (which we will assume in this chapter), Ω is given by equation 6.25:

$$\Omega = 0.400 \cdot \left[\ln \left(\frac{11 \cdot Y}{y_r} \right) \right]^{-1}, \tag{9.7}$$

where y_r is the local effective height of bed-roughness elements. Using equation 9.1, we can write the Chézy equation for discharge as

$$Q = \Omega^{-1} \cdot g^{1/2} \cdot W \cdot Y^{3/2} \cdot S_0{}^{1/2}. \tag{9.8}$$

Although we have seen that the Chézy equation is preferable on theoretical grounds, the Manning equation (equation 6.40c) is commonly assumed to be the uniform-flow equation:

$$U = u_M \cdot n_M{}^{-1} \cdot Y^{2/3} \cdot S_0{}^{1/2}, \tag{9.9}$$

where u_M is a unit-conversion factor (section 6.8.1), and n_M is the local resistance factor called Manning's n (section 6.8.2; note that we are assuming a wide channel, so hydraulic radius $R = Y$). This relation can also be written in terms of discharge:

$$Q = u_M \cdot n_M{}^{-1} \cdot W \cdot Y^{5/3} \cdot S_0{}^{1/2}. \tag{9.10}$$

As noted in section 9.1.2, the friction loss is the energy loss due to the boundary. We define the local **friction slope**, S_{fi}, as

$$S_{fi} = \frac{\mathcal{M}_i - \mathcal{M}_{i-1}}{X_i - X_{i-1}} = \frac{\Delta \mathcal{M}_{i,i-1}}{X_i - X_{i-1}}. \tag{9.11}$$

A critical assumption in gradually varied flow computations is that the uniform-flow resistance relation applies when local channel slope S_{0i} is replaced by the **friction slope**, S_{fi}. Thus, we assume that one of the following relations applies at each cross section:

$$\text{Chézy: } Q = \Omega_i{}^{-1} \cdot g^{1/2} \cdot W_i \cdot Y_i{}^{3/2} \cdot S_{fi}{}^{1/2} \tag{9.12C}$$

or

$$\text{Manning: } Q = u_M \cdot n_{Mi}{}^{-1} \cdot W_i \cdot Y_i{}^{5/3} \cdot S_{fi}{}^{1/2} \tag{9.12M}$$

9.2 Water-Surface Profiles: Classification

9.2.1 Normal Depth and Critical Depth

9.2.1.1 Normal Depth

As noted above, water-surface computations are done for a specified design discharge in the reach of interest; thus, Q is a specified value. For a given discharge in a given reach,[1] the **normal depth**, Y_n, is defined as the depth of a uniform flow. Thus, using the Chézy equation, the normal depth is computed from equation 9.12C as

$$Y_n = \left(\frac{\Omega \cdot Q}{g^{1/2} \cdot W \cdot S_0{}^{1/2}} \right)^{2/3}, \tag{9.13C}$$

and using the Manning equation, from equation 9.12M as

$$Y_n = \left(\frac{n_M \cdot Q}{u_M \cdot W \cdot S_0{}^{1/2}} \right)^{3/5}. \tag{9.13M}$$

Note that for a given discharge, normal depth depends on channel resistance, width, and slope.

Recall that uniform flow represents the condition in which the driving and resisting forces balance, and that turbulent resistance increases as the square of velocity. Thus, if the actual depth is above or below the normal depth, the driving and resisting forces are not in balance. If the local flow depth is greater than the normal depth for the discharge, the velocity will be lower than for uniform flow, the driving forces will exceed the resisting forces, and the flow will tend to accelerate until a balance is achieved. Conversely, if the depth if less than the normal depth, velocity and hence resistance will be greater than required to balance the driving force, and the "excess" resistance will tend to slow the flow until the forces again balance. As a parcel of water moves through a succession of reaches, changing conditions of slope, roughness, geometry, and discharge (due to tributary and groundwater inflows) continually modify the normal depth, but the flow is continuously driven toward the uniform-flow condition.

9.2.1.2 Critical Depth

As defined in section 8.1.2.2, **critical depth**, Y_c, is the depth of critical flow (i.e., flow with Froude number $Fr = 1$). For a given discharge in a channel of a given width, Y_c is found from equation 8.14:

$$Y_c \equiv \left(\frac{Q}{g^{1/2} \cdot W} \right)^{2/3}. \tag{9.14}$$

Note that, for a given discharge, critical depth depends only on width (not on resistance or slope).

Subcritical flow encountering a sudden drop in bed elevation, such as a weir or waterfall, accelerates and may pass through the critical state. Flow can also be forced to change from subcritical to critical if it passes through a sudden width contraction, such as a bridge opening, or encounters a sudden increase in slope or decrease in resistance. The marked decrease in elevation accompanying the subcritical-to-supercritical transition is called a **hydraulic drop**. Conversely, supercritical flows may be forced into the subcritical state by conditions that produce sudden decreases in velocity, such as encountering an obstacle like a dam, a channel widening, a decrease in slope, or an increase in resistance. The supercritical-to-subcritical transition is marked by a sudden increase in water-surface elevation called a **hydraulic jump**. The surface elevations before and after hydraulic drops and jumps are the sequent depths discussed in section 8.2.2. In these rapid changes in flow configuration and geometry, one cannot assume uniform-flow conditions and hydrostatic pressure, so they are not gradually varied flows. These rapidly varied flows are discussed in chapter 10.

9.2.2 Mild and Steep Reaches

Consider a reach of a natural channel with a particular width, slope, and resistance and transmitting a particular discharge. The depths Y_n and Y_c can be computed via equations 9.13 and 9.14, respectively, and shown as lines parallel to the channel bottom (figure 9.2).

> If $Y_n > Y_c$, a uniform flow would be subcritical, and the reach slope is said to be **mild**.
>
> If $Y_n < Y_c$, a uniform flow would be supercritical, and the reach slope is said to be **steep**.

Although it is possible for $Y_n = Y_c$, this precise condition (called a **critical slope**) is unlikely. We should also note that the local channel slope could be zero (**horizontal slope**) or even negative (**adverse slope**), but these conditions are very rare over any distance in natural channel reaches. Thus, we will consider only mild and steep reaches here; Chow (1959) treats the other possibilities in some detail.

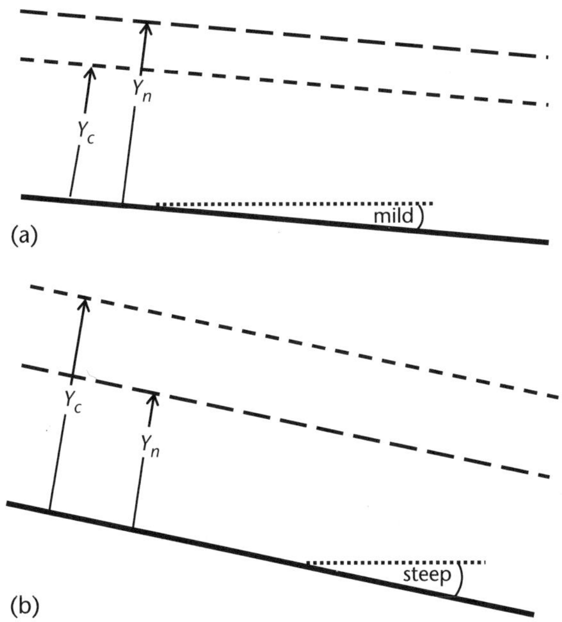

Figure 9.2 Relations between normal depth Y_n (long-dashed line) and critical depth Y_c (short-dashed line) for uniform flows on (a) mild and (b) steep slopes.

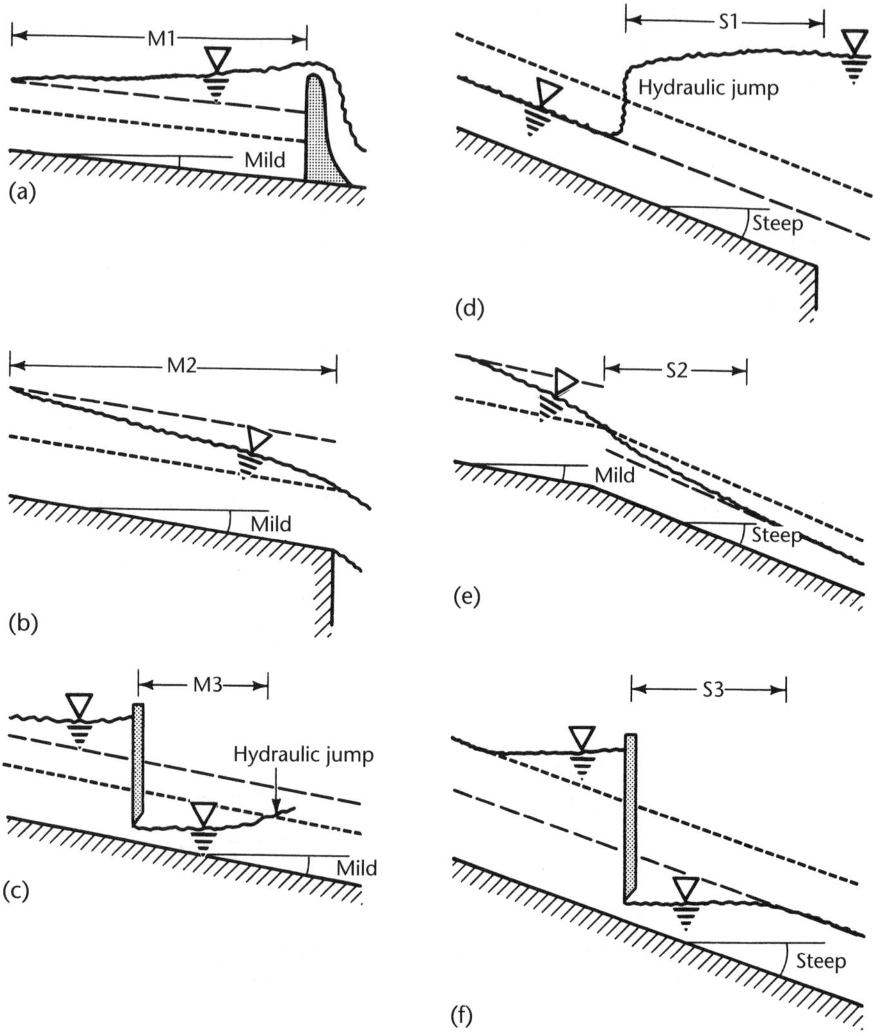

Figure 9.3 Typical situations associated with the most common types of water-surface profiles. Long-dashed lines represent normal depth; short-dashed lines represent critical depth. For details, see table 9.2. After Daily and Harleman (1966).

9.2.3 Profile Classification

Flow profiles are classified according to two criteria: 1) whether the channel slope is mild or steep, and 2) the relation of the actual depth to the normal depth and the critical depth. The classification is illustrated in figure 9.3 and summarized in table 9.1. The letters "M" for mild and "S" for steep specify whether a uniform flow in the reach would be subcritical or supercritical, respectively. Profiles lying above both Y_n and Y_c are designated "1," those lying between Y_n and Y_c are designated "2," and those lying

Table 9.1 Classification of flow profiles in natural channels.[a]

Designation	Depth relations	Type	Flow state	Figure
Mild slopes ($Y_n > Y_c$)				
M1	$Y > Y_n > Y_c$	Backwater; $\frac{dY}{dX} > 0$	Subcritical	9.3a
M2	$Y_n > Y > Y_c$	Drawdown; $\frac{dY}{dX} < 0$	Subcritical	9.3b
M3	$Y_n > Y_c > Y$	Backwater; $\frac{dY}{dX} > 0$	Supercritical	9.3c
Steep slopes ($Y_c > Y_n$)				
S1	$Y > Y_c > Y_n$	Backwater; $\frac{dY}{dX} > 0$	Subcritical	9.3d
S2	$Y_c > Y > Y_n$	Drawdown; $\frac{dY}{dX} < 0$	Supercritical	9.3e
S3	$Y_c > Y_n > Y$	Backwater; $\frac{dY}{dX} > 0$	Supercritical	9.3f

[a]Typical situations inducing the various profile types are shown in figure 9.3.

below Y_n and Y_c are designated "3." Profiles in which depth increases downstream are called **backwater profiles**; those in which depth decreases downstream are called **drawdown profiles**.

Because most natural-channel flows are subcritical, by far the most common profile types encountered are M1 and M2.

9.3 Controls

As can be seen in equation 9.13, the normal depth for a given discharge is determined by the local channel width, slope, and resistance. Thus, a spatial change in one or more of these factors produces a change in depth as the flow seeks to achieve the new normal depth. A **control** is a portion of a channel in which a relatively marked change occurs in one or more of the factors controlling normal depth such that it determines the depth associated with a given discharge for some distance along the channel—upstream, downstream, or both. More succinctly, "A control [is] any channel feature, natural or man-made, which fixes a relationship between depth and discharge in its neighborhood" (Henderson 1966, p. 174).

A change in depth can be viewed as a positive or negative gravity wave that travels along the channel at the celerity C_{gw} given by equation 6.4:

$$C_{gw} = (g \cdot Y)^{1/2}. \tag{9.15}$$

The wave celerity is its velocity with respect to the water velocity. Thus, if the flow is subcritical, $C_{gw} > U$ and the depth change can be transmitted both upstream and downstream. However, if the flow is supercritical, $C_{gw} < U$, and the "information" about the new normal depth cannot be transmitted upstream; that is, "the water doesn't know what's happening downstream" (Henderson 1966, p. 40).

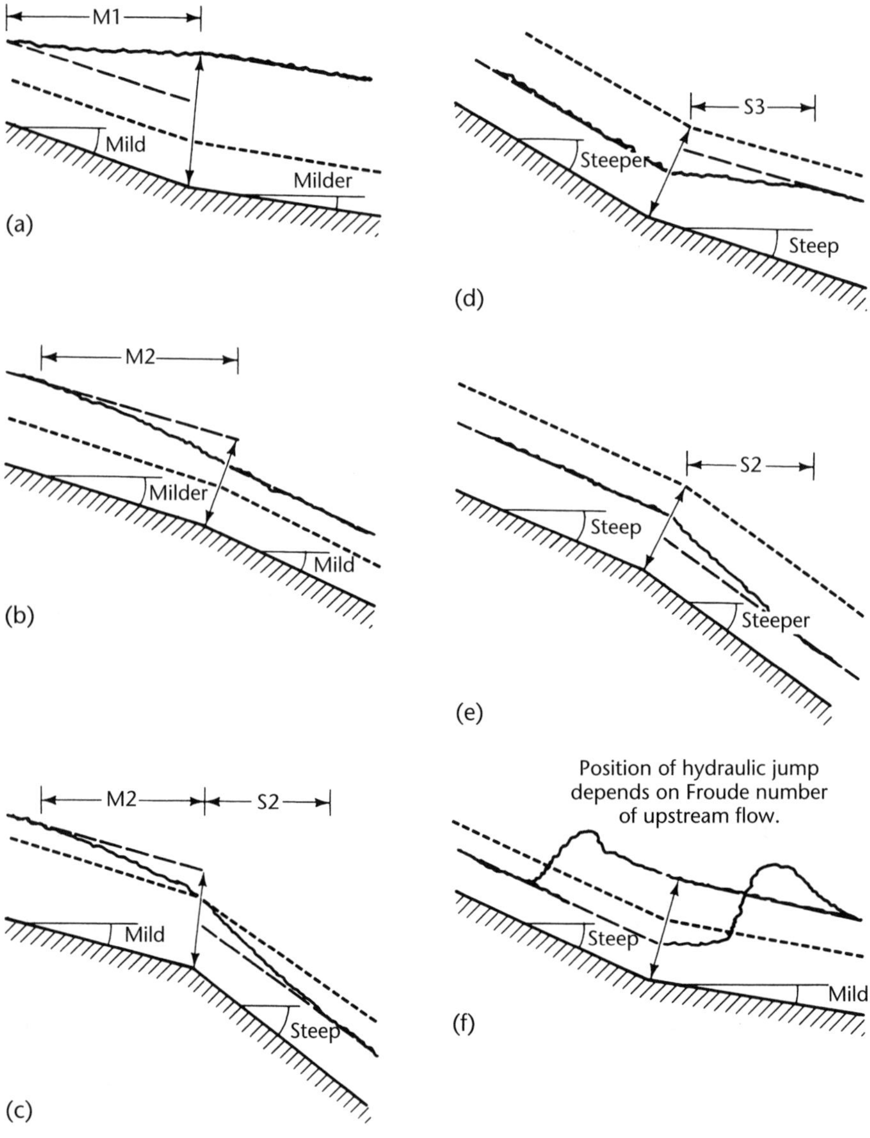

Figure 9.4 Water-surface profiles associated with controls exerted by changes in slope. Abrupt changes in width and/or resistance produce similar effects. Long-dashed lines represent normal depth; short-dashed lines represent critical depth. Vertical arrows indicate the control section.

Figure 9.4 shows how abrupt changes in channel slope act as controls; changes in width and/or resistance have similar effects. In figure 9.4a–c, the flow upstream of the control is subcritical and the control therefore determines the depth to the next control upstream. In figure 9.4c the flow changes from subcritical to supercritical, so the influence of the control extends both upstream and downstream. In figure 9.4d and e,

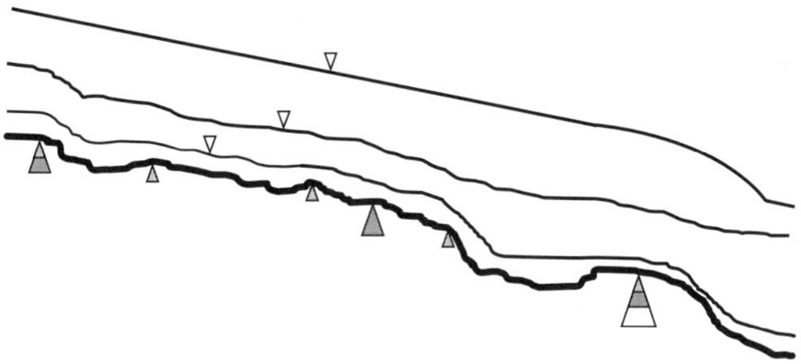

Figure 9.5 Diagram illustrating partial section controls. The lowest line is the channel-bottom profile; the other lines represent water surfaces at successively higher discharges. The smallest triangles indicate section controls effective over short distances at low flows; the successively larger triangles indicate section controls effective over successively longer distances at successively higher flows.

the upstream flow is supercritical, so the control cannot affect the upstream situation and only determines the depth for a distance downstream. In figure 9.4f, the transition from supercritical to subcritical flow is marked by a highly turbulent standing wave—the hydraulic jump—whose exact position and form are determined by the Froude number of the upstream flow and the channel slopes (section 10.1).

In natural channels, the changes in slope, width, or resistance that produce a control may occur within a relatively short, distinct reach, in which case they are called **section controls**. More diffuse changes that take place over longer distances are **channel controls** (Corbett 1945). Section controls are good places to establish discharge-measurement stations, because the depth-discharge relation immediately upstream tends to be stable. However, sections that act as controls at relatively low discharges may be "drowned out" at higher discharges if more profound controls downstream extend their influence over longer distances; these are called **partial controls** (figure 9.5).

The various types of weirs and flumes discussed in chapter 10 are **artificial controls** designed to provide stable, precise relations between depth and discharge for accurate flow measurement.

9.4 Water-Surface Profiles: Computation

If the flow in a given channel is uniform, the depth corresponding to a given discharge can be computed via the Chézy (or Manning) equation. Natural channels, however, are highly variable in geometry and bed material, and as indicated in section 6.2.2.1 and suggested by figures 9.3 and 9.4, the uniform-flow condition is more realistically considered to be an asymptotic condition rarely exactly achieved. Here, we examine the methodology for computing depths, and hence water-surface profiles, for these asymptotic situations.

First, section 9.4.1 presents a theoretical development using continuous mathematics that provides some physical and mathematical insight to water-surface profiles and the classification introduced in table 9.1 and figure 9.3. In section 9.4.2 we develop a discrete-mathematics approach that is the basis for the methodology incorporated in the computer models that are widely used for determining flood-prone areas. Both the theoretical and practical approaches are based on one-dimensional macroscopic versions of the three fundamental physical discussed in section 9.1:

1. The continuity relation
2. The energy equation
3. A uniform-flow resistance relation

Both approaches arrive at equations for computing the spatial rate-of-change of depth in a given channel at a given discharge, and they both require that computation begin at a cross section where the depth is known.

Most texts and conventional engineering practice adopt the Manning equation to express uniform-flow relations, but as discussed in chapter 6, the Chézy equation has a firmer theoretical basis. Thus, in the theoretical development we will use both equations, but in the practical methodology we will use only the traditional Manning equation.

Following presentation of the continuous and discrete mathematical approaches to profile computation, we conclude with a discussion of the some of the practical aspects of profile computation (section 9.4.2.3).

9.4.1 Theoretical Basis

Consider a channel carrying a steady flow of specified discharge Q. To simplify the development, assume the energy coefficient $\alpha = 1$ and hydrostatic pressure distribution with $\cos \theta_0 = 1$.

From the definition of specific head (section 8.1.2.1), the total energy per weight of flowing water, $\mathcal{H}$, at a given cross section can be written as the sum of the elevation head Z and the specific head, $\mathcal{H}_S$:

$$\mathcal{H} = Z + \mathcal{H}_S. \tag{9.16}$$

Taking the derivative of $\mathcal{H}$ relative to the downstream direction X,

$$\frac{d\mathcal{H}}{dX} = \frac{dZ}{dX} + \frac{d\mathcal{H}_S}{dX}. \tag{9.17}$$

We can now substitute the definition of the channel slope, S_0, from equation 7.11 and of the friction slope, S_f, from equation 8B2.2 and write

$$\frac{d\mathcal{H}_S}{dX} = S_0 - S_f. \tag{9.18}$$

Noting that

$$\frac{d\mathcal{H}_S}{dX} = \frac{d\mathcal{H}_S}{dY} \cdot \frac{dY}{dX}, \tag{9.19}$$

we can substitute 9.19 into 9.18 and solve for dY/dX:

$$\frac{dY}{dX} = \frac{S_0 - S_f}{d\mathcal{H}_S/dX}. \tag{9.20}$$

BOX 9.1 Derivation of the Downstream Rate-of-Change-of-Depth Relation (Equation 9.21)

We saw in equation 8.13 that

$$\frac{d\mathcal{H}_S}{dY} = 1 - \frac{Q^2}{g \cdot W^2 \cdot Y^3}, \tag{9B1.1}$$

and substituting equation 9B1.1 into equation 9.20 gives

$$\frac{dY}{dX} = \frac{S_0 - S_f}{1 - \frac{Q^2}{g \cdot W^2 \cdot Y^3}}. \tag{9B1.2}$$

This expression can be simplified by recalling from equation 8.14 that

$$\frac{Q^2}{g \cdot W^2} = Y_c^3, \tag{9B1.3}$$

so equation 9B1.2 can be written as

$$\frac{dY}{dX} = \frac{S_0 - S_f}{1 - \left(\frac{Y_c}{Y}\right)^3}. \tag{9B1.4}$$

We can write the numerator of equation 9B1.4 in a form similar to the denominator by noting that $S_0 - S_f = S_0 \cdot (1 - S_f/S_0)$:

$$\frac{dY}{dX} = S_0 \cdot \frac{1 - \left(\frac{S_f}{S_0}\right)}{1 - \left(\frac{Y_c}{Y}\right)^3} \tag{9B1.5}$$

Then, following the steps in box 9.1, we can express the downstream rate of change of depth as

$$\frac{dY}{dX} = S_0 \cdot \left[\frac{1 - (S_f/S_0)}{1 - (Y_c/Y)^3}\right]. \tag{9.21}$$

Our next goal is to develop an expression for dY/dX as a function of the normal, critical, and actual depths. To do this, we invoke a uniform-flow relation—either the Chézy equation (the theoretically preferred approach) or the Manning equation (the traditional approach). For both relations, 1) the normal depth Y_n is related to the channel slope, S_0, directly from the uniform-flow relations; and 2) the actual depth Y is related to the friction slope, S_f, assuming that the uniform-flow relations are applicable to gradually varied flow.

For the Chézy equation, the relation between channel slope and normal depth is given by equation 9.13C:

$$Y_n = \left(\frac{\Omega \cdot Q}{g^{1/2} \cdot W \cdot S_0^{1/2}}\right)^{2/3}, \tag{9.22}$$

which is rearranged to give

$$S_0 = \frac{\Omega^2 \cdot Q^2}{g \cdot W^2 \cdot Y_n^3}. \quad (9.23)$$

On the assumption that the uniform-flow relation applies to gradually varied flow, we substitute S_f for S_0 and Y for Y_n in equation 9.23 to give

$$S_f = \frac{\Omega^2 \cdot Q^2}{g \cdot W^2 \cdot Y^3}. \quad (9.24)$$

Then, from equations 9.23 and 9.24 we see that

$$\frac{S_f}{S_0} = \left(\frac{Y_n}{Y}\right)^3. \quad (9.25C)$$

Using the Manning equation, the relation between slope and normal depth is given by equation 9.13M, and we find

$$\frac{S_f}{S_0} = \left(\frac{Y_n}{Y}\right)^{10/3}. \quad (9.25M)$$

Now substituting equations 9.25C and 9.25M into equation 9.21 yields the expressions we sought:

Chézy:

$$\frac{dY}{dX} = S_0 \cdot \left[\frac{1-(Y_n/Y)^3}{1-(Y_c/Y)^3}\right] \quad (9.26C)$$

Manning:

$$\frac{dY}{dX} = S_0 \cdot \left[\frac{1-(Y_n/Y)^{10/3}}{1-(Y_c/Y)^3}\right] \quad (9.26M)$$

These expressions can be directly related to the profile classifications in table 9.1 and figure 9.3. To see this, define

$$N \equiv 1 - \left(\frac{Y_n}{Y}\right)^3 \quad (9.27C)$$

if the Chézy equation is used or

$$N \equiv 1 - \left(\frac{Y_n}{Y}\right)^{10/3} \quad (9.27M)$$

if the Manning equation is used, and

$$D \equiv 1 - \left(\frac{Y_c}{Y}\right)^3. \quad (9.28)$$

Now we see that if $Y_n < Y, N > 0$; if $Y_n > Y, N < 0$; and if $Y_c < Y, D > 0$; if $Y_c > Y$, $D < 0$. Then, the sign of the ratio N/D determines the sign of dY/dX, that is, whether the depth increases or decreases in the downstream direction. The various possibilities are shown in table 9.2.

Table 9.2 Relation of water-surface profile classification (table 9.1, figure 9.3) to equations 9.26–9.28, assuming $S_0 > 0$.

Depth relations	N	D	N/D, $\frac{dY}{dX}$	Profile type
Mild slopes ($Y_n > Y_c$)				
$Y > Y_n > Y_c$	$1 > N > 0$	$1 > D > 0$	>0	M1, backwater
$Y_n > Y > Y_c$	$N < 0$	$1 > D > 0$	<0	M2, drawdown
$Y_n > Y_c > Y$	$N < 0$	$D < 0$	>0	M3, backwater
Steep slopes ($Y_c > Y_n$)				
$Y > Y_c > Y_n$	$1 > N > 0$	$1 > D > 0$	>0	S1, backwater
$Y_c > Y > Y_n$	$1 > N > 0$	$D < 0$	<0	S2, drawdown
$Y_c > Y_n > Y$	$N < 0$	$D < 0$	>0	S3, backwater

Two other implications of equation 9.26 are of interest. When $Y = Y_n$, $dY/dX = 0$, consistent with the fact that depth does not change in a reach with uniform flow. However, when $Y = Y_c$, the change in depth is not defined. This reflects the fact that water surfaces are unstable when flows are near critical, as discussed in section 6.2.2.2. This instability is also suggested in the specific-head diagram (see figure 8.6), which has a very steep slope in the vicinity of the critical depth. This means that when the flow is near the critical regime, a small change in its energy leads to a relatively large depth change. In natural channels, there are ubiquitous small variations in slope, width, and resistance that affect the energy of the flow, so when the flow is near critical, the surface is wavy and irregular (figure 9.6). Under these conditions, the flow is rapidly varied rather than gradually varied, and the assumptions of uniform flow are no longer valid.

Equation 9.26 can be rearranged to

$$dY = S_0 \cdot \left[\frac{1 - (Y_n/Y)^3}{1 - (Y_c/Y)^3}\right] \cdot dX \tag{9.29}$$

(the exponent in the numerator = 10/3 if the Manning relation is used). We see from equations 9.13 and 9.14 that, in general, Y_n and Y_c are functions of distance along the channel, X. Thus, we can integrate equation 9.29 between a location X_i where the depth is Y_i and a location X_{i+1} where the depth is Y_{i+1}:

$$\int_{Y_i}^{Y_{i+1}} dY = \int_{X_i}^{X_{i+1}} S_0 \cdot \left[\frac{1 - [Y_n(X)/Y(X)]^3}{1 - [Y_c(X)/Y(X)]^3}\right] \cdot dX$$

$$Y_{i+1} = Y_i + \int_{X_i}^{X_{i+1}} S_0 \cdot \left[\frac{1 - [Y_n(X)/Y(X)]^3}{1 - [Y_c(X)/Y(X)]^3}\right] \cdot dX. \tag{9.30}$$

(Again, the exponent in the numerator = 10/3 if the Manning relation is used.) If, for a given discharge, we know 1) the depth at a starting location ($i = 0$), 2) the bottom elevation and channel geometry at successive locations along the channel, and 3) information required for estimating resistance (Ω or n_M) at successive locations,

Figure 9.6 A high flow in a small New England stream. The extremely uneven surface is characteristic of flows that are close to critical. Photo by the author.

equation 9.30 can be integrated numerically to provide successive depths and water-surface elevations. As noted above, if the flow is subcritical, the integration proceeds in the upstream direction, and if supercritical, it proceeds in the downstream direction.

Chow (1959) described various mathematical approaches to integrating equation 9.30. However, in practice, the integration is usually carried out by a finite-difference approach, called the standard step method, described in the following subsection. This method is incorporated, with many elaborations, in computer programs for calculating water-surface profiles, such as the widely used U.S. Army Corps of Engineers' Hydrologic Engineering Center River Analysis System (HEC-RAS; Brunner 2001a).

9.4.2 The Standard Step Method

9.4.2.1 Basic Approach

From equation 8.7, the total mechanical energy per unit weight (head) at cross section i, $\mathcal{H}_i$, can be written as the sum of the potential-energy head, $\mathcal{H}_{\text{PE}i}$, and the kinetic-energy (velocity) head, $\mathcal{H}_{\text{KE}i}$:

$$\mathcal{H}_i = \mathcal{H}_{\text{PE}i} + \mathcal{H}_{\text{KE}i}. \tag{9.31}$$

The potential-energy head represents the elevation of the water surface above a datum (see figure 8.5) (assuming, as we will throughout this section, that $\cos\theta_0 = 1$). Thus, from equation 8.8b, we can write the energy equation between an upstream section (designated by subscript i) and a downstream section (designated by subscript $i-1$) as

$$\mathcal{H}_{\text{PE}i} + \mathcal{H}_{\text{KE}i} = \mathcal{H}_{\text{PE}i-1} + \mathcal{H}_{\text{KE}i-1} + \Delta\mathcal{H}, \tag{9.32}$$

where $\Delta\mathcal{H}$ is the total head loss between the two sections. For subcritical flow we compute in the upstream direction, so the working form of equation 9.32 is

$$\mathcal{H}_{\text{PE}i} = \mathcal{H}_{\text{PE}i-1} + \mathcal{H}_{\text{KE}i-1} + \Delta\mathcal{H} - \mathcal{H}_{\text{KE}i}. \tag{9.33a}$$

For supercritical flow, we solve for the downstream water-surface elevation:

$$\mathcal{H}_{\text{PE}i-1} = \mathcal{H}_{\text{PE}i} + \mathcal{H}_{\text{KE}i} - \Delta\mathcal{H} - \mathcal{H}_{\text{KE}i-1}. \tag{9.33b}$$

Since subcritical flow is by far the more common, subsequent developments here will use only equation 9.33a.

Following the discussion in section 8.3, the total head loss between sections is usually divided into two parts:

$$\Delta\mathcal{H} = \Delta\mathcal{M} + \Delta\mathcal{H}_{\text{eddy}}, \tag{9.34}$$

where $\Delta\mathcal{M}$ represents the energy loss due to friction with the flow boundary (**friction loss),** and $\Delta\mathcal{H}_{\text{eddy}}$ represents the energy losses due to flow expansion or contraction (**eddy loss** or **shock loss**). The friction loss is computed from the average friction slope $\bar{S}_f$, which is computed from the selected uniform-flow equation at the upstream and downstream sections:

$$\frac{\Delta\mathcal{M}}{\Delta X} \equiv \bar{S}_f,$$

$$\Delta\mathcal{M} = \bar{S}_f \cdot \Delta X. \tag{9.35}$$

The eddy loss is usually estimated via equation 9.5b as

$$\Delta H_{\text{eddy}} = k_{\text{eddy}} \cdot \left| \left(\frac{U_i^2}{2 \cdot g} - \frac{U_{i\text{-}1}^2}{2 \cdot g} \right) \right|, \tag{9.36}$$

where k_{eddy} is estimated as described in table 9.3.

Combining equations 9.33a and 9.34, the basic working equation for computing water-surface profiles in subcritical flows is

$$\mathcal{H}_{\text{PE}i} = \mathcal{H}_{\text{PE}i-1} + \mathcal{H}_{\text{KE}i-1} - \mathcal{H}_{\text{KE}i} + \Delta\mathcal{M} + \Delta\mathcal{H}_{\text{eddy}}, \tag{9.37}$$

from which the upstream depth, Y_i, is calculated as the difference between the potential head and the bed elevation, Z_i:

$$Y_i = \mathcal{H}_{\text{PE}i} - Z_i \tag{9.38}$$

Table 9.3 Values of the eddy-loss coefficient k_{eddy} for subcritical flows (after Brunner 2001b).

Nature of width transition	Contraction ($W_U > W_D$)	Expansion ($W_U < W_D$)
None to very gradual	0.0	0.0
Gradual	0.1	0.3
Typical bridge sections	0.3	0.5
Abrupt	0.6	0.8

Table 9.4 Example of water-surface profile computations.

1	2	3	4	5	6	7	8	9	10	11
Sect., i	Distance, X (m)	Bed elev., Z_0 (m)	Width, W (m)	Manning n_M	Est. depth, $\hat{Y}$ (m)	Area, A (m^2)	Velocity, U (m/s)	Froude no., Fr	Vel. head, $\mathcal{H}_{KE}$ (m)	Fric. slope, S_f
0	0	843.14	147.0	0.043	10.21	1500.85	2.67	0.27	4.71E−01	5.93E−04
1	100	843.25	137.2	0.036	10.16	1393.95	2.87	0.29	5.46E−01	4.85E−04
2	200	843.00	152.4	0.039	10.73	1624.58	2.46	0.24	4.02E−01	3.93E−04
3	400	844.05	162.8	0.037	9.67	1574.28	2.54	0.26	4.28E−01	4.29E−04
4	600	844.57	162.0	0.045	9.26	1500.12	2.67	0.28	4.71E−01	7.43E−04
5	1000	845.81	167.3	0.040	8.48	1418.70	2.82	0.31	5.27E−01	7.36E−04
6	1500	846.74	128.2	0.051	8.36	1071.75	3.73	0.41	9.23E−01	2.14E−03
7	2000	847.23	150.3	0.038	9.05	1360.22	2.94	0.31	5.73E−01	6.27E−04
8	2500	849.00	161.0	0.043	7.45	1199.45	3.34	0.39	7.37E−01	1.41E−03

See text for discussion. The computed profile is plotted in figure 9.8. Pot., potential.

9.4.2.2 Detailed Steps and Example Calculation

Here we describe the details and show the results of an example computation using the standard step method. The procedure used here, based on computations carried out via the spreadsheet program WSPROFILE.XLS (available at the book's website, http://www.oup.com/us/fluvialhydraulics; see appendix D), is a much-simplified version of the approach incorporated in such programs as the U.S. Army Corps of Engineers' HEC-RAS (Brunner 2001a, 2001b) or the U.S. Geological Survey's WSPRO (for Water-Surface Profile) program (Shearman 1990). HEC-RAS is a very elaborate but user-friendly program that is widely used by practitioners for calculating water-surface profiles.

The computations can be followed in table 9.4. In this (fictitious) example, we calculate the water-surface profile for a river upstream of its entrance into a reservoir when the discharge is 4,000 m^3/s. The channel characteristics determined by survey and observation are entered in columns 2–5. The depth at the downstream end (Y_0) is fixed by the known reservoir elevation, which has been entered in the first row of column 6. At all sections, we assume a rectangular channel with $\alpha = 1.3$ and $k_{eddy} = 0.3$ for expanding sections and 0.1 for contracting sections. We specify a tolerance of $\Delta Y = 0.02$ m as the maximum acceptable difference between the initial trial depth and the computed depth.

Once Y_0 is entered, the other quantities for that section (except slope) are calculated in other columns. Beginning at the next upstream section ($i = 1$), computation proceeds via the following steps. Quantities that have been predetermined by survey, observation, or estimation are shown in boldface:

Column 6. Enter a trial depth $\hat{Y}_i$.

Column 7. The cross-sectional area of the flow is computed as $A_i = \hat{Y}_i \cdot \mathbf{W}_i$.

12	13	14	15	16	17	18	19	20	21
Avg. slope, $\bar{S}_f$	Frict. loss, $\Delta\mathcal{M}$ (m)	Eddy loss, $\Delta\mathcal{H}_{\text{eddy}}$ (m)	Total head, $\mathcal{H}$ (m)	Pot. head, $\mathcal{H}_{\text{PE}}$ (m)	Pot. head OK?	Calc. depth, Y (m)	Depth OK?	Normal depth, Y_n (m)	Critical, Y_c (m)
			853.82			10.21		8.48	4.23
5.39E−04	5.39E−02	7.50E−03	853.88	853.40	yes	10.15	yes	7.95	4.43
4.39E−04	4.39E−02	4.32E−02	853.97	853.65	yes	10.65	yes	6.12	4.13
4.11E−04	8.22E−02	7.82E−03	854.06	853.72	yes	9.67	yes	4.56	3.95
5.85E−04	1.17E−01	4.33E−03	854.18	853.83	yes	9.26	yes	6.35	3.96
7.38E−04	2.95E−01	1.67E−02	854.49	854.28	yes	8.47	yes	5.57	3.88
1.44E−03	7.18E−01	3.96E−02	855.25	855.08	yes	8.34	yes	8.71	4.63
1.37E−03	6.99E−01	1.05E−01	856.05	856.29	yes	9.06	yes	8.05	4.16
1.04E−03	5.19E−01	4.92E−02	856.62	856.45	yes	7.45	yes	5.66	3.98

Column 8. The cross-sectional average velocity is computed as $U_i = \boldsymbol{Q}/A_i$.

Column 9. The Froude number is computed as $Fr_i = U_i/(\boldsymbol{g}\cdot\hat{Y}_i)^{1/2}$. This provides a check that the flow is subcritical ($Fr_i < 1$) so that computations can proceed in the upstream direction.

Column 10. The velocity head is computed as $\mathcal{H}_{\text{KE}i} = \alpha_i \cdot U_i^2/(2\cdot\boldsymbol{g})$.

Column 11. The friction slope S_{fi} is computed. Here we use the Manning equation, so $S_{fi} = \boldsymbol{n}_{Mi}^2 \cdot \boldsymbol{Q}^2/(\boldsymbol{u}_m^2 \cdot \boldsymbol{W}_i^2 \cdot \hat{Y}_i^{10/3})$. We assume a single value of $\boldsymbol{n}_{Mi}$ at each cross section, but in many sections the resistance varies significantly as a function of width, especially if floodplains are included, and this variation must be accounted for. Box 9.2 and figure 9.7 describe the general approach for doing this. (The example assumes no overbank flow or cross-sectional variation of resistance.)

Column 12. The average friction slope for adjacent sections, $\bar{S}_{fi}$, is determined as the arithmetic mean of the slope at the current section and the adjacent downstream section: $\bar{S}_{fi} = (S_{fi-1} + S_{fi})/2$.[2]

Column 13. The friction loss between sections i and $i-1$, $\Delta\mathcal{M}_{i,i-1}$, is calculated as the product of $\bar{S}_{fi}$ and the distance between the two sections: $\Delta\mathcal{M}_{i,i-1} = \bar{S}_{fi}\cdot(\boldsymbol{X}_i - \boldsymbol{X}_{i\text{-}1})$.

Column 14. The eddy loss is computed as $\Delta\mathcal{H}_{\text{eddy}:i,i-1} = \boldsymbol{k}_{\text{eddy}i}|[U_i^2/(2\cdot\boldsymbol{g}) - U_{i-1}{}^2/(2\cdot\boldsymbol{g})]|$.

Column 15. The total head $\mathcal{H}_i$ is computed as $\mathcal{H}_i = \mathcal{H}_{i-1} + \Delta\mathcal{H}_{i,i-1} = \mathcal{H}_{i-1} + \Delta\mathcal{M}_{i,i-1} + \Delta\mathcal{H}_{\text{eddy}:i,i-1}$. To satisfy the second law of thermodynamics, it must be true that $\mathcal{H}_i > \mathcal{H}_{i-1}$ where section i is upstream of section $i-1$.

Column 16. The potential head $\mathcal{H}_{PEi}$ is computed as $\mathcal{H}_{\text{PE}i} = \mathcal{H}_{i-1} - \mathcal{H}_{\text{KE}i} + \Delta\mathcal{M}_{i,i-1} + \Delta\mathcal{H}_{\text{eddy}:i,i-1}$.

Column 17. Here we check that $\mathcal{H}_{\text{PE}i} > \mathcal{H}_{\text{PE}i-1}$, which must be true in order for flow to occur. "No" appears in this column if this condition is not satisfied.

BOX 9.2 Accounting for Resistance Variations in Channel Cross Sections

Marked variations in resistance in various parts of a cross section are common. These can occur within the channel where the roughness height, y_r, changes significantly and will almost always be present if the floodplain, which typically contains trees and/or brush, is included in the section. Failure to account for such changes can lead to large errors in computed water-surface profiles.

The general resistance relation can be written as

$$\mathcal{Q} = \mathcal{K} \cdot S_f{}^{1/2}, \tag{9B2.1}$$

where $\mathcal{K}$ is called the **conveyance**:

$$\mathcal{K} \equiv \frac{\mathcal{Q}}{S_f{}^{1/2}}. \tag{9B2.2}$$

Thus, if the Chézy equation (equation 9.8) is used,

$$\mathcal{K} = \Omega^{-1} \cdot g^{1/2} \cdot W \cdot Y^{3/2}; \tag{9B2.3-C}$$

if the Manning equation (equation 9.10) is used,

$$\mathcal{K} = n_M{}^{-1} \cdot u_M \cdot W \cdot Y^{5/3}. \tag{9B2.3-M}$$

Cross-channel resistance changes at a given cross section are accounted for by assuming that the friction slope S_f is constant across the section and computing it via equation 9B2.2:

$$S_f = \left(\frac{\mathcal{Q}}{\sum_{j=1}^{m} \mathcal{K}_j} \right)^2, \tag{9B2.4}$$

where Q is the discharge and $\mathcal{K}_j$ are the conveyances for segments j = 1, 2, ... , m of the section (figure 9.7). The $\mathcal{K}_j$ values are computed as

$$\text{Chézy: } \mathcal{K}_j = \Omega_j{}^{-1} \cdot g^{1/2} \cdot W_j \cdot Y_j^{3/2}; \tag{9B2.5-C}$$

$$\text{Manning: } \mathcal{K}_j = n_{Mj}{}^{-1} \cdot u_M \cdot W_j \cdot Y_j^{5/3}; \tag{9B2.5-M}$$

where the subwidths W_j are determined by survey.

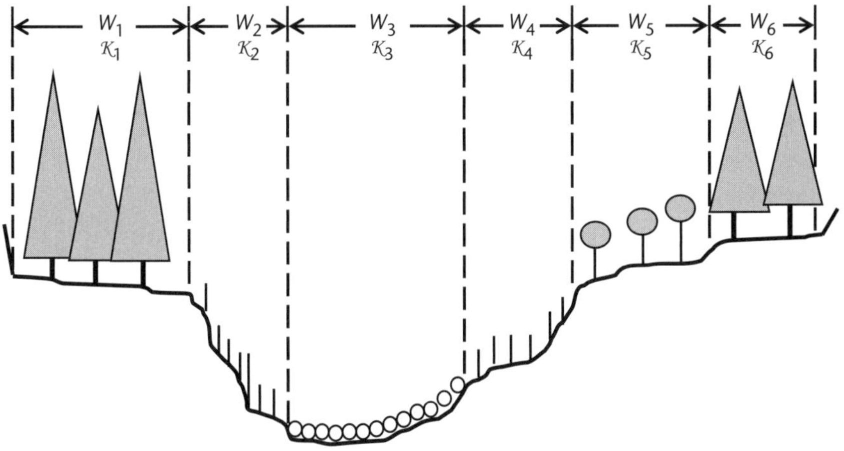

Figure 9.7 Division of a cross section into $m = 6$ segments of differing resistance for computation of conveyance (see box 9.2).

Column 18. The depth Y_i (i.e., the pressure head) is calculated as $Y_i = \mathcal{H}_{\mathrm{PE}i} - \mathbf{Z}_i$.

Column 19. The calculated depth Y_i is compared to the trial value of step 1, $\hat{Y}_i$. If the two depths differ by an amount greater than a prespecified **tolerance ΔY**, "NO" appears in this column, the computations for the section are invalid, and we return to column 6 and assume a new trial depth.

Column 20. The normal depth is calculated as $\mathbf{Y}_{ni} = [\mathbf{n}_{Mi} \cdot \mathbf{Q}/(u_M \cdot S_{0i}{}^{1/2} \cdot \mathbf{W}_i)]^{3/5}$. This value is computed for comparison with the calculated depth. $Y_i > \mathbf{Y}_{ni}$ in reaches with M1 profiles; $Y_i < \mathbf{Y}_{ni}$ in reaches with M2 profiles. (No value is shown if the slope is adverse.)

Column 21. The critical depth is calculated as $\mathbf{Y}_{ci} = [\mathbf{Q}/(\mathbf{g}^{1/2} \cdot \mathbf{W}_i)]^{2/3}$. This value is computed for comparison with the calculated depth. $Y_i > \mathbf{Y}_{ci}$ for reaches with subcritical flow, which was assumed in the calculations here.

The computed profile for this example is shown in figure 9.8.

9.4.2.3 Factors Affecting Accuracy

Accuracy of the computed water-surface profile for a specified design discharge in an actual channel segment depends fundamentally on 1) the degree to which the assumptions of steady gradually varied flow are appropriate, 2) the accuracy to which the channel-bed elevation is measured, and 3) the fidelity with which the geometry and resistance of the segment are captured in the computations. Although complex but user-friendly computer programs for computing water-surface profiles such as HEC-RAS (Brunner 2001a, 2001b) and WSPRO (Shearman 1990) are readily available, successful application of the methods described here requires accurate field measurements and considerable experience and judgment.

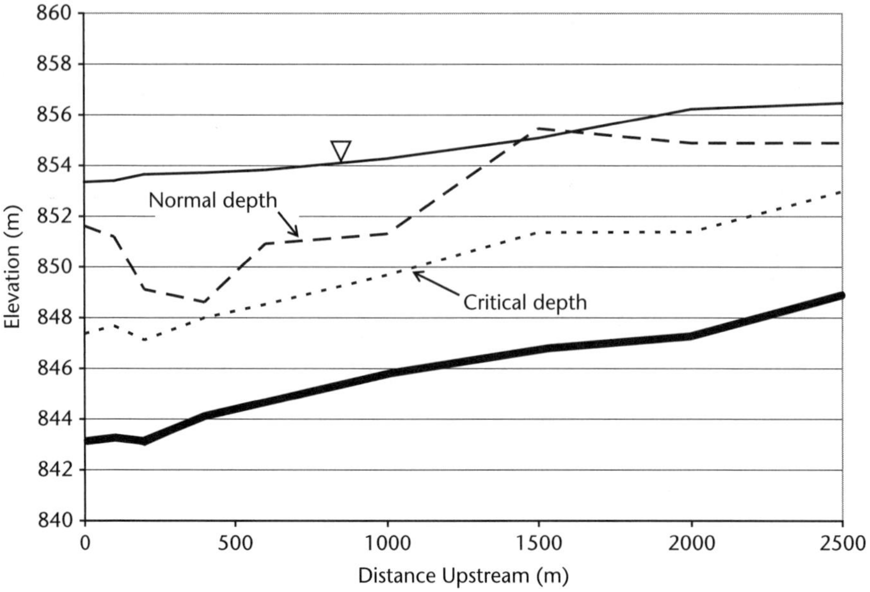

Figure 9.8 Computed water-surface profile for the example in table 9.4.

The major specific issues affecting the representation of hydraulic conditions are as follows:

1. Location and spacing of the surveyed cross sections. Cross sections should be representative of the reach between them and located so that the energy, water-surface, and bed slopes are as parallel possible. To help assure this, Davidian (1984) recommended locating sections at

   a. Major breaks in bed profile
   b. Points of minimum and maximum cross-sectional areas
   c. Shorter intervals in expanding regions and bends
   d. Shorter intervals where there are rapid changes of width, depth, and/or resistance
   e. Shorter intervals in streams with very low slopes
   f. At or near control sections (section 9.3) and at shorter intervals near control sections
   g. Upstream and downstream of large tributary junctions

   However, the accuracy of a finite-difference computation such as the standard step method depends critically on the spacing of cross sections, and one should not hesitate to insert cross sections even though the additional sections do not reflect major changes in geometry or resistance. The location of cross sections is more important than exact shape and area of the cross section for properly defining the energy loss, and the U.S. Army Corps of Engineers (1969) stated that the cross sections should not necessarily be restricted to the actual surveyed cross sections that are available. For large rivers where the cross sections are fairly uniform and slopes are approximately = 0.002, cross sections may be spaced up to one mile (1.6 kilometers) apart. For small streams on very steep slopes, five or more cross sections per mile may be required. Additional cross sections should be added when the cross-sectional area changes appreciably, when a change in roughness occurs, or when a marked change in bottom slope occurs.

2. The accuracy with which the resistance of the channel and floodplain is represented. In a study to evaluate factors that affect the accuracy of computed water-surface profiles, the U.S. Army Corps of Engineers (1986) found that the error in computed profiles increases significantly with decreased reliability of the estimate of channel resistance (Manning's n_M) and can be several times the error resulting from typical errors in surveying cross-section geometry. The study also showed that even experienced hydraulic engineers can differ widely in their estimate of n_M for a given reach, when that estimate is based only on the use of expedient methods (i.e., verbal descriptions and photographs; see table 6.3 and figure 6.22). The study results emphasize the importance of obtaining reliable determinations of resistance via field measurement, as shown in box 6.9.

3. The accuracy of surveying of cross-section geometry, including floodplains. The U.S. Army Corps of Engineers (1986) study found that on-site measurements of cross-section geometry by standard field techniques (see Harrelson et al. 1994) introduced little error into profile computations. Determining cross-section geometry from spot elevations measured from aerial photographs produced relatively small typical profile-elevation errors, ranging from 0.02 to 0.2 ft, depending on the contour interval. However, determining geometry from conventional topographic maps produced typical profile-elevation errors from 0.1 to more than 1 ft, again depending on contour interval. New techniques are now becoming available that make use of digital elevation models consisting of closely spaced elevations determined by airborne laser altimetry (LIDAR). These techniques show much promise for combining with water-surface profile programs to provide automated approaches to generating profiles and mapping flood-inundation areas (e.g., Noman et al. 2001; Bates et al. 2003; Omer et al. 2003).

4. Precision to which the depth at the initial section is known. As noted above, profile computations must begin at a section where the water-surface elevation or depth is known for the discharge(s) of interest. This is typically a gaging station where the rating curve (stage-discharge relation) has been established by standard field measurements. Other possible starting points are at a weir, dam, or channel constriction where the flow becomes critical (see chapter 10) or at the inflow to a lake or reservoir where the water-surface elevation is known. Where no known elevation is available, one can begin the computations with an assumed depth at a point downstream (assuming subcritical flow) from the reach where the profile is needed. If the starting point is far enough downstream and the assumed elevation is not too different from the true value, the computed profile will converge to the correct profile as you approach the reach of interest. Bailey and Ray (1966) give equations for estimating the distance X^* required:

M1 profiles:

$$X^* = (0.860 - 0.640 \cdot Fr^2) \cdot \left(\frac{Y_n}{S_0}\right) \tag{9.39a}$$

M2 profiles:

$$X^* = (0.568 - 0.788 \cdot Fr^2) \cdot \left(\frac{Y_n}{S_0}\right), \tag{9.39b}$$

where Fr is the Froude number, Y_n is the normal depth, and S_0 is the channel slope. Equation 9.39, a and b, assumes that the starting depth is between 0.75 and 1.25 times the true depth.

10

Rapidly Varied Steady Flow

10.0 Introduction and Overview

Rapidly varied flow is flow in which the spatial rates of change of velocity and depth are large enough to make the assumptions of uniform and gradually varied flow inapplicable. Such flow occurs at relatively abrupt changes in channel geometry (bed elevation, width, slope, curvature, resistance) and is quite common in natural streams, particularly cascade and step-pool mountain streams (see figure 2.14, table 2.4) and flows over pronounced bedforms (see section 6.6.4.2, table 6.2). Rapidly varied flow is also common at engineered structures such as bridges, culverts, weirs, and flumes.

In rapidly varied flow, the nature of the flow changes is determined by 1) the geometry of the stream bed or structure and 2) the flow regime. Recall from sections 6.2.2.2 and 8.1.2 that the flow regime is determined by the value of the Froude number, Fr:

$$Fr \equiv \frac{U}{(g \cdot Y)^{1/2}}, \tag{10.1}$$

where U is average velocity, g is gravitational acceleration, and Y is depth. The Froude number is the ratio between the flow velocity and the celerity of a shallow-water gravity wave. When $Fr = 1$, the flow is critical; when $Fr < 1$, the flow regime is subcritical; and when $Fr > 1$, the flow regime is supercritical.

Recall also from equation 9.14 (section 9.2.1.2) that in a channel of specified width W and discharge Q, the critical depth Y_c is given by

$$Y_c \equiv \left(\frac{Q}{g^{1/2} \cdot W} \right)^{2/3}. \tag{10.2}$$

Box 10.1 and figure 10.1 show that the flow regime can also be expressed in terms of the ratio of the actual depth to the critical depth, Y/Y_c:

$$Fr = \left(\frac{Y_c}{Y}\right)^{3/2}; \qquad (10.3)$$

when $Y > Y_c$, the flow is subcritical; when $Y < Y_c$, the flow is supercritical.

The following features distinguish rapidly varied flow from gradually varied flow (Chow 1959):

- The rapid changes in flow configuration produce eddies, rollers, and zones of flow separation resulting in velocity distributions that cannot be characterized by the Prandtl-von Kármán or other regular distributions discussed in chapter 5.
- The curvature of the streamlines is pronounced, and the pressure distribution cannot be assumed to be hydrostatic (see figure 4.5).

BOX 10.1 Relation between Y/Y_c and Froude Number

Here we show that the ratio Y/Y_c has a one-to-one relation to the Froude number and hence is an alternate way of expressing the flow regime.

To derive the relation between Y/Y_c and Fr, we begin with the definition of specific head, $\mathcal{H}_S$, from section 8.1.2.1 (continuing to assume that $\alpha = 1$):

$$\mathcal{H}_S \equiv Y + \frac{U^2}{2 \cdot g} \qquad (10B1.1a)$$

Rearranging equation 10B1.1a,

$$\frac{U^2}{2 \cdot g} = \mathcal{H}_S - Y. \qquad (10B1.1b)$$

Then, using the definition of Fr (equation 10.1), we can write equation 10B1.1b as

$$Fr^2 = \frac{2 \cdot (\mathcal{H}_S - Y)}{Y} = 2 \cdot \left(\frac{\mathcal{H}_S}{Y} - 1\right), \qquad (10B1.2a)$$

which can also be written as

$$Fr^2 = 2 \cdot \left(\frac{\mathcal{H}_S/Y_c}{Y/Y_c} - 1\right). \qquad (10B1.2b)$$

Using the conservation-of-mass relation $U = Q/(W \cdot Y)$, equation 10B1.1a can be written as

$$\mathcal{H}_S \equiv Y + \frac{Q^2}{2 \cdot g \cdot W^2 \cdot Y^2}, \qquad (10B1.3)$$

and dividing this by Y_c gives

$$\frac{\mathcal{H}_S}{Y_c} \equiv \frac{Y}{Y_c} + \frac{Q^2}{2 \cdot g \cdot W^2 \cdot Y^2 \cdot Y_c}. \qquad (10B1.4a)$$

Now using equation 10.2, equation 10B1.4a becomes

$$\frac{\mathcal{H}_S}{Y_C} \equiv \frac{Y}{Y_C} + \frac{1}{2} \cdot \left(\frac{Y_C}{Y}\right)^2. \qquad (10B1.4b)$$

Finally, we substitute equation 10B1.4b into 10B1.2b, and after some algebraic manipulation, we have the relation between *Fr* and Y/Y_C:

$$Fr = \left(\frac{Y_C}{Y}\right)^{3/2}. \qquad (10B1.5)$$

This is the relation shown in figure 10.1.

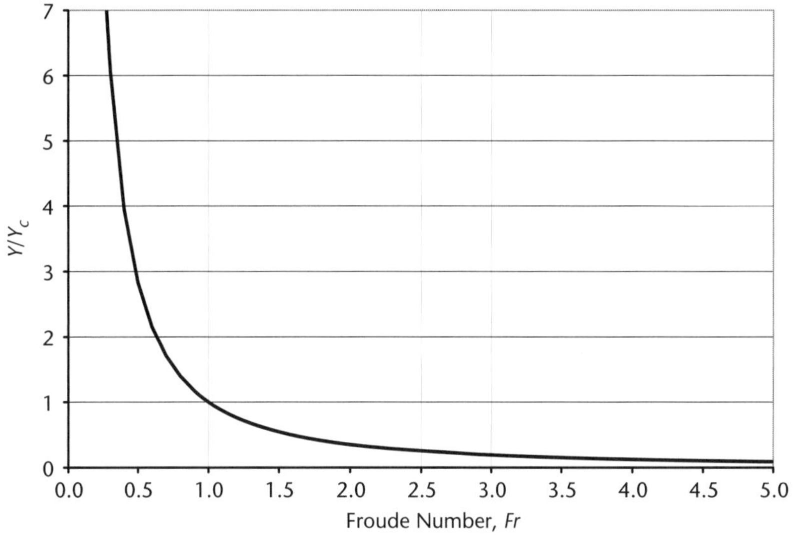

Figure 10.1 Relationship between Y/Y_C and *Fr* (equation 10.3).

- The changes in flow configuration take place in a relatively short reach; this means that boundary friction is commonly of negligible magnitude compared to other forces, particularly those associated with convective acceleration.
- The velocity-distribution coefficients for energy (α) and momentum (β) (see box 8.1) are typically considerably greater than 1 and are difficult to determine.

These characteristics of rapidly varied flow make the derivation of applicable equations from basic physics applicable in only the simplest situations. As a consequence, rapidly varied flow is generally treated by considering various typical situations as isolated cases, applying the basic principles of conservation of mass and of momentum and/or energy as a starting point, and placing heavy reliance on dimensional analysis (section 4.8.2) and empirical relations established in laboratory experiments. In most cases, the analysis is not applied to the region of rapidly varied flow itself, but to cross sections immediately upstream and downstream where gradually varied flow exists.

This chapter discusses the three broad cases of rapidly varied flow that are of primary interest to surface-water hydrologists:

1. **Hydraulic jumps**, which are standing waves that mark a sudden transition from supercritical to subcritical flow
2. **Abrupt transitions** in channel elevation or width, which are further subdivided into 1) transitions without energy loss and 2) transitions with energy loss, which include structures such as bridges
3. **Discharge measurement structures** designed for the measurement of discharge, including weirs and flumes, which usually involve a transition from subcritical to supercritical flow

10.1 Hydraulic Jumps

Natural reaches containing bank-to-bank supercritical flows are uncommon, but they do occur in steep bedrock channels and in meltwater channels on glaciers (figure 10.2), where the channel provides very low resistance. Local or partial supercritical flows are common in step-pool and cascade mountain streams (see table 2.4, figure 2.14) where the flow plunges over a bank-to-bank step or an individual boulder (figure 10.3) (Grant 1997; Comiti and Lenzi 2006; Vallé and Pasternack 2006), and are common in engineered structures such as spillways (figure 10.4a) and artificial channels (figure 10.4b). A change from supercritical to subcritical flow may be brought about by gradual deceleration due to frictional energy loss or by more abrupt decreases in channel slope, increases in resistance, or changes in bed elevation or width that force an increase in depth and/or a decrease in velocity, as discussed in section 10.2.

Whether such changes are abrupt or gradual, the location at which a supercritical flow becomes critical ($Fr = 1$) is commonly marked by an abrupt increase in depth and a relatively short reach of very high turbulence and an irregular to undulating surface. This phenomenon, clearly visible in figure 10.4, is called a **hydraulic jump**. Hydraulic jumps are standing waves that are stationary relative to an observer on the river bank, but are traveling upstream at a celerity (speed relative to the water) equal to the flow velocity. The physical cause of hydraulic jumps is epitomized in the specific-head and specific-force diagrams (figures 8.6 and 8.11): for a given discharge in a given channel, there are two depths that satisfy the specific-head and specific-force equations (equations 8.12 and 8.37b), and the flow jumps from the depth corresponding to supercritical flow to that corresponding to subcritical flow.

Flow within a jump is highly turbulent, so there is much energy loss due to eddies. Downstream from the jump, the flow gradually reestablishes as quasi-uniform or gradually varied subcritical flow at a higher depth and lower velocity.[1] The aspects of hydraulic jumps that are of most interest to hydraulic engineers, geomorphologists, and surface-water hydrologists are their physical characteristics, especially the associated depth and velocity changes and their downstream lengths,

Figure 10.2 A channel eroded in ice in central Alaska. The very low resistance of the ice boundary induces supercritical flow even at moderate slopes. Note the irregular water surface, which is typical of supercritical flow. The channel is about 0.5 m wide. Photo by the author.

and the energy loss that occurs within them. The discussion here begins with a qualitative classification of jumps, and then develops the conservation-of-momentum principle to provide tools for obtaining quantitative descriptions of those aspects.

Note that most of the information on hydraulic jumps has been published in the engineering literature and is based on data from flumes with fixed beds. Only a few studies have investigated jumps in mobile-bed settings that are more applicable to natural streams (Kennedy 1963; Comiti and Lenzi 2006).

10.1.1 Classification

Chow (1959) describes empirical studies showing that hydraulic jumps on fixed beds have characteristic forms that depend on the upstream Froude number, Fr_U

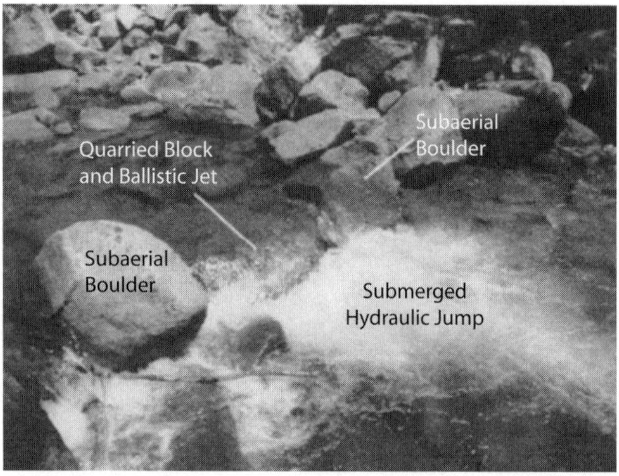

Figure 10.3 Local supercritical flow ("ballistic jet") over a stone block with a submerged hydraulic jump downstream. From Vallé and Pasternack (2006); reproduced with permission of Elsevier.

(figures 10.5 and 10.6). In most natural streams Froude numbers rarely exceed 2, so only the undular and weak jumps are likely to be observed; oscillating, steady, and strong jumps may occur in association with various engineering works. The Froude-number limits shown in figure 10.5 are not strict; for example, undular jumps have been reported at Fr_U as high as 3.6, and there is evidence that the limit is affected by the width/depth ratio and the Reynolds number (Comiti and Lenzi 2006).

In many cases in natural streams, the water-surface elevation immediately downstream of a jump, which is determined by conditions farther downstream, is higher than the amplitude of the jump. In these cases the jump is said to be **submerged** (figure 10.7), and the distinct water-surface rise that occurs in unsubmerged jumps of figures 10.5 and 10.6 is not observed.

10.1.2 Sequent Depths and Jump Heights

Recall from equation 8.37 (section 8.2.2) that the specific force, F_S, at any cross section is given by

$$F_S \equiv \frac{Y^2 \cdot W}{2} + \frac{Q^2}{g \cdot W \cdot Y}, \tag{10.4}$$

where Y is average depth, W is width, Q is discharge, and g is gravitational acceleration, and that the specific-force diagram (see figure 8.11) relates the depths upstream and downstream (the **sequent depths**) of a hydraulic jump to the specific force. Thus, if Q and W are specified, one of the major questions concerning hydraulic jumps can be answered simply by constructing such a curve. It is not practicable to construct a dimensionless version of the specific-force curve, so using this approach requires constructing a separate curve for each problem.

(a)

(b)

Figure 10.4 Hydraulic jumps at engineering structures: (a) Irregular jump at the base of a spillway; (b) undular jump in a stone-lined canal. Flow is from right to left; the V-shape is due to the cross-channel velocity gradient. Note the jump profile on the far wall left by a previous higher flow. Photos by the author.

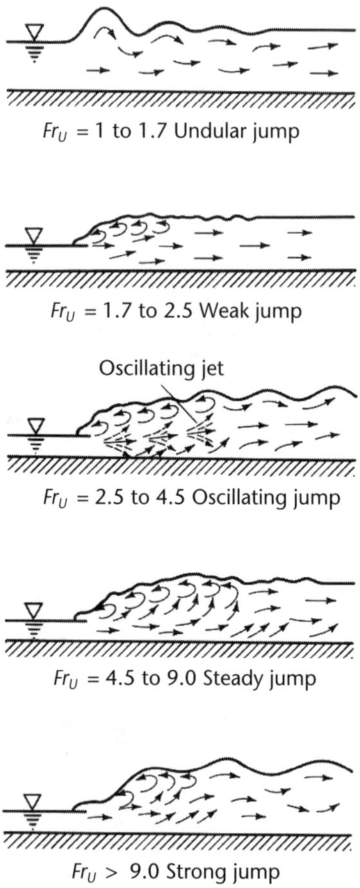

Figure 10.5 Types of hydraulic jumps and their associations with upstream Froude number, Fr_U. From Chow (1959).

However, we can develop a general approach to determining sequent depths by applying the principle of conservation of momentum to the situation depicted in figure 10.8. To simplify the development and emphasize the principles involved, we make the following assumptions: 1) the channel is horizontal, so that gravitational forces are not considered; 2) the distance L_J is small enough that we can neglect boundary frictional force; 3) the channel is rectangular with constant width; 4) the discharge is constant through the jump; and 5) the momentum coefficient (see box 8.1) $\beta = 1$. Many engineering-oriented texts (e.g., Chow 1959; French 1985) extend the analysis of hydraulic jumps to account for sloping and nonprismatic channels.

Equation 4.22 gave the time rate of change of momentum through a channel segment of infinitesimal length dX as

$$\frac{\mathrm{dM}}{\mathrm{d}t} = \rho \cdot Q \cdot \frac{\mathrm{d}U}{\mathrm{d}X} \cdot \mathrm{d}X, \tag{10.5}$$

(a)

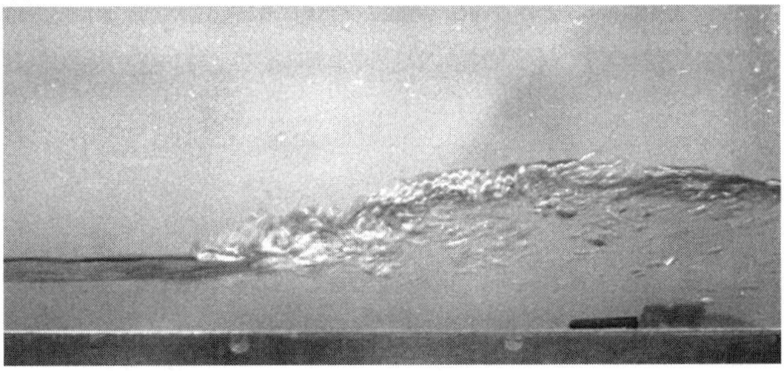

(b)

(c)

(d)

Figure 10.6 Hydraulic jump types in a laboratory flume: (a) weak; (b) oscillating, (c) steady, (d) strong. Compare with figure 10.5. Photos by the author.

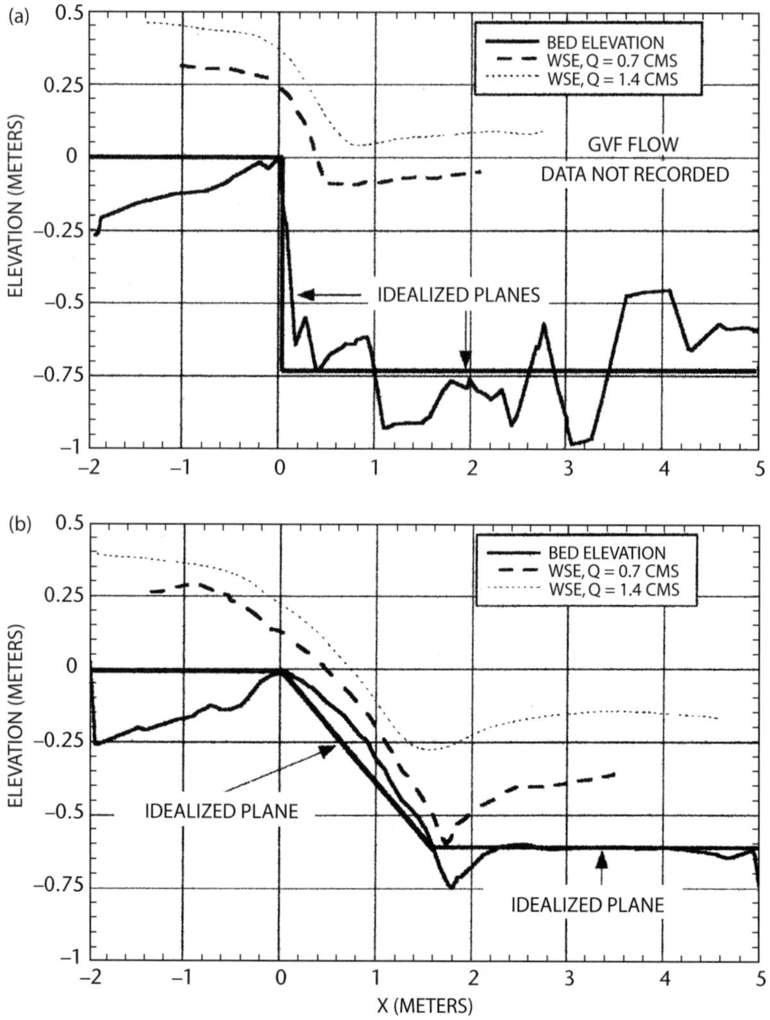

Figure 10.7 Centerline water-surface profiles through (a) a submerged jump region and (b) an unsubmerged jump region for lower ($Q = 0.7\,\mathrm{m}^3/\mathrm{s}$, dashed line) and higher ($Q = 1.4\,\mathrm{m}^3/\mathrm{s}$, dotted line) discharges in a mountain stream. Straight lines are idealized planes drawn through each jump for modeling purposes. (CMS = cubic meters per second). From Vallé and Pasternack (2006); reproduced with permission of Elsevier.

where M is momentum, ρ is mass density of water, and U is average velocity.[2] From the principle of conservation of momentum, the time rate of change of momentum is equal to the net force acting on the water. Because we have assumed that gravity forces and frictional forces are negligible, the only force acting on the water in the jump is the pressure force, F_P. As shown in equation 4.25, this net force

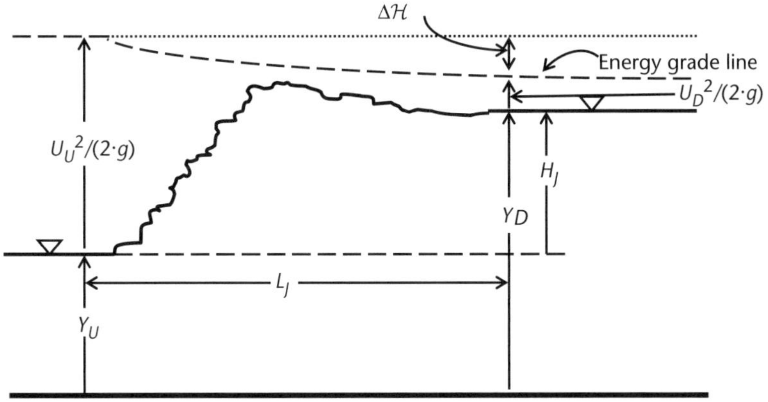

Figure 10.8 Definitions of terms for analyzing hydraulic jumps. L_J is the jump length; The jump height $H_J = (Y_D - Y_U)$. $\Delta\mathcal{H}_J$ is the energy loss through the jump.

is given by

$$F_P = -\gamma \cdot W \cdot Y \cdot \frac{\mathrm{d}Y}{\mathrm{d}X} \cdot \mathrm{d}X, \tag{10.6}$$

where γ is the weight density of water. Equating equations 10.6 and 10.5,

$$\rho \cdot Q \cdot \frac{\mathrm{d}U}{\mathrm{d}X} = -\gamma \cdot W \cdot Y \cdot \frac{\mathrm{d}Y}{\mathrm{d}X}. \tag{10.7}$$

To apply equation 10.7 to figure 10.8, we write it in finite-difference form. To do this, we express $\mathrm{d}U$ as $(U_D - U_U)$, $\mathrm{d}Y$ as $(Y_D - Y_U)$, and Y as $(Y_U + Y_D)/2$, so that

$$Q \cdot (U_U - U_D) = \left(\frac{1}{2}\right) \cdot g \cdot W \cdot (Y_D^2 - Y_U^2), \tag{10.8}$$

where $g = \gamma/\rho$. Then, following the steps in box 10.2, we arrive at

$$\left(\frac{Y_D}{Y_U}\right)^2 + \frac{Y_D}{Y_U} - 2 \cdot Fr_U^2 = 0. \tag{10.9}$$

Equation 10.9 is a quadratic equation in Y_D/Y_U, with one positive root and one negative root. The negative root is of no physical significance; the positive root is

$$\frac{Y_D}{Y_U} = \frac{(1 + 8 \cdot Fr_U^2)^{1/2} - 1}{2}, \tag{10.10}$$

which is valid for $Fr_U > 1$.

Equation 10.10 is the dimensionless universal equation for computing sequent depths that we have been seeking; its graph is shown in figure 10.9. If we are given the depth and velocity (or depth, discharge, and width) of the flow just upstream of the jump, we can compute Fr_U, find Y_D/Y_U from equation 10.10, and then compute the sequent depth Y_D.[3]

BOX 10.2 Derivation of Dimensionless Expression for Sequent Depths

Defining $\mathcal{Q} \equiv Q/W$, dividing equation 10.8 through by Y_U^2, and rearranging yields

$$\frac{Y_D^2}{Y_U^2} - 1 = \frac{2 \cdot \mathcal{Q} \cdot U_U}{g \cdot Y_U^2} - \frac{2 \cdot \mathcal{Q} \cdot U_D}{g \cdot Y_U{}^2}. \qquad (10B2.1)$$

From the conservation of mass,

$$\mathcal{Q} = U_U \cdot Y_U = U_D \cdot Y_D. \qquad (10B2.2)$$

We can use equation 10B2.2 to rewrite equation 10B2.1 as

$$\left(\frac{Y_D}{Y_U}\right)^2 - 1 = \frac{2 \cdot U_U^2}{g \cdot Y_U} - \frac{2 \cdot \mathcal{Q}^2}{g \cdot Y_U^2 \cdot Y_D}. \qquad (10B2.3)$$

Multiplying both sides of 10B2.3 by Y_D/Y_U and using the definition of the Froude number (equation 10.1), we obtain

$$\left[\left(\frac{Y_D}{Y_U}\right)^2 - 1\right] \cdot \left(\frac{Y_D}{Y_U}\right) = 2 \cdot Fr_U^2 \cdot \left(\frac{Y_D}{Y_U}\right) - 2 \cdot Fr_U^2 = 2 \cdot Fr_U^2 \cdot \left(\frac{Y_D}{Y_U} - 1\right). \qquad (10B2.4)$$

Dividing both sides of equation 10B2.4 by $(Y_D/Y_U - 1)$ and rearranging yields

$$\left(\frac{Y_D}{Y_U}\right)^2 + \frac{Y_D}{Y_U} - 2 \cdot Fr_U^2 = 0. \qquad (10B2.5)$$

The **jump height**, H_J, is defined as $H_J \equiv Y_D - Y_U$; this value can also be expressed in dimensionless form as a function of the upstream Froude number:

$$\frac{H_J}{\mathcal{H}_{SU}} = \frac{(1 + 8 \cdot Fr_U{}^2)^{1/2} - 3}{Fr_U{}^2 + 2}, \qquad (10.11)$$

where $\mathcal{H}_{SU}$ is the upstream specific head (Chow 1959). This relation is also plotted on figure 10.9.

10.1.3 Jump Length

The length, L_J, of a hydraulic jump is defined the distance from the front face of the jump to the point where a constant downstream depth is established. Jump lengths have been investigated experimentally and, like the general jump form and height, have been found to be determined by the entering Froude number Fr_U (Chow 1959). The relationship can be expressed in dimensionless form as a plot of L_J/Y_D versus Fr_U; this relation is shown on figure 10.10.

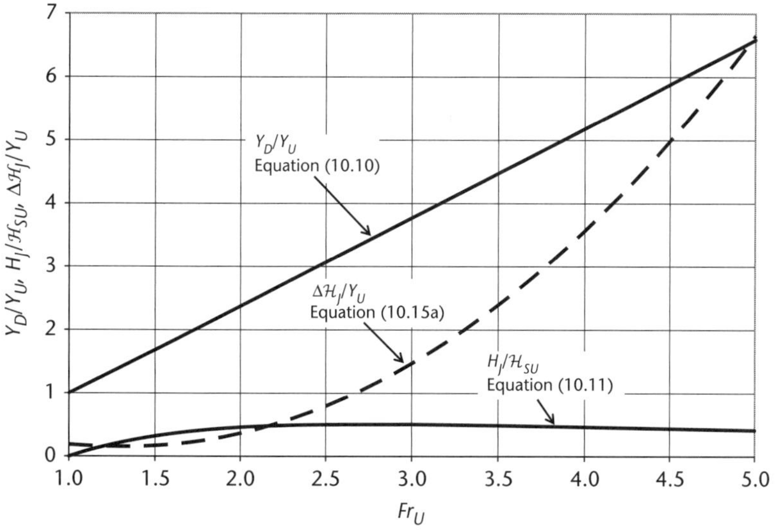

Figure 10.9 Jump conditions as a function of upstream Froude number, Fr_U. Curves show ratio of sequent depths Y_D/Y_U (equation 10.10), the ratio of jump height to upstream specific head $H_J/\mathcal{H}_{SU}$ (equation 10.11), and the ratio of energy loss through a jump to upstream depth $\Delta\mathcal{H}_J/Y_U$ (equation 10.15a).

10.1.4 Characteristics of Waves in Undular Jumps

Several investigators have studied the amplitudes and lengths of the waves in undular jumps and have related these characteristics to the upstream Froude number (Comiti and Lenzi 2006); Reinauer and Hager (1995) found in fixed-bed flume studies that the distance between the first and second wave crests in an undular jump, λ_{12}, is related to the entering Froude number as

$$\frac{\lambda_{12}}{Y_U} = 6.5 + 3.25 \cdot (Fr_U - 1). \tag{10.12}$$

Comiti and Lenzi (2006) found a very similar relation for jumps formed downstream of abrupt drops (**sills**) in channels with mobile beds. Equation 10.12 is shown in figure 10.10.

Andersen (1978) related the amplitude $\mathcal{A}_J$ (vertical distance between trough and crest) of the first wave of an undular jump on a fixed bed to Fr_U as

$$\mathcal{A}_J/Y_c = 1.48 \cdot (Fr_U - 1)^{1.03}, \tag{10.13}$$

where Y_c is the critical depth (figure 10.10). For mobile-bed channels, Comiti and Lenzi (2006) found that $\mathcal{A}_J/Y_c$ values centered around 1, with considerable scatter. Other studies (Chanson 2000) found a strong relation between amplitude and the ratio Y_D/W.

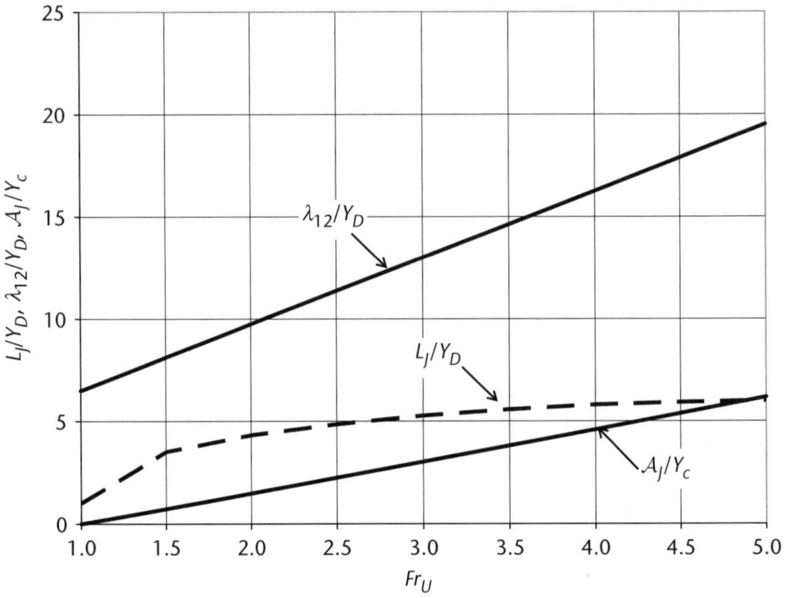

Figure 10.10 More jump conditions as a function of upstream Froude number, Fr_U. Curves show ratio of jump length to downstream depth L_J/Y_D, the ratio of wavelength of first wave of an undular jump to upstream depth λ_{12}/Y_U (equation 10.12), and the ratio of wave amplitude of first wave of an undular jump to critical depth $\mathcal{A}_J/Y_c$ (equation 10.13).

10.1.5 Energy Loss

Given channel width, discharge, and upstream depth or Froude number, the downstream depth and hence velocity can be obtained from equation 10.10. The head loss through the jump, $\Delta\mathcal{H}_J$, can then be computed via the energy equation:

$$\Delta\mathcal{H}_J = Y_U + \frac{U_U{}^2}{2\cdot g} - Y_D - \frac{U_D{}^2}{2\cdot g} \qquad (10.14)$$

This energy loss can be expressed in dimensionless form by using an approach similar to that described in box 10.2 to arrive at

$$\frac{\Delta\mathcal{H}_J}{Y_U} = \frac{(1+8\cdot Fr_U{}^2)}{16} - \frac{(1+8\cdot Fr_U{}^2)^{1/2}}{2} - \frac{1}{2\cdot(1+8\cdot Fr_U{}^2)-2} + \frac{19}{16}, \qquad (10.15a)$$

or, in terms of Y_D/Y_U,

$$\frac{\Delta\mathcal{H}_J}{Y_U} = \frac{1}{4}\cdot\left(\frac{Y_D}{Y_U}\right)^2 - \frac{3}{4}\cdot\left(\frac{Y_D}{Y_U}\right) - \frac{1}{4}\left(\frac{Y_U}{Y_D}\right) + \frac{3}{4}. \qquad (10.15b)$$

Equation 10.15a is shown in figure 10.9. Note that energy losses are relatively small in jumps at Froude numbers <2, which is the range that would typically occur in natural streams.

10.2 Abrupt Channel Transitions with No Energy Loss

The methods for determining the changes in depth and velocity through abrupt changes in channel elevation and width are based on the principles of conservation of mass, energy, and momentum and the concept of specific energy. In this section, we apply these principles with the simplifying assumption that the total head does not change through the transition. This assumption is acceptable when 1) the transition occurs over a distance that is short enough to make the boundary-friction loss negligible, and 2) the energy losses due to expansion and contraction (the "eddy losses" discussed in section 9.1.2) are negligible. Transitions with energy losses often occur at structures such as bridges and culverts, which are discussed in section 10.3. Energy losses are usually significant when the change in channel elevation or width forces a change in flow regime.

10.2.1 Elevation Transitions

10.2.1.1 Basic Approach

We assume that the width W and discharge Q are specified and constant through the transition and that the change in bottom elevation is specified. We are given the depth Y_U at a section just upstream of the transition and want to calculate the depth Y_D and velocity U_D at a section just downstream from it.[4]

To solve this problem we invoke the principles of conservation of mass and conservation of energy. The conservation-of-mass relation for this situation (assuming constant density) is

$$Q = W \cdot Y_U \cdot U_U = W \cdot Y_D \cdot U_D. \tag{10.16}$$

If we assume negligible energy loss between sections U and D, the energy equation (equation 8.8b) is

$$Z_U + Y_U + \frac{\alpha_U \cdot U_U{}^2}{2 \cdot g} = Z_D + Y_D + \frac{\alpha_D \cdot U_D{}^2}{2 \cdot g}, \tag{10.17}$$

where Z is channel-bottom elevation, α is energy coefficient, and g is gravitational acceleration. To simplify the development, we assume henceforth that $\alpha_U, \alpha_D \approx 1$.

If we take the channel elevation on the upstream side as the elevation datum so that $Z_U = 0$, we can further simplify equation 10.17 to

$$Y_U + \frac{U_U{}^2}{2 \cdot g} = Z_D + Y_D + \frac{U_D{}^2}{2 \cdot g}, \tag{10.18}$$

where Y_U, U_U, and Z_D are known and Y_D and U_D are to be determined. We can make use of equation 10.16 to write equation 10.18 as

$$Y_D + \frac{Q^2}{2 \cdot g \cdot W^2 \cdot Y_D^2} = Y_U + \frac{Q^2}{2 \cdot g \cdot W \cdot Y_U^2} - Z_D, \tag{10.19}$$

where there is now a single unknown, Y_D, and the velocity head is expressed in terms of discharge, width, and depth.

One way to solve equation 10.19 is by trial and error. However, if we recall the definition of specific head, $\mathcal{H}_S$, as the sum of the pressure head and the velocity head (section 8.1.2), we see that

$$\mathcal{H}_S = Y + \frac{Q^2}{2 \cdot g \cdot W^2 \cdot Y^2} \tag{10.20}$$

and can write equation 10.19 as

$$\mathcal{H}_{SD} = \mathcal{H}_{SU} - Z_D. \tag{10.21}$$

The value of $\mathcal{H}_{SU}$ is determined from the specified values of Q, W, and Y_U, and we can make use of a specific-head diagram to find Y_D, as explained in the following subsections.

10.2.1.2 Elevation Drops

For an abrupt channel drop, the elevation change Z_D is a negative number, and equation 10.21 can be written as

$$\mathcal{H}_{SD} = \mathcal{H}_{SU} + |Z_D|. \tag{10.22}$$

The nature of the change in depth through an abrupt channel drop is determined by whether the upstream flow is subcritical (figure 10.11a) or supercritical (figure 10.11b). The upper portions of this figure are the specific-head curves for the specified discharge. The known value of $\mathcal{H}_{SU}$ is plotted on the horizontal axis, and the corresponding depth Y_U, on the vertical axis. Then, $\mathcal{H}_{SD}$ is found via equation 10.22 and plotted on the horizontal axis. If the upstream flow is subcritical, the downstream depth Y_D is found where the vertical line drawn from $\mathcal{H}_{SD}$ intersects the upper limb of the specific-head curve. If the upstream flow is supercritical, the lower limb of the curve is used to find Y_D. The changes induced by the abrupt drop in channel elevation are summarized below and in the top two rows of table 10.1:

> At an abrupt drop in channel-bed elevation a subcritical flow becomes deeper, slower, and "more subcritical" (i.e., the Froude number decreases), whereas a supercritical flow becomes shallower, faster, and "more supercritical" (i.e., the Froude number increases).

10.2.1.3 Elevation Rises

The same approach is used to determine the changes induced by an abrupt increase in bed elevation (figure 10.12). In this case, however, $Z_D > 0$, so from equation 10.21, $\mathcal{H}_{SD} < \mathcal{H}_{SU}$, and we move to the left on the appropriate arm of the specific-head curve, that is, toward the critical point at the "nose" of the curve. The changes induced by the abrupt rise in channel elevation are summarized below and in the bottom two rows of table 10.1:

> At an abrupt rise in channel-bed elevation a subcritical flow becomes shallower, faster, and "less subcritical" (i.e., the Froude number increases),

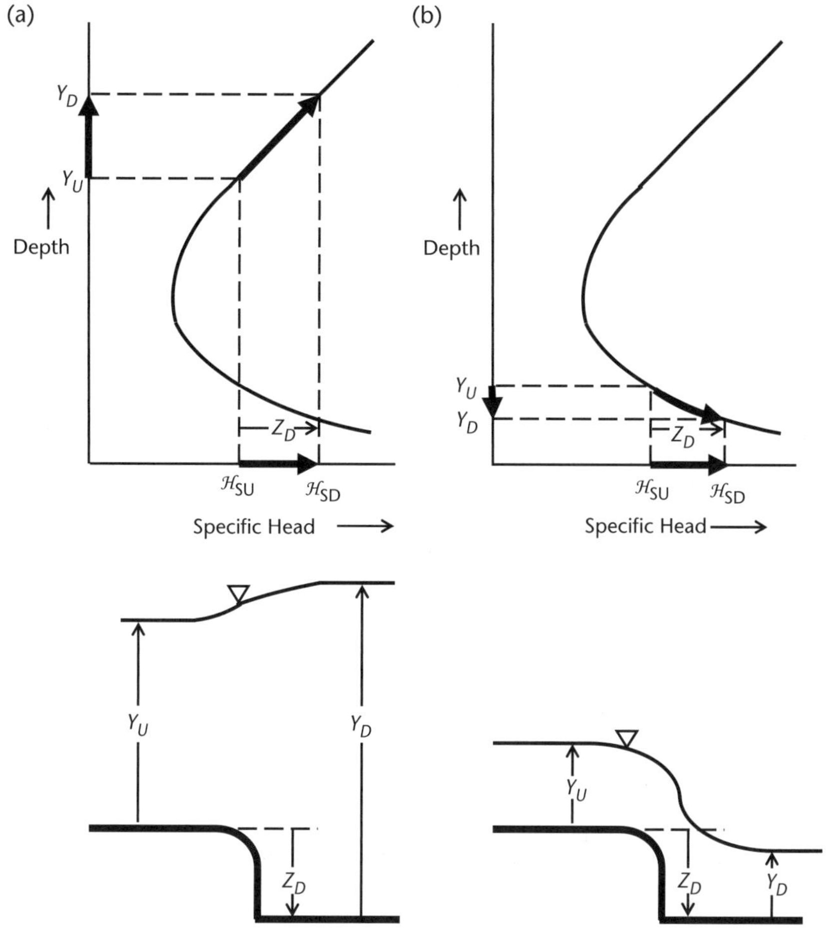

Figure 10.11 Definition diagrams (lower) and specific-head diagrams (upper) for calculating energy relations and depth changes due to an abrupt decrease in channel elevation, assuming no energy loss: (a) subcritical flow; (b) supercritical flow.

whereas a supercritical flow becomes deeper, slower, and "less supercritical" (i.e., the Froude number decreases).

However, the "nose" of the specific-head curve represents an important constraint in applying this approach to abrupt channel rises: We cannot move leftward of the critical point where the specific head is at its minimum value. This reflects the fact that critical flow represents an instability that produces significant energy losses in the form of a marked contraction of streamlines (subcritical to supercritical transition) or a highly turbulent hydraulic jump (supercritical to subcritical transition). These energy losses violate the assumptions of the above analysis.

Table 10.1 Depth and velocity changes induced by abrupt drops and rises in channel-bed elevation under the assumption of no energy loss (figures 10.11 and 10.12).

Elevation change	Z_D	Downstream flow regime	Upstream flow regime	Change in Froude no.	Change in depth	Change in velocity
Drop	<0	Subcritical ($Fr_D < 1$)	Subcritical ($Fr_U < 1$)	↓	↑	↓
		Supercritical ($Fr_D > 1$)	Supercritical ($Fr_U > 1$)	↑	↓	↑
Rise	>0	Subcritical ($Fr_D < 1$)	Subcritical ($Fr_U < 1$)[a]	↑[a]	↓[a]	↑[a]
		Supercritical ($Fr_D > 1$)	Supercritical ($Fr_U > 1$)[a]	↓[a]	↑[a]	↓[a]

Upward (downward) arrows indicate increases (decreases). See examples in box 10.3.
[a]If Z_D is large enough to induce the flow to pass through the critical point, the upstream depth and velocity cannot be determined under the assumption of negligible energy loss. If the flow changes from supercritical downstream to subcritical upstream, the rise acts as a weir (section 10.4.1); if the flow changes from subcritical downstream to supercritical upstream, the rise induces a hydraulic jump.

To quantify this constraint, note that the value of the minimum specific head, $H_{S\min}$, is given by

$$\mathcal{H}_{S\min} = Y_c + \frac{U_c^2}{2 \cdot g}, \tag{10.23}$$

where Y_c is critical depth, and U_c is the velocity at critical depth. From equation 10.1, $U_c^2 = g \cdot Y_c$ at critical flow ($Fr = 1$), so we can also write

$$\mathcal{H}_{S\min} = Y_c + \frac{Y_c}{2} = 1.5 \cdot Y_c, \tag{10.24}$$

where Y_c can be found via equation 10.2. Thus, we see that when

$$Z_D \leq \mathcal{H}_{SU} - \mathcal{H}_{S\min} = \mathcal{H}_{SU} - 1.5 Y_c, \tag{10.25}$$

the flow is forced through the critical point and the downstream conditions cannot be determined using this approach.

10.2.1.4 Dimensionless Specific-Head Curve

Because specific head is a function of discharge and width, application of the methods described in sections 10.2.1.2 and 10.2.1.3 requires constructing separate curves for each discharge and width of interest. To avoid this requirement, we can make use of a universal **dimensionless specific-head diagram**. Such a curve is constructed by dividing equation 10.20) by the critical depth Y_c:

$$\frac{\mathcal{H}_S}{Y_c} = \frac{Y}{Y_c} + \frac{Q^2}{2 \cdot g \cdot W^2 \cdot Y_c \cdot Y^2}, \tag{10.26a}$$

which is simplified by substituting equation 10.2 to give

$$\frac{\mathcal{H}_S}{Y_c} = \frac{Y}{Y_c} + \left(\frac{1}{2}\right) \cdot \left(\frac{Y_c}{Y}\right)^2. \tag{10.26b}$$

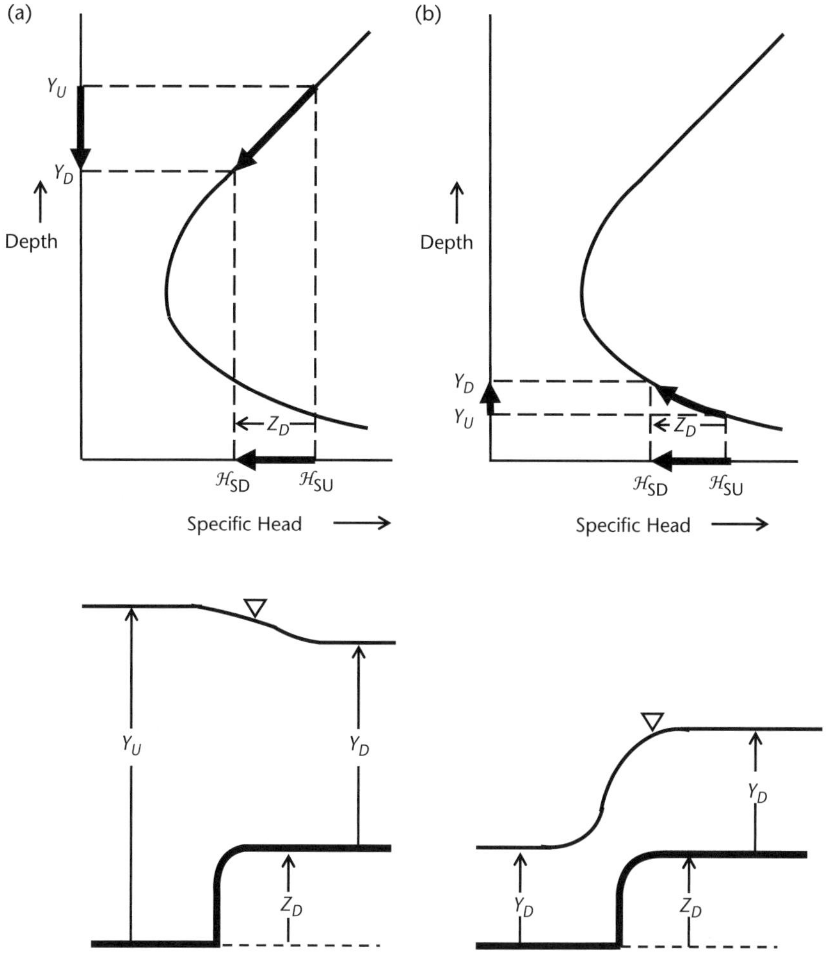

Figure 10.12 Definition diagrams (lower) and specific-head diagrams (upper) for calculating energy relations and depth changes due to an abrupt increase in channel elevation, assuming no energy loss: (a) subcritical flow; (b) supercritical flow.

Figure 10.13 shows a plot of the dimensionless specific-head curve, and box 10.3 gives examples of its application in computing depth and velocity changes through abrupt changes in channel-bed elevation.

10.2.1.5 Implications for Flow over Bedforms

When threshold shear-stress values are exceeded in sand-bed streams, bed-load transport begins and a typical sequence of bedforms develops as shear stress increases (see section 6.6.4.2, table 6.2). Dunes are the large bedforms that occur in flows with high but still subcritical Froude numbers; antidunes (see figure 6.19) occur in supercritical flows.

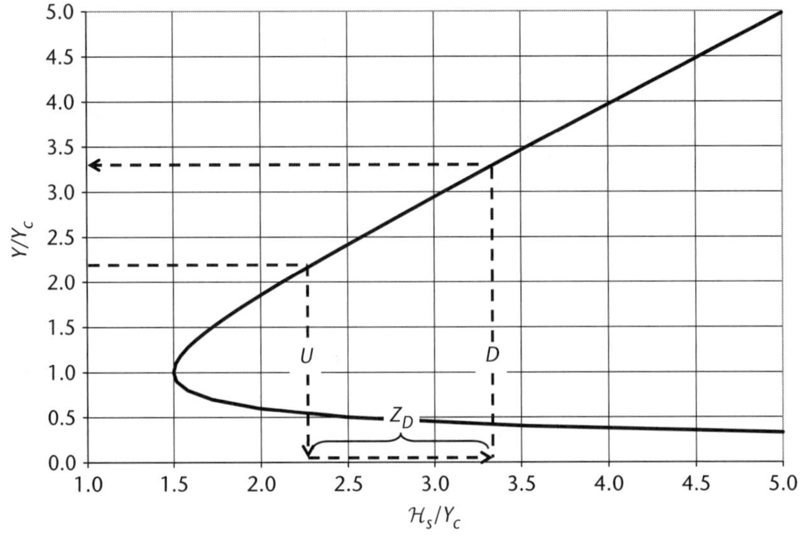

Figure 10.13 Dimensionless specific-head diagram. The dashed lines show the computations for example 1 of box 10.3; *D* denotes downstream values; *U*, upstream values.

BOX 10.3 Example Calculations of Abrupt Channel-Elevation Changes

Example 1: Channel Drop

Specified Values

Quantity	Width, W (m)	Elevation change, Z_D (m)	Discharge, Q (m^3/s)	Upstream depth, Y_U (m)
Value	5.0	−0.80	10.0	1.60

Computation of Other Upstream Quantities

The upstream velocity U_U is found from equation 10.16 as

$$U_U = \frac{10.0\ \text{m}^3/\text{s}}{5.0\ \text{m} \cdot 1.60\ \text{m}} = 1.25\ \text{m/s}.$$

The critical depth Y_c is found from equation 10.2 as

$$Y_c = \left(\frac{(10.0\ \text{m}^3/\text{s})^2}{(9.81\ \text{m/s}^2) \cdot (5.00\ \text{m})^2} \right)^{1/3} = 0.74\,\text{m}.$$

The critical depth is less than the actual depth, so the upstream flow is subcritical; the upstream Froude number is

$$Fr_U = \frac{1.25\ \text{m/s}}{(9.81\ \text{m/s}^2 \cdot 1.60\ \text{m})^{1/2}} = 0.32.$$

To use figure 10.13, we first compute $Y_U/Y_c = 1.60/0.74 = 2.16$. Entering figure 10.13 (or using equation 10.26) with this value gives $\mathcal{H}_{SU}/Y_c = 2.27$. We then find $\mathcal{H}_{SU} = 2.27 \times 0.74\text{ m} = 1.68\text{ m}$.

To Find Downstream Values

Use of the dimensionless specific-head diagram for this example is shown on figure 10.13. Applying equation 10.21,

$$\mathcal{H}_{SD} = 1.68\text{ m} - (-0.80\text{ m}) = 2.48\text{ m}.$$

Thus, $\mathcal{H}_{SD}/Y_c = 2.48\text{ m}/0.74\text{ m} = 3.35$, and from figure 10.13, the corresponding value of $Y_D/Y_c = 3.30$.

The downstream values are thus

$$Y_D = 3.30 \times 0.74\text{ m} = 2.45\text{ m};$$

$$U_D = \frac{10.0\text{ m}^3/\text{s}}{5.0\text{ m} \cdot 2.45\text{ m}} = 0.82\text{ m/s};$$

$$Fr_D = \frac{0.82\text{ m/s}}{(9.81\text{ m/s}^2 \cdot 2.45\text{ m})^{1/2}} = 0.17.$$

Example 2: Channel Rise

Specified Values

Quantity	Width, W (m)	Elevation change, Z_D (m)	Discharge, Q (m^3/s)	Upstream depth, Y_U (m)
Value	12.0	0.80	50	2.70

Computation of Other Upstream Quantities

The upstream velocity U_U is found from equation 10.16 as

$$U_U = \frac{50\text{ m}^3/\text{s}}{12.0\text{ m} \cdot 2.70\text{ m}} = 1.54\text{ m/s}.$$

The critical depth Y_c is found from equation 10.2 as

$$Y_c = \left(\frac{(50\text{ m}^3/\text{s})^2}{(9.81\text{ m/s}^2) \cdot (2.70\text{ m})^2}\right)^{1/3} = 1.21\text{ m}.$$

The critical depth is less than the actual depth, so the upstream flow is subcritical; the upstream Froude number is

$$Fr_U = \frac{1.54\text{ m/s}}{(9.81\text{ m/s}^2 \cdot 2.70\text{ m})^{1/2}} = 0.30.$$

To use figure 10.13, we first compute $Y_U/Y_c = 2.70/1.21 = 2.23$. Entering figure 10.13 (or using equation 10.26) with this value gives $\mathcal{H}_{SU}/Y_c = 2.33$. We then find $\mathcal{H}_{SU} = 2.33 \times 1.21\text{ m} = 2.82\text{ m}$.

(Continued)

BOX 10.3 *Continued*

To Find Downstream Values

Applying equation 10.21,

$$\mathcal{H}_{SD} = 2.82\,\text{m} - 0.80\,\text{m} = 2.02\,\text{m}.$$

Thus, $\mathcal{H}_{SD}/Y_c$ = 2.02 m/1.21 m = 1.67, and from figure 10.13, the corresponding value of Y_D/Y_c = 1.43. The downstream values are thus

$$Y_D = 1.43 \times 1.21\,\text{m} = 1.73\,\text{m};$$

$$U_D = \frac{50\,\text{m}^3/\text{s}}{12.0\,\text{m} \cdot 1.73\,\text{m}} = 2.41\,\text{m/s};$$

$$Fr_D = \frac{2.41\,\text{m/s}}{(9.81\,\text{m/s}^2 \cdot 1.73\,\text{m})^{1/2}} = 0.58.$$

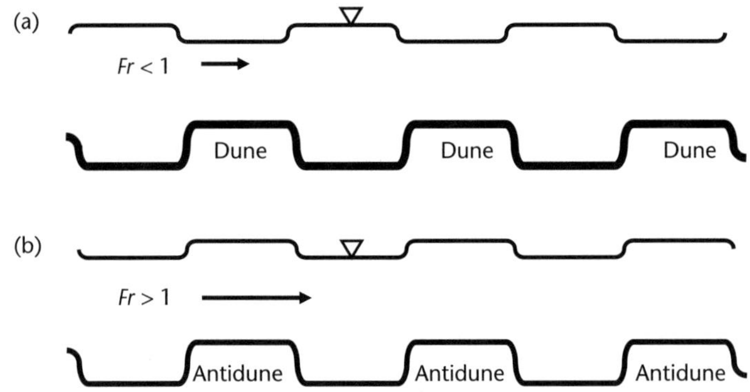

Figure 10.14 Idealized diagram of the form of the water surface over the bedforms often seen in sand-bed streams. The surface configuration can be explained by its response to abrupt rises and drops of bed elevation as shown in figures 10.11 and 10.12: The water surface is (a) out of phase with dunes that form in subcritical flows (compare figure 6.18a) and (b) in phase with the antidunes that form in supercritical flows (compare figure 6.19).

Based on the discussions in sections 10.2.1.2 and 10.2.1.3, figure 10.14 schematically represents bedforms as a succession of abrupt changes in bed elevation and the accompanying changes in water-surface elevation that occur when the flow is subcritical (figure 10.14a) and supercritical (figure 10.14b): The water surface over dunes is out of phase with the bed topography (compare figure 6.18a); the water surface over antidunes is in phase with the bed topography (compare figure 6.19).

10.2.2 Width Transitions

The typical problem is that the width, discharge, and depth are specified at a section immediately upstream or downstream of a specified abrupt change in width, and we want to compute the depth and velocity downstream or upstream. This problem can be approached by making use of the dimensionless specific-head curve if we assume negligible energy change through the transition (which is not the case if the flow is forced through a subcritical/supercritical transition) and that the bottom elevation is constant. Note that the assumption of no energy loss would often be inappropriate at, for example, a typical bridge opening, as discussed in section 10.3.3.

The assumptions of negligible energy loss and a horizontal channel bed allow us to equate the specific heads at the upstream and downstream sections:

$$\mathcal{H}_{SD} = \mathcal{H}_{SU}, \tag{10.27}$$

from which

$$Y_D + \frac{Q^2}{2 \cdot g \cdot W_D{}^2 \cdot Y_D{}^2} = Y_U + \frac{Q^2}{2 \cdot g \cdot W_U{}^2 \cdot Y_U^2}, \tag{10.28}$$

and all the quantities on the right-hand side are known. We can also compute the critical depths at each section from the given information via equation 10.2:

$$Y_{cU} = \left(\frac{Q^2}{g \cdot W_U{}^2} \right)^{1/3} \tag{10.29a}$$

$$Y_{cD} = \left(\frac{Q^2}{g \cdot W_D{}^2} \right)^{1/3} \tag{10.29b}$$

The value of $\mathcal{H}_{SD}/Y_{cD}$ can now be determined from equations 10.28 and 10.29b. Entering the horizontal axis of figure 10.13 with that value (assuming $\mathcal{H}_{SD}/Y_{cD} > 1.5$), we can find Y_D/Y_{cD} on the vertical axis and compute Y_D.

As in the case of changes of bed elevations, the computations are valid only if there is no change in flow regime through the transition. Table 10.2 summarizes changes induced by width transitions, and box 10.4 provides example calculations. The following section provides a theoretical analysis that includes cases in which the flow regime changes through the transition and which allows estimation of energy losses due to contractions and expansions.

10.3 Abrupt Transitions with Energy Loss

This section begins the discussion of energy losses in abrupt channel transitions with a theoretical analysis, and then provides an introduction to the effects of bridges on flows. The analyses of channel transitions here are limited to the simplest cases; engineering texts on open-channel flow (e.g., Chow 1959; Henderson 1961; French 1985) should be consulted for approaches to more complex situations. The use of abrupt width constrictions to measure discharge is discussed later in the chapter (section 10.4.3).

Table 10.2 Depth and velocity changes induced by abrupt width contractions and expansions under the assumption of no energy loss.

Width change	Downstream flow regime	Upstream flow regime	Change in Froude no.	Change in depth	Change in velocity
Contraction	Subcritical ($Fr_D < 1$)	Subcritical ($Fr_U < 1$)[a]	↑[a]	↓[a]	↑[a]
	Supercritical ($Fr_D > 1$)	Supercritical ($Fr_U > 1$)	↓[a]	↑[a]	↓[a]
Expansion	Subcritical ($Fr_D < 1$)	Subcritical ($Fr_U < 1$)	↓	↑	↓
	Supercritical ($Fr_D > 1$)	Supercritical ($Fr_U > 1$)[a]	↑	↓	↑

Upward (downward) arrows indicate increases (decreases). See examples in box 10.4.
[a]If the contraction is severe enough to induce the flow to pass through the critical point, the upstream depth and velocity cannot be determined from the assumption of negligible energy loss.

BOX 10.4 Example Calculation of Abrupt Width Changes

Example 1: Width Contraction

Specified Values

Quantity	Upstream width, W_U (m)	Downstream width, W_D (m)	Discharge, Q (m^3/s)	Upstream depth, Y_U (m)
Value	4.20	3.80	2.00	0.39

Computation of Other Upstream Quantities

The upstream velocity U_U is found from equation 10.16 as

$$U_U = \frac{2.00\ \text{m}^3/\text{s}}{4.20\ \text{m} \cdot 0.39\ \text{m}} = 1.22\ \text{m/s}.$$

The critical depth Y_{cU} is found from equation 10.2 as

$$Y_{cU} = \left(\frac{(2.00\ \text{m}^3/\text{s})^2}{(9.81\ \text{m/s}^2) \cdot (4.20\ \text{m})^2} \right)^{1/3} = 0.28\ \text{m}.$$

The critical depth is less than the actual depth, so the upstream flow is subcritical; the upstream Froude number is

$$Fr_U = \frac{1.22\ \text{m/s}}{(9.81\ \text{m/s}^2 \cdot 0.39\ \text{m})^{1/2}} = 0.62.$$

To use figure 10.13, we first compute $Y_U/Y_{cU} = 0.39/0.28 = 1.37$. Entering figure 10.13 (or using equation 10.26) with this value gives $\mathcal{H}_{SU}/Y_{cU} = 1.64$. We then find $\mathcal{H}_{SU} = 1.64 \times 0.28\ \text{m} = 0.47\ \text{m}$.

To Find Downstream Values

The critical depth Y_{cD} is found from equation 10.2 as

$$Y_{cD} = \left(\frac{(2.00\ \text{m}^3/\text{s})^2}{(9.81\ \text{m/s}^2) \cdot (3.80\ \text{m})^2} \right)^{1/3} = 0.30\ \text{m}.$$

From equation 10.27, $\mathcal{H}_{SD} = \mathcal{H}_{SU} = 0.47\ \text{m}$, so $\mathcal{H}_{SD}/Y_{cD} = 0.47\ \text{m}/0.30\ \text{m} = 1.53$. Entering figure 10.13 with this value, we find $Y_D/Y_{cD} = 1.16$. Therefore, $Y_D = 1.16 \times 0.30\ \text{m} = 0.35\ \text{m}$. This depth is greater than the critical depth, so the flow remains subcritical and the computations are valid.

The downstream values are thus

$$Y_D = 0.35\ \text{m};$$

$$U_D = \frac{2.00\ \text{m}^3/\text{s}}{3.80\ \text{m} \cdot 0.35\ \text{m}} = 1.49\ \text{m/s};$$

$$Fr_D = \frac{1.49\ \text{m/s}}{(9.81\ \text{m/s}^2 \cdot 0.35\ \text{m})^{1/2}} = 0.80.$$

Example 2: Width Expansion

Specified Values

Quantity	Upstream width, W_U (m)	Downstream width, W_D (m)	Discharge, Q (m^3/s)	Upstream eepth, Y_U (m)
Value	4.00	5.00	10.0	0.93

Computation of Other Upstream Quantities

The upstream velocity U_U is found from equation 10.16 as

$$U_U = \frac{10.0\ \text{m}^3/\text{s}}{4.00\ \text{m} \cdot 0.93\ \text{m}} = 2.69\ \text{m/s}.$$

The critical depth Y_{cU} is found from equation 10.2 as

$$Y_{cU} = \left(\frac{(10.0\ \text{m}^3/\text{s})^2}{(9.81\ \text{m/s}^2) \cdot (4.00\ \text{m})^2} \right)^{1/3} = 0.86\ \text{m}.$$

The critical depth is less than the actual depth, so the upstream flow is subcritical; the upstream Froude number is

$$Fr_U = \frac{2.69\ \text{m/s}}{(9.81\ \text{m/s}^2 \cdot 0.39\ \text{m})^{1/2}} = 0.89.$$

To use figure 10.13, we first compute $Y_U/Y_{cU} = 0.93/0.86 = 1.08$. Entering figure 10.13 (or using equation 10.26) with this value gives $\mathcal{H}_{SU}/Y_{cU} = 1.51$. We then find $\mathcal{H}_{SU} = 1.51 \times 0.86\ \text{m} = 1.30\ \text{m}$.

(Continued)

To Find Downstream Values

The critical depth Y_{cD} is found from equation 10.2 as

$$Y_{cD} = \left(\frac{(10.0\ \mathrm{m}^3/\mathrm{s})^2}{(9.81\ \mathrm{m/s}^2) \cdot (5.00\ \mathrm{m})^2} \right)^{1/3} = 0.74\ \mathrm{m}.$$

From equation 10.27, $\mathcal{H}_{SD} = \mathcal{H}_{SU} = 1.30$ m, so $\mathcal{H}_{SD}/Y_{cD} = 1.30\ \mathrm{m}/0.74\ \mathrm{m} = 1.75$. Entering figure 10.13 with this value, we find $Y_D/Y_{cD} = 1.54$. Therefore, $Y_D = 1.54 \times 0.74\ \mathrm{m} = 1.14$ m. This depth is greater than the critical depth, so the flow remains subcritical and the computations are valid.

The downstream values are thus

$$Y_D = 1.14\ \mathrm{m};$$

$$U_D = \frac{10.0\ \mathrm{m}^3/\mathrm{s}}{5.00\ \mathrm{m} \cdot 1.14\ \mathrm{m}} = 1.75\ \mathrm{m/s};$$

$$Fr_D = \frac{1.75\ \mathrm{m/s}}{(9.81\ \mathrm{m/s}^2 \cdot 1.14\ \mathrm{m})^{1/2}} = 0.52.$$

10.3.1 General Theoretical Approach

The basic approach to computing the energy losses associated with abrupt transitions employs the strategy alluded to in section 8.3: The changes in depth (and velocity) induced by the transition are determined by applying the momentum principle, and the results of that analysis are used to calculate the energy losses via the energy equation.

10.3.1.1 Momentum Equation

The macroscopic momentum equation was given in equation 8.32 as

$$\rho \cdot Q \cdot (\beta_D \cdot U_D - \beta_U \cdot U_U) = F_G + F_P - F_T, \tag{10.30a}$$

where ρ is the mass density of water; Q is the discharge (constant through the transition); U_D and U_U are the average velocities at the gradually varied sections immediately downstream and upstream of the transition, respectively; β_D and β_U are the momentum coefficients at the respective sections; and F_G, F_P, and F_T are the net forces on the water between the two sections due to gravity, pressure, and turbulent resistance, respectively. To simplify the development, we again make the assumptions that 1) β_D, $\beta_U = 1$, 2) the channel bed is horizontal so that $F_G = 0$, and 3) the distance between the two sections is short enough to justify assuming $F_T = 0$. Thus,

$$\rho \cdot Q \cdot (U_D - U_U) = F_P. \tag{10.30b}$$

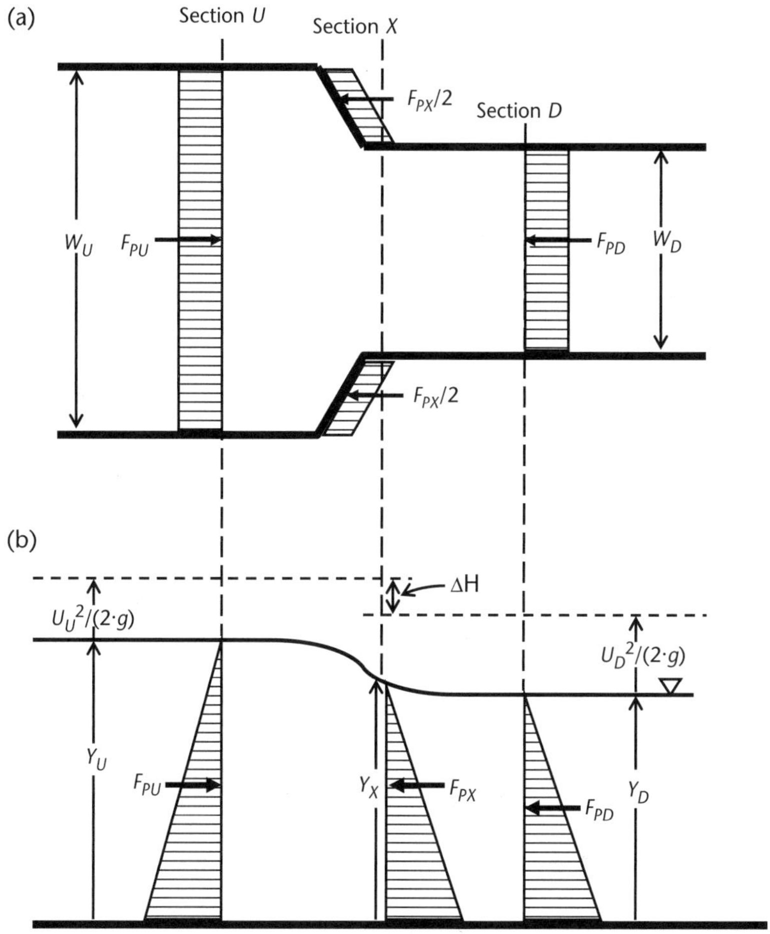

Figure 10.15 Definition diagram for analysis of a width contraction: (a) plan view; (b) longitudinal profile. See text for discussion. After Chow (1959).

Following the analysis of Chow (1959), we here apply this approach to the width contraction depicted in figure 10.15. The net pressure force on the water between the two sections is calculated as

$$F_P = F_{PU} - F_{PX} - F_{PD}, \tag{10.31}$$

where F_{PU} is the pressure force at the upstream section, F_{PX} is the pressure force exerted by the walls forming the contraction, and F_{PD} is the pressure force at the downstream section. These forces are calculated by applying equation 7.17 at the respective sections:

$$F_{Pi} = \frac{\gamma \cdot W_i \cdot Y_i^2}{2}, \tag{10.32}$$

where γ is the weight density of water, W_i is the channel width at section i, and Y_i is the average depth at section i.

Now making the additional assumption that the depth at the transition, Y_X, equals the downstream depth Y_D, we can combine equations 10.30b, 10.31, and 10.32 to write

$$\rho \cdot Q \cdot (U_D - U_U) = \frac{\gamma \cdot W_U \cdot {Y_U}^2}{2} - \frac{\gamma \cdot (W_U - W_D) \cdot {Y_D}^2}{2} - \frac{\gamma \cdot W_D \cdot {Y_D}^2}{2}. \quad (10.33)$$

Equation 10.33 can be manipulated (box 10.5) to derive a dimensionless expression that relates the upstream Froude number, Fr_U, to the ratios of depths and widths at the upstream and downstream sections:

$${Fr_U}^2 = \frac{(Y_D/Y_U) \cdot [(Y_D/Y_U) - 1]}{2 \cdot [(Y_D/Y_U) - (W_U/W_D)]}. \quad (10.34)$$

This relation is plotted in figure 10.16a, where Y_D/Y_U is plotted against Fr_U for various values of $W_D/W_U \leq 1$. The same approach can be applied to width expansions; this yields

$${Fr_U}^2 = \frac{(Y_D/Y_U) \cdot [1 - (Y_D/Y_U)^2]}{2 \cdot (W_U/W_D) \cdot [(W_U/W_D) - (Y_D/Y_U)]}, \quad (10.35)$$

which is plotted on figure 10.16b for various values of $W_D/W_U \geq 1$ (Chow 1959).

The upstream flow is, of course, subcritical for $Fr_U < 1$ and supercritical for $Fr_U > 1$. It can be shown (box 10.5) that the ratio of the downstream to upstream Froude numbers is given by

$$\frac{{Fr_D}^2}{{Fr_U}^2} = \frac{(W_U/W_D)^2}{(Y_D/Y_U)^3}; \quad (10.36)$$

therefore, critical flow at the downstream section ($Fr_D = 1$) occurs when ${Fr_U}^2 = (Y_D/Y_U)^3/(W_U/W_D)^2$. The curve defined by this equality and the line defined by $Fr_U = 1$ define four fields that reflect the flow regimes of the upstream and downstream flows, as shown on figure 10.16.

10.3.1.2 Energy Equation

To determine the energy loss through an abrupt width transition, the upstream and downstream widths, the discharge, and the upstream depth (or velocity) are specified. This allows us to compute the upstream Froude number; entering figure 10.16a

BOX 10.5 Derivation of Equations 10.34 and 10.36

Equation 10.34

Noting that $\gamma/\rho = g$, equation 10.33 can be written as

$$\left(\frac{Q}{g}\right)\cdot(U_D - U_U) = \frac{W_U\cdot {Y_U}^2}{2} - \frac{(W_U - W_D)\cdot {Y_D}^2}{2} - \frac{W_D\cdot {Y_D}^2}{2},$$

which reduces to

$$\left(\frac{Q}{g}\right)\cdot(U_D - U_U) = \frac{W_U\cdot {Y_U}^2}{2} - \frac{W_U\cdot {Y_D}^2}{2}. \tag{10B5.1}$$

Since

$$Q = W_U\cdot Y_U\cdot U_U = W_D\cdot Y_D\cdot U_D, \tag{10B5.2}$$

equation 10B5.1 can be written as

$$\left(\frac{1}{g}\right)\cdot(U_D - U_U) = \left(\frac{1}{2}\right)\cdot\left(\frac{Y_U}{U_U} - \frac{{Y_D}^2}{U_U\cdot Y_U}\right)$$

or

$$\frac{U_U\cdot U_D}{g} - \frac{{U_U}^2}{g} = \left(\frac{1}{2}\right)\cdot\left(Y_U - \frac{{Y_D}^2}{Y_U}\right). \tag{10B5.3}$$

Again using equation 10B5.2, equation 10B5.3 becomes

$$\left(\frac{{U_U}^2}{g}\right)\cdot\left(\frac{W_U\cdot Y_U}{W_D\cdot Y_D} - 1\right) = \left(\frac{1}{2}\right)\cdot\left(Y_U - \frac{{Y_D}^2}{Y_U}\right). \tag{10B5.4}$$

We now divide equation 10B5.4 by Y_U to yield

$$\left(\frac{{U_U}^2}{g\cdot Y_U}\right)\cdot\left(\frac{W_U\cdot Y_U}{W_D\cdot Y_D} - 1\right) = \left(\frac{1}{2}\right)\cdot\left(1 - \frac{{Y_D}^2}{{Y_U}^2}\right). \tag{10B5.5}$$

Since $U_U^2/(g\cdot Y_U) \equiv Fr_U^2$, equation 10B5.5 becomes

$${Fr_U}^2 = \frac{1 - (Y_D/Y_U)^2}{2\cdot[(W_U/W_D)\cdot(Y_U/Y_D) - 1]}, \tag{10B5.6}$$

which, when multiplied by −1 and Y_D/Y_U, yields equation 10.34.

Equation 10.36

The ratio of downstream to upstream Froude numbers is

$$\frac{{Fr_D}^2}{{Fr_U}^2} = \frac{{U_D}^2/(g\cdot Y_D)}{{U_U}^2/(g\cdot Y_U)} = \frac{{U_D}^2\cdot Y_U}{{U_U}^2\cdot Y_D}. \tag{10B5.7}$$

From equation 10B5.2, $U_i = Q/(W_i\cdot Y_i)$, so equation 10B5.7 is equivalently

$$\frac{{Fr_D}^2}{{Fr_U}^2} = \frac{(W_U/W_D)^2}{(Y_D/Y_U)^3}. \tag{10B5.8}$$

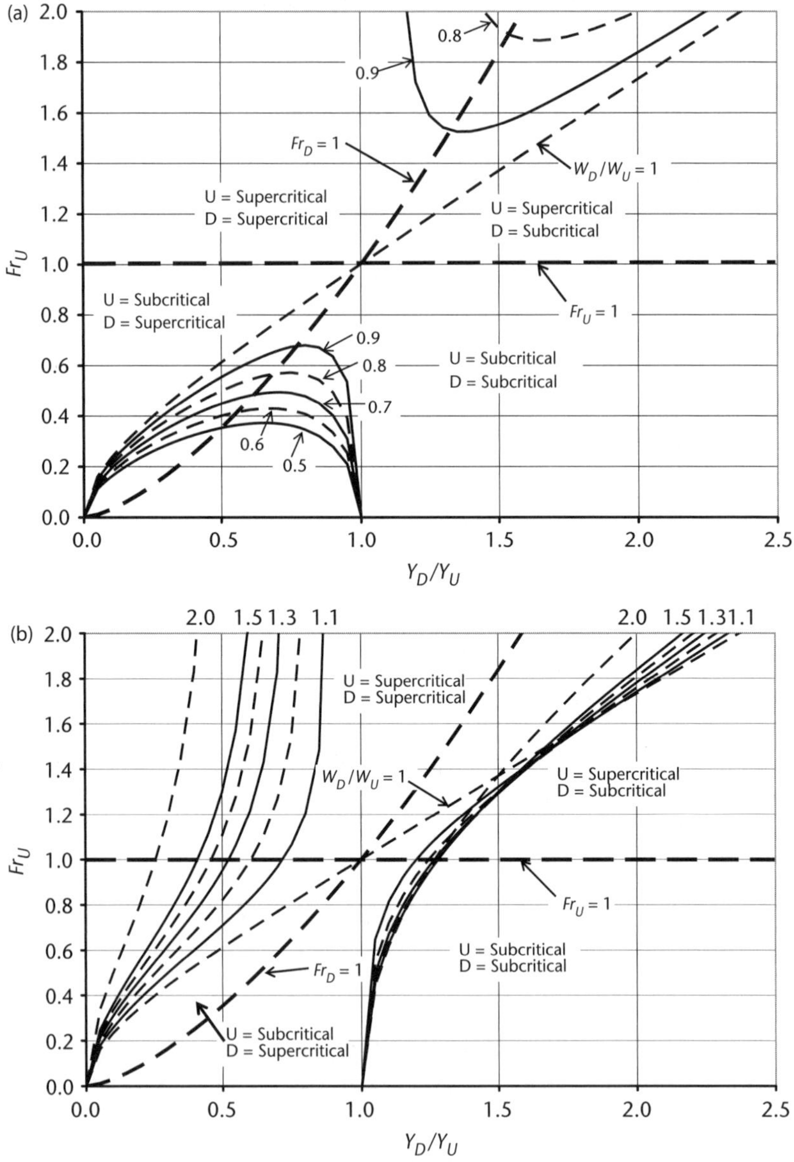

Figure 10.16 Ratio of downstream to upstream depth Y_D/Y_U (x-axis) as a function of width ratio W_D/W_U (contours on graph) and upstream Froude number Fr_U (y-axis) for (a) contractions ($W_D/W_U \leq 1$) (equation 10.34) and (b) expansions ($W_D/W_U \geq 1$) (equation 10.35). The long-dashed lines indicate when Froude numbers upstream (Fr_U) and downstream (Fr_D) = 1 and divide the graph into fields that indicate when upstream (U) and downstream (D) flows are subcritical or supercritical.

(for contractions) or 10.16b (for expansions) allows us to determine the ratio Y_D/Y_U, and hence Y_D and U_D, for the specified width ratio W_D/W_U. The head loss, $\Delta\mathcal{H}$, is then computed from the energy equation:

$$\Delta\mathcal{H} = Y_U + \frac{U_U{}^2}{2 \cdot g} - Y_D - \frac{U_D{}^2}{2 \cdot g}, \tag{10.37a}$$

or, in dimensionless form,

$$\frac{\Delta\mathcal{H}}{Y_U} = 1 + \frac{Fr_U{}^2}{2} - \left[\frac{Y_D}{Y_U} + \frac{Fr_U{}^2}{2 \cdot (Y_D/Y_U) \cdot (W_D/W_U)}\right], \tag{10.37b}$$

where we continue to assume that the energy coefficients $\alpha_U = \alpha_D = 1$.

In using this approach, it is important to note that many of the flow solutions given by equations 10.34 and 10.35 and indicated on figure 10.16 cannot actually occur because using the theoretical values they provide in equation 10.37 results in a negative energy loss ($\Delta\mathcal{H} < 0$), which violates the law of conservation of energy. Equation 10.37b can be used to identify situations that are energetically possible, but as Chow (1959) pointed out, the energy loss in transitions is typically very small and can readily be changed from negative to positive by a slight change in the terms in the equation. This also means that some theoretical solutions that appear impossible may actually be possible, because the real flow situation may not conform to the simplifications incorporated in the theoretical analysis (horizontal bed, no friction loss, $Y_X = Y_D$, and uniform velocity distribution).

Thus, although the analysis just described provides a theoretical framework for understanding flows through transitions, in practice hydraulic engineers usually refer to experimental results as described in the following section.

10.3.2 Experimental Results

In practice, the energy losses through transitions are treated separately for subcritical and supercritical flows. Referring to experimental work of Formica (1955), Chow (1959) reported that energy loss for subcritical flows through abrupt width contractions and expansions can be calculated as follows:

$$\text{Contractions:} \Delta\mathcal{H} = k_{\text{con}} \cdot \frac{U_D{}^2}{2 \cdot g}, \tag{10.38a}$$

$$\text{Expansions:} \Delta\mathcal{H} = k_{\text{exp}} \cdot \frac{(U_U - U_D)^2}{2 \cdot g}, \tag{10.38b}$$

where typical values of the loss coefficients are $0.06 \leq k_{\text{con}} \leq 0.10$ and $0.44 \leq k_{\text{exp}} \leq 0.82$, increasing with the abruptness of the transition. Note that equation 10.38b is of the same form as equations 9.5 and 9.36 used for computing eddy losses in gradually varied flow, and that the coefficient values cited here are consistent with those given in table 9.3.

Transitions in supercritical flows are accompanied by cross waves that originate at the walls where the width changes and are reflected off the channel walls downstream. Chow (1959) and Henderson (1961) provided analyses of these situations that emphasize the design of channels to minimize the height and downstream extent of the surface disturbances. Irregular and complex cross waves are observed in supercritical reaches of natural channels, which most often occur in steep bedrock channels.

10.3.3 Constrictions (Bridge Openings)

Constrictions create a single-opening width contraction of limited downstream extent (figure 10.17). They may occur naturally where local resistant geological formations are present or where entering tributaries, landslides, or debris flows deposit large amounts of coarse sediment. However, by far the most common occurrences of constrictions are at bridge openings, and a principal concern is determining their effects on water-surface profiles. Thus, profile-computation programs such as HEC-RAS and WSPRO (see section 9.4) contain algorithms for computing these effects. This section introduces approaches to estimating the water-surface profile effects and associated energy losses of constrictions. The use of constrictions in measuring streamflow (discharge) is discussed in section 10.4.3.

Figure 10.17 shows the four possible cases of rapidly varied flow induced by constrictions. In figure 10.17, a and b, the entering flow is subcritical; in 10.17a it remains subcritical through the constriction, whereas in 10.17b a short reach of supercritical flow occurs within and just downstream, followed by a return to subcritical flow via a hydraulic jump. In both of these cases a backwater effect (M1 profile; see figure 9.3) is induced that typically extends a considerable distance upstream. In figure 10.17, c and d, the entering flow is supercritical; in 10.17c supercritical flow is maintained in the constriction, whereas in 10.17d a hydraulic jump is induced upstream and a somewhat longer reach of subcritical flow (S1 profile) forms.

Here, we determine the backwater effect induced by constrictions to subcritical flows. Referring to figure 10.18, we again consider the simplest situation, with a horizontal channel of constant width upstream and downstream of the constriction ($W_U = W_D$) and uniform velocity distributions ($\alpha = 1$, $\beta = 1$) at all sections. The backwater effect is $\Delta Y \equiv Y_U - Y_D$, and we assume that Y_D is known from water-surface profile computations proceeding in the upstream direction.

As noted by Henderson (1961), the most elementary approach to determining ΔY would be to equate the energy at sections U and O ($\mathcal{H}_U = \mathcal{H}_O$) and the momentum at sections O and D($\mathsf{M}_O = \mathsf{M}_D$). However, this is not appropriate because 1) unless the constriction ratio $w \equiv W_O/W_U < 0.5$, the velocity distribution at section O will not be quasi uniform, and 2) more important, there typically will be significant energy loss between sections U and O. A second possible approach would estimate the friction loss $\Delta\mathcal{M}$ between sections U and D in the constriction and use the momentum equation $\mathsf{M}_U - \mathsf{M}_D = \Delta\mathcal{M}$ to find ΔY. This is a valid approach but requires experimental data on which to base the estimate of $\Delta\mathcal{M}$.

Because experimental data are required in any case, the most straightforward approach to determining the backwater effect is to use the experimental results of Yarnell (1934). Based on dimensional analysis (section 4.8.2) and measurements on

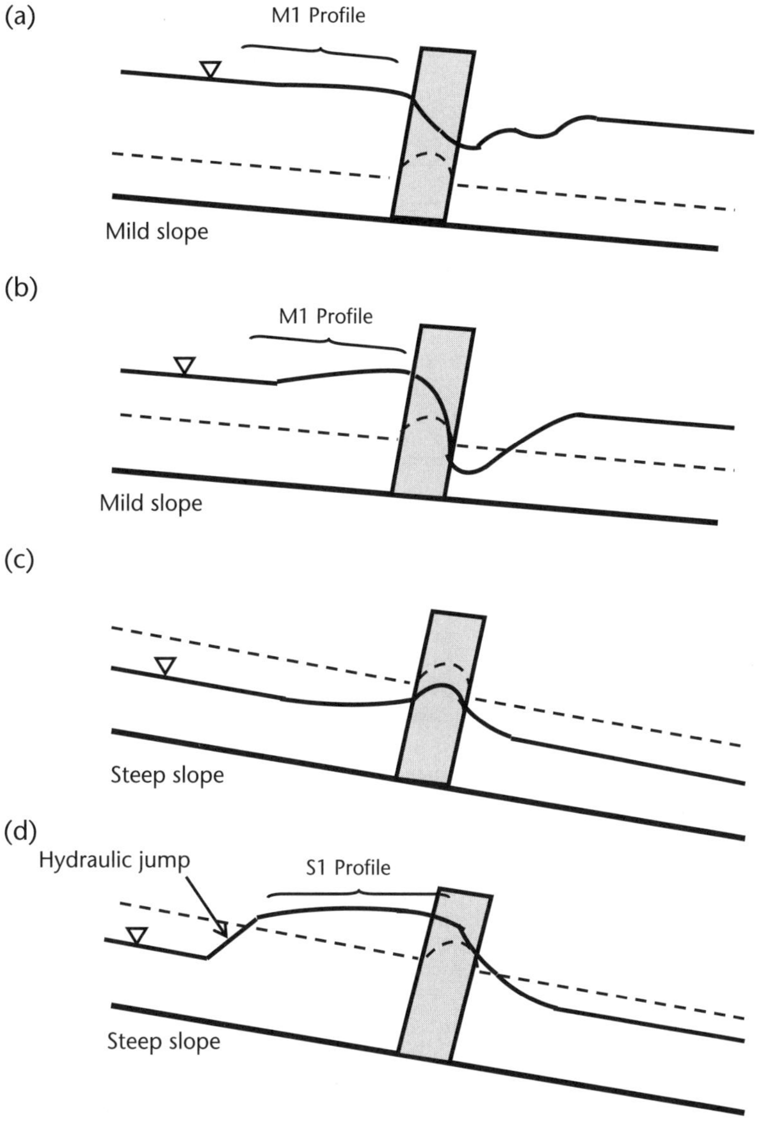

Figure 10.17 Four cases of rapidly varied flow induced by a constriction. Dashed line is critical depth. (a) subcritical flow throughout; (b) supercritical flow induced in constriction with hydraulic jump downstream; (c) supercritical flow throughout; (d) subcritical flow induced in constriction, producing a hydraulic jump upstream. After Chow (1959).

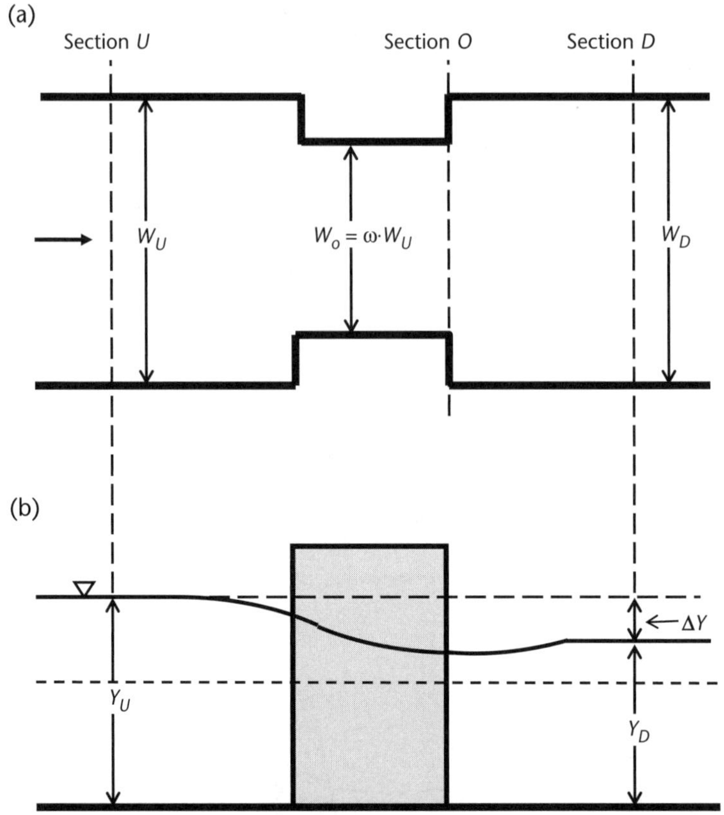

Figure 10.18 Definition diagram for computing the backwater effect ΔY due to a subcritical flow through a width constriction (equation 10.39): (a) plan view; (b) longitudinal profile. The short-dashed line is the critical-depth line. After Henderson (1966).

scale models of bridge piers with varying geometries, Yarnell (1934) found that ΔY can be directly estimated as

$$\frac{\Delta Y}{Y_D} = k_B \cdot Fr_D{}^3 \cdot (k_B + 5 \cdot Fr_D{}^2 - 0.6) \cdot [(1 - w) + 15 \cdot (1 - w)^4], \qquad (10.39)$$

where k_B is a coefficient that depends on the shape of the bridge pier (table 10.3). Figure 10.19 plots the values of $\Delta Y/Y_D$ as a function of Fr_D and w as given by equation 10.39 for $k_B = 1$; it shows that the backwater effect increases with the downstream Froude number and with the narrowness of the opening.

If discharge, upstream width, and other factors are constant, the Froude number of the flow in a constriction increases as the opening narrows (i.e., as w decreases). It is of interest to determine the point at which the flow is forced through the critical point; this is the condition called **choking**. Chow (1959) approached this problem via the energy equation, defining Y_{min} as the depth and U_{min} as the velocity at the section

Table 10.3 Values of shape factor, k_B, in equation 10.39 for various bridge-pier shapes determined by Yarnell (1934), as cited in Henderson (1961).

Shape	$k_B{}^a$
Semicircular nose and tail	0.9
Lens-shaped nose and tail	0.9
Twin-cylinder with connecting diaphragm	0.95
Twin-cylinder	1.05
90°-triangle nose and tail	1.05
Square nose and tail	1.25

[a]These values are for piers with lengths equal to four times their width ($L_P = 4 \cdot W_P$) and oriented parallel to the flow. Yarnell (1934) obtained slightly lower values for longer piers parallel to flow. For piers at an angle to the flow direction, Henderson (1961) states that the effective width W'_P equals the projected width; that is, $W'_P = L_P \cdot \sin\theta$, where θ is the angle between the pier axis and the flow direction. This effect may be large: For $\theta = 20°$, the backwater effect is 2.3 times the value for $\theta = 0°$.

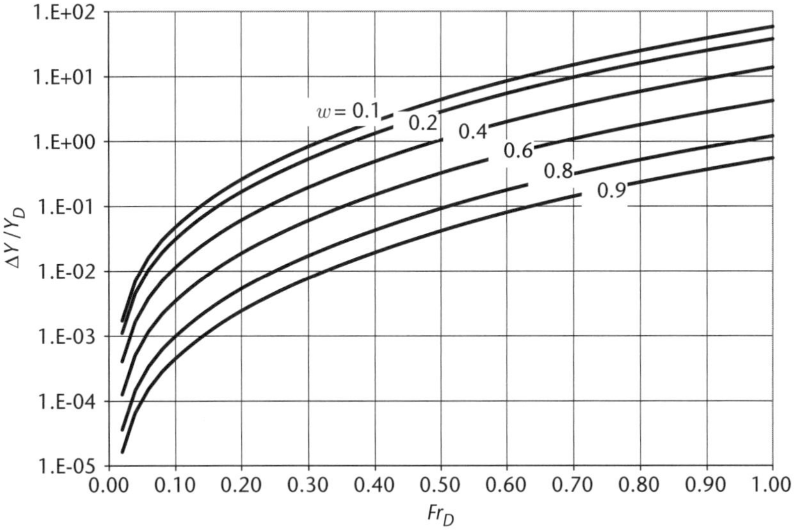

Figure 10.19 Relative backwater effect $\Delta Y/Y_D$ (logarithmic scale) as a function of downstream Froude number Fr_D for various constriction ratios $w \equiv W_O/W_U$ as given by equation 10.39, with $k_B = 1$.

with minimum depth and writing

$$\varepsilon_{\mathcal{H}} \cdot \left(Y_{\min} + \frac{U_{\min}{}^2}{2 \cdot g}\right) = \left(Y_D + \frac{U_D{}^2}{2 \cdot g}\right), \tag{10.40}$$

where $\varepsilon_{\mathcal{H}}$ is the fractional energy loss between the section with minimum depth and the downstream section. Using this relation, the definition of the Froude number, and

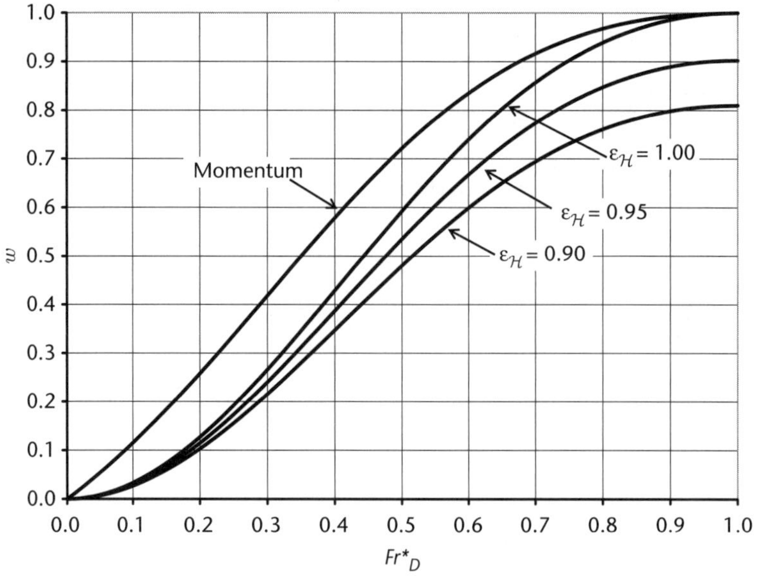

Figure 10.20 Critical value of downstream Froude number, $Fr_D{}^*$, as a function of width-constriction ratio w. Curves labeled with values of the energy-loss ratio $\varepsilon_{\mathcal{H}}$ are given by equation 10.41(b), derived from the energy equation. Curve labeled "Momentum" is derived from the momentum equation (equation 10.42).

the continuity relation $Q = W_O \cdot Y_{\min} \cdot U_{\min} = W_D \cdot Y_D \cdot U_D$ leads to

$$w^2 = \frac{\varepsilon_{\mathcal{H}}{}^3 \cdot Fr_D{}^2 \cdot (2 + Fr_{\min}{}^2)^3}{Fr_{\min}{}^2 \cdot (2 + Fr_D{}^2)^3}, \tag{10.41a}$$

where w is the constriction ratio. When $Fr_{\min} = 1$, the flow at the location of minimum depth becomes critical; substituting that value in equation 10.41a yields the expression for the critical value of the downstream Froude number, $Fr_D{}^*$, as a function of the constriction ratio and $\varepsilon_{\mathcal{H}}$:

$$\frac{Fr_D{}^{*2}}{(2 + Fr_D{}^{*2})^3} = \frac{w^2}{27 \cdot \varepsilon_{\mathcal{H}}{}^3} \tag{10.41b}$$

This relation is plotted in figure 10.20 for $\varepsilon_{\mathcal{H}} = 0.90, 0.95$, and 1.00.

In an alternative approach to the determination of $Fr_D{}^*$, Henderson (1961) equated the momentum at the opening to the downstream momentum ($\mathsf{M}_O = \mathsf{M}_D$) and derived

$$\frac{Fr_D{}^{*4}}{(1 + 2 \cdot Fr_D{}^{*2})^3} = \frac{w}{(2 + 1/w)^3}. \tag{10.42}$$

This relation is also plotted in figure 10.20. Note that equation 10.42 predicts that the critical Froude number for a given constriction ratio is smaller than predicted by equation 10.41. This more conservative value is probably more correct and more useful, because it does not require any estimate of the energy loss ($\varepsilon_{\mathcal{H}}$) (Henderson 1961).

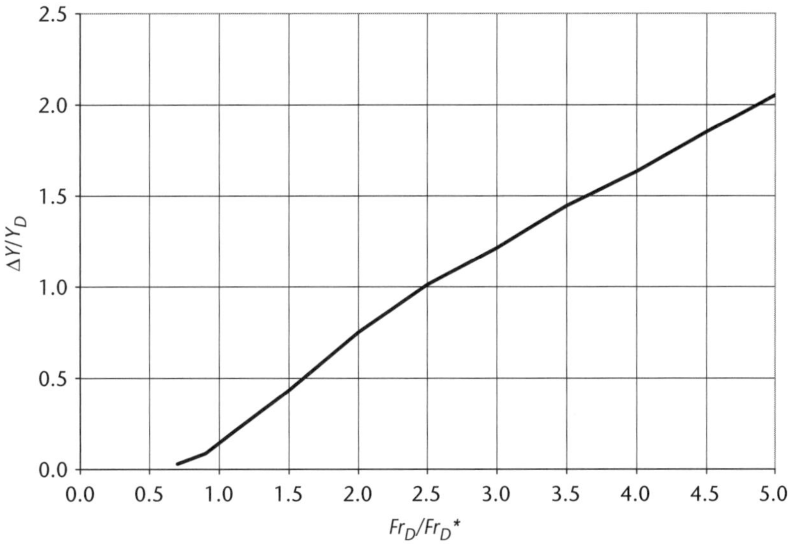

Figure 10.21 Graph for determining relative backwater effect $\Delta Y/Y_D$ for supercritical flow ($Fr_D > 1$) through a width constriction when downstream Froude number Fr_D is known and the value of $Fr_D{}^*$ has been determined from figure 10.20.

For a given opening, the flow is choked and becomes supercritical when the downstream Froude number exceeds the value, $Fr_D{}^*$, that satisfies equation 10.41b or 10.42. (This is the case shown in figure 10.17b, in which a hydraulic jump forms downstream from the constriction.) The value of $Fr_D{}^*$ can be determined for a given constriction ratio and the appropriate curve in figure 10.20. Then, given the actual downstream Froude number, Fr_D, the backwater effect, ΔY, can be found by entering the graph shown in figure 10.21 with the applicable value of $Fr_D{}^*/Fr_D$ (Yarnell 1934).

Once ΔY is determined from equation 10.39 or figure 10.21, the energy loss $\Delta \mathcal{H}$ is readily calculated from the energy equation:

$$\Delta \mathcal{H} = \Delta Y + \frac{Q^2}{2 \cdot g \cdot W_D{}^2} \cdot \left[\frac{1}{(\Delta Y + Y_D)^2} - \frac{1}{Y_D{}^2} \right], \tag{10.43}$$

where Q is the discharge.

10.4 Artificial Controls for Flow Measurement

10.4.1 Weirs

Weirs are damlike barriers constructed across channels in order to measure flow rates (discharge). They are of particular interest to hydrologists because they are generally the most practical means for continuous measurement where high accuracy and precision are required, such as on research watersheds. As discussed in section 2.5.3, weirs provide this accuracy by assuring a consistent relation between the elevation of the water surface (**stage**) and the discharge. The basic aspects of the stage-discharge

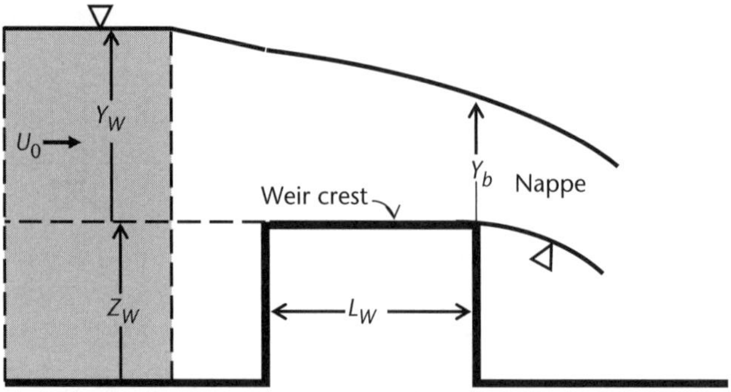

Figure 10.22 Definition of terms for describing flow over weirs. The shaded region is the approach section in which flow is assumed uniform. Z_W is the weir height, Y_W is the weir head, U_0 is the approach velocity, L_W is the weir length, and Y_b is the brink depth.

relation are determined by applying the conservation-of-energy principle, with empirically based modifications to account for the rapidly varied flow.

Figure 10.22 defines the basic terms characterizing weir geometry. The top surface of the weir is the **weir crest**, and the opening through which the water issues is the **weir notch**. The shape of the notch when viewed from upstream or downstream may be rectangular, triangular, or some other regular geometric form. In the **approach section**, well upstream of the crest, the flow is assumed to be uniform with hydrostatic pressure distribution, and the average approach velocity is designated U_0. The surface (and streamline) curvature increases as the flow accelerates toward the weir crest, and the pressure distribution increasingly deviates from hydrostatic. The flow velocity passes through the critical point near (usually slightly upstream of) the weir crest. The jet of water exiting the weir is called the **nappe**.[5] The free-falling nappe contracts and reaches a minimum cross-sectional area some distance beyond the crest. Concomitantly, the average velocity is a maximum at that point. The **weir length**, L_w, is the streamwise dimension of the weir; the **weir height**, Z_W, is the elevation of the crest above the weir floor (assumed horizontal); W_W is the **weir width** (cross-channel distance of a rectangular opening), and the vertical distance of the water surface in the approach section upstream of the weir crest is the **weir head**, Y_W.

Weirs are described in terms of 1) their relative "thickness," that is, the ratio Y_W/L_W; and 2) the shape of the notch. If Y_W/L_W is less than about 1.6–9, the weir is **broad crested**; if $Y_W/L_W > 1.6$, the flow springs free from the upstream edge of the weir and the weir is described as **sharp crested**. Broad-crested weirs usually present a horizontal surface extending across the stream width. Sharp-crested weirs with **rectangular**, **triangular**, or **trapezoidal** notches (or combinations of these shapes) are the types usually installed for the specific purpose of discharge measurement.

The remainder of this section introduces the basic hydraulics of weirs and flumes and the more important practical aspects of measuring discharge at such structures. The books by Ackers et al. (1978) and Herschy (1999a, 1999b) should be consulted for more detailed discussions of flow measurement with flumes.

10.4.1.1 Sharp-Crested Weirs

Basic Hydraulics An actual flow over a sharp-crested weir is shown in figure 10.23, and figure 10.24 defines terms characterizing the flow over an ideal rectangular sharp-crested weir. Note that the pressure at all surfaces of the nappe is atmospheric; that is, the gage pressure = 0. The pressure head and velocity head at the notch are indicated in the figure; friction losses are assumed to be negligible. Following Henderson (1966), the velocity head in the flow at the notch equals the vertical distance from the surface to the total head line, so the velocity at an arbitrary level "A" is $u_A = (2 \cdot g \cdot h_A)^{1/2}$. Thus, if the curvature of the surface is ignored, the discharge per unit width through the notch, $\mathcal{Q} \equiv Q/W_W$, where W_W is the width of the notch, is

$$\mathcal{Q} = \int_{U_0^2/2\cdot g}^{Y_w+U_0^2/2\cdot g} (2\cdot g\cdot h)^{1/2}\cdot dh = \frac{2}{3}\cdot(2\cdot g)^{1/2}\cdot\left[\left(\frac{U_0{}^2}{2\cdot g}+Y_W\right)^{3/2} - \left(\frac{U_0{}^2}{2\cdot g}\right)^{3/2}\right]. \tag{10.44a}$$

To account for the surface curvature and other effects (e.g., surface tension and friction losses), a **contraction coefficient**, C_{cR}, is introduced so that

$$\mathcal{Q} = \frac{2}{3}\cdot C_{cR}\cdot(2\cdot g)^{1/2}\cdot\left[\left(\frac{U_0{}^2}{2\cdot g}+Y_W\right)^{3/2} - \left(\frac{U_0{}^2}{2\cdot g}\right)^{3/2}\right]. \tag{10.44b}$$

This coefficient depends on the ratio Y_W/Z_W.

Equation 10.44b is more compactly written as

$$\mathcal{Q} = \frac{2}{3}\cdot C_{sR}\cdot(2\cdot g)^{1/2}\cdot Y_W{}^{3/2}, \tag{10.45a}$$

or, in terms of discharge,

$$Q = \frac{2}{3}\cdot C_{sR}\cdot(2\cdot g)^{1/2}\cdot W_W\cdot Y_W{}^{3/2}, \tag{10.45b}$$

Figure 10.23 Flow over a rectangular sharp-crested weir in a laboratory flume. Photo by the author.

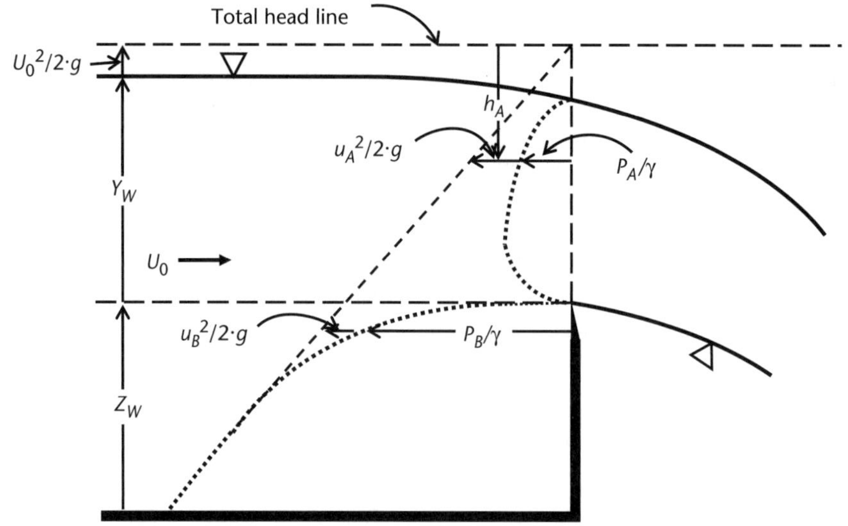

Figure 10.24 Definition diagram for flow over a sharp-crested weir, leading to equation 10.46. Z_W is the weir height, Y_W is the weir head, and U_0 is the approach velocity. The sloping short-dashed line is the total head at the exit section; the dotted lines show the pressure heads at two arbitrary levels A (P_A/γ) within the opening and B (P_B/γ) below the opening; $u_A^2/2\cdot g$ and $u_B^2/2\cdot g$ are the velocity heads at the corresponding levels. The velocity $u_A = (2\cdot g\cdot h_A)^{1/2}$. After Henderson (1966).

where C_{sR} is a **discharge coefficient** for a sharp-crested rectangular weir equal to

$$C_{sR} = C_{cR}\cdot\left[\left(\frac{U_0{}^2}{2\cdot g\cdot Y_W}+1\right)^{3/2}-\left(\frac{U_0{}^2}{2\cdot g\cdot Y_W}\right)^{3/2}\right]. \qquad (10.46)$$

Note that if the approach velocity U_0 is negligible, $C_{sR} = C_{cR}$. Thus, we can conclude that C_{sR} also depends essentially on Y_W/Z_W; the relation has been found by experiment to be

$$C_{sR} = 1.06+\left(1+\frac{Z_W}{Y_W}\right)^{3/2},\ \frac{Z_W}{Y_W}<0.05\left(\frac{Y_W}{Z_W}>20\right); \qquad (10.47a)$$

$$C_{sR} = 0.611+0.08\cdot\frac{Y_W}{Z_W},\ \frac{Z_W}{Y_W}>0.15\left(\frac{Y_W}{Z_W}<6.67\right). \qquad (10.47b)$$

Figure 10.25 plots equation 10.47, a and b, with a smooth curve (supported by modeling studies) connecting the curves for the two ranges ($0.05 < Z_W/Y_W < 0.15$; $6.67 < Y_W/Z_W < 20$). Note that when $Z_W/Y_W = 0$, the weir crest disappears, and there is a free overfall.

The presence of side walls, or **contractions**, on the notch opening also determines the degree of contraction of the nappe. Kindsvater and Carter (1959) conducted

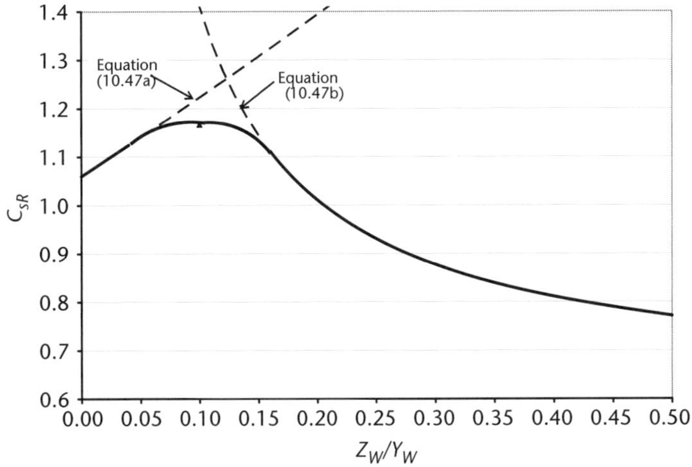

Figure 10.25 Solid curve shows discharge coefficient for sharp-crested rectangular weirs, C_{sR}, as a function of the ratio of weir height Z_W to weir head Y_W. After Daily and Harleman (1966).

a series of experiments on rectangular sharp-crested weirs to determine the effect of the relative opening width on the discharge coefficient. Figure 10.26 shows their results and indicates the combined effects of Y_W/Z_W and W_W/W on C_{sR}. Clearly, the presence of contractions causes C_{sR} to decrease, and for highly contracted weirs, C_{sR} decreases, rather than increases, with Y_W/Z_W.

Sharp-crested weirs with triangular openings, or **V-notch weirs** (figure 10.27), are commonly used for discharge measurement because they provide higher relative sensitivity at low flows than do rectangular weirs. To find the relation for discharge through a triangular notch, note from figure 10.28 that the cross-sectional area of flow through a triangular opening A_T is related to the weir head and the vertex angle θ_T as

$$A_T = {Y_W}^2 \cdot \tan(\theta_T/2). \tag{10.48}$$

Using this relation, Henderson (1966) showed that applying the approach that led to equation 10.45 to a triangular notch gives

$$Q = \frac{8}{15} \cdot C_{sT} \cdot (2 \cdot g)^{1/2} \cdot \tan(\theta_T/2) \cdot {Y_W}^{5/2}, \tag{10.49}$$

where the applicable coefficient is designated C_{sT}. For $\theta = 90°$, a common value for measurement weirs, $C_{sT} = 0.585$. However, as noted in the following section, weir coefficients should be determined by calibration.

It is important to note that the theoretical relations and the experimental results described below all assume that the nappe is completely aerated such that atmospheric pressure is maintained over all of its surface. Because the flow over the weir tends to entrain and deplete the air beneath the nappe, a vent pipe may be required to continually replenish the air (see French 1985, pp. 344–347).

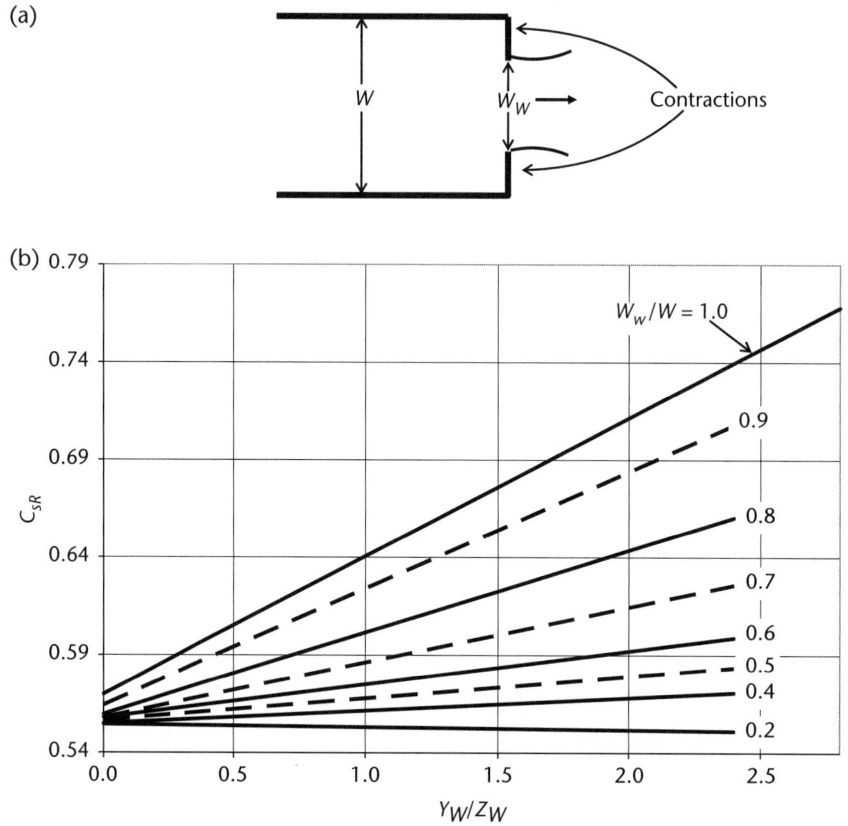

Figure 10.26 (a) Plan view of contracted rectangular sharp-crested weir. W_W/W is the contraction ratio. (b) Weir coefficient C_{sR} as a function of Y_W/Z_W and contraction ratio from experiments by Kindsvater and Carter (1959).

Practical Considerations Practical forms of the weir equations 10.45 and 10.49 can be presented in simplified form as follows:

Rectangular weirs:

$$Q = C_{WR}(W_W/W, Y_W/Z_W) \cdot W \cdot {Y_W}^{3/2} \tag{10.50R}$$

Triangular weirs:

$$Q = C_{WT}(Y_W/Z_W) \cdot \tan(\theta/2) \cdot {Y_W}^{5/2} \tag{10.50T}$$

The weir coefficients C_{WR} and C_{WT} have dimensions $[L^{1/2}T]$ and hence vary with the unit system. For any given weir, W and W_W/W (rectangular) or θ (triangular), and Z_W (both) will be constant so that the weir coefficient

(a)

(b)

(c)

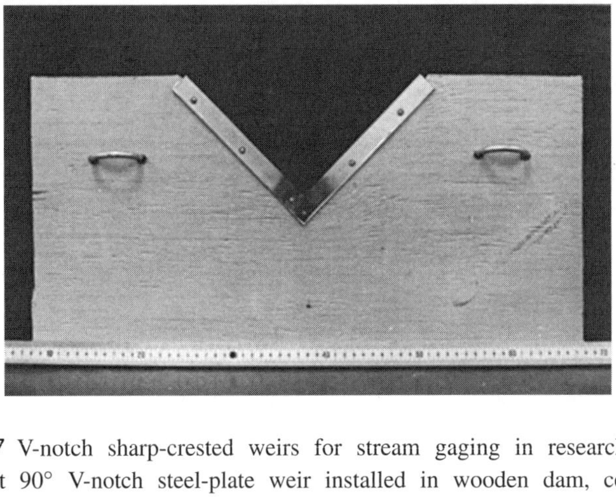

Figure 10.27 V-notch sharp-crested weirs for stream gaging in research watersheds. (a) Permanent 90° V-notch steel-plate weir installed in wooden dam, central Alaska. (b) Permanent 120° V-notch concrete weir, northeastern Vermont. (c) Portable 90° metal V-notch weir of plywood (scale in centimeters).

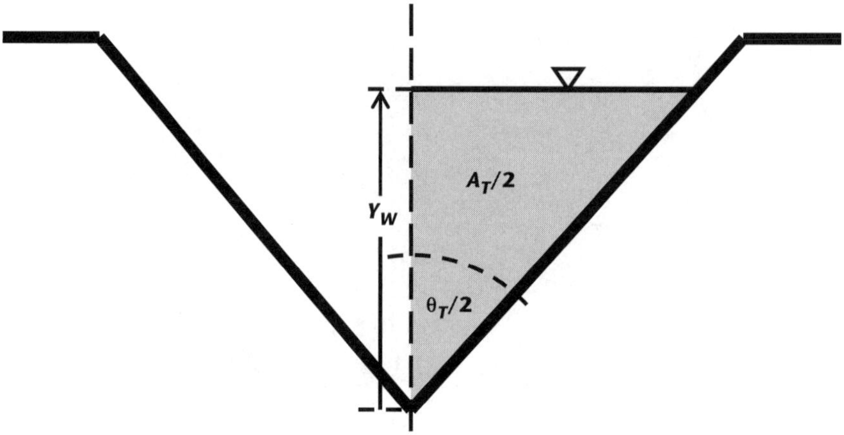

Figure 10.28 Definition diagram for deriving the equation for discharge through a V-notch weir (equations 10.48 and 10.49).

varies only as a function of water level (discharge). Thus, equation 10.50 can be further simplified as follows:

Rectangular weirs:

$$Q = C^*_{WR}(Y_W/Z_W) \cdot {Y_W}^{3/2} \tag{10.51R}$$

Triangular weirs:

$$Q = C^*_{WT}(Y_W/Z_W) \cdot {Y_W}^{5/2} \tag{10.51T}$$

The coefficients with asterisks also have dimensions $[L^{1/2}T]$.

Although, as we have seen, general values for the coefficients have been obtained by experiment, measurement weirs should be individually calibrated. Of special concern are the coefficient values at very low flows, because these are strongly influenced by irregularities in the construction and surface condition of the notch. Figure 10.29 shows the results of calibration for the weir in figure 10.27a: The weir coefficient C_{sT} (equation 10.49) decreases rapidly with Y_W/Z_W below $Y_W/Z_W = 0.3$ and is effectively constant at $C_{sT} = 0.57$ above that level. Note that this latter value is substantially below the commonly accepted value of $C_{sT} = 0.585$ noted above.

Other practical aspects of flow measurement with sharp crested weirs should be noted:

1. The range of discharge values that can be measured by a given weir depends on the vertical extent of the notch, so careful consideration must be given to the expected discharge range. The range can be extended by combining a triangular notch with a small angle and a larger-angle notch, either in the same weir plate (figure 10.30a) or separately (figure 10.30b).
2. Care must be taken to assure that all the flow to be measured is directed to the notch; this may involve installing wing-wall barriers to prevent surface and subsurface flow from bypassing the weir.

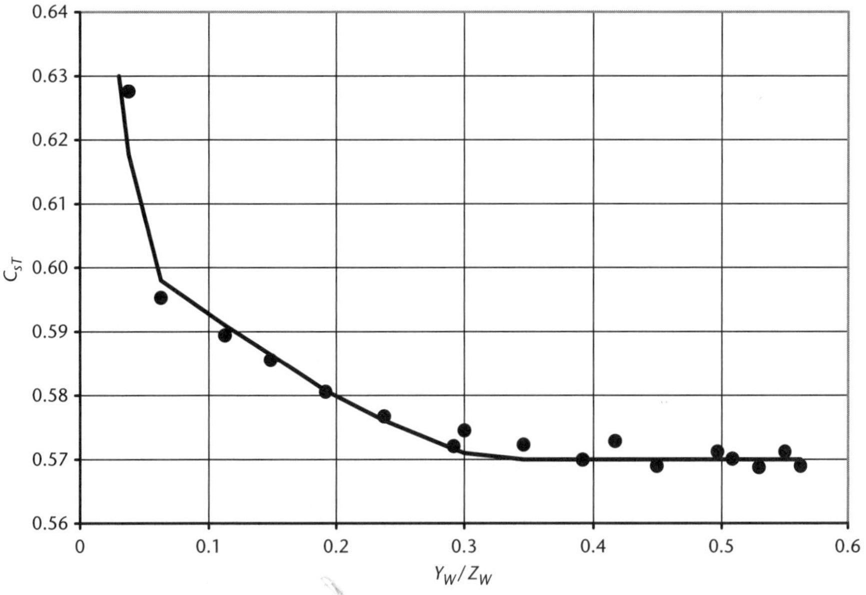

Figure 10.29 Weir coefficient C_{sT} as a function of relative weir head Y_W/Z_w as determined by laboratory calibration of the 90° V-notch weir shown in figure 10.27a.

3. The theoretical weir equations assume that the weir head, Y_w, is measured upstream of where the surface is affected by curvature; this requires that the measurement be made at an upstream distance at least two-times the vertical dimension of the notch. The head may also be measured on the upstream face of the weir plate as far from the notch as possible.
4. Every attempt should be made to reduce the approach velocity U_0 to near zero. If $U_0 \approx 0$, the weir head will approximate the total head.
5. Because the approach velocity is small, sediment tends to settle in the weir pool. If it builds up sufficiently, the value of Z_W and hence the ratio Y_W/Z_W will change, which will alter the weir coefficient and the calibration. Thus, periodic cleaning of the approach pool may be required—and may provide a useful way of measuring sediment yield (see section 12.2.2).

10.4.1.2 Broad-Crested Weirs

Basic Hydraulics We saw in section 10.2.1.3 that when a subcritical flow encounters an abrupt rise in the channel bottom, its depth decreases and its velocity increases (figure 10.12a). If the rise Z_D is large enough, the flow will be forced through the critical point at which (from equation 10.2)

$$Q = g^{1/2} \cdot W_W \cdot Y_c^{3/2}, \tag{10.51a}$$

(a)

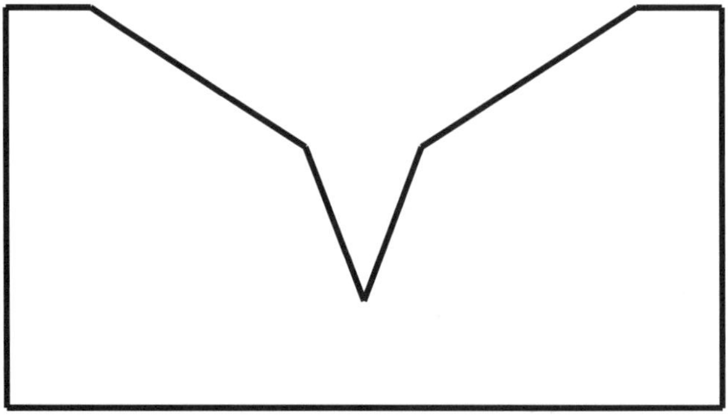

(b)

Figure 10.30 Combination V-notch weirs. (a) Diagram of compound weir plate. The small-angle notch increases precision at low flows, and the wide-angle notch increases weir capacity. (b) The same effect can be achieved by installing separate wide-angle and small-angle (lower right) V-notch weirs, as at this gaging station on a research watershed in Vermont. Photo by the author.

where Y_c is critical depth or, in terms of $\mathcal{Q} \equiv Q/W_W$ (assuming a horizontal surface across the weir),

$$\mathcal{Q} = g^{1/2} \cdot Y_c{}^{3/2}. \tag{10.51b}$$

At critical flow, the specific head $\mathcal{H}_S = (3/2) \cdot Y_c$ (equation 10.24), and assuming hydrostatic pressure distribution and no head loss due to friction, this relation can be substituted into equation 10.51b to yield

$$\mathcal{Q} = \left(\frac{2}{3}\right)^{3/2} \cdot g^{1/2} \cdot Y_W{}^{3/2} = 0.544 \cdot g^{1/2} \cdot Y_W{}^{3/2}, \tag{10.52}$$

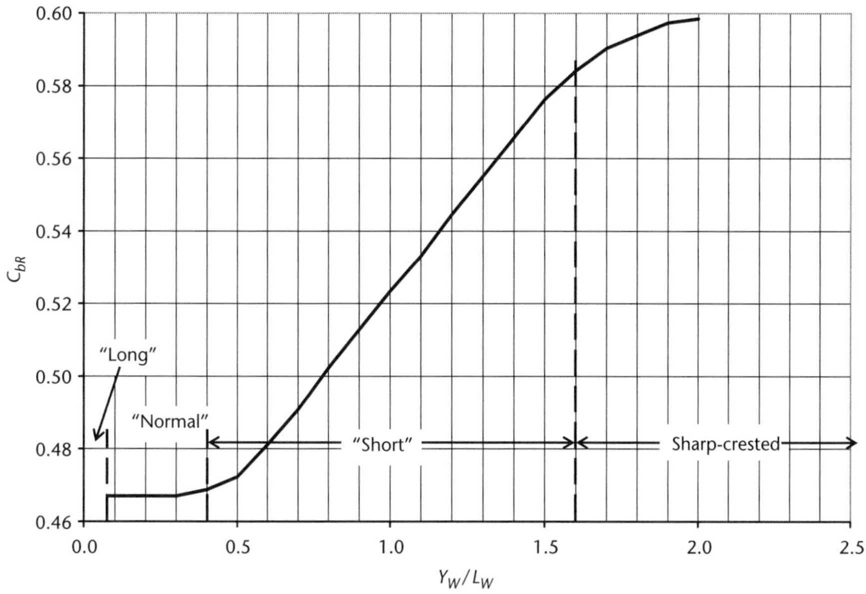

Figure 10.31 Weir coefficient C_{bR} for rectangular broad-crested weirs as a function of relative weir height Y_W/L_W. Data from Tracy (1957).

where the weir head Y_W is defined as in figures 10.22 and 10.24.

Equation 10.52 is the basic discharge relation for a rectangular broad-crested weir. However, it applies only when the assumptions of hydrostatic pressure distribution and negligible friction loss are met. Because these assumptions are generally more-or-less violated in actual situations, it is appropriate to write the discharge relation for a rectangular broad-crested weir as

$$Q = C_{bR} \cdot g^{1/2} \cdot {Y_W}^{3/2}. \tag{10.53}$$

Experiments and literature review by Tracy (1957) showed how the weir coefficient C_{bR} varies as a function of the ratio of weir head to weir thickness, Y_W/L_W (figure 10.31), and the following terminology is used:

Long weir, $Y_W/L_W < 0.08$ (figure 10.32a): The flow over the weir crest is long enough to create a significant turbulent boundary layer (see figure 3.28), such that friction losses become significant and the above hydraulic analysis is not appropriate. However, such a weir can be used for flow measurement if calibrated. If there is a free overfall, the depth at the brink, $Y_b = 0.715 \cdot Y_c$, can be measured, in which case discharge per unit width, Q, can be determined as $Q = 1.65 \cdot g^{1/2} \cdot {Y_b}^{3/2}$ (Henderson 1966).

Normal weir, $0.08 < Y_W/L_W < 0.4$ (figure 10.32b): The flow over the weir crest is long enough to permit a quasi-horizontal water surface but short enough to keep

(a)

(b)

(c)

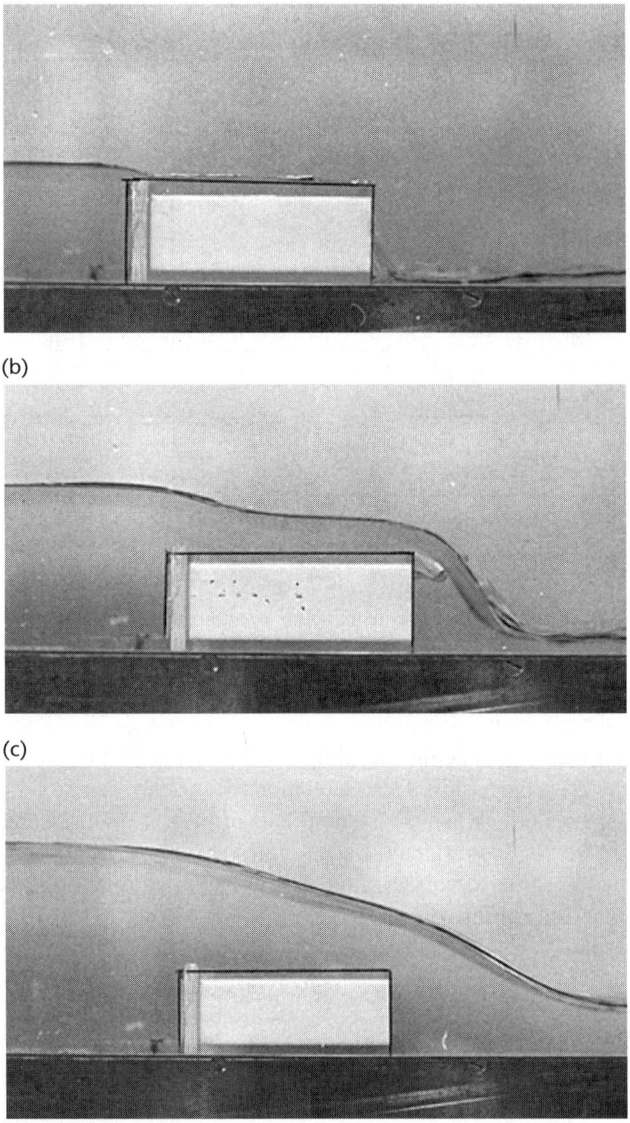

Figure 10.32 Flows over a rectangular broad-crested weir in a laboratory flume: (a) "long," (b) "normal," and (c) "short." Photos by the author.

frictional effects small. This situation conforms most closely to the theoretical hydraulic analysis above (equation 10.52), and the weir coefficient does not vary significantly with discharge. However, the actual value of the weir coefficient differs from the theoretical value due to frictional effects, water-surface curvature, and other deviations from the ideal situation.

Short weir, $0.4 < Y_W/L_W <\approx 1.6$ (figure 10.32c): The water surface is curvilinear over the entire crest length, so the assumption of hydrostatic pressure is violated. However, the flow still goes through the critical point, and the weir can be used for measurement, although the weir coefficient changes as the degree of curvature changes with discharge.

Sharp-crested weir, $1.6 < Y_W/L_W$: In this range, the flow separates from the upstream edge of the weir, and it acts as a sharp-crested weir.

Practical Considerations The practical considerations listed in section 10.4.1.1 for sharp-crested weirs apply equally for broad-crested weirs. Tracy (1957) summarized studies showing the effects on the weir coefficient of degree of nappe aeration, submergence, rounding of the upstream face, boundary roughness, and shape of the upstream and downstream faces of broad-crested weirs. However, as with sharp-crested weirs, broad-crested weirs used for measurement should be calibrated.

10.4.2 Flumes

A **flume** is an artificial channel, usually designed to convey water at an accelerated velocity. As noted, one disadvantage of using weirs for discharge measurement is that the low approach velocities induce sediment accumulation. To avoid this problem, hydrologists often install measurement flumes. The most commonly used type, at least in the United States, is the **Parshall flume**, designed by R.L. Parshall in the 1920s. Parshall flumes are a form of critical-depth flume that forces the flow to become supercritical by a combination of width constriction and local steepening in a throat section.

Parshall flumes are constructed in a range of sizes, following the general design shown in figure 10.33. The various dimensions denoted by letters in that figure are given in tables (e.g., French 1985). Note that the weir head, Y_W, is measured a prescribed distance upstream of the throat; the relation between weir head and discharge has been established by careful calibration studies, and for flumes with throat widths, W_T, of 1–8 ft is

$$Q = 4 \cdot W_T \cdot Y_W^{1.522 \cdot W_T{}^{0.026}}, \qquad (10.54)$$

where Q is in ft^3/s, and W_T and Y_W are in ft (Henderson 1966). The standard rating relations such as equation 10.54 are valid as long as the water surface downstream of the throat is not high enough to submerge the hydraulic jump in the exit section. Correction factors must be used when submergence occurs.

The principal practical considerations in using Parshall flumes are 1) properly sizing the flume for the range of discharges to be expected, 2) installing the weir so that the converging section is horizontal, and 3) installing wing walls or other means to ensure that all the flow to be measured passes through the flume. Small Parshall flumes are portable and are commercially manufactured. Further details are given by Herschy (1999a) and Dingman (2002).

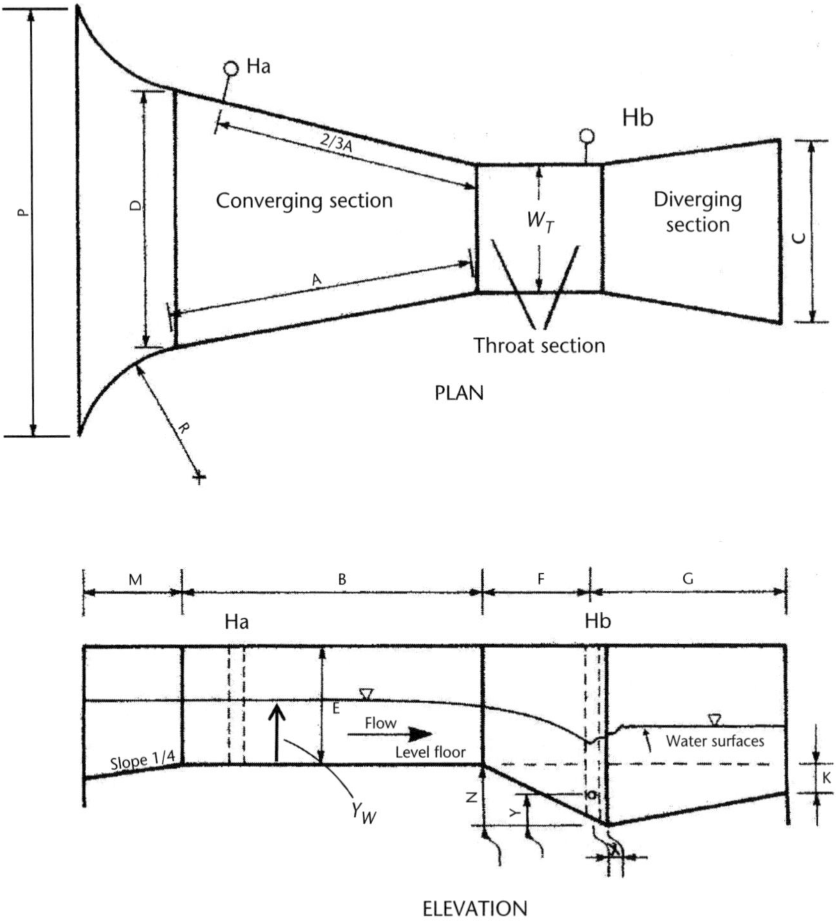

Figure 10.33 Plan and elevation of a Parshall flume. The letters indicate the various dimensions that have a prescribed relation to the throat width W_T. The weir head Y_W is measured in a stilling well at location Ha on the plan. The submergence depth is measured at Hb. See French (1985) and Herschy (1999b) for details. After Herschy (1999b).

10.4.3 Flows through Width Constrictions

Constrictions such as bridge openings can be used for estimating the discharge of a past flood peak if marks recording the water-surface configuration at the maximum discharge are apparent through the constriction. This is a form of slope-area measurement, introduced in section 6.10.2.

Section 10.3.3 discussed flows through bridge openings and derived relations for estimating the backwater effect (equation 10.39) and energy loss (equation 10.43). Here, we derive relations that allow computation of discharge from measurements of bridge-opening geometry, channel characteristics, and high-water marks. These relations are based on the continuity equation, the energy equation, and a resistance relation.

10.4.3.1 Conceptual Approach

In principle, discharge through a typical bridge constriction can be found by solving equation 10.43 for Q:

$$Q = \left[\frac{2 \cdot g \cdot W_D{}^2 \cdot (\Delta \mathcal{H} - \Delta Y)}{\dfrac{1}{(\Delta Y + Y_D)^2} - \dfrac{1}{Y_D{}^2}} \right]^{1/2}, \tag{10.55}$$

where $\Delta \mathcal{H}$ is the head loss through the constriction, and the other terms are defined in figure 10.18. The geometric terms would be determined by field survey, while the estimate of the energy loss $\Delta \mathcal{H}$ between an upstream and a downstream cross section could be based on the assumption that all energy loss is due to boundary friction, that is, that

$$\Delta \mathcal{H} = S_f \cdot \Delta X = \Delta \mathcal{M}, \tag{10.56}$$

where S_f is the friction slope (head loss due to boundary friction) and ΔX is the distance between sections. Estimation of the friction slope, in turn, requires the assumption of a resistance relation, typically the Manning equation (see section 6.8),

$$Q = \frac{u_M \cdot A \cdot Y^{2/3} \cdot S_f{}^{1/2}}{n_M}, \tag{10.57a}$$

from which

$$S_f = \frac{n_M{}^2 \cdot Q^2}{u_M^2 \cdot A^2 \cdot Y^{4/3}}, \tag{10.57b}$$

where A is cross-section area, and n_M would be estimated using one of the techniques described in table 6.3.

There are two difficulties with this approach: 1) We would need a way of averaging A, Y, and n_M for the reach between the two sections; and 2) it requires an iterative solution, because computing the value of S_f from 10.57b requires specifying a value of Q. The following section describes the approach developed by Matthai (1967) to get around these difficulties.

10.4.3.2 Approach of Matthai (1967)

Referring to figure 10.34, when a subcritical flow enters a constriction, the "live stream" contracts to a minimum area and then expands as it leaves the constriction. The energy equation can be written between an upstream approach section (designated by subscript U) and a downstream contracted section (designated by subscript C):

$$Y_U + \frac{\alpha_U \cdot U_U{}^2}{2 \cdot g} = Y_C + \frac{\alpha_C \cdot U_C{}^2}{2 \cdot g} + \Delta \mathcal{H}, \tag{10.58}$$

where $\Delta \mathcal{H}$ is the energy loss between the two sections. From the continuity relation, $U_U = Q/A_U$ and $U_C = Q/A_C$, where A_U is the area at the upstream section, and A_C is

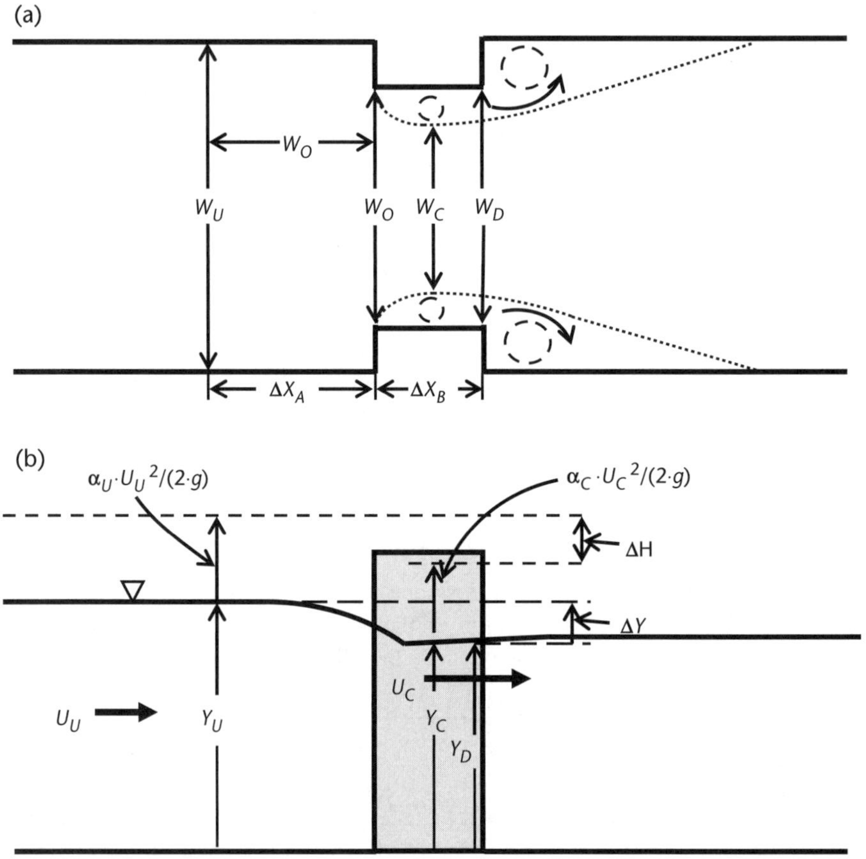

Figure 10.34 Definition diagram for derivation of equations 10.59 and 10.64: (a) plan view and (b) profile view. The upstream section is at a distance equal to one opening (W_O) upstream of the constriction. The "live" flow contracts to a minimum width (W_C) within the constriction. The downstream section is located at or upstream of the bridge-opening exit, depending on bridge geometry. Short-dashed lines are energy-grade lines.

the area of the live stream at the contracted section. In practice, A_C is not known, so it is replaced by $A_C = C_d \cdot A_D$, where C_d is a discharge coefficient (discussed further below), and the area A_D is the downstream area, measured at a prescribed location that depends on the detailed geometry of the bridge abutments. Incorporating these relations, equation 10.58 can be written as

$$Q = (2 \cdot g)^{1/2} \cdot C_d \cdot A_D \cdot \left(\Delta Y - \frac{\alpha_U \cdot Q^2}{2 \cdot g \cdot A_U{}^2} - \Delta \mathcal{H} \right)^{1/2}, \tag{10.59}$$

where ΔY is the difference between the upstream and downstream water-surface elevations as revealed by the high-water marks; that is, $\Delta Y \equiv Y_U - Y_D$.

The next step in Matthai's development was to invoke the concept of **conveyance**, $\mathcal{K}$ (see box 9.2), defined as

$$\mathcal{K} \equiv \frac{u_M \cdot A \cdot Y^{2/3}}{n_M}, \tag{10.60}$$

so that the resistance relation 10.57b can be written as

$$S_f = \frac{Q^2}{\mathcal{K}^2}, \tag{10.61}$$

and, using equation 10.56,

$$\Delta\mathcal{H} = \frac{Q^2}{\mathcal{K}^2} \cdot \Delta X. \tag{10.62}$$

Matthai then divided the distance between the approach section and the downstream section into two segments and replaced equation 10.62 with

$$\Delta\mathcal{H} = \left(\frac{Q^2}{\mathcal{K}_U \cdot \mathcal{K}_D}\right) \cdot \Delta X_A + \left(\frac{Q^2}{\mathcal{K}_D^2}\right) \cdot \Delta X_B, \tag{10.63}$$

where ΔX_A is the distance from the upstream approach section to the bridge opening, ΔX_B is the distance from the opening to the downstream section, and $\mathcal{K}_U$ and $\mathcal{K}_D$ are the conveyances of the upstream and downstream sections, respectively.

Finally, substituting equation 10.63 into 10.59 and solving for Q yields the working relation:

$$Q = (2 \cdot g)^{1/2} \cdot C_d \cdot A_D \cdot \left(\frac{\Delta Y}{1 - \alpha_U \cdot C_d^2 \cdot \left(\frac{A_D}{A_U}\right)^2 + 2 \cdot g \cdot C_d{}^2 \cdot \left(\frac{A_D}{\mathcal{K}_D}\right)^2 \cdot \left(\Delta X_B + \frac{\Delta X_A \cdot \mathcal{K}_D}{\mathcal{K}_U}\right)}\right)^{1/2} \tag{10.64}$$

All the quantities on the right-hand side of equation 10.64 can be determined by field measurement and observation, as described in detail by Matthai (1967). The upstream section is located a distance of one bridge-opening width upstream of the opening (i.e., $\Delta X_A = W_O$). The downstream section is located within the bridge opening or at its exit, depending on the geometry of the bridge opening. The conveyances and α_U are determined by field survey of the areas and depths and application of the conventional empirical approach described in box 8.2.

The discharge coefficient C_d accounts for 1) the degree of contraction, 2) the eddy losses associated with the contraction, and 3) the kinetic-energy coefficient at the contracted section, α_C. Dimensional analysis reveals that C_d depends on a number of aspects of the geometry of the bridge opening and abutments, the most important of which are 1) the degree of contraction imposed by the bridge opening, and 2) the ratio of bridge-opening width to the length of the opening, W_O/X_B. Much of Matthai's report presents graphs for estimating C_d for bridge openings of various geometries.

11

Unsteady Flow

11.0 Introduction and Overview

This chapter focuses on one-dimensional flows and is concerned with changes in the downstream direction only. In general, the average downstream velocity, U, is a function of space (downstream location, X) and time, t, that is,

$$U = U(X, t), \tag{11.1}$$

and the definition of acceleration given in equation 4.11 simplifies to

$$\frac{dU}{dt} = \underbrace{\frac{\partial U}{\partial t}}_{\text{local acceleration}} + \underbrace{\frac{\partial U}{\partial X}}_{\text{convective acceleration}} \cdot U. \tag{11.2}$$

Thus, for one-dimensional flows, the definition of **unsteady flow** given in section 4.2.1.2 becomes "flow in which $|\partial U/\partial t| > 0$." It is essential to note that temporal changes in velocity always involve concomitant changes in depth and so can be viewed as wave phenomena. In fact, most unsteady-flow situations in natural channels are produced by natural or human-caused depth disturbances, including the following:

1. Flood waves produced by watershed-wide increases in streamflow due to rain or snowmelt
2. Waves due to landslides or debris avalanches into lakes or streams
3. Waves generated by the failure of natural or artificial dams
4. Waves produced by tidal fluctuations (tidal bores)
5. Waves produced by the operation of engineering structures, such as starting or stopping turbines or pumps, or opening or closing control gates or navigation locks

Some of the most important applications of the principles of open-channel flow are in the prediction and modeling of the depth and speed of travel of these waves.

The objective of this chapter is to provide a basic understanding of unsteady-flow phenomena, and we begin by applying the by-now familiar principles of conservation of mass and conservation of momentum to derive the basic equations for one-dimensional unsteady flow.

11.1 The Saint-Venant Equations: The Basic Equations of Unsteady Gradually Varied Flow

As with the relations for steady gradually varied open-channel flows, the basic relations for analysis of unsteady flows are 1) the conservation-of-mass equation, and 2) a dynamic relation that can be derived from either the conservation of energy or of momentum. Because we are now dealing with spatial and temporal changes, these relations take the form of partial-differential equations. The dynamic relation can be incorporated into a resistance relation to show how discharge is determined by the various forces that influence open-channel flows.

The conservation-of-mass equation and the dynamic equation were first developed by Jean-Claude Barré de Saint-Venant (1797–1886) in France in 1848 and are known as the **Saint-Venant equations**.

11.1.1 Conservation of Mass Equation (Continuity)

Referring to figure 11.1, we can derive the conservation-of-mass equation for one-dimensional (macrosopic) open-channel flow as in section 4.3.2 to arrive at

$$q_L - U \cdot \frac{\partial A}{\partial X} - A \cdot \frac{\partial U}{\partial X} = \frac{\partial A}{\partial t}, \tag{11.3a}$$

where U is cross-sectional average velocity [L T^{-1}], A is cross-sectional area [L^2], and q_L is the net rate of lateral inflow (which might include rainfall and seepage into or out of the channel) per unit channel distance [L^2 T^{-1}]. Since the discharge $Q = U \cdot A$, we can use the rules of derivatives to note that $U \cdot (\partial A/\partial X) + A \cdot (\partial U/\partial X) = \partial Q/\partial X$ and write equation 11.3 more compactly as

$$q_L - \frac{\partial Q}{\partial X} = \frac{\partial A}{\partial t} \tag{11.3b}$$

or, in the absence of lateral inflow,

$$-\frac{\partial Q}{\partial X} = \frac{\partial A}{\partial t}. \tag{11.3c}$$

Note that equation 11.3c makes logical sense if we imagine a wave traveling through a channel, as in figure 2.33: In the channel downstream (upstream) of the peak, discharge decreases (increases) in the downstream direction, so $\partial Q/\partial X < 0$ (> 0), but the discharge and hence the cross-sectional area are increasing (decreasing) with time, so $\partial A/\partial t > 0$ (< 0). Thus, the two rates of change must have opposite signs.

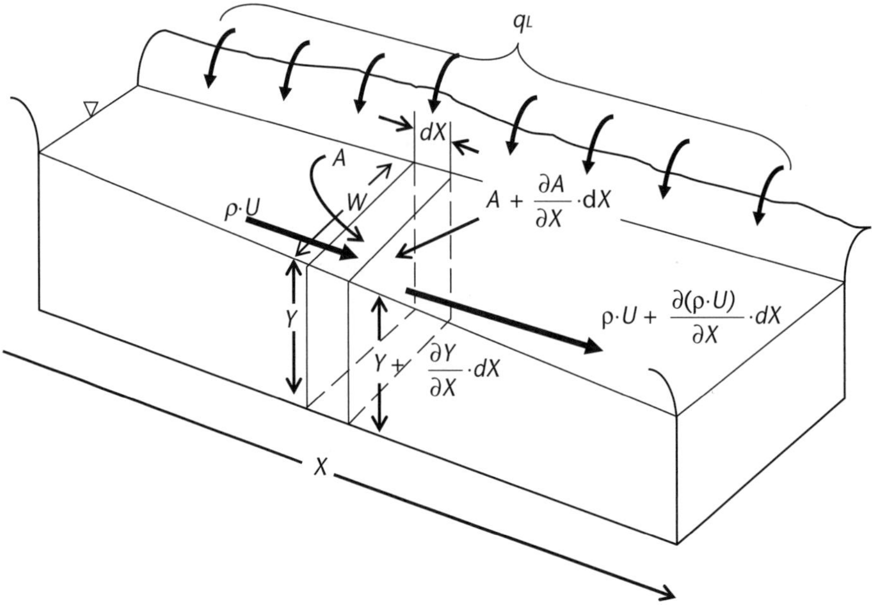

Figure 11.1 Definition diagram for derivation of macroscopic continuity equation (equation 11.3) and macroscopic conservation-of-energy equation (equation 11.6). The area of the upstream and downstream faces of the control volume are A and $A + (\partial A/\partial X) \cdot \mathrm{d}X$, respectively.

11.1.2 Dynamic Equation (Momentum/Energy)

11.1.2.1 Derivation

If we assume hydrostatic pressure distribution and uniform velocity distribution, the one-dimensional energy equation for steady flow between an upstream cross section (subscript i) and a downstream cross section (subscript $i-1$) is

$$Z_i + Y_i + \frac{U_i^2}{2 \cdot g} = Z_{i-1} + Y_{i-1} + \frac{U_{i-1}^2}{2 \cdot g} + \Delta\mathcal{H}_{i,i-1}, \tag{11.4a}$$

where Z is the channel-bottom elevation, g is gravitational acceleration, and $\Delta\mathcal{H}_{i,i-1}$ is the energy loss between section i and section $i-1$. (Equation 11.4a is identical to equation 8.8b.)

Again referring to figure 11.1, if we consider a small increment of channel length $\mathrm{d}X$ and define $\mathrm{d}Z \equiv Z_{i-1} - Z_i$ and similarly for $\mathrm{d}Y$, $\mathrm{d}(U^2/2 \cdot g)$, and $\mathrm{d}\mathcal{H}$, we can rewrite 11.4a in differential form:

$$Z + Y + \frac{U^2}{2 \cdot g} = (Z + \mathrm{d}Z) + (Y + \mathrm{d}Y) + \left[\frac{U^2}{2 \cdot g} + \mathrm{d}\left(\frac{U^2}{2 \cdot g}\right)\right] + \mathrm{d}\mathcal{H}, \tag{11.4b}$$

which reduces immediately to

$$d\mathcal{H} = -\left[\mathrm{d}Z + \mathrm{d}Y + \left(\frac{1}{2 \cdot g}\right) \cdot \mathrm{d}(U^2)\right]. \tag{11.5a}$$

Since $d(U^2) = 2 \cdot U \cdot dU$, we write equation 11.5a as

$$d\mathcal{H} = -\left[dZ + dY + \left(\frac{1}{g}\right) \cdot U \cdot dU\right]. \tag{11.5b}$$

Now if we divide equation 11.5b by dX, we have an expression for the downstream rate of change of total head for steady nonuniform flow:

$$\frac{d\mathcal{H}}{dX} = -\left[\frac{dZ}{dX} + \frac{dY}{dX} + \frac{U}{g} \cdot \frac{dU}{dX}\right] \tag{11.6}$$

Recalling the discussion in section 7.1, equation 11.6 reflects the force balance as written in equation 7.4:

$$a_V + a_T = a_G + a_P - a_X, \tag{11.7}$$

where the terms represent the forces per unit mass (accelerations), and the subscripts denote the viscous (V), turbulent (T), gravitational (G), pressure (P), and convectional (X) accelerations. These accelerations have the following correspondences to the gradients in equation 11.6:

$$a_V + a_T \leftrightarrow \frac{d\mathcal{H}}{dX}$$

$$a_G \leftrightarrow -\frac{dZ}{dX}$$

$$a_P \leftrightarrow -\frac{dY}{dX}$$

$$a_X \leftrightarrow -\frac{U}{g} \cdot \frac{dU}{dX}$$

In unsteady flows, velocity changes with time, so there is local acceleration, a_t, as well as convective acceleration, where

$$a_t \equiv \frac{\partial U}{\partial t}. \tag{11.8}$$

The expression for head loss due to local acceleration is developed by invoking Newton's second law,

$$F_t = \rho \cdot V \cdot \frac{\partial U}{\partial t}, \tag{11.9}$$

where ρ is mass density, and F_t is the force exerted on the volume of water V undergoing the local acceleration. The work done, or energy expended, in accelerating this volume is the force times the downstream distance dX, so

$$dE_t = \rho \cdot V \cdot \frac{\partial U}{\partial t} \cdot dX, \tag{11.10}$$

where dE_t is the energy expended as a result of the local acceleration. Dividing this energy loss by the weight of the volume of water, $\gamma \cdot V$, where γ is weight density, gives the corresponding head loss, $d\mathcal{H}_t$:

$$d\mathcal{H}_t = \frac{\rho \cdot V}{\gamma \cdot V} \cdot \frac{\partial U}{\partial t} \cdot dX = \frac{1}{g} \cdot \frac{\partial U}{\partial t} \cdot dX \tag{11.11}$$

The downstream rate of energy loss due to local acceleration is thus

$$\frac{d\mathcal{H}_t}{dX} = \frac{1}{g} \cdot \frac{\partial U}{\partial t}. \tag{11.12}$$

Now including the term for local acceleration (which corresponds to $-a_t$ in equation 7.5) and using partial-differential notation to reflect changes with respect to both space and time, the complete **dynamic equation for unsteady flow**[1] is

$$\frac{d\mathcal{H}}{dX} = -\left[\frac{\partial Z}{\partial X} + \frac{\partial Y}{\partial X} + \frac{U}{g} \cdot \frac{\partial U}{\partial X} + \frac{1}{g} \cdot \frac{\partial U}{\partial t}\right]. \tag{11.13}$$

It is useful to write equation 11.13 incorporating the following identities:

$$\frac{d\mathcal{H}}{dX} \equiv S_e, \tag{11.14}$$

$$\frac{\partial Z}{\partial X} \equiv -S_0, \tag{11.15}$$

where S_e and S_0 are the energy slope and the channel slope, respectively. With these substitutions, equation 11.13 becomes

$$S_e = S_0 - \frac{\partial Y}{\partial X} - \frac{U}{g} \cdot \frac{\partial U}{\partial X} - \frac{1}{g} \cdot \frac{\partial U}{\partial t} \tag{11.16a}$$

or

$$S_0 - S_e = \frac{\partial Y}{\partial X} + \frac{U}{g} \cdot \frac{\partial U}{\partial X} + \frac{1}{g} \cdot \frac{\partial U}{\partial t}. \tag{11.16b}$$

In deriving equations 11.13 and 11.16, we have not considered the effect of the lateral-inflow rate q_L on the energy/momentum balance. These inflows/outflows could be due to in-falling rain, evaporation, overland flow from the banks, or seepage into or out of the channel ($q_L < 0$ for lateral outflow). Their contribution to the acceleration in the X-direction would be equal to $U_L \cdot q_L/A$, where U_L is the component of the velocity of the inflow in the downstream direction. In virtually all natural situations, inflow would enter perpendicularly to the downstream direction and with a very small velocity, so U_L will be negligible, and we are justified in leaving the term out.

11.1.2.2 Incorporation in Resistance Relations

The general resistance relation (equation 6.19) can be written as

$$U = \Omega^{-1} \cdot g^{1/2} \cdot Y^{1/2} \cdot S_e^{1/2}, \tag{11.17}$$

where Ω is resistance and S_e is the energy slope. In terms of discharge, Q, this becomes

$$Q = \Omega^{-1} \cdot g^{1/2} \cdot A \cdot Y^{1/2} \cdot S_e^{1/2}, \tag{11.18}$$

where A is cross-sectional area. Substituting equation 11.16a gives

Forces: viscous + turbulent resistance; gravitational; pressure; convectional; local

$$Q = \Omega^{-1} \cdot g^{1/2} \cdot A \cdot Y^{1/2} \cdot \left(S_0 - \frac{\partial Y}{\partial X} - \frac{U}{g} \cdot \frac{\partial U}{\partial X} - \frac{1}{g} \cdot \frac{\partial U}{\partial t} \right)^{1/2} \quad (11.19)$$

Flow types: steady uniform (kinematic) [S_0]; quasi-uniform (diffusive) [through $\partial Y/\partial X$]; steady nonuniform [through $(U/g)\cdot\partial U/\partial X$]; unsteady nonuniform (complete dynamic) [through $(1/g)\cdot\partial U/\partial t$]

In equation 11.19, we have identified the terms that represent the influences of various forces and the terms that are included to characterize steady uniform, steady nonuniform, and unsteady nonuniform flows. Equation 11.19 is central to later discussion of the application of unsteady-flow concepts. In section 7.5 (see figure 7.14), we compared the typical magnitudes of the various forces in natural open-channel flows. We found that the viscous resistance was almost always negligible and that in straight reaches the turbulent-resistance force is balanced by gravitational, pressure, convective-acceleration, and local-acceleration forces, generally in that order of importance. In formulating solutions to various unsteady-flow problems, we are justified in simplifying the mathematics by dropping the dynamic terms that are of negligible relative magnitude, and we will employ this strategy in subsequent analyses.

11.1.3 Solution of the Saint-Venant Equations

The Saint-Venant equations involve two dependent variables (U or Q and Y) and two independent variables (X and t). General solutions to these equations cannot be obtained by analytical methods; they can only be solved by numerical techniques that approximate the partial-differential equations with algebraic difference equations. There are many varieties of numerical technique, and there is an extensive literature on numerical solution of the Saint-Venant equations; reviews include those of Strelkoff (1970), Price (1974), Lai (1986), Fread (1992), and Chaudhry (1993). In all numerical techniques, the space and time continuums are discretized into a grid system, and solutions are found for specific points in space, separated by a distance ΔX, and instants in time, separated by Δt (figure 11.2).

Detailed discussion of numerical solution of the Saint-Venant equations is beyond the scope of this text. However, to illustrate the general approach, we describe the **explicit finite-difference scheme** used by Ragan (1966). This is not usually the best numerical technique, but it is the most straightforward and is thus appropriate for purposes of illustration here. In explicit techniques, there is the possibility that

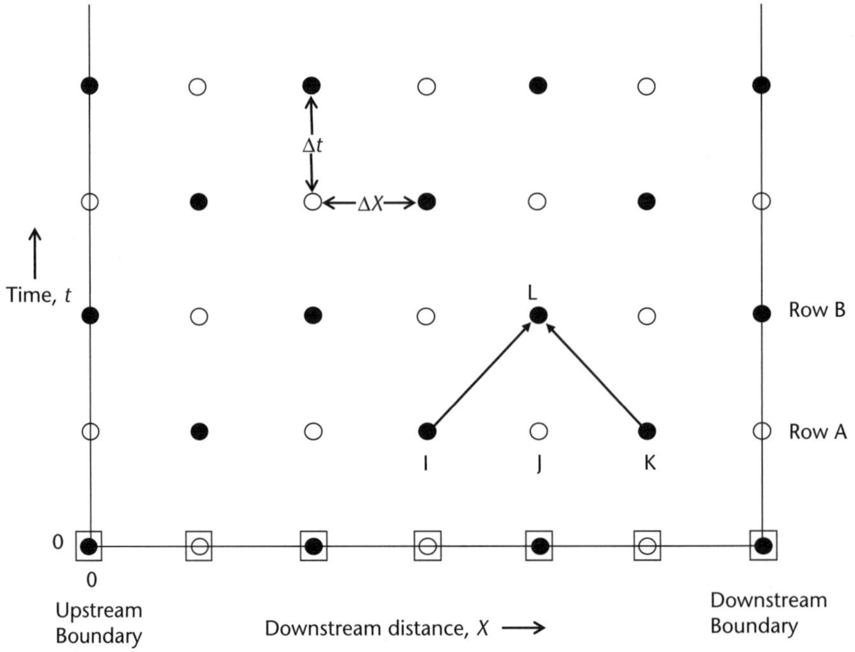

Figure 11.2 Definition diagram for discretization of the Saint-Venant equations. Depths and velocities are computed for grid points represented by dark circles; open circles are intermediate points used in computation. Depths and velocities at grid points marked with squares are specified initial conditions. See text. After Ragan (1966).

computations will become unstable and the results deviate markedly from physical reality if Δt is too large. To avoid this, the **Courant condition** is imposed; this requires that $\Delta t < \Delta X/U$; more detailed discussion of numerical stability issues was given by Fread (1992) and Chaudhry (1993).

To simplify the development here, we consider a rectangular channel of constant width W, so that we can write the continuity relation (equation 11.3b) as

$$\frac{\partial(U \cdot Y)}{\partial X} + \frac{\partial Y}{\partial t} = \frac{q_L}{W}, \qquad (11.20a)$$

which is discretized as

$$\frac{\Delta(U \cdot Y)}{\Delta X} + \frac{\Delta Y}{\Delta t} = \frac{q_L}{W}; \qquad (11.20b)$$

the dynamic equation (equation 11.16b) is

$$g \cdot \frac{\partial Y}{\partial X} + U \cdot \frac{\partial U}{\partial X} + \frac{\partial U}{\partial t} - g \cdot (S_0 - S_e) = 0, \qquad (11.21a)$$

discretized as

$$g \cdot \frac{\Delta Y}{\Delta X} + U \cdot \frac{\Delta U}{\Delta X} + \frac{\Delta U}{\Delta t} - g \cdot (S_0 - S_e) = 0. \qquad (11.21b)$$

Because the differential equations are written in terms of spatial and temporal rates of change, the values of depths and velocities at all locations at the initial instant ($t = 0$) must be specified; these are called the **initial conditions**. Similarly, we must specify the upstream and downstream boundary conditions at all values of time: the depth and velocity at the upstream end of channel; the relation between depth, velocity, and discharge at the downstream end; and the lateral input rate (for further discussion, see Ragan 1966).

In figure 11.2, the dark circles represent the points for which a solution is obtained; the open circles are intermediate points needed in the computations. A typical computation step uses the depths and velocities at the points in row A ($t = t_A$) to compute the depths and velocities at row B ($t = t_B$). This requires that the depths and velocities at all points in row A be known either from the preceding step or as initial conditions.

The computations for an interior grid point L proceed by writing the space and time derivatives as

$$\frac{\Delta U}{\Delta X} = \frac{U_K - U_I}{2 \cdot \Delta X} \tag{11.22}$$

and

$$\frac{\Delta Y}{\Delta t} = \frac{Y_L - Y_J}{\Delta t}. \tag{11.23}$$

The channel slope S_0 and the resistance Ω are determined from field or laboratory measurements, and the energy slope S_e is calculated from the resistance relation, so that

$$S_e = \frac{U^2 \cdot \Omega^2}{g \cdot Y}, \tag{11.24}$$

and at point L

$$S_{eL} = 0.5 \cdot (S_{eK} + S_{eI}) \tag{11.25}$$

and

$$q_{LL} = 0.5 \cdot (q_{LK} + q_{LI}), \tag{11.26}$$

where q_{Li} is the lateral-inflow rate at point i. Then, substituting equations 11.23 and 11.26 into equation 11.20b,

$$Y_L = Y_J - \frac{\Delta t}{2 \cdot \Delta X} \cdot (Y_K \cdot U_K - Y_I \cdot U_I) + \frac{1}{2} \cdot \frac{(q_{LK} + q_{LI})}{W} \cdot \Delta t, \tag{11.27}$$

and equation 11.22, 11.23, and 11.25 into equation 11.21b,

$$U_L = U_J - \frac{U_J \cdot \Delta t}{2 \cdot \Delta x} \cdot (U_K - U_I) - \frac{g \cdot \Delta t}{2 \cdot \Delta x} \cdot (Y_K - Y_I) - \frac{g}{2} \cdot (S_{eK} + S_{eI}) \cdot \Delta t. \tag{11.28}$$

Computations at upstream and downstream boundary points require a somewhat different approach, as explained in Ragan (1966).

11.1.4 Tests of the Saint-Venant Equations

Laboratory experiments by Ragan (1966) provided an excellent test of the ability of the Saint-Venant equations to model open-channel flows with lateral inputs. These experiments were conducted in a 20-cm-wide, 22-m-long tiltable flume in which water was continually supplied at the upper end and additional water could be supplied from a series of lateral-inflow pipes distributed along the channel, representing runoff contributions from a watershed (figure 11.3). The Manning equation (section 6.8, equation 6.40c) was used as the resistance relation, and the relation between resistance and discharge for the flume was determined by measurements of steady uniform flows prior to the main experimental runs. Figure 11.4 shows the close correspondence of the hydrographs computed by numerical solution of the Saint-Venant equations and the measured hydrographs at the downstream end of the flume for four spatial distributions of lateral inflow.

In a field test of the Saint-Venant equations, Morgali (1963) modeled runoff from a rainstorm on a 9.2-ha watershed in Wisconsin. In this case the Saint-Venant equations were applied twice, to simulate first the overland flow with rainfall constituting the lateral inflow, and then the flow in the channel with the overland flow as lateral inflow. As shown in figure 11.5, the modeled hydrograph closely matched the measured flow.

11.2 Hydraulic Geometry

Recall from section 2.6.3 that the at-a-station hydraulic geometry functions relate values of the hydraulic variables width (W), depth (Y), and velocity (U) to discharge (Q) in a given reach, and that these functions are usually given as simple power-law equations:

Width–discharge:

$$W = a \cdot Q^b \tag{11.29}$$

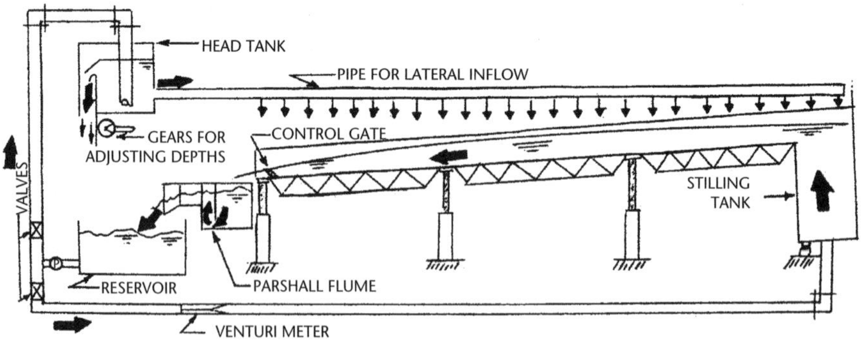

Figure 11.3 Flume arrangement used by Ragan (1966) for tests of the Saint-Venant equations. From Ragan (1966).

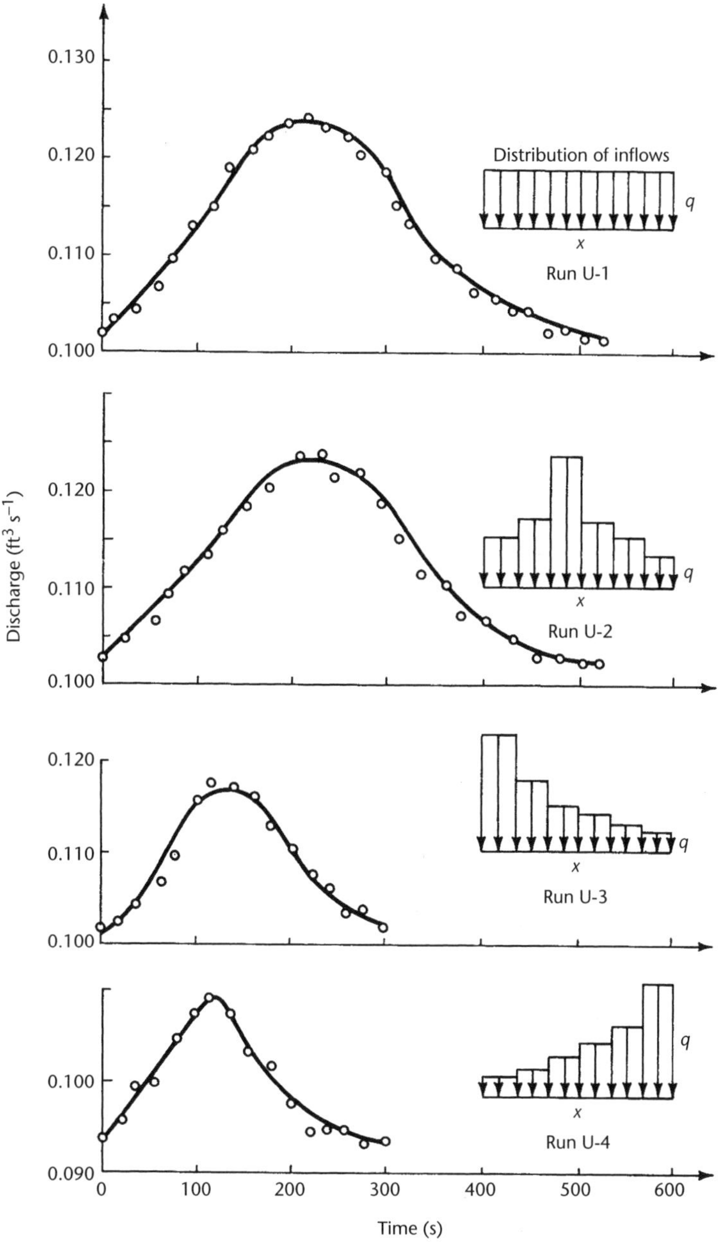

Figure 11.4 Ragan's (1966) comparisons of measured hydrographs (circles) and hydrographs simulated by solution of the Saint-Venant equations (lines) for different spatial patterns of lateral inflows (insets). From Ragan (1966).

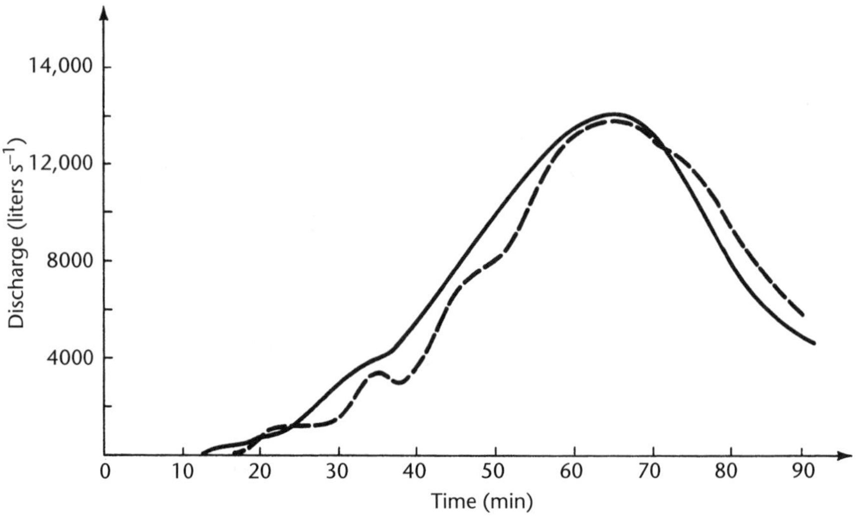

Figure 11.5 Comparison of measured hydrograph (solid line) and hydrograph simulated by numerical solution of the Saint-Venant equations (dashed line) for a storm on a 9.2-ha watershed in Wisconsin. After Morgali (1963).

Average depth–discharge:

$$Y = c \cdot Q^f \tag{11.30}$$

Average velocity–discharge:

$$U = k \cdot Q^m \tag{11.31}$$

The ranges of values of the exponents b, f, and m reported in a number of field studies were shown in figure 2.41. There is wide variation from reach to reach, but there is a tendency for the exponent values to center on $b \approx 0.11, f \approx 0.44, m \approx 0.45$. However, although the coefficients and exponents in equations 11.29–11.31 vary from reach to reach, because $Q = W \cdot Y \cdot U$, it must be true that

$$b + f + m = 1 \tag{11.32}$$

and

$$a \cdot c \cdot k = 1. \tag{11.33}$$

The analysis summarized in box 2.4 shows that the exponents depend only on the exponent r in the general cross-section-shape relation (equations 2.20 and 2B4.2) and the depth exponent p in the general hydraulic relation (equation 2B4.3). The effects of channel shape and different values of p on the exponents can be clearly seen in figure 2.41. Box 2.4 also shows the theoretical relations for the coefficients, which can take on a wide range of values depending on the channel dimensions, conductance, and slope as well as on r and p.

It can be shown from equations 11.29–11.31 that $\mathrm{d}W/W = b \cdot (\mathrm{d}Q/Q)$, $\mathrm{d}Y/Y = f \cdot (\mathrm{d}Q/Q)$, and $\mathrm{d}U/U = m \cdot (\mathrm{d}Q/Q)$. Thus, the at-a-station hydraulic geometry relations give information on how small changes in discharge are allocated among changes in width, depth, and velocity in a reach. For example, if $b = 0.23, f = 0.46$, and $m = 0.31$, a 10% increase in discharge is accommodated by a 2.3% increase in width, a 4.6% increase in depth, and a 3.1% increase in velocity.

Thus, the at-a-station hydraulic geometry relations contain important information about unsteady-flow relations for a particular reach, and can be thought of as empirical hydraulic relations.[2] For example, we can show from equations 11.29–11.31 that velocity can be related to depth as

$$U = \frac{k}{c^{m/f}} \cdot Y^{m/f}, \tag{11.34}$$

which is an empirical version of the basic resistance relation of equation 11.17 in which $p = m/f$; and that discharge can be related to depth as

$$Q = \frac{1}{c^{1/f}} \cdot Y^{1/f}, \tag{11.35}$$

which is an empirical version of equation 11.18. We can also show that

$$W = \frac{a}{c^{b/f}} \cdot Y^{b/f}, \tag{11.36}$$

which is an empirical representation of cross-section geometry in which $r = f/b$. And, because cross-sectional area $A = W \cdot Y$,

$$A = a \cdot c \cdot Q^{b+f} = \frac{a}{c^{b/f}} \cdot Y^{(b+f)/f}. \tag{11.37}$$

Equations 11.34–11.37 are useful because they relate all the hydraulic variables of interest to depth and can be used to relate changes in those variables to changes in depth. We will make use of these relations later in this chapter.

11.3 Waves

11.3.1 Basic Characteristics

As noted above, unsteady flow in open channels is essentially a wave phenomenon. For our purposes, a **wave** is a surface disturbance (i.e., a relatively abrupt change in surface elevation) that travels (propagates) with respect to a water body. At a given cross section or reach, variations in water-surface elevation are equivalent to variations in the maximum depth, Ψ. Recalling the general cross-section geometry formula introduced in section 2.4.3.2, we can relate the maximum depth to the average depth as

$$Y = \left(\frac{r}{r+1}\right) \cdot \Psi = \mathcal{R} \cdot \Psi, \tag{11.38}$$

where r is the exponent that reflects the cross-section shape in equation 2.20 and figure 2.25, and we have defined $\mathcal{R} \equiv r/(r+1)$. Now cross-section shape can be compactly expressed as the value of $\mathcal{R}$ ($\mathcal{R} = 1/2$ for triangle, $\mathcal{R} = 2/3$ for a parabola, $\mathcal{R} = 1$ for a rectangle), and $\mathcal{R} \cdot \Psi$ may be substituted for Y in equations 11.34–11.37. However, to simplify the notation and some of the mathematical derivations in the

Table 11.1 Qualitative characteristics of waves due to various causes.

Cause	Addition/ Displacement	Solitary/ Periodic	Translatory/ Oscillatory	Dynamic/ Kinematic
Wind	Displacement	Periodic	Oscillatory[a]	Dynamic
Seiches	Displacement	Periodic	Oscillatory	Dynamic
Tides	Displacement	Periodic	Translatory	Dynamic
Earthquake tsunami	Displacement	Solitary	Translatory	Dynamic
Landslide	Displacement	Solitary	Translatory	Dynamic
Dam failure	Addition	Solitary	Translatory	Kinematic and dynamic
Tidal bores	Addition	Solitary	Translatory	Dynamic
Engineering operations	Displacement or addition	Solitary	Translatory	Kinematic and dynamic
Flood waves	Addition	Solitary	Translatory	Kinematic and dynamic

See text for definitions of terms.
[a]Wind waves become translatory as they approach the shore.

remainder of this chapter, we will assume a rectangular channel, so that $\mathcal{R} = 1$ and $Y = \Psi$.

Table 11.1 lists the principal types of waves that occur in natural water bodies and their qualitative characteristics. Some wave types are due to the addition of water, whereas others are generated by the displacement of a constant volume of water. Most of the wave types of practical concern in streams are **solitary waves**; wind waves, seiches,[3] and tides are periodically repeating waves. Waves that involve the net movement of water in the direction of wave movement are **translatory**; **oscillatory waves** involve no net water movement. As we will explore further in later sections of this chapter, the characteristics of **dynamic waves** are deduced from energy or momentum principles as well as conservation of mass, whereas those of **kinematic waves** can be deduced from the conservation-of-mass principle alone.

The essence of a surface wave is a functional relation between water-surface elevation, or depth Y; streamwise location, X; the wave speed relative to the water, which is called the **celerity**, C_W; and time, t. This relation can be stated in general form as

$$Y = \Phi_w(X - C_W \cdot t), \tag{11.39}$$

where $\Phi_w(.)$ denotes a **wave function**.

The **wave velocity**, U_W, is the speed of the wave relative to a stationary observer. The relation between celerity and wave velocity is

$$U_W = C_W \pm U, \tag{11.40}$$

where U and U_W are positive in the downstream direction; the plus applies to a wave traveling downstream, and the minus to a wave traveling upstream. The form of equation 11.39 reflects the fact that, to an observer moving along the stream bank at a velocity equal to U_W, the surface elevation will appear to remain constant.

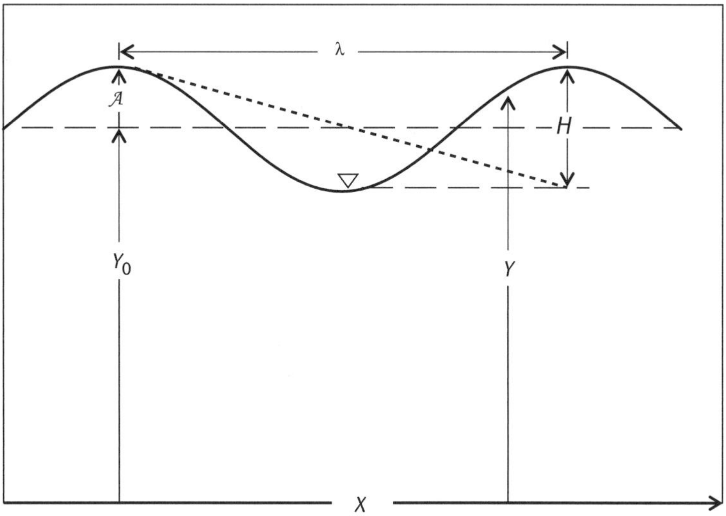

Figure 11.6 A sinusoidal wave (equation 11.41). The heavy dashed line is the equilibrium level; Y_0 is the undisturbed depth, and the actual depth Y is a function of location, X, and time, t. λ is wavelength, $\mathcal{A}$ is wave amplitude, $H \equiv 2 \cdot \mathcal{A}$ is wave height. Wave steepness $S_w \equiv H/\lambda$ is represented by the dotted line.

In classical wave theory, the wave function $\Phi_w(.)$ is sinusoidal (figure 11.6):

$$Y = Y_0 + \mathcal{A} \cdot \sin\left[\frac{2 \cdot \pi}{\lambda} \cdot (X - C_w \cdot t)\right], \tag{11.41}$$

where Y_0 is the undisturbed depth, $\mathcal{A}$ is the wave **amplitude** (maximum vertical displacement of the surface), and λ is the **wavelength** (distance between successive peaks or troughs). Waves are also described in terms of their **period**, T_w, which is the time interval required for two successive peaks (or troughs) to pass a fixed point:

$$T_w \equiv \frac{\lambda}{C_w}; \tag{11.42}$$

or their **frequency**, f_w, which is the number of peaks or troughs passing a fixed point per unit time:

$$f_w \equiv \frac{C_w}{\lambda} = \frac{1}{T_w}. \tag{11.43}$$

Waves are also described in terms of their **height**, H, where

$$H \equiv 2 \cdot \mathcal{A}, \tag{11.44}$$

and their **steepness**, S_w, where

$$S_w \equiv \frac{H}{\lambda}. \tag{11.45}$$

Whatever the cause or type of wave, when a disturbance is produced in a water surface, two **restoring forces** that tend to reduce the magnitude of the disturbance immediately come into play: surface tension and gravity. The disturbance displaces the wave medium (the water) from its equilibrium position, and the restoring forces cause the medium to "overshoot" on either side of the equilibrium position. The resulting alternating displacement and restoration produce the wave motion.

We begin the exploration of waves by introducing classical wave theory, which was developed for oscillatory waves.[4]

11.3.2 Classical Theory of Oscillatory Waves

Accounting for the two restoring forces of gravity and surface tension, Sir George Airy (1801–1892) derived in 1845 the general relation between celerity and wavelength for water-surface waves of small amplitude:

$$C_w = \left[\left(\frac{g \cdot \lambda}{2 \cdot \pi} + \frac{2 \cdot \pi \cdot \sigma}{\rho \cdot \lambda}\right) \cdot \tanh\left(\frac{2 \cdot \pi \cdot Y_0}{\lambda}\right)\right]^{1/2}, \tag{11.46}$$

where g is gravitational acceleration, σ is surface tension, ρ is mass density of water, and Y_0 is undisturbed depth (Henderson 1966). In equation 11.46, "tanh(ξ)" denotes the **hyperbolic tangent** function of a quantity ξ, which is defined as

$$\tanh(\xi) \equiv \frac{\exp(\xi) - \exp(-\xi)}{\exp(\xi) + \exp(-\xi)}. \tag{11.47}$$

A graph of this function is shown in figure 11.7; it has the interesting properties that for $\xi \leq 0.3$, $\tanh(\xi) \approx \xi$; for $\xi \geq 3$, $\tanh(\xi) \approx 1$. Clearly, the value of the argument in equation 11.46 depends on the ratio of depth to wavelength, Y_0/λ, and we see that when $(Y_0/\lambda) > 0.5$, $\tanh(2 \cdot \pi \cdot Y_0/\lambda) \approx 1$ and

$$C_w \approx \left(\frac{g \cdot \lambda}{2 \cdot \pi} + \frac{2 \cdot \pi \cdot \sigma}{\rho \cdot \lambda}\right)^{1/2}. \tag{11.48}$$

Thus, the celerity of waves in situations where the depth exceeds one-half the wavelength is given by equation 11.48. Using typical values of mass density and surface tension (see sections 3.3.1 and 3.3.2), we show in figure 11.8 the dependency of C_w on λ for such waves. The minimum value of $C_w = 0.23$ m/s occurs at $\lambda = 0.017$ m; this is taken as the boundary between shorter **capillary waves**, for which surface tension is the principal restoring force, and longer **gravity waves**. Capillary waves are always present; they can be important in laboratory situations, particularly in small-scale hydraulic models, but can generally be ignored in natural streams.

Now neglecting surface tension, equation 11.46 becomes

$$C_{gw} = \left[\left(\frac{g \cdot \lambda}{2 \cdot \pi}\right) \cdot \tanh\left(\frac{2 \cdot \pi \cdot Y_0}{\lambda}\right)\right]^{1/2}, \tag{11.49}$$

which is the general equation relating celerity, C_{gw}; wavelength, λ; and depth, Y_0, for gravity waves.

We have seen that $\tanh(2 \cdot \pi \cdot Y/\lambda) \approx 1$ when $(Y/\lambda) > 0.5$. Thus, waves in water with a depth exceeding one-half the wavelength are called **deep-water waves**, and

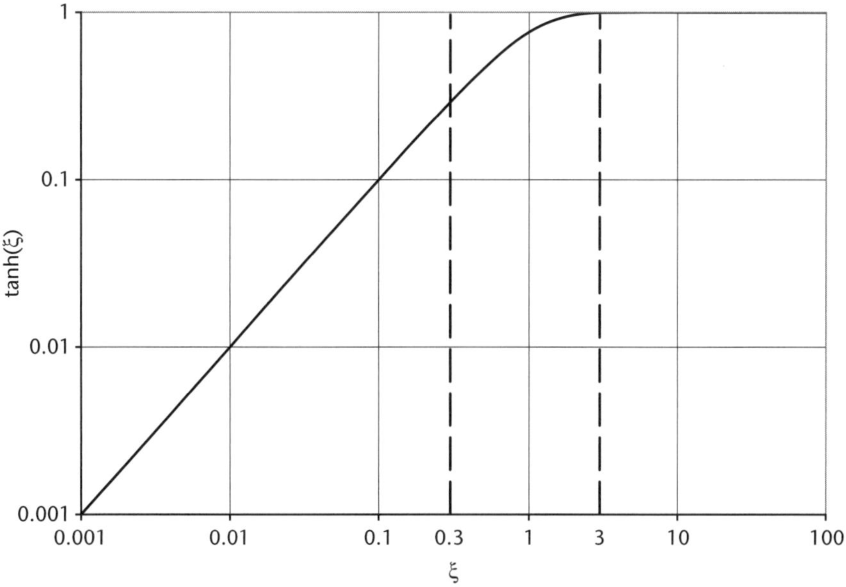

Figure 11.7 The hyperbolic-tangent function (equation 11.47). For $\xi \leq 0.3$, $\tanh(\xi) \approx \xi$; for $\xi \geq 3$, $\tanh(\xi) \approx 1$.

we conclude from equation 11.49 that the celerity of deep-water gravity waves, C_{gwD}, is a function of wavelength only:

$$C_{gwD} \approx \left(\frac{g \cdot \lambda}{2 \cdot \pi}\right)^{1/2} \tag{11.50}$$

As noted above, when $2 \cdot \pi \cdot Y_0/\lambda \leq 0.3$, $\tanh(2 \cdot \pi \cdot Y_0/\lambda) \approx 2 \cdot \pi \cdot Y_0/\lambda$. This occurs when $Y_0/\lambda \leq 0.05$. Thus, waves in water with a depth less than 1/20th the wavelength are called **shallow-water waves**, and we see that the celerity of shallow-water gravity waves, C_{gwS}, is a function of depth only:

$$C_{gwS} = \left[\left(\frac{g \cdot \lambda}{2 \cdot \pi}\right) \cdot \left(\frac{2 \cdot \pi \cdot Y_0}{\lambda}\right)\right]^{1/2} = (g \cdot Y_0)^{1/2}. \tag{11.51}$$

Virtually all the waves of practical interest in open-channel flows are shallow-water waves, and equation 11.51 is consistent with equation 6.4 and the discussion of surface waves in section 6.2.2.2.

We can summarize the relations of oscillatory gravity waves in useful dimensionless form by writing equation 11.49 as

$$\frac{C_{gw}}{(g \cdot Y_0)^{1/2}} = \left[\left(\frac{\lambda}{2 \cdot \pi \cdot Y_0}\right) \cdot \tanh\left(\frac{2 \cdot \pi \cdot Y_0}{\lambda}\right)\right]^{1/2}, \tag{11.52}$$

as shown in figure 11.9.

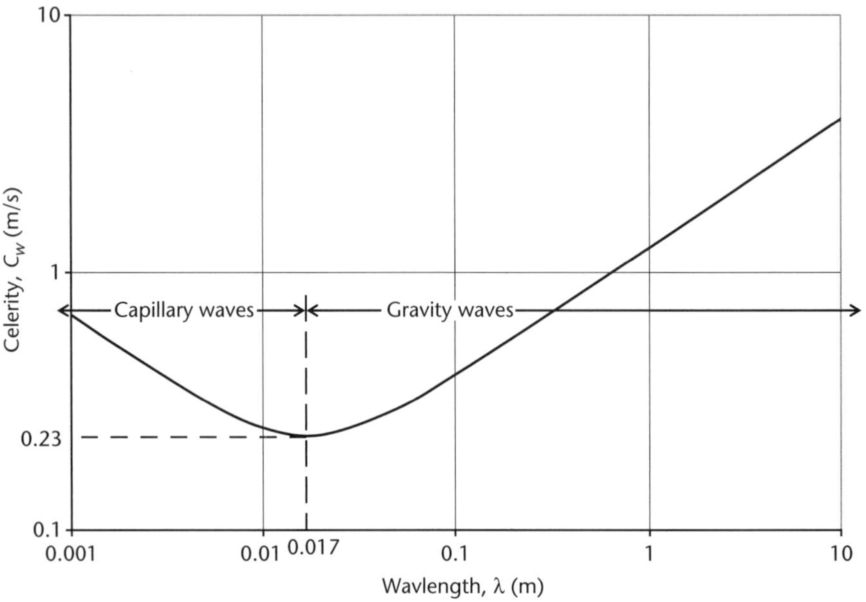

Figure 11.8 Wave celerity C_w as a function of wavelength for deep-water waves (equation 11.48). The curve minimum at $C_w = 0.23$ m/s and $\lambda = 0.017$ m defines the boundary between capillary and gravity waves.

For ideal sinusoidal waves, equation 11.41 describes the motion of the surface. Beneath the surface, water particles move in orbital paths as successive surface waves pass (figure 11.10). In deep-water waves (figure 11.10c), the paths are circles whose diameters decrease exponentially with depth to become negligible at a depth of $\lambda/2$. Thus, there is no net transport of water in deep-water oscillatory waves.

If the depth is less than $\lambda/2$, the friction of the bottom affects the movement, and the particle paths become ellipses (figure 11.10b). When the depth is less than about $\lambda/20$ (i.e., shallow-water waves), the ellipses are nearly completely flattened, and the oscillatory displacement becomes nearly independent of depth. As the depth decreases relative to wavelength (i.e., as the waves approach the shore), the ideal oscillatory waves become increasingly translatory.

As noted above, the Airy wave equation was derived for sinusoidal waves in which the amplitude is small relative to the depth. For water waves with amplitudes that are a significant fraction of the wavelength, the shape is not truly sinusoidal, the orbits of water particles are not closed, and there is some transport of water in the direction of wave movement. Such waves have celerities larger than given by equations 11.46, 11.50, and 11.51, as shown in figure 11.9, and section 11.4.2 shows how amplitude affects celerity in the case of a simple shallow-water translatory wave.

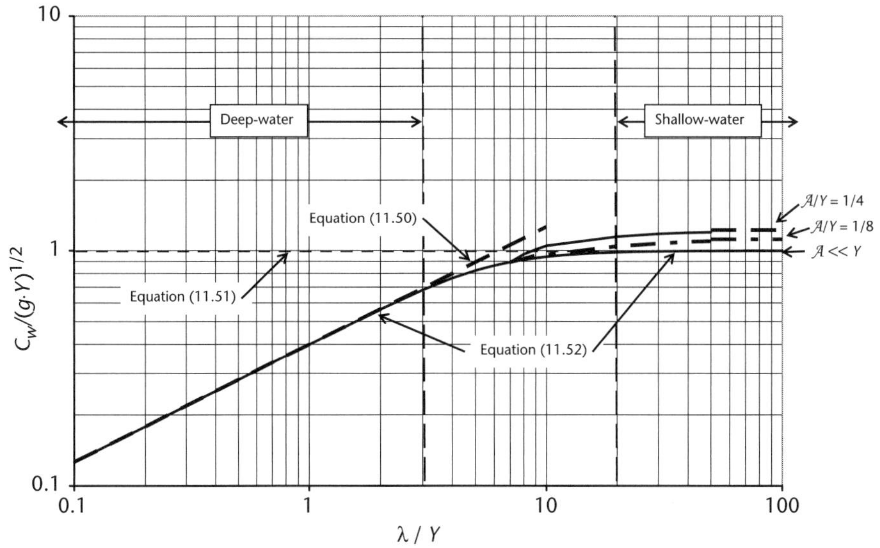

Figure 11.9 Dimensionless wave celerity $C_w/(g \cdot Y)^{1/2}$ as a function of λ/Y. The heavy solid line is equation 11.52, for waves with small amplitude ($\mathcal{A} << Y$) (the Airy wave equation). Equation 11.50 gives the celerity for deep-water waves ($\lambda/Y < 3$); equation 11.51 gives the celerity for shallow-water waves ($\lambda/Y > 20$). The curves above the heavy line in the range $\lambda/Y > 7$ show the effect of amplitude in increasing wave celerity for $\mathcal{A}/Y = 1/8$ and 1/4.

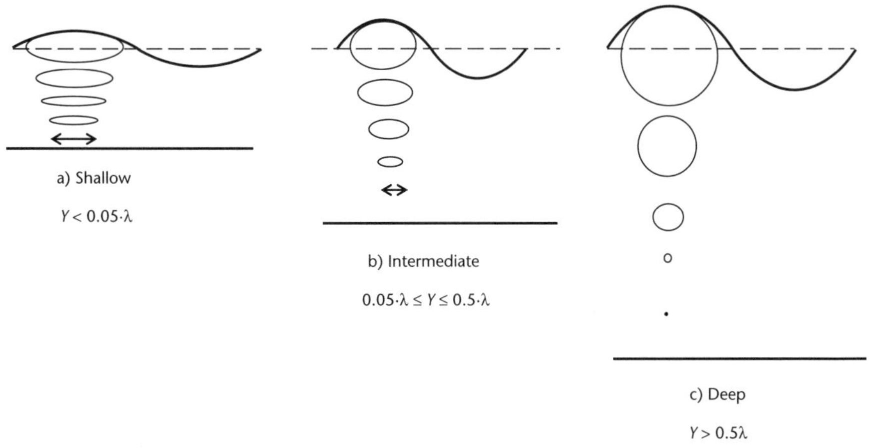

Figure 11.10 Schematic (not to scale) showing orbital paths of water parcels beneath (a) shallow-water, (b) intermediate, and (c) deep-water waves. Y is depth, λ is wavelength.

11.4 Gravity Waves in Open Channels

11.4.1 Simple Gravity Waves

Figure 11.11 shows wave patterns created by dropping a stone into a body of water. The waveform is approximately sinusoidal, the wavelength is proportional to the size of the stone, and the waves travel with a celerity determined by their wavelength and the water depth (equations 11.49–11.51). The velocity of the waves relative to a stationary observer is given by equation 11.40. In the case where $U > C_{gw}$ (figure 11.11d), the upstream wavefront forms an angle θ where

$$\theta = 2 \cdot \sin^{-1}\left(\frac{C_{gw}}{U}\right). \tag{11.53}$$

These waves gradually dissipate as they spread.

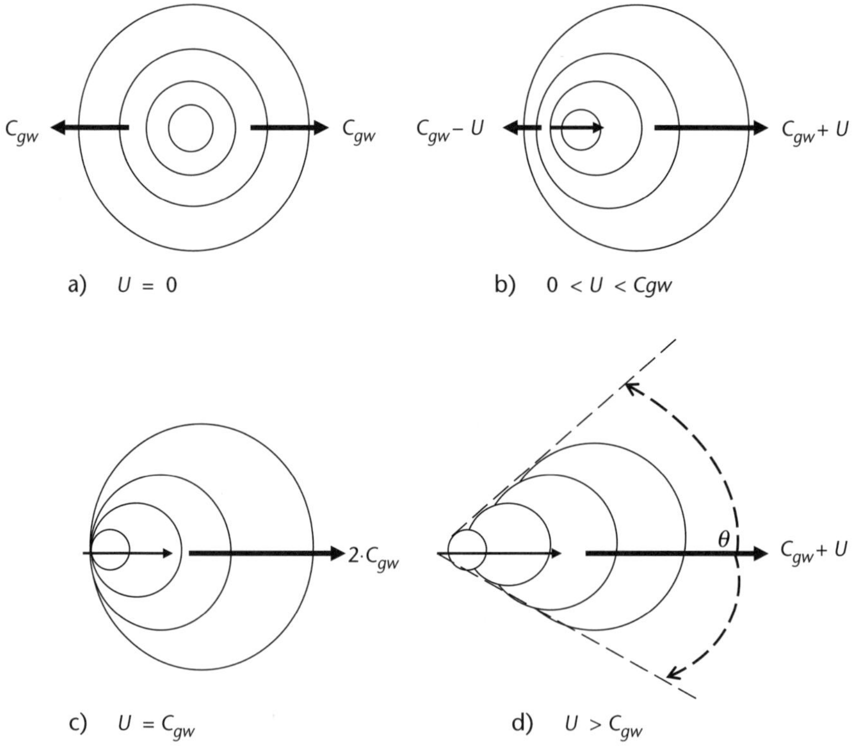

Figure 11.11 Propagation of gravity waves created by dropping a stone into water. The heavier arrow indicates the wave velocity, U_w; the lighter arrow, the water velocity, U. (a) When $U = 0$, the wave crests travel at $U_w = C_{gw}$ in all directions. (b) When $0 < U < C_{gw}$, wave crests travel upstream at $U_w = C_{gw} - U$ and downstream at $U_w = C_{gw} + U$. (c) When $U = C_{gw}$, waves travel only downstream at $U_w = C_{gw} + U = 2 \cdot C_{gw}$. (d) When $U > C_{gw}$, waves travel only downstream at $U_w = C_{gw} + U > 2 \cdot C_{gw}$, and the upstream wavefront forms an angle θ given by equation 11.53.

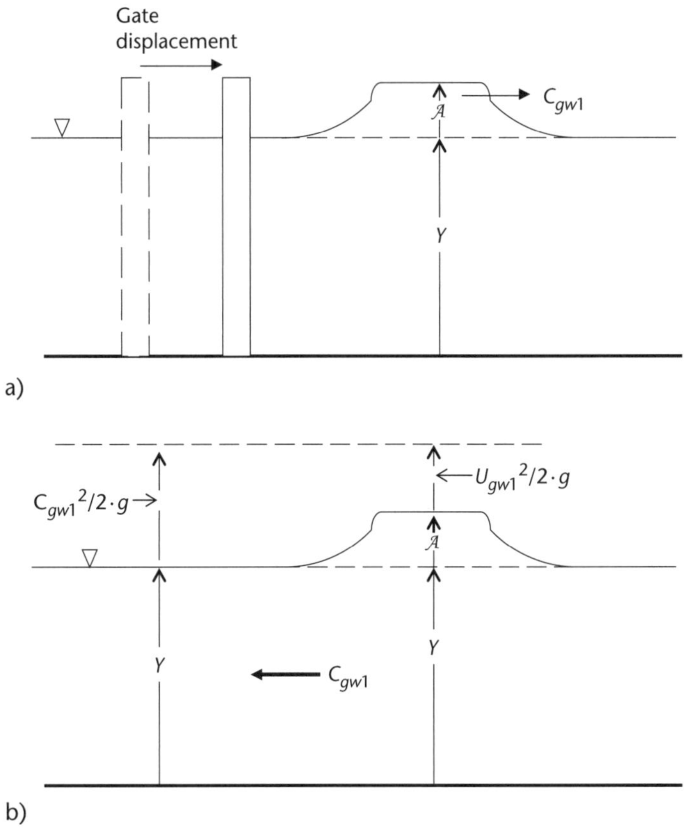

Figure 11.12 The solitary wave generated by displacement of a gate. (a) Unsteady-flow view of wave to a stationary observer. (b) Steady-flow view to an observer moving with the wave. After Chow (1959).

11.4.2 The Soliton

The **soliton** (also called the **solitary wave**) is a shallow-water gravity wave consisting of an elevation without an associated depression (figure 11.12). Such a wave can be created by a sudden horizontal movement of a gate, the movement of a barge in a shallow canal, or by sudden natural displacements caused by earthquakes or landslides. As described by Chow (1959, p. 537), "The wave lies entirely above the normal water surface and moves smoothly and quietly without turbulence at any place along its profile. In a frictionless channel the wave can travel an infinite distance without change of shape or velocity, but in an actual channel the height of the wave is gradually reduced by the effects of friction."

Solitary waves were first studied in canals in England by John Scott Russell (1808–1882). He created these waves by suddenly stopping a towed barge, and found that, even in real channels with friction, solitons can travel long distances with very little change of form. This feature was noted by Scales and Snieder (1999, p. 739): "In solitons, the wave spreading by dispersion is exactly (and miraculously) offset by the nonlinear steepening of the wave, so that a solitary wave maintains its identity." We will discuss the conditions under which flood waves spread or steepen in section 11.5.3.

Russell made very accurate measurements of soliton velocity, from which he concluded (Russell 1844) that the celerity C_{gw1} depends on wave amplitude $\mathcal{A}$ as well as depth:

$$C_{gw1} = [g \cdot (Y_0 + \mathcal{A})]^{1/2}. \tag{11.54}$$

Subsequent investigators have attempted to derive expressions for the celerity of solitons; the detailed analysis by Dean and Dalrymple (1991) yields

$$C_{gw1} = (g \cdot Y_0)^{1/2} \cdot \left(1 + \frac{\mathcal{A}}{2 \cdot Y_0}\right). \tag{11.55}$$

Clearly, the above expressions for the celerity of a soliton reduce to the shallow-water value given by equation 11.51 when wave amplitude $\mathcal{A}$ is very small relative to depth Y_0. We see in figure 11.13 that equations 11.54 and 11.55 give similar values.

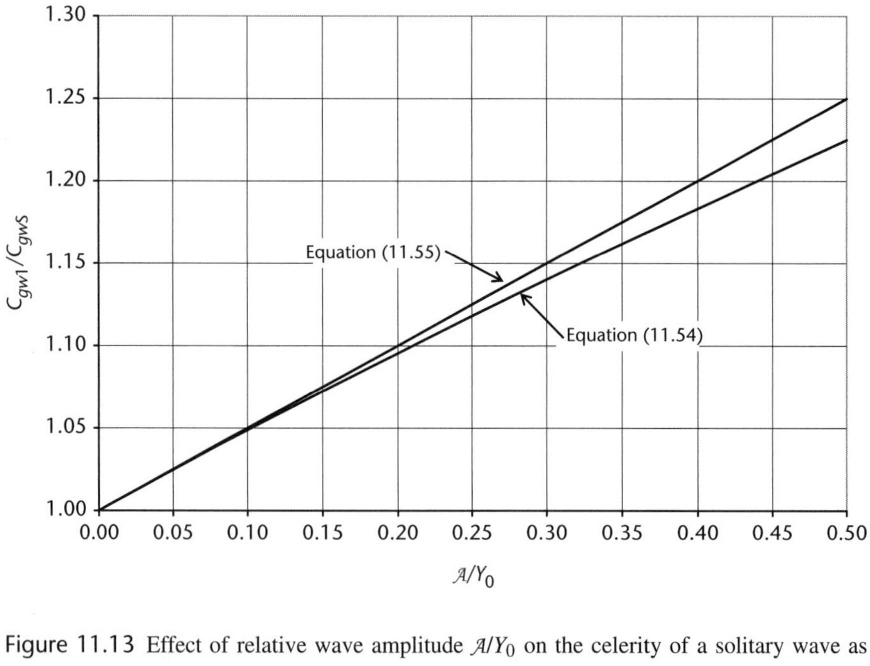

Figure 11.13 Effect of relative wave amplitude $\mathcal{A}/Y_0$ on the celerity of a solitary wave as given by the experiments of Russell (1844) (equation 11.54) and the analysis of Dean and Dalrymple (1991) (equation 11.55). C_{gw1}/C_{gwS} is the ratio of the solitary-wave velocity to the small-amplitude shallow-water celerity $(g \cdot Y_0)^{1/2}$ (equation 11.51).

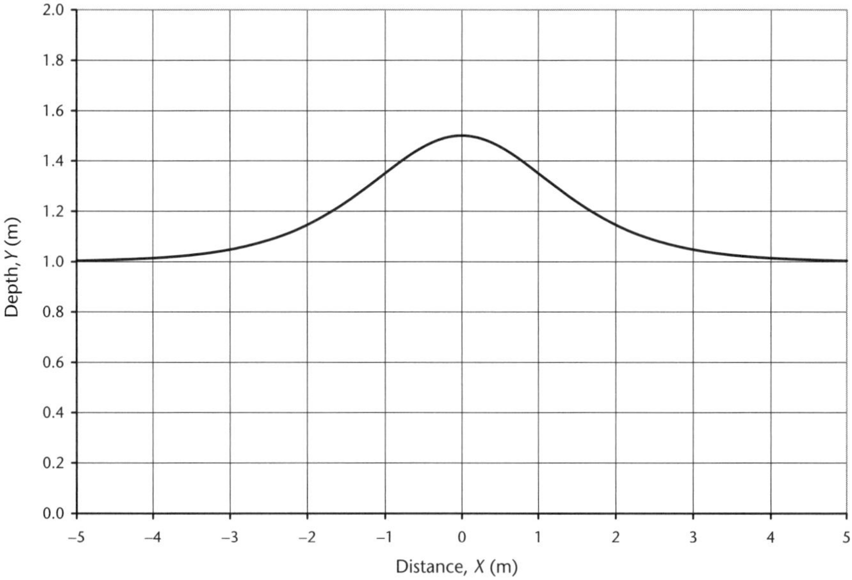

Figure 11.14 Profile of a solitary wave with an amplitude of $\mathcal{A} = 0.5$ m in water with an undisturbed depth of $Y_0 = 1$ m (equation 11.56). Note the approximately threefold vertical exaggeration. The theoretical profile extends to infinity in both directions, but 95% of the wave volume is contained within ±3 m (equation 11.57). This wave would have a celerity of 3.84 m/s (equation 11.54).

The soliton profile is given by

$$Y = Y_0 + \mathcal{A} \cdot \text{sech}^2 \left[\left(\frac{3 \cdot \mathcal{A}}{4 \cdot Y_0^3} \right)^{1/2} \cdot (X - C_{w1} \cdot t) \right], \tag{11.56}$$

where sech(ξ) is the **hyperbolic secant** function of the quantity ξ: $\text{sech}(\xi) \equiv 2/[\exp(\xi) + \exp(-\xi)]$; this form is shown in figure 11.14. Theoretically, the profile extends to infinity in both directions, but as shown by Dean and Dalrymple (1991), 95% of the volume of the wave is contained within a distance $X_{0.95}$, where

$$X_{0.95} = \frac{2.12 \cdot Y}{(\mathcal{A}/Y)^{1/2}}; \tag{11.57}$$

thus, for a wave with an amplitude equal to half the depth ($\mathcal{A}/Y = 0.5$), 95% of the wave volume is contained in a distance equal to only about six times the depth.

11.5 Flood Waves

11.5.1 Qualitative Aspects

Flood waves are usually represented as **discharge hydrographs** (graphs of discharge vs. time at a measurement station) but, for our present purposes, are better shown

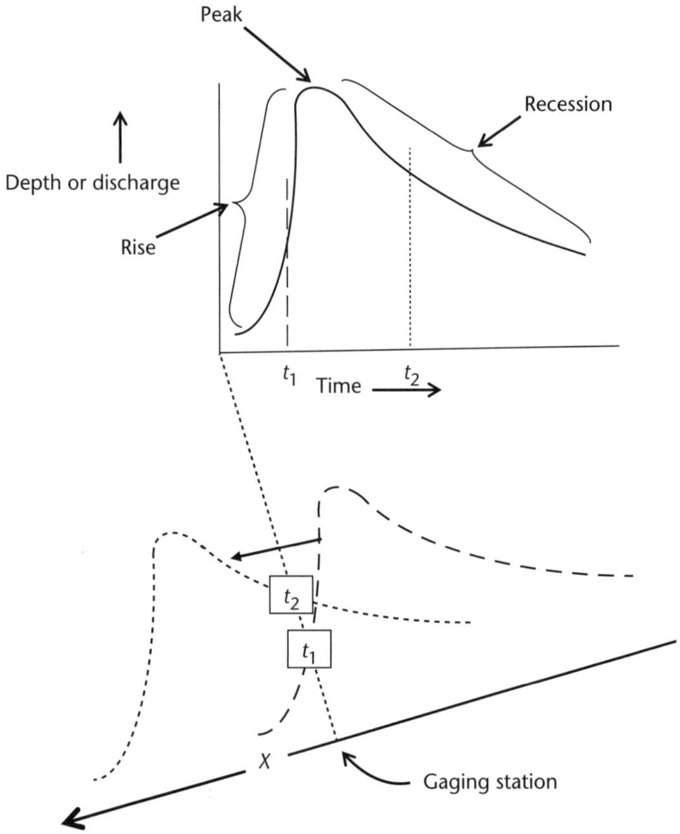

Figure 11.15 Time-space relations for a typical flood wave. The lower diagram shows the physical flood wave passing a gaging station at successive times t_1 (dashed wave) and t_2 (dotted wave). The upper graph shows the depth (or stage) hydrograph recorded at the gaging station.

as **depth** (or **stage** [water-surface elevation]) **hydrographs** (figure 11.15). The connection between discharge hydrographs and depth hydrographs is the **depth-** (or **stage-**) **discharge relation**, or **rating curve**, which is an aspect of the at-a-station hydraulic geometry relations discussed in section 2.6.3.1.

The hydrograph records the passage of the wave through the measurement location. The typical form of a flood wave has a relatively steep leading limb (the **hydrograph rise**) rising to a peak, followed by a less steep trailing limb (the **hydrograph recession**). This means that the water-surface slope downstream of the peak is steeper than that upstream of the peak; we will explore the implications of this slope change later in this section.

Flood waves are produced by relatively rapid accumulations of water in the channel system due to 1) significant rain or snowmelt on a watershed entering the stream system (section 2.5.5) or 2) the opening or breach of a natural or artificial dam. As flood waves travel downstream, the peak discharge tends to decrease, and the wave tends to lengthen and dissipate, or spread, because 1) deeper portions of the wave travel with higher velocities than shallower portions (equation 11.17), 2) pressure forces act to accelerate the flow downstream of the peak and decelerate it upstream of the peak, 3) channel friction differentially retards portions of the flow, and 4) the rising water tends to spread laterally to fill channel irregularities, cover the adjacent floodplain, and/or enter ground-water storage in the banks. However, the tendency for downstream-decreasing peak flow may be reversed by lateral inflows and inputs from tributaries. We will discuss the spreading of flood waves more fully in section 11.5.3.

Figure 11.16 shows typical depth and discharge hydrographs resulting from a watershed-wide rainfall event. In a rain or snowmelt event, the channel system receives watershed-wide lateral inputs from ground or surface water (see figure 2.32), and the wave tends to grow in discharge as it moves downstream. However, as noted above, the dissipation due to pressure forces, friction, and storage operates to lengthen the wave and diminish the peak flow per unit watershed area, as shown in figure 2.34.

Flood waves caused by rain or snowmelt are, of course, of central interest in hydrology and fluvial hydraulics. However, to better set the stage for exploring the nature of flood waves, we first examine a simpler flood wave generated by a sudden input of water at a single location, which is the case shown in figure 11.17. Clearly, the square-wave form of the initial release pulse dissipated and changed to the typical hydrograph shape as it traveled. The analysis in box 11.1 shows that 91% of the water in the original release was present at the downstream site, so only 9% was "lost" to storage; thus, most of modification of the wave form was because the water "parcels" were differentially affected by pressure and friction forces and traveled at different speeds. Most interesting, this simple flood wave traveled at a velocity much lower than that of a gravity wave, but greater than the water velocity. The analysis in the following section will show why that is the case.

11.5.2 Kinematic Waves

The American engineer James Seddon (1900) made the first observations of flood waves on the Mississippi and Missouri rivers moving at speeds that were greater than the actual water velocity but slower than shallow-water gravity waves. The mathematics of the phenomenon had previously been explored by the Frenchman M. Kleitz (1877); however, the first comprehensive treatment of the subject was by two English mathematicians, M.J. Lighthill and G.B. Witham (1955). They stated that such waves are a general occurrence that arises in any flow in which there is a functional relationship between 1) the flow rate (discharge) and 2) the amount of flowing substance in a segment of the flow (reach cross-sectional area or average depth). As we shall see, the basic relationships for such waves can be derived without invoking force (dynamic) relations, so Lighthill and Witham called the phenomenon the **kinematic wave**.[5]

Interestingly, Lighthill and Witham (1955) showed that kinematic waves occur in automobile traffic and devoted the second part of their seminal paper to a discussion of traffic flow. In traffic the flow rate is inversely, rather than directly, related to the amount of flowing substance (as your own experience will no doubt verify), and

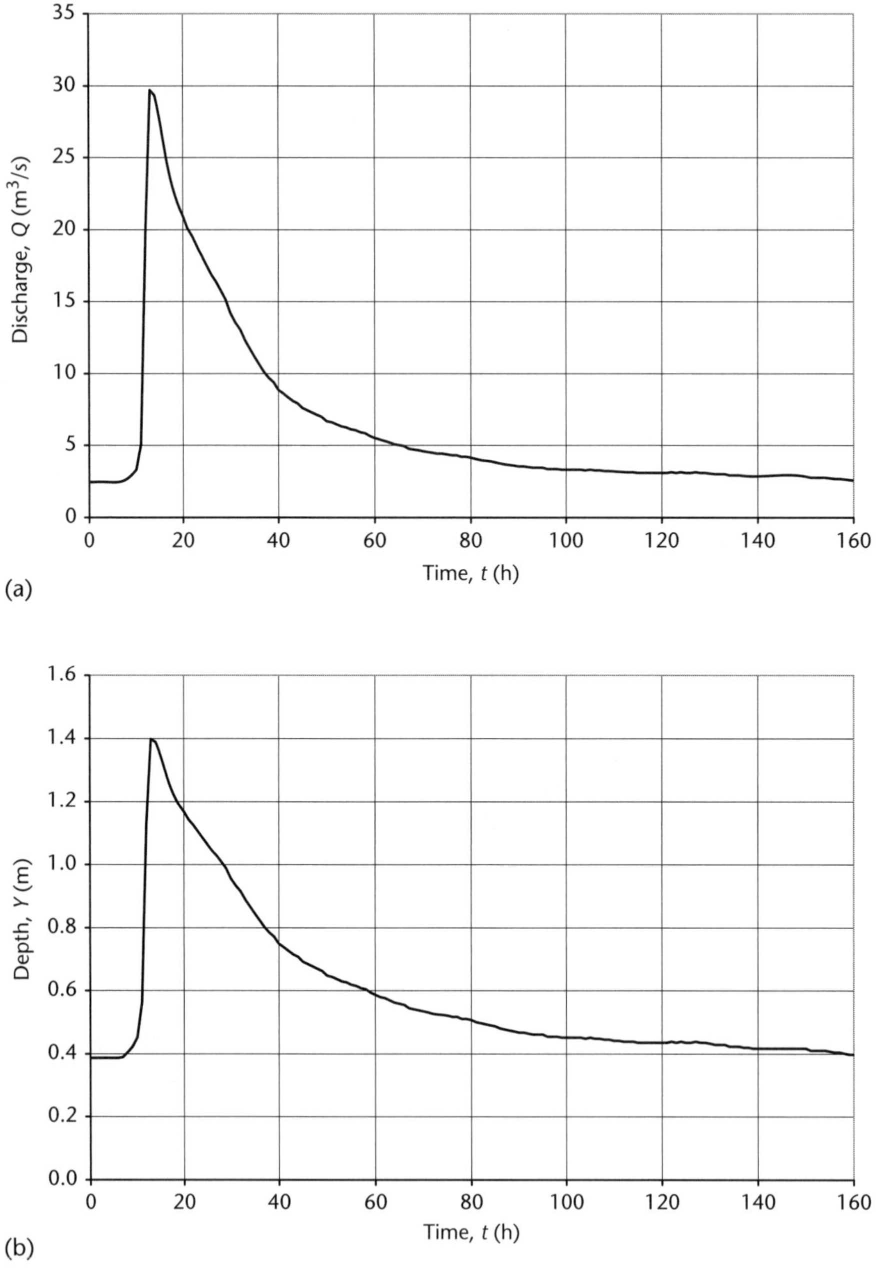

Figure 11.16 Hydrographs of the Diamond River near Wentworth Location, New Hampshire, in response to an intense rainstorm on 23 July 2004. (a) Discharge hydrograph. (b) Depth hydrograph. Data courtesy of Ken Toppen, U.S. Geological Survey, Pembroke, New Hampshire.

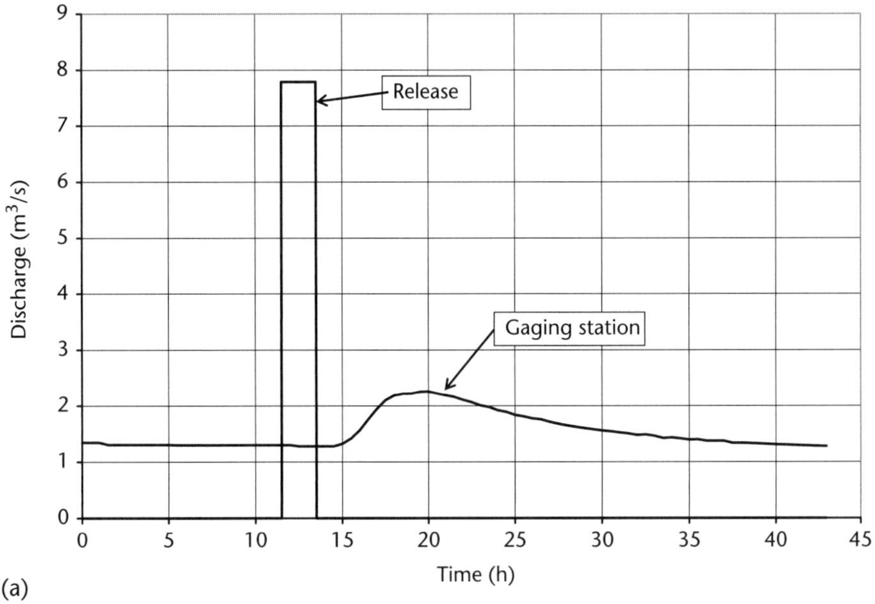

(a)

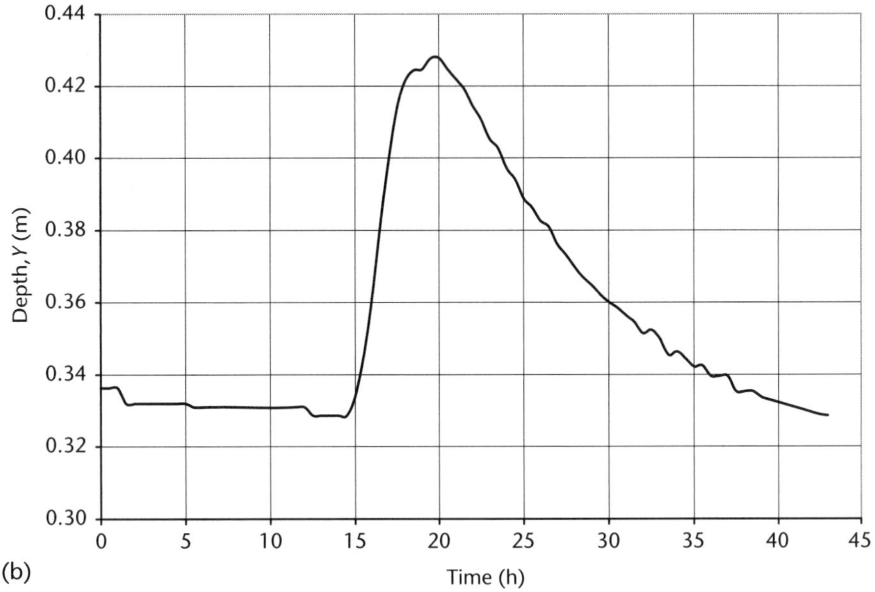

(b)

Figure 11.17 (a) Hydrographs showing sudden release of 7.79 m^3/s for 2.0 h from Jackman Hydroelectric Dam on the North Branch of the Contoocook River, New Hampshire, and arrival of the wave at the gaging station 12.6 km downstream. (b) Depth hydrograph at gaging station. (See box 11.1.) Data courtesy of Walter Carlson, New Hampshire Department of Environmental Services.

BOX 11.1 Contoocook River Flood Wave

Figure 11.17 shows the hydrograph of the flood wave recorded at the U.S. Geological Survey gage on the Contoocook River near Henniker, New Hampshire, resulting from the sudden release of a constant discharge of 7.79 m^3/s for 2.0 h from Jackman Hydroelectric Dam on the North Branch of the Contoocook River, New Hampshire, 12.6 km downstream. (The releases are controlled automatically.)

The travel time from midpoint of the release to the peak flow at the gage was 7.5 h (27,000 s), so the wave velocity $U_w = 0.465$ m/s. From examination of the hydraulic geometry relations based on measurements at the gaging station, the average depth Y at the gage was about 0.38 m, and the average velocity U was about 0.14 m/s. Thus, the wave velocity was about 3.3 times the water velocity.

Using the average depth, the celerity of a gravity wave is $U_{gw} = (9.81 \cdot 0.38)^{1/2} = 1.93$ m/s. Thus, the velocity of a gravity wave would be about $1.93 + 0.14 = 2.07$ m/s, about 4.5 times faster than the actual wave velocity. Thus, we conclude that the wave was not a gravity wave.

As described later in the text, we would expect the velocity of a kinematic wave to be about 1.5–2 times the water velocity. The actual ratio was somewhat higher at $0.465/0.14 = 3.3$. It is possible that the higher ratio is due to higher water velocities in reaches upstream of the gage, and it seems reasonable to assume that this wave traveled as a kinematic wave.

The total release was 42,100 m^3, and the total flow increment in the hydrograph at the gage was 38,400 m^3. This is 91.2% of the release; the "missing" 8.8% presumably entered relatively long-term channel storage between the dam and the gage.

because of this, kinematic waves in traffic travel upstream rather than downstream as in rivers.

11.5.2.1 Kinematic-Wave Velocity

As discussed in section 11.3.1, the essence of a wave is that an observer moving with the wavefront at the wave velocity U_w sees a steady discharge Q (figure 11.18). Thus, to this observer, $dQ = 0$, and since $Q = f(X, t)$, we can write

$$dQ = \frac{\partial Q}{\partial X} \cdot dX + \frac{\partial Q}{\partial t} \cdot dt = 0. \tag{11.58}$$

Then, starting with equation 11.58 and invoking the one-dimensional conservation-of-mass equation (equation 11.3c), we find via the derivation in box 11.2 that

$$U_{kw} = \frac{\partial Q}{\partial A}, \tag{11.59a}$$

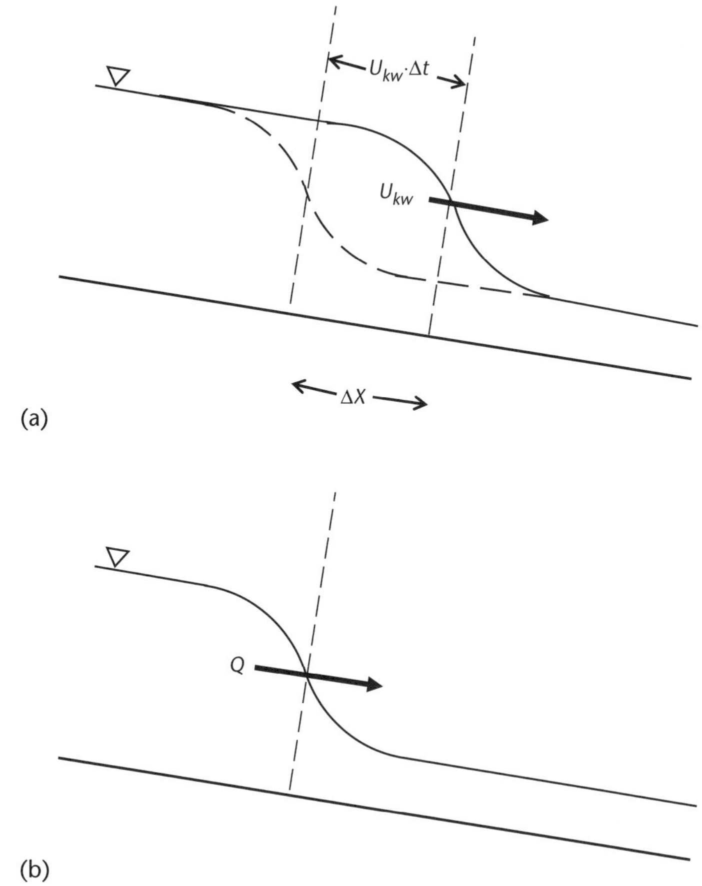

Figure 11.18 Definition diagram for uniformly progressive flow (monoclinal rising wave). (a) View of a stationary observer (unsteady flow): The wavefront moves a distance ΔX in time Δt, and the wave velocity $U_{\text{kw}} = \Delta X/\Delta t$. (b) View of observer moving with the wavefront at velocity U_{kw} (steady flow). The observer sees a constant discharge Q; that is, $\mathrm{d}Q(X,t) = 0$.

where U_{kw} is the kinematic-wave velocity, and A is cross-sectional area. For a rectangular channel, the width is constant, and the relation becomes

$$U_{\text{kw}} = \frac{1}{W} \cdot \frac{\partial Q}{\partial Y}. \tag{11.59b}$$

We see from equation 11.59b that the wave velocity is essentially determined by the slope of the depth-discharge relation, or rating curve.

We can relate the kinematic-wave velocity to the water velocity U by first generalizing the basic resistance relation (equation 6.19) to

$$U = \Omega^{-1} \cdot g^{1/2} \cdot S_e^{1/2} \cdot Y^p, \tag{11.60}$$

BOX 11.2 Derivation of Equation 11.59: Kinematic-Wave Velocity

Equation 11.58 can be rearranged to give

$$\frac{dX}{dt} = -\frac{\partial Q/\partial t}{\partial Q/\partial X}, \qquad (11B2.1a)$$

where Q is discharge, X is downstream distance, and t is time. Because dX/dt is the velocity of the observer and the flood wave, U_{kw}, we can also write

$$U_{kw} = -\frac{\partial Q/\partial t}{\partial Q/\partial X}. \qquad (11B2.1b)$$

From the properties of derivatives,

$$\frac{\partial Q}{\partial t} = \frac{\partial Q}{\partial A} \cdot \frac{\partial A}{\partial t}, \qquad (11B2.2)$$

where A is cross-sectional area.

Now we see from the conservation-of-mass equation (equation 11.3c) that

$$\frac{\partial A}{\partial t} = -\frac{\partial Q}{\partial X}, \qquad (11B2.3)$$

and substituting 11B2.3 into equation 11B2.2 yields

$$\frac{\partial Q}{\partial t} = -\frac{\partial Q}{\partial X} \cdot \frac{\partial Q}{\partial A}. \qquad (11B2.4)$$

Now replacing the numerator of equation 11B2.1b with equation 11B2.4 gives equation 11.59:

$$U_{kw} = -\frac{-\partial Q/\partial X \cdot \partial Q/\partial A}{\partial Q/\partial X} = \partial Q/\partial A. \qquad (11B2.5)$$

where Ω is resistance, and S_e is energy slope. The exponent $p = 1/2$ for the Chézy relation and 2/3 for the Manning relation and, more generally, can be related to the exponents in the hydraulic geometry relations (equations 11.30 and 11.31) as

$$p = \frac{m}{f} \qquad (11.61)$$

(see box 2.4). Then, with equation 11.60, the discharge in a rectangular channel is given by

$$Q = \Omega^{-1} \cdot g^{1/2} \cdot S_e^{1/2} \cdot W \cdot Y^{p+1}, \qquad (11.62)$$

from which

$$U_{kw} = \frac{1}{W} \cdot \frac{\partial Q}{\partial Y} = (p+1) \cdot \Omega^{-1} \cdot g^{1/2} \cdot S_e^{1/2} \cdot Y^p = (p+1) \cdot U. \qquad (11.63)$$

Because $p > 0$, we see that the velocity of a kinematic wave is always greater than the water velocity. Assuming that the Chézy equation approximately applies (i.e., $p \approx 1/2$), the wave velocity will be on the order of 1.5 times the water velocity.

The derivations in boxes 11.3 and 11.4 explore the relation between U and U_{kw} in more detail. Box 11.3 shows that in a rectangular channel U_{kw} exceeds $1.5 \cdot U$ by an amount that increases with slope and depth (equation 11B3.5) and with resistance (equation 11B3.6, figure 11.19a). Box 11.4 explores the significance of equation 11.59 from the point of view of hydraulic geometry, showing that the ratio U_{kw}/U increases toward 1.5 as the channel shape approaches a rectangle (figure 11.19b).

Equation 11.63 shows that, in a channel with constant slope and resistance, kinematic-wave velocity increases with depth. This implies that the deeper portions of a flood wave will move faster than the shallower portions and that the wave will tend to steepen as it travels downstream (figure 11.20). However, pressure force, which is proportional to the downstream depth gradient (equation 7.20), opposes this tendency to steepen and may cancel it altogether. We quantitatively explore the conditions under which flood waves steepen or dissipate in section 11.5.3.

BOX 11.3 Kinematic-Wave Velocity and Resistance (Rectangular Channel)

In rough turbulent flow, resistance is

$$\Omega = 0.4 \cdot \left[\ln\left(\frac{11 \cdot Y}{y_r}\right)\right]^{-1}, \qquad (11B3.1)$$

where y_r is the effective height of boundary roughness elements (equation 6.25). Thus, the Chézy-Keulegan resistance relation (equation 6.26) can be written as

$$U = 2.5 \cdot g^{1/2} \cdot S_e^{1/2} \cdot \ln\left(\frac{11 \cdot Y}{y_r}\right) \cdot Y^{1/2}, \qquad (11B3.2)$$

where g is gravitational acceleration, and S_e is energy slope. Because $Q = W \cdot Y \cdot U$, we can write 11B3.2 for discharge as

$$Q = 2.5 \cdot g^{1/2} \cdot S_e^{1/2} \cdot W \cdot \ln\left(\frac{11 \cdot Y}{y_r}\right) \cdot Y^{3/2}. \qquad (11B3.3)$$

From equations 11B3.1 and 11B3.3,

$$U_{\mathrm{kw}} = \frac{1}{W} \cdot \frac{\partial Q}{\partial Y} = 2.5 \cdot g^{1/2} \cdot S_e^{1/2} \cdot \left[\frac{3}{2} \cdot \ln\left(\frac{11 \cdot Y}{y_r}\right) \cdot Y^{1/2} + Y^{1/2}\right]. \qquad (11B3.4)$$

From equation 11B3.2, equation 11B3.4 can also be written as

$$U_{\mathrm{kw}} = \frac{3}{2} \cdot U + 2.5 \cdot g^{1/2} \cdot S_e^{1/2} \cdot Y^{1/2}. \qquad (11B3.5)$$

From equations 11B3.2 and 11B3.5, the ratio U_{kw}/U is

$$\frac{U_{\mathrm{kw}}}{U} = \frac{3}{2} + \frac{2.5 \cdot u_*}{U} = \frac{3}{2} + 2.5 \cdot \Omega = \frac{3}{2} + \left[\ln\left(\frac{11 \cdot Y}{y_r}\right)\right]^{-1}, \qquad (11B3.6)$$

where u_* is friction velocity ($\equiv (g \cdot Y \cdot S)^{1/2}$).

BOX 11.4 Kinematic-Wave Velocity and Hydraulic Geometry

We saw in equation 11.37 that

$$A = a \cdot c \cdot Q^{b+f}, \tag{11B4.1}$$

so we can write

$$Q = \left(\frac{1}{a \cdot c}\right)^{1/(b+f)} \cdot A^{1/(b+f)}. \tag{11B4.2}$$

Thus,

$$U_{kw} = \frac{\partial Q}{\partial A} = \left(\frac{1}{b+f}\right) \cdot \left(\frac{1}{a \cdot c}\right)^{1/(b+f)} \cdot A^{(1-b-f)/(b+f)}. \tag{11B4.3}$$

We can show from the basic hydraulic geometry relations that

$$A = \frac{1}{k^{1/m}} \cdot U^{(1-m)/m}, \tag{11B4.4}$$

and if equation 11B4.4 is substituted into equation 11B4.3 and simplified (noting that $b + f + m = 1$ and $a \cdot c \cdot k = 1$), we find that

$$U_{kw} = \frac{1}{1-m} \cdot U. \tag{11B4.5a}$$

Since $m < 1$, $U_{kw}/U > 1$. Note that $1 - m = b + f$, and $b = 0$ for a rectangular channel, so the kinematic-wave velocity in a rectangular channel is

$$U_{kw} = \frac{1}{f} \cdot U. \tag{11B4.5b}$$

The ratio U_{kw}/U depends on channel geometry. We saw in box 2.4 that the value of m is given by

$$m = \frac{r \cdot p}{1 + r + r \cdot p}, \tag{11B4.6}$$

where r is an exponent that reflects channel cross-section form ($r = 1$ for a triangle, $r = 2$ for a parabola, successively higher values of r reflect channels with successively steeper sides, and $r \to \infty$ reflects a rectangle), and p is the depth exponent in the resistance relation. Equations 11B4.5 and 11B4.6 are used in plotting figure 11.19b, with $p = 1/2$ as given by the Chézy relation (equation 11B3.2).

11.5.2.2 Effects of Overbank Flow on Kinematic-Wave Velocity

The above analysis strictly applies to within-bank flows. A simple analysis by Gray and Wigham (1970) shows, at least qualitatively, how overbank flow affects kinematic-wave velocity. Figure 11.21 is a cross section of a channel with a flow spilling over to floodplains on either side. Treating all three portions of the channel as rectangular,

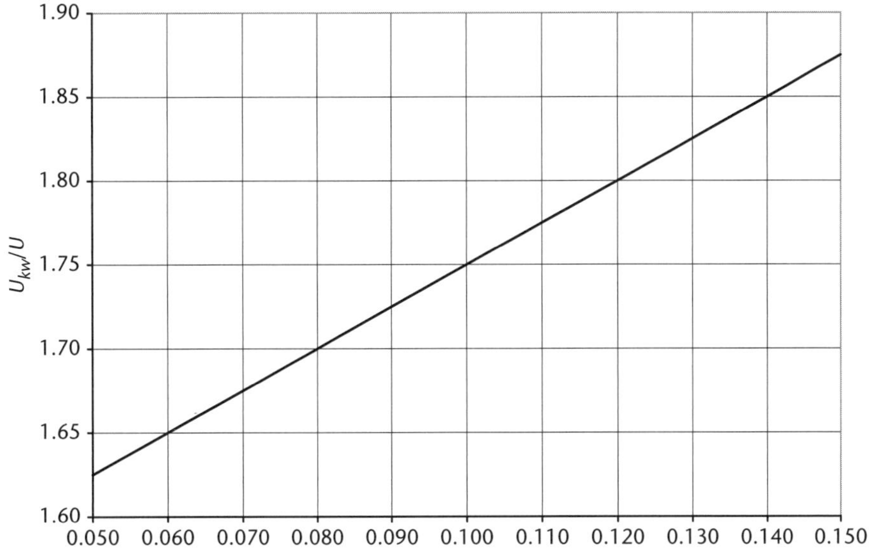

(a)

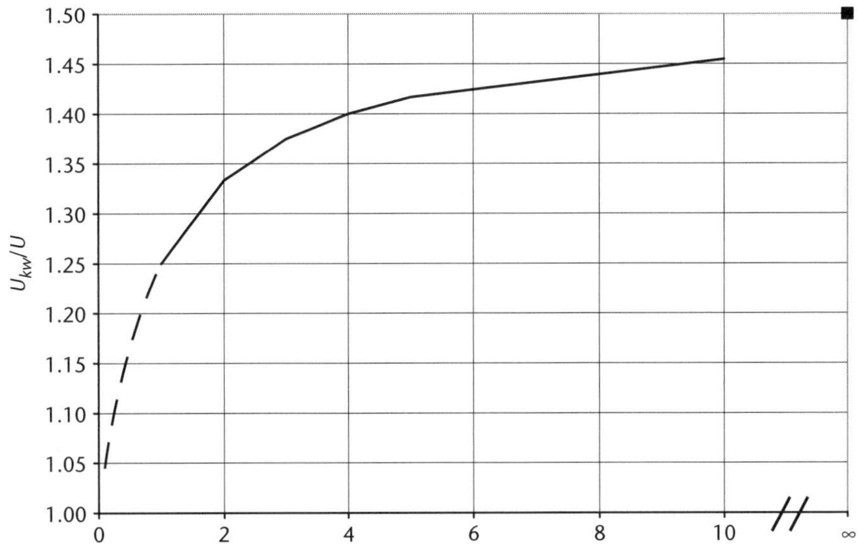

(b)

Figure 11.19 (a) Ratio of kinematic-wave velocity to water velocity, U_{kw}/U, as a function of resistance as given by equation 11B3.6. (b) U_{kw}/U as a function of cross-section geometric form deduced from hydraulic geometry relations (equations 11B4.5 and 11B4.6). $r = 1$ for a triangle, and $r = 2$ for a parabola; successively higher values of r reflect channels with successively steeper sides, and $r \to \infty$ reflects a rectangle.

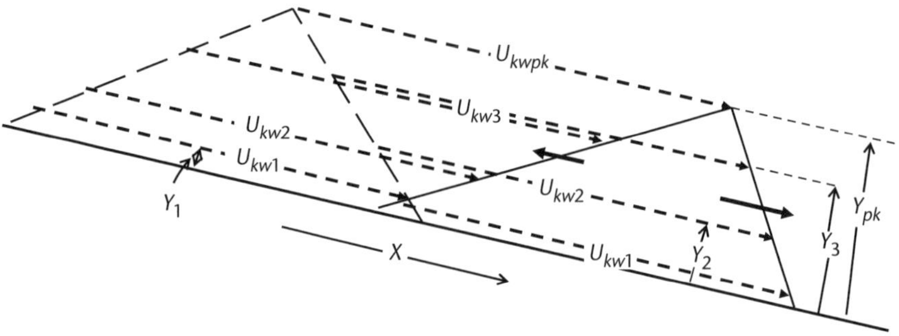

Figure 11.20 Schematic diagram illustrating steepening of kinematic wave as it travels. The dashed triangle is the wave at time t_1, and the solid triangle is the wave at a later time t_2. Wave velocity U_{kw} increases with depth Y, and if slope and resistance are constant, the distances between the rising limb and recession limb at each level remain constant, but the higher levels move a greater distance in each time increment, so the rising-limb slope increases while the recession-limb slope decreases. However, the difference in pressure force (bold arrows) between the steeper downstream face and the upstream face acts to reduce this tendency to steepen.

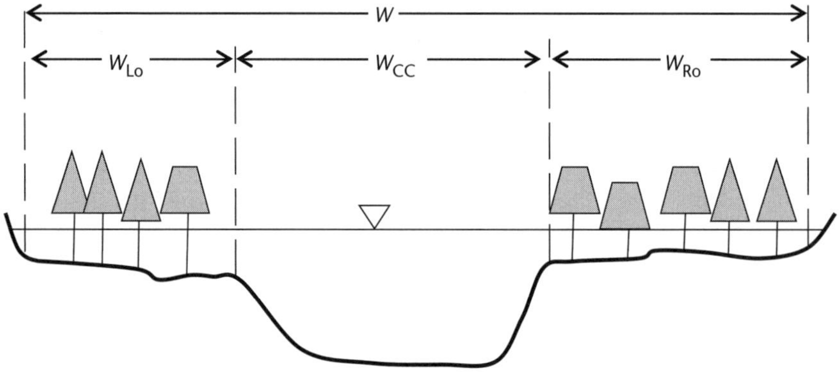

Figure 11.21 Definitions of terms used in estimating the effects of overbank flow on flood-wave velocity (equations 11.64 and 11.65). After Gray and Wigham (1970).

the average velocity of the flow U is approximately

$$U = \frac{W_{\mathrm{Lo}} \cdot U_{\mathrm{Lo}} + W_{\mathrm{cc}} \cdot U_{\mathrm{cc}} + W_{\mathrm{Ro}} \cdot U_{\mathrm{Ro}}}{W}, \tag{11.64a}$$

where the subscripts denote the left overbank (Lo), channel (cc), and right overbank (Ro), and W is the entire flow width. If we assume that the velocities on the floodplains are negligible due to the high resistance typically offered by brush and trees,

$$U = \frac{W_{\mathrm{cc}} \cdot U_{\mathrm{cc}}}{W}, \tag{11.64b}$$

and the kinematic-wave velocity is

$$U_{kw} = (p+1) \cdot U = (p+1) \cdot \frac{W_{cc} \cdot U_{cc}}{W}. \tag{11.65}$$

From equation 11.65, we see that $U_{kw} < U$ when $W_{cc}/W < 1/(p+1)$. Thus, this analysis indicates that for $p = 1/2$, the flood-wave velocity will be less than the water velocity when $W_{cc}/W < 2/3$. While this analysis is only approximate, it indicates that, by slowing the velocities of overbank flows, floodplains tend to reduce the velocity of a flood wave.

11.5.2.3 Relations between Kinematic Waves and Gravity Waves

The celerity of a simple shallow-water gravity wave or a solitary wave is given by equation 11.51, and combining this with equation 11.40 gives the downstream velocity of such waves, U_{gw}:

$$U_{gw} = U + (g \cdot Y)^{1/2}, \tag{11.66}$$

where U is the water velocity. Equating 11.66 and 11.63, we see that for a rectangular channel, $U_{kw} = U_{gw}$ when

$$U + p \cdot U = U + (g \cdot Y)^{1/2},$$

and, from the definition of the Froude number, when

$$Fr = \frac{1}{p}. \tag{11.67}$$

Thus, assuming $p = 1/2$, we see that

$$\begin{aligned} U_{gw} &> U_{kw} \text{ for when } Fr < 2; \\ U_{gw} &= U_{kw} \text{ for } Fr = 2; \\ U_{gw} &< U_{kw} \text{ for } Fr > 2. \end{aligned} \tag{11.68}$$

Flows in natural streams are almost always subcritical, that is, $Fr < 1$, so we conclude that gravity waves almost always travel faster than do kinematic waves.

Henderson (1966, p. 368) summarizes the relation between dynamic (gravity) waves and kinematic waves as follows:

> Evidently both types of wave movement—kinematic and dynamic [gravity]—may be present in any natural flood wave. The bed [channel] slope S_0 is usually by far the most important term [in equation 11.19] even if the other three terms are not negligible; the main bulk of the flood wave therefore moves substantially as a kinematic wave .... In particular, the speed of the main flood wave may be expected to approximate that of the kinematic wave, given by [equation 11.59], and this result was in fact proved by Lighthill and Witham's study .... But unless the other slope terms are absolutely negligible (which they seldom are) they will produce dynamic wave fronts also, moving at speeds $[U \pm (g \cdot Y)^{1/2}]$ in front of and behind the main body of the flood wave.

Lighthill and Witham (1955) showed that gravity waves attenuate rapidly due to friction and disappear quickly, whereas kinematic waves dissipate slowly and hence dominate even in flows with $Fr < 2$. After summarizing the basic qualities of kinematic waves, we will quantitatively explore the kinematic and dynamic aspects of flood waves in section 11.5.3.

11.5.2.4 Kinematic Waves: Summary

- The motion of most flood waves is approximated by the kinematic wave, which is a translatory shallow-water wave with a single wavefront that moves downstream (only) with a constant velocity.
- For in-channel flows, the ratio of kinematic-wave velocity to water velocity is $p+1$, which is always greater than 1.
- The ratio of kinematic-wave velocity to water velocity increases with resistance and decreases with relative submergence.
- The ratio of kinematic-wave velocity to water velocity increases as the channel cross-section form approaches a rectangular shape.
- In a given reach, kinematic-wave velocity increases with discharge (and depth).
- When overbank flow occurs, the kinematic-wave velocity may be less than the water velocity.

11.5.3 Quantitative Analysis of Flood Waves

We saw from equation 11.19 that the dynamic equation for unsteady flow can be incorporated into a resistance relation and written as

Forces: viscous + turbulent resistance ($\Omega^{-1} \cdot g^{1/2} \cdot A \cdot Y^{1/2}$); gravitational ($S_0$); pressure ($\partial Y/\partial X$); convectional ($\frac{U}{g} \cdot \frac{\partial U}{\partial X}$); local ($\frac{1}{g} \cdot \frac{\partial U}{\partial t}$)

$$Q = \Omega^{-1} \cdot g^{1/2} \cdot A \cdot Y^{1/2} \cdot \left(S_0 - \frac{\partial Y}{\partial X} - \frac{U}{g} \cdot \frac{\partial U}{\partial X} - \frac{1}{g} \cdot \frac{\partial U}{\partial t} \right)^{1/2} \qquad (11.69)$$

Flow types: steady uniform (kinematic) — terms through S_0; quasi-uniform (diffusive) — terms through $\partial Y/\partial X$; steady nonuniform — terms through $\frac{U}{g} \cdot \frac{\partial U}{\partial X}$; unsteady nonuniform (complete dynamic) — all terms.

Chapter 7 explores the relative magnitudes of the various terms in natural channels; the results are summarized in figure 7.14. Recalling these results:

- The gravitational-force term due to the channel slope S_0 is usually the dominant driving force.
- The pressure force due to the spatial gradient of depth $(\partial Y/\partial X)$ may often be of comparable magnitude to the gravitational force.
- The convective-acceleration term $(U/g) \cdot (\partial U/\partial X)$ is usually of lesser importance than the gravitational and/or pressure terms and may often be negligible.
- The local-acceleration term $(1/g) \cdot (\partial U/\partial t)$ is usually of negligible relative magnitude.

Following Julien (2002), it is possible to further compare the magnitudes of the terms in equation 11.69 by expressing the convective and local accelerations in terms of the spatial gradient of depth ($\partial Y/\partial X$). Doing this will give insight into the conditions under which flood waves tend to steepen or dissipate when lateral inflow is negligible.

We begin by writing the continuity equation (equation 11.3c) in the form

$$\frac{\partial A}{\partial t} = -\frac{\partial Q}{\partial X} = -U \cdot \frac{\partial A}{\partial X} - A \cdot \frac{\partial U}{\partial X}. \tag{11.70}$$

Julien (2002) showed that for a rectangular channel with no lateral inflow,

$$\frac{\partial U}{\partial X} = p \cdot \frac{U}{Y} \cdot \frac{\partial Y}{\partial X} \tag{11.71}$$

and

$$\frac{\partial U}{\partial t} = -p \cdot (p+1) \cdot \frac{U^2}{Y} \cdot \frac{\partial Y}{\partial X}, \tag{11.72}$$

where p is the depth exponent in the basic resistance relation. Now substituting equations 11.71 and 11.72 into equation 11.69 and using the definition of the Froude number, $Fr \equiv U/(g \cdot Y)^{1/2}$, equation 11.69 can be written in terms of channel slope and the depth gradient alone:

$$Q = \Omega^{-1} \cdot g^{1/2} \cdot W \cdot Y^{p+1} \cdot \left[S_0 - \left(1 - p^2 \cdot Fr^2\right) \cdot \frac{\partial Y}{\partial X} \right]^{1/2}. \tag{11.73a}$$

The value of p is determined by the applicable resistance relation. Recall that $p = 1/2$ for the Chézy relation, $p = 2/3$ for the Manning relation, and more generally, $p = m/f$, where m is the velocity exponent and f the depth exponent in the hydraulic geometry relations. For the Chézy relation (equation 11.17) with $p = 1/2$,

$$Q = \Omega^{-1} \cdot g^{1/2} \cdot W \cdot Y^{3/2} \cdot \left[S_0 - \left(1 - \frac{Fr^2}{4}\right) \cdot \frac{\partial Y}{\partial X} \right]^{1/2}. \tag{11.73b}$$

Equation 11.73 gives us considerable insight into the behavior of flood waves in the absence of significant lateral inflow. To see this, it is useful to define the **dimensionless flood-wave diffusivity**, D_{fw}, as

$$D_{\text{fw}} \equiv 1 - p^2 \cdot Fr^2, \tag{11.74a}$$

so that with the Chézy relation,

$$D_{\text{fw}} \equiv 1 - \frac{Fr^2}{4}. \tag{11.74b}$$

Now we can relate the energy slope S_e to the channel slope S_0 and the depth gradient as

$$S_e = S_0 - D_{\text{fw}} \cdot \frac{\partial Y}{\partial X} \tag{11.75}$$

and see from equation 11.63 that the flood-wave velocity U_{fw} is given by

$$U_{\text{fw}} = (p+1) \cdot \Omega^{-1} \cdot g^{1/2} \cdot Y^{1/2} \cdot \left[S_0 - D_{\text{fw}} \cdot \frac{\partial Y}{\partial X} \right]^{1/2}. \tag{11.76}$$

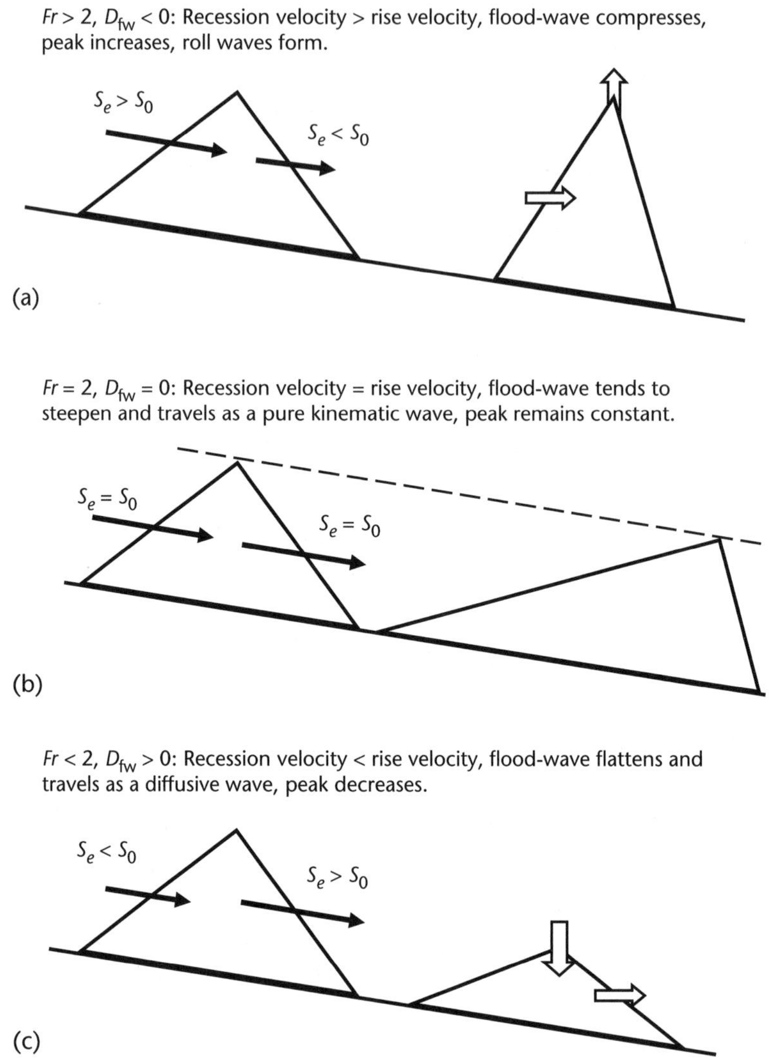

Figure 11.22 Schematic diagram illustrating how Froude number Fr affects flood-wave diffusivity D_{fw} (equation 11.76). After Julien (2002).

Now, referring to figure 11.22, we can identify the following cases (in which we assume constant channel slope S_0, geometry, resistance, and no lateral inflow):

Case 1: $Fr > 1/p(Fr > 2)$;	$D_{\mathrm{fw}} < 0$
Recession ($\partial Y/\partial X > 0$):	$S_e > S_0$
Rise ($\partial Y/\partial X < 0$):	$S_e < S_0$

Therefore, recession $U_{fw}(Y) >$ rise $U_{fw}(Y)$, so the flood wave tends to compress and the peak discharge is amplified. This produces the surface instabilities discussed in section 6.2.2.2, in which the flow forms pulses or surges called **roll waves**.

Case 2: $Fr = 1/p (Fr = 2)$; $D_{fw} = 0$
Recession ($\partial Y/\partial X > 0$): $S_e = S_0$
Rise ($\partial Y/\partial X < 0$): $S_e = S_0$

Therefore, recession $U_{fw}(Y) =$ rise $U_{fw}(Y)$. For a given depth, the velocity is equal for the rise and the recession but because the velocity is a function of depth, the flood wave tends to steepen (as in figure 11.20). However, the peak discharge is constant. This is the **pure kinematic wave**.

Case 3: $Fr < 1/p (Fr < 2)$; $D_{fw} > 0$
Recession ($\partial Y/\partial X > 0$): $S_e < S_0$
Rise ($\partial Y/\partial X < 0$): $S_e > S_0$

Therefore, recession $U_{fw}(Y) <$ rise $U_{fw}(Y)$, so the peak discharge decreases and the wave tends to attenuate. This is a **diffusive wave**.

Because Fr is almost always <1 in natural channels, we conclude from the above analysis that most flood waves are diffusive. However, from equation 11.76 we see that the degree of attenuation of a flood wave depends on the magnitude of $D_{fw} \cdot (\partial Y/\partial X)$ relative to the magnitude of S_0. From the derivation in box 11.5, we see that

$$\left|\frac{D_{fw}}{S_0} \cdot \frac{\partial Y}{\partial X}\right| = \frac{1-p^2}{(p+1)^2 \cdot g \cdot S_0 \cdot Y} \cdot \frac{\partial q}{\partial t} = \frac{1-p^2}{(p+1)^2 \cdot g \cdot S_0 \cdot Y \cdot W} \cdot \frac{\partial Q}{\partial t}, \qquad (11.77a)$$

BOX 11.5 Derivation of Equation 11.77

From equation 11.74a,

$$\frac{D_{fw}}{S_0} \cdot \frac{\partial Y}{\partial X} = \frac{1-p^2 \cdot Fr^2}{S_0} \cdot \frac{\partial Y}{\partial X}. \qquad (11B5.1)$$

Julien (2002, table 5.1) shows that

$$\frac{\partial Y}{\partial X} = -\frac{Y^2}{(p+1)^2 \cdot q^2} \cdot \frac{\partial q}{\partial t}, \qquad (11B5.2)$$

where $q \equiv Q/W$ (discharge per unit width) $= U \cdot Y$. Substituting equation 11B5.2 and the definition of the Froude number $Fr \equiv U/(g \cdot Y)^{1/2}$ into equation 11B5.1 yields

$$\left|\frac{D_{fw}}{S_0} \cdot \frac{\partial Y}{\partial X}\right| = \frac{1-p^2}{(p+1)^2 \cdot g \cdot S_0 \cdot Y} \cdot \frac{\partial q}{\partial t}. \qquad (11B5.3)$$

where $\mathcal{Q} \equiv Q/W$. With $p = 1/2$,

$$\left|\frac{D_{\mathrm{fw}}}{S_0} \cdot \frac{\partial Y}{\partial X}\right| = \frac{1}{3 \cdot g \cdot S_0 \cdot Y \cdot W} \cdot \frac{\partial Q}{\partial t}. \tag{11.77b}$$

Thus we see that the relative importance of flood-wave diffusivity (i.e., the tendency for the flood wave to attenuate) decreases with slope and depth and increases with the time rate of increase of discharge.

Again, note that the above analysis applies in the absence of tributary inputs, lateral inflows, and overbank flows. Lateral inflows and tributary contributions act to reduce the dissipation of the flood wave, and overbank flow and lateral outflows tend to accelerate the dissipation.

11.6 Flood-Wave Routing

11.6.1 Overview

Flood-wave routing is the general term for mathematical procedures for forecasting the magnitude, shape, and speed of flood waves as a function of time at one or more cross sections in a channel or channel network (Fread 1992). As noted above, such waves may be generated by watershed-wide rainfall or snowmelt, by the operation of engineering works (locks, reservoirs), or by catastrophic events such as landslides or the failure of dams or levees. Such forecasts are essential for the design and operation of engineering and land-use planning measures to reduce flood damages and for implementing emergency procedures when floods threaten.

As noted in section 11.2, the most complete physical descriptions of the movement of flood waves are given by numerical solutions to the Saint-Venant equations. We have seen that such solutions can provide excellent predictions but are data intensive, requiring information about channel geometry and resistance at many cross sections that are incorporated into elaborate mathematical procedures requiring computer implementation. Fread (1992), Moussa and Bocquillon (1996), Moramarco and Singh (2000), and Wang and Chen (2003) provide useful overviews of various approaches to routing procedures based on the Saint-Venant equations and guidance in selecting the appropriate procedure.

In many cases, the full Saint-Venant equations will be the method of choice. However, before the widespread accessibility of high-speed computers, hydraulic engineers and scientists had developed simpler approaches to flood-wave routing, and these may still be satisfactory where the availability of data, the accuracy requirements, and the resources available for developing the prediction are limited.

A major class of these simpler methods is called **hydrologic routing procedures**. As with the Saint-Venant equations, these procedures are based on 1) a continuity equation, and 2) a dynamic relation. However, in hydrologic routing, the dynamic equation is not developed from basic energy or momentum considerations, but is based on heuristic[6] relations involving only depth, discharge, and storage volumes.

Hydrologic routing generally gives satisfactory results only in cases where the rate of hydrograph rise (dQ/dt) is not too large, where backwater effects caused by constrictions (bridges) are not present (i.e., where local and convectional accelerations are negligible), and where the flow is subcritical. Thus, hydrologic routing is not suitable for predicting dam-break floods and similar phenomena.

With the goal of further developing an intuitive understanding of how flood waves move through channels, we explore the most widely used hydrologic routing approach in the following section.

11.6.2 Hydrologic Routing: The Muskingum Method

11.6.2.1 Basic Development

The Muskingum routing method was first developed by the U.S. Army Corps of Engineers for design of flood-control measures in the Muskingum River basin in Ohio in the late 1930s. Although considerably simpler than the numerical solutions of the Saint-Venant equations described in section 11.1.3, the method is conceptually similar in that 1) the routing equations are derived from the principles of conservation of mass (continuity) and "dynamic" hydraulic relations, and 2) the space and time continua are divided into discrete increments and solutions are found for successive increments. We describe the Muskingum method as applied to a single reach, but the method can be applied to any number of successive reaches, where the outflow from an upstream reach is the inflow to a downstream reach.

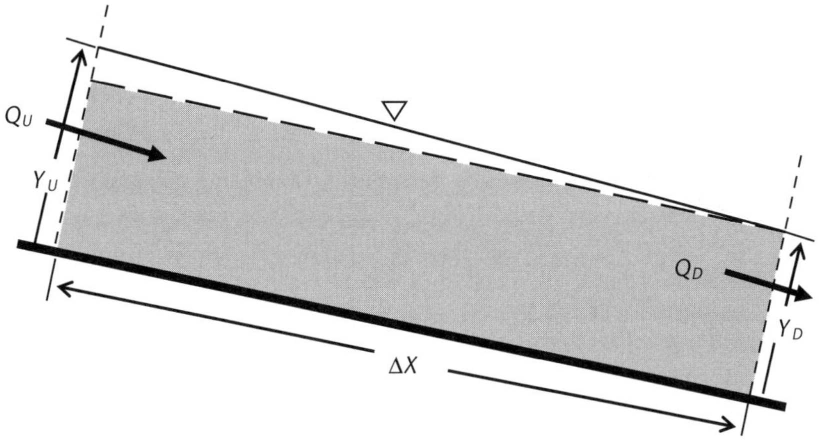

Figure 11.23 Definition diagram for the Muskingum routing procedure. The volume stored in the reach, V, is the area under the water surface times the channel width. This volume is divided into **prism storage**, with its upper surface parallel to the channel bed (shaded area), and **wedge storage**, the portion between the upper surface of the prism and the water surface (unshaded area). The prism storage is a function of the downstream depth Y_D; the wedge storage is a function of the upstream depth Y_U. Q_U is the discharge entering the reach (the input hydrograph), which is known; Q_D is the discharge leaving the reach (the output hydrograph), which is to be predicted.

Referring to figure 11.23, the basic continuity equation for a reach is

$$Q_U - Q_D = \frac{dV}{dt}, \tag{11.78}$$

where Q_U is the **inflow rate** (the instantaneous discharge entering the reach at the upstream end [$L^3\,T^{-1}$]), Q_D is the **outflow rate** (the instantaneous discharge leaving the reach at the downstream end [$L^3\,T^{-1}$]), and V is the **storage** (the instantaneous volume of water stored in the reach [L^3]). The graphs of Q_U, Q_D, and V versus time are the upstream, downstream, and storage hydrographs, respectively (figure 11.24). For a particular reach, the general problem is to predict the downstream hydrograph given the upstream hydrograph. Note that lateral inflow is not included in equation 11.78, and the volumes of water in the upstream and downstream hydrographs (areas under the respective hydrographs) are equal; we will see later that lateral inflow can be accounted for in the method.

In place of the dynamic relation used in the Saint-Venant formulation (equation 11.19), the Muskingum method relates the inflow and outflow rates to upstream and downstream depths Y_U and Y_D, respectively, via simple heuristic power-law functions, as in the hydraulic geometry relations of section 11.2:

$$Q_U = a \cdot Y_U{}^{b}, \tag{11.79a}$$

$$Q_D = a \cdot Y_D{}^{b}, \tag{11.79b}$$

where a is an empirical coefficient and b an empirical exponent. The reach storage is similarly modeled by first defining an "upstream storage" V_U and a "downstream storage" V_D:

$$V_U = c \cdot Y_U{}^{d}, \tag{11.80a}$$

$$V_D = c \cdot Y_D{}^{d}, \tag{11.80b}$$

where c and d are again empirically determined. Combining equations 11.79 and 11.80, we can write

$$V_U = c \cdot \left(\frac{Q_U}{a}\right)^{d/b} \tag{11.81a}$$

and

$$V_D = c \cdot \left(\frac{Q_D}{a}\right)^{d/b}. \tag{11.81b}$$

At any instant the discharges at the upstream and downstream ends of the reach differ, and the actual reach storage at any instant, V, is expressed as a weighted average of the values given by equations 11.81a and 11.81b:

$$V = X \cdot V_U + (1 - X) \cdot V_D, \tag{11.82}$$

where X is the weighting factor. If the stages in a reach are determined solely by a control at the downstream end, as at the spillway of a level-pool reservoir, $X = 0$. If there is **prism storage** (figure 11.23), $X > 0$; however, as will be shown below,

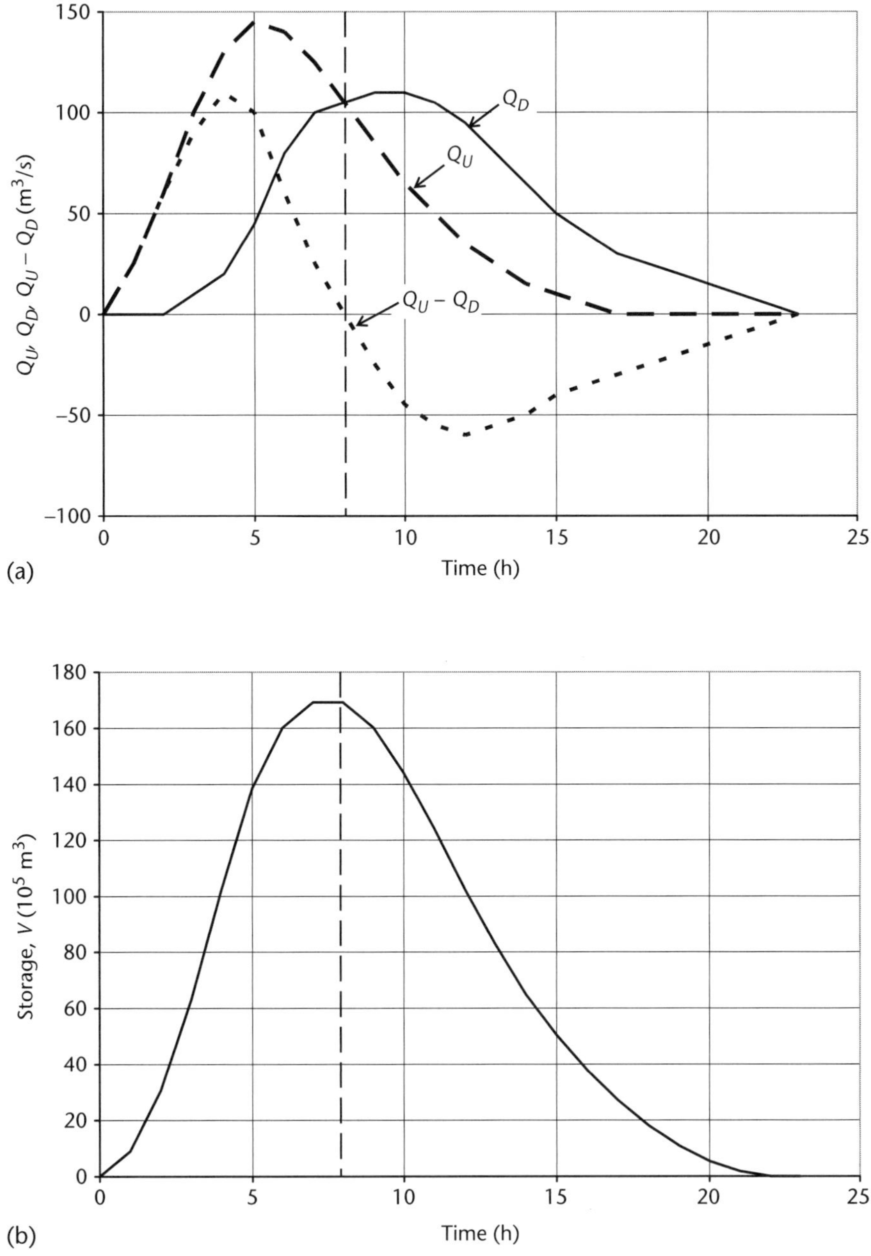

Figure 11.24 Hydrographs illustrating the Muskingum routing procedure. (a) Hydrographs of inflow, Q_U (long-dash line), outflow Q_D (solid line), and rate of storage accumulation $Q_U - Q_D$ (short-dash line). (b) Hydrograph of volume of water in storage. Note that storage is maximum when $Q_U = Q_D$.

it must be true that $0 \leq \mathcal{X} \leq 0.5$. Now substituting equations 11.81 into 11.82, we have

$$V = T^* \cdot [\mathcal{X} \cdot Q_U{}^{\mathcal{d}/\mathcal{b}} + (1 - \mathcal{X}) \cdot Q_D{}^{\mathcal{d}/\mathcal{b}}], \tag{11.83a}$$

where $T^* \equiv c/a^{\mathcal{d}/\mathcal{b}}$.

We now consider the values of the exponents $\mathcal{d}/\mathcal{b}$ and the coefficient T^*. Examining the basic resistance relation (equation 11.19), we would expect $\mathcal{b} = 1.5$, whereas the hydraulic geometry relation (equation 11.35) with a typical value of $f = 0.44$ suggests that $\mathcal{b} \approx 1/0.44 = 2.28$. In a prismatic channel, it is reasonable to assume that $\mathcal{d} \approx 1$ (although it could be greater if water spreads out over a floodplain), so the value of $\mathcal{d}/\mathcal{b}$ might be expected to be in the range 1/2.28 to 1/1.5, or $0.44 \leq \mathcal{d}/\mathcal{b} \leq 0.67$. However, V has the dimensions [L^3] and Q the dimensions [$L^3\,T^{-1}$], so for equations 11.81, 11.82, and 11.83a to be dimensionally correct, it should be true that $\mathcal{d}/\mathcal{b} = 1$ and that the dimensions of T^* be [T]. In the Muskingum method, the dimensional considerations prevail (and are mathematically convenient), and it is assumed that

$$V = T^* \cdot [\mathcal{X} \cdot Q_U + (1 - \mathcal{X}) \cdot Q_D]. \tag{11.83b}$$

We will see in section 11.6.2.3 that T^* is the time it takes the flood wave to travel through the reach.

11.6.2.2 Discretization

For practical application, equation 11.78 must be discretized by writing the derivative as the difference in storage at successive time increments t and $t+1$,

$$\frac{dV}{dt} \approx \frac{V(t+1) - V(t)}{\Delta t}. \tag{11.84}$$

where Δt is the duration of the time increment (i.e., $t+1 = t + \Delta t$). Equation 11.83b is then used to relate this rate-of-change of storage to the inflow and outflow discharges at successive time increments:

$$\frac{V(t+1) - V(t)}{\Delta t} = \frac{T^* \cdot \{[\mathcal{x} \cdot Q_U(t+1) + (1-\mathcal{x}) \cdot Q_D(t+1)] - [\mathcal{x} \cdot Q_U(t) + (1-\mathcal{x}) \cdot Q_D(t)]\}}{\Delta t}, \tag{11.85}$$

One can then derive the **Muskingum routing equation** from equation 11.85 as

$$Q_D(t+1) = C_1 \cdot Q_U(t+1) + C_2 \cdot Q_U(t) + C_3 \cdot Q_D(t). \tag{11.86}$$

The **routing coefficients** C_1, C_2, and C_3 are given by

$$C_1 = \frac{\Delta t - 2 \cdot T^* \cdot \mathcal{x}}{2 \cdot T^* \cdot (1 - \mathcal{x}) + \Delta t}, \tag{11.87}$$

$$C_2 = \frac{\Delta t + 2 \cdot T^* \cdot \mathcal{x}}{2 \cdot T^* \cdot (1 - \mathcal{x}) + \Delta t}, \tag{11.88}$$

$$C_3 = \frac{2 \cdot T^* \cdot (1 - \mathcal{x}) - \Delta t}{2 \cdot T^* \cdot (1 - \mathcal{x}) + \Delta t}, \tag{11.89}$$

and

$$C_1 + C_2 + C_3 = 1. \tag{11.90}$$

If lateral inflow is important, it is incorporated into a fourth routing coefficient as

$$C_4 = \frac{q_L \cdot \Delta t \cdot \Delta X}{2 \cdot T^* \cdot (1 - x) + \Delta t}, \tag{11.91}$$

where q_L is the lateral-inflow rate per unit channel length $[L^2T^{-1}]$, and C_4 is added to the right-hand side of the routing equation 11.86 (Fread 1992).

To apply the method to a reach of length ΔX, one must know the values of the upstream (input) hydrograph $Q_U(t)$ for all time increments, an initial value of the downstream (output) hydrograph $Q_D(0)$, the lateral-inflow rate q_L for all time increments, and appropriate values of the routing parameters T^*, X, and Δt. Equation 11.86 is then applied at each time step to generate successive values of $Q_D(t+1)$. Methods for determining the values of the routing parameters are described in box 11.6.

BOX 11.6 Determination of Parameter Values for Muskingum Routing

A Posteriori Determination from Inflow and Outflow Hydrographs

If inflow and outflow hydrographs for past floods in the reach of interest are available, the value of Δt can be selected as a convenient time interval providing that

$$\Delta t \leq \frac{T_R}{5}, \tag{11B6.1}$$

where T_R is the **time of rise** of the inflow hydrograph.

The appropriate value of X can be determined graphically by plotting successive values of storage,

$$V(t+1) = V(t) + \frac{Q_U(t) + Q_U(t+1)}{2} - \frac{Q_D(t) + Q_D(t+1)}{2}, \tag{11B6.2}$$

against values of discharge $\hat{Q}$ estimated as

$$\hat{Q} = X \cdot Q_U + (1 - X) \cdot Q_D \tag{11B6.3}$$

using trial values of X. For each value of X, the plot will trace out a loop as in figure 11.25, and the appropriate value of X is the one for which the loop is "tightest," that is, closest to a straight line. The appropriate value of T^* is then determined as the slope of the straight line that best fits the tightest loop. (In the plot of figure 11.25, the storage values are plotted on the abscissa, so T^* is given by the inverse of the slope on that graph.)

(Continued)

BOX 11.6 *Continued*

In an alternative approach applicable in the absence of lateral inflow, McCuen (2005) stated that the values of the three routing coefficients can be determined most accurately by solving three simultaneous equations for C_1, C_2, and C_3:

$$C_1 \cdot \Sigma[Q_U(t+1)]^2 + C_2 \cdot \Sigma[Q_U(t) \cdot Q_U(t+1)] + C_3 \cdot \Sigma[Q_D(t) \cdot Q_U(t+1)] = \Sigma[Q_D(t+1) \cdot Q_U(t+1)], \quad (11B6.4a)$$

$$C_1 \cdot \Sigma[Q_U(t) \cdot Q_U(t+1)] + C_2 \cdot \Sigma[Q_U(t)]^2 + C_3 \cdot \Sigma[Q_D(t) \cdot Q_U(t)] = \Sigma[Q_D(t+1) \cdot Q_U(t)], \quad (11B6.4b)$$

$$C_1 \cdot \Sigma[Q_D(t) \cdot Q_U(t+1)] + C_2 \cdot \Sigma[Q_D(t) \cdot Q_U(t)] + C_3 \cdot \Sigma[Q_D(t)]^2 = \Sigma[Q_D(t) \cdot Q_D(t+1)]. \quad (11B6.4c)$$

Then, $\mathcal{X}$ and T^* are found as

$$\mathcal{X} = \frac{C_2 - C_1}{2 \cdot (1 - C_1)} \quad (11B6.5)$$

and

$$T* = \frac{\Delta t \cdot (1 - C_1)}{C_1 + C_2}. \quad (11B6.6)$$

A Priori Determination of Parameter Values

For situations in which inflow and outflow hydrographs are not available for the reach of interest, the routing parameters can be estimated from basic hydraulic considerations.

As above, the time interval Δt is found from equation 11B6.1. Assuming that the reach length ΔX is given, the travel time through the reach T^* is found via equation 11.93 where the flood-wave velocity U_{fw} can be estimated from reach characteristics via equations 11.63, 11B3.4, or 11B3.5.

Two approaches have been suggested for a priori estimates of $\mathcal{X}$. Cunge (1969) derived

$$\mathcal{X} = 0.5 - \frac{Q}{2 \cdot U_{fw} \cdot W \cdot S_0 \cdot \Delta X} = 0.5 - \frac{Y \cdot U}{2 \cdot U_{fw} \cdot S_0 \cdot \Delta X}, \quad (11B6.7)$$

where Q is the time- and space-averaged discharge, and the other quantities are their values at that discharge. Dooge et al. (1982) derived

$$\mathcal{X} = 0.5 - 0.3 \cdot \left(1 - p^2 \cdot Fr^2\right) \cdot \frac{Y}{S_0 \cdot \Delta X}, \quad (11B6.8)$$

where p is the depth exponent in the resistance relation (equation 11.60). Note the similarity between 11B6.8 and the definition of flood-wave diffusivity D_{fw} in equation 11.74.

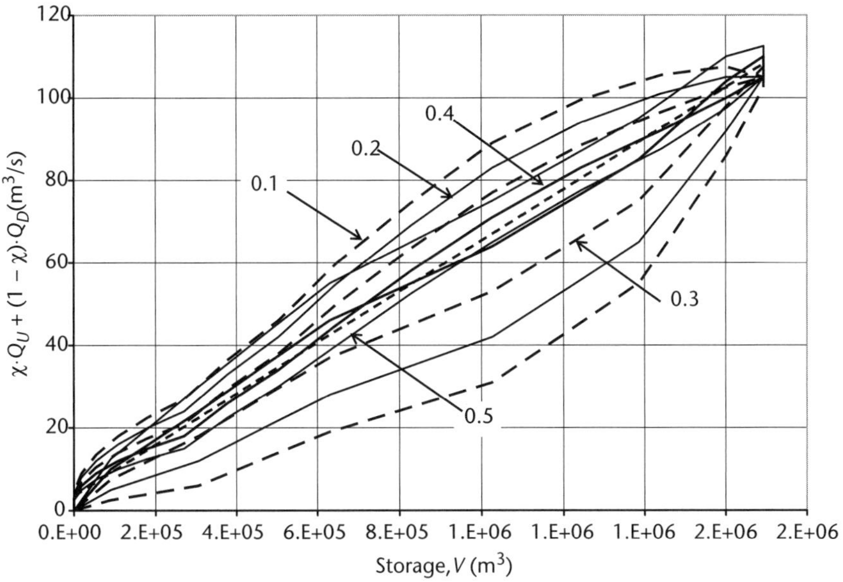

Figure 11.25 A posteriori graphical determination of Muskingum routing parameters X and T^* (box 11.6) for the hydrographs of figure 11.24 (box 11.7). Each curve is a plot of $\chi \cdot Q_U + (1-\chi)\cdot Q_D$ versus storage, V, using trial values of X (curve labels). The value $X = 0.4$ (heavy line) is selected as the one giving the plot nearest a straight line. The appropriate value of T^* is the inverse of the slope of the straight (heavy dashed) line that best fits the selected X loop.

BOX 11.7 Muskingum Routing Example

This example uses the a posteriori methods described in box 11.6 to determine the appropriate Muskingum routing parameters for the (fictitious) case shown in figure 11.24.

Δt: The time of rise, T_R, of the input hydrograph is about 5 h, so following equation 11B6.1, we select a time increment of $\Delta t = 1$ h.

X: Using $\Delta t = 1$ h, we plot storage calculated via equation 11B6.2 versus $[X \cdot Q_U + (1 - X] \cdot Q_D)$ (equation 11B6.3) for values of $X = 0.1, 0.2, 0.3, 0.4,$ and 0.5 (figure 11.25). The loop for $X = 0.4$ is tightest, so we select that value.

T^*: Using regression analysis (section 4.8.3.1), we determine the slope of the plot of $V(t+1)$ versus $Q(t+1)$ to be 16158 s so that $T^* = 16158/3600 = 4.49$ h.

(Continued)

BOX 11.7 *Continued*

Routing Coefficients

Now using these parameter values in equations 11.87–11.89, we compute the routing coefficients

$$C_1 = -0.406,$$

$$C_2 = 0.719,$$

$$C_3 = 0.687.$$

Routing Procedure

With these values, the routing computations proceed as in table 11B7.1.

Table 11B7.1

Time (h)	Input, Q_U (m^3/s)	Computation	Predicted output, $\hat{Q}_D$ (m^3/s)	Actual output, Q_D (m^3/s)
0	0		0	0
1	25	$-0.406 \times 25 + 0.719 \times 0 + 687 \times 0 =$	−10.1	0
2	60	$-0.406 \times 60 + 0.719 \times 25 + 0.687 \times -10.1 =$	−6.4	0
3	100	$-0.406 \times 100 + 0.719 \times 60 + 0.687 \times -6.4 =$	2.6	10
4	130	$-0.406 \times 130 + 0.719 \times 100 + 0.687 \times 2.6 =$	26.0	20
...	...	...	...	
9	85	$-0.406 \times 85 + 0.719 \times 105 + 0.687 \times 115.9 =$	113.1	110
10	65	$-0.406 \times 65 + 0.719 \times 85 + 0.687 \times 113.1 =$	110.3	110
11	50	$-0.406 \times 50 + 0.719 \times 65 + 0.687 \times 110.3 =$	102.0	105
12	35	$-0.406 \times 35 + 0.719 \times 50 + 0.687 \times 102.0 =$	93.9	95
...	...	...	...	
20	0	$-0.406 \times 0 + 0.719 \times 0 + 0.687 \times 17.2 =$	13.7	15
21	0	$-0.406 \times 0 + 0.719 \times 0 + 0.687 \times 13.7 =$	10.3	10
22	0	$-0.406 \times 0 + 0.719 \times 0 + 0.687 \times 10.3 =$	6.9	5
23	0	$-0.406 \times 0 + 0.719 \times 0 + 0.687 \times 6.9 =$	3.4	0

The predicted and actual output hydrographs are compared in figure 11.26.

Box 11.7 derives the parameter values and coefficients for the case plotted in figure 11.24 and shows how the routing equation is used to generate successive values of $Q(t+1)$ via equation 11.86. Figure 11.26 compares the measured and predicted outflow hydrographs; the estimated values are quite close to the actual, but the predicted peak is about 5% higher and occurs about 1.5 h earlier. Note that the method gives negative discharges for the first two time steps; these are, of course, not physically possible, and Q_D for those times would be considered zero. The occurrence of negative values early in the predicted hydrograph is a common

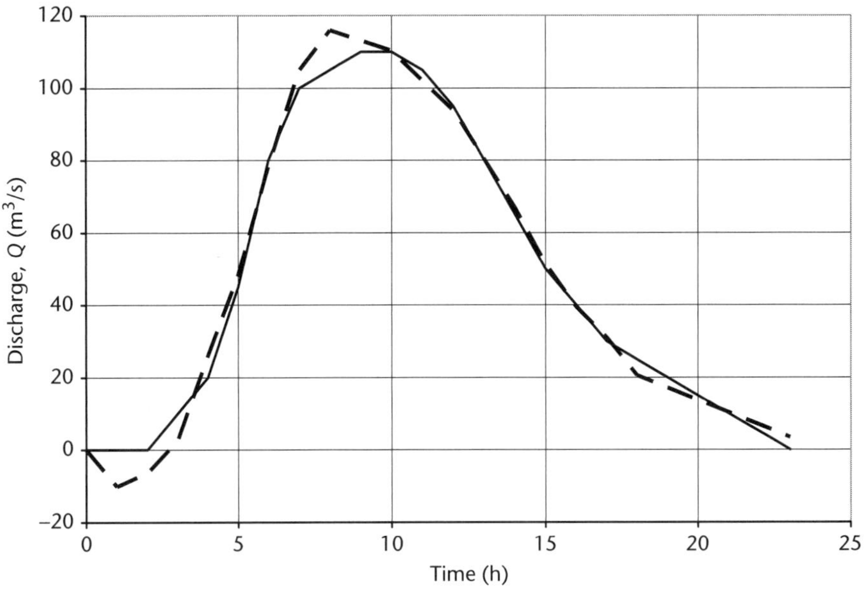

Figure 11.26 Comparison of measured (solid) and predicted (dashed) output hydrographs for the example in box 11.7. The predicted peak is slightly higher and occurs earlier than the measured peak. The negative discharge values in the earliest time steps of the predicted hydrograph are an artifact that commonly occurs in Muskingum routing.

artifact of the Muskingum method; reducing ΔX by dividing the reach of interest into shorter subreaches or reducing Δt may eliminate the problem.

11.6.2.3 Significance of Routing Parameters

If we assume for the moment that $X = 0$, we see from equation 11.83b that

$$T^* = \frac{V}{Q_D}; \tag{11.92}$$

that is, T^* is the total volume of storage in the reach divided by the rate of input. This is the definition of **residence time**, which is the average length of time that a "parcel" of water is in the reach. Thus, T^* is the time it takes a flood wave to travel through the reach, and we can write

$$T^* = \frac{\Delta X}{U_{\text{fw}}}, \tag{11.93}$$

where ΔX is the reach length. Note that T^* is also the time lag between the peaks of the input and output hydrographs.

As noted above, X is a weighting factor that determines the degree to which reach storage is controlled by upstream (wedge storage) or downstream (prism storage) discharge (equations 11.81 and 11.82). If $X = 0$, there is no wedge storage, and the

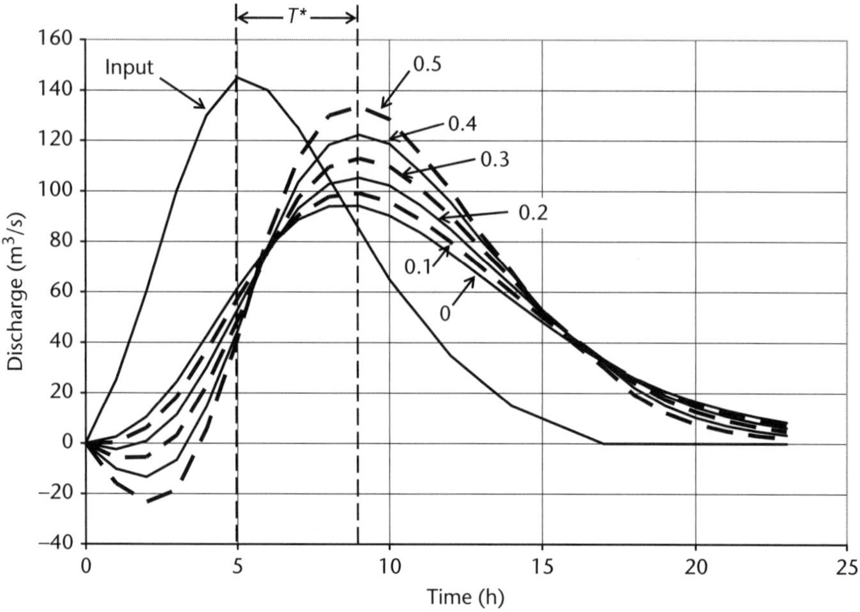

Figure 11.27 Effects of routing parameter X on hydrograph attenuation for the input hydrograph of box 11.7. Curve labels are values of X; decreasing values of χ produce more attenuated downstream hydrographs with decreasing peaks. The time lag between the peak of the inflow hydrograph and the peak of the outflow hydrographs is T^* and is the same for all values of X.

reach storage depends only on the downstream discharge; if $X = 0.5$, the storage depends equally on upstream and downstream discharge. In this case, the output discharge at a given time step is essentially equal to the input discharge of the previous time step, and the input hydrograph has simply been translated through the reach with little change in form or decrease in peak discharge. Successively smaller values of X reflect the increasing effects of reach storage in attenuating the input hydrograph and reducing the peak, as shown in figure 11.27. Thus, X is an inverse measure of flood-wave diffusivity.

A value of $X = 0.5$ approximates the case of a pure kinematic flood wave. McCuen (2005) noted that $X \approx 0.2$ for reaches with large floodplains and $X \approx 0.4$ for most natural reaches. The value of X should not exceed 0.5, because this leads to amplification of the downstream hydrograph and increasing problems with negative discharges.

11.7 Unsteady Flow: Summary

Unsteady flow is flow in which temporal changes in velocity are significant. The fundamental equations describing unsteady flow in open channels were first formulated by J.C.B. de Saint-Venant in 1848. These equations reflect 1) the basic principle of conservation of mass and 2) dynamic (force) considerations that can be derived from the conservation of momentum or energy. In its complete form, the dynamic equation accounts for forces associated with gravity (bed slope), pressure (depth gradient), and convective and local accelerations and can be incorporated in hydraulic relations that give discharge or velocity as a function of resistance and net driving force. The Saint-Venant equations cannot be solved analytically; we introduced approaches to developing numerical solutions and showed that such solutions can provide useful predictions of unsteady-flow phenomena.

The at-a-station hydraulic geometry relations introduced in chapter 2 reflect the interrelations among temporal changes in hydraulic quantities. These can be viewed as empirical hydraulic relations that can be incorporated in unsteady-flow analysis to help assess the relative importance of dynamic terms that influence flood-wave movement.

Temporal changes in velocity are always accompanied by temporal and spatial changes in depth; thus, unsteady-flow phenomena are waves. Classical wave theory deals with oscillatory waves in which the primary restoring force is gravity. In water that is deeper than one-half the wavelength (deep-water waves), the celerity of such waves depends only on the wavelength; in water shallower than 1/20th the wavelength (shallow-water waves), the celerity depends only on the depth. Virtually all the waves of practical importance in open-channel flows are shallow-water waves. However, shallow-water gravity waves are of secondary importance in open-channel flow because they tend to dissipate rapidly as they travel. (Solitons are an exception to this and may travel long distances without dissipation; however, they are usually generated by engineering structures or activities and are not common phenomena.)

Flood waves are shallow-water waves produced by relatively rapid accumulations of water in the channel system due to significant rain or snowmelt on a watershed or the opening or breach of a natural or artificial dam and are of central interest in hydrology and open-channel hydraulics. Flood waves typically travel at a velocity much lower than that of a gravity wave, but greater than the water velocity. As they travel downstream, the peak discharge tends to decrease and the wave tends to lengthen and dissipate because deeper portions of the wave travel with higher velocities than shallower portions, pressure forces act to accelerate the flow downstream of the peak and decelerate it upstream of the peak, channel friction differentially retards portions of the flow, and the rising water tends to enter into storage in the channel and the adjacent floodplain and banks.

The motion of most flood waves is approximated by the kinematic wave, which is a translatory shallow-water wave with a single wavefront that moves downstream (only) with a constant velocity that is usually faster than the velocity of the water itself. Although the Saint-Venant equations describe kinematic waves, their essential properties can be derived solely from the local relation between discharge and cross-sectional area, that is, from the slope of the rating curve. In a given reach, kinematic-wave velocity increases with discharge (and depth).

Using the continuity relation—or, alternatively, the hydraulic geometry relations—and the rules of derivatives, the convective and local accelerations as well as the pressure forces can be related to the depth gradient in a reach. From this analysis, we find that in the absence of lateral inflow or outflow, flood waves tend to steepen and form roll waves when the Froude number exceeds about 2, tend to steepen but travel as pure kinematic waves with constant peak discharge when Froude number equals 2, and tend to flatten (diffuse) as they travel when the Froude number is less than 2. Because Fr is almost always <1 in natural channels, we conclude that most flood waves are diffusive. The tendency for flattening is inversely related to the Froude number and directly related to the ratio of the depth gradient to the channel slope. However, the presence of lateral inflows or tributary inputs may reverse the tendency for downstream attenuation of the peak.

Flood-wave routing is the process of forecasting the magnitude, shape, and speed of flood waves as a function of time at one or more cross sections in a channel or channel network. Numerical solutions based on discretization of the Saint-Venant equations provide the complete physical basis for such forecasts but require extensive data on reach characteristics and elaborate computer models. Simpler hydrological routing methods are often used; these are based on conservation-of-mass considerations and simple heuristic relations among reach storage, discharge, and depth. The Muskingum routing method is perhaps the most widely used hydrological method. Like the Saint-Venant equations, it requires discretization of time but typically can be applied to the entire reach of interest. In addition, it requires determination of two routing parameters, one of which reflects the kinematic-wave velocity and the other of which reflects the tendency for flood-wave dissipation. The routing parameters can be estimated either a posteriori from measured inflow and outflow hydrographs for the reach of interest or a priori from knowledge of reach characteristics.

12

Sediment Entrainment and Transport

12.0 Introduction and Overview

As noted in section 2.3, most natural streams are **alluvial**; that is, their channels are made of particulate sediment that is subject to entrainment and transport by the water flowing in them. This sediment is entrained during sporadic periods of higher flows and deposited as discharge subsides. It may be entrained, transported, deposited, and stored as part of the channel or floodplain many times over and for widely varying time periods as it moves inexorably seaward. In the long geological view, this sediment movement is a link in the geologic/tectonic cycle of the destruction and construction of continents.

Understanding the processes by which and conditions under which entrainment and transport of particulate sediment occur is clearly of immense scientific and practical import. We begin our development of this understanding by defining some of the basic terminology and the techniques used to measure sediment in streams. We then explore empirical relations between sediment transport and streamflow and how these relations are used to estimate continental denudation rates and to understand some fundamental aspects of geomorphic processes. Next, we formulate the basic physics of the forces that act on sediment particles in suspension and on the stream bed to provide an essential foundation for understanding entrainment and transport processes. One central scientific question, touched on in section 2.4, concerns the shape of the channel cross section that an alluvial stream constructs through the processes of entrainment and deposition. This question is complex and not completely answered, but we can apply force-balance principles to provide useful insight.

We then explore the question of stream competence: the maximum size of sediment that can be entrained by a given flow. This, too, involves force considerations but, as in many hydraulic problems we have seen, ultimately requires experimental observations to answer. Understanding the question of competence also provides insights to the conditions under which various bedforms (introduced in section 6.6.4.2) occur.

Finally, we focus our understanding of the physics of sediment entrainment on a central problem of stream hydraulics: the prediction of sediment load, and the capacity of a given flow to transport the material constituting the channel bed.

12.1 Definitions and Measurement

12.1.1 Definitions

Streams transport organic matter and inorganic earth materials in dissolved and particulate forms, as well as dissolved gases (figure 12.1). For each component, **concentration** is the amount (weight or volume) of sediment component per amount (weight or volume) of water-plus-sediment mixture, and may be reported in several ways, as described in box 12.1. Unless otherwise specified, we will state sediment concentrations as weight of sediment per unit volume of water plus sediment ($[\mathrm{F\,L^{-3}}]$; C_{wv} in box 12.1).

The **load** (also called **sediment discharge**) is the mass-rate of transport (weight per unit time; $[\mathrm{F\,T^{-1}}]$). The load L_i at a given instant is the product of its concentration, C_i, and the instantaneous discharge, Q (volume of water plus sediment mixture per unit time; $[\mathrm{L^3\,T^{-1}}]$):

$$L_i = C_i \cdot Q, \tag{12.1a}$$

where i denotes a particular component. The usual units for concentration are milligrams per liter (mg/L), and for load are tons per day (T/day). For these units and discharge in m^3/s,

$$L_i = 0.0864 \cdot C_i \cdot Q. \tag{12.1b}$$

As noted in box 12.1, the discharge measured by the standard techniques described in section 2.5.3 and denoted Q here and throughout this text is actually the volume flow rate of water plus its contained sediment, but unless the sediment concentration is greater than about 130,000 mg/L (a very high value), the sediment makes up less than 5% of the total flow volume.

The inorganic solid load is of primary relevance for geological processes; it has the following constituents:

> **Dissolved load** is in the form of ions except for silica (quartz; SiO_2), which is carried in nonionic form. Inorganic dissolved constituents are of molecular and atomic sizes, less than 5×10^{-5} mm (Hem 1970). The uptake, transport, and deposition of these constituents are not determined by hydraulic conditions and so are not discussed further here. However, they usually constitute a significant portion of the total solid load.

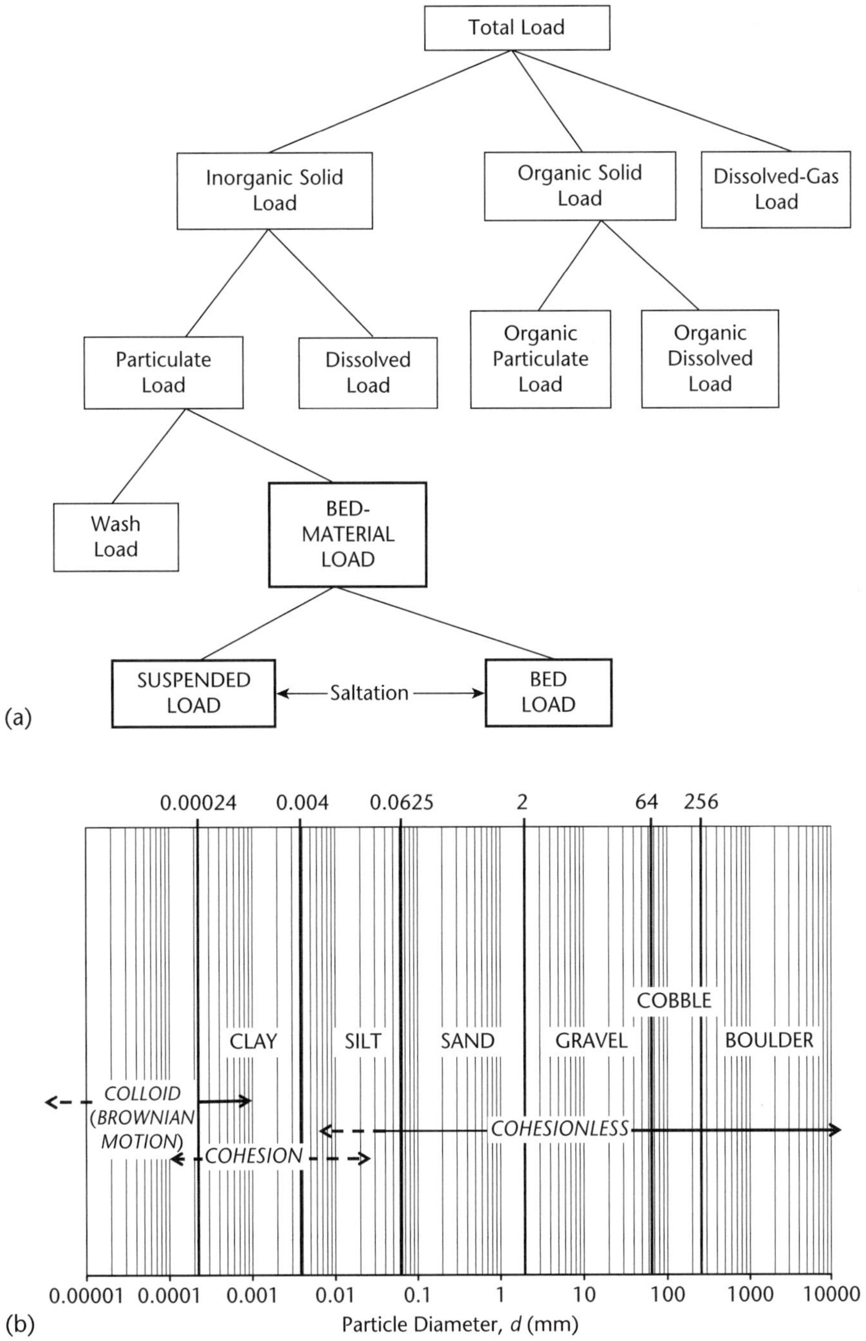

Figure 12.1 (a) Classification of solid loads transported by streams. See text for definitions. Only bed-material load depends on hydraulic conditions; this is the focus of this chapter. (b) Sediment-texture terms and behavior as related to particle diameter.

BOX 12.1 Sediment Concentration

For geomorphological computations, sediment concentration is expressed most usefully as **weight of sediment per unit volume of water**, C_{wv} [F L^{-3}]:

$$C_{wv} \equiv \frac{\text{weight of sediment}}{\text{volume of (water + sediment)}} \qquad (12B1.1)$$

The standard units for this quantity are milligrams per liter (mg/L), and

$$C_{\text{mg/L}} = 10^3 \cdot C_{wv}, \qquad (12B1.2)$$

when C_{wv} is in kg/m^3.

The **weight concentration**, C_{ww} [F F^{-1}], is

$$C_{ww} \equiv \frac{\text{weight of sediment}}{\text{weight of (water + sediment)}}; \qquad (12B1.3)$$

this is usually expressed as parts per million (ppm), C_{ppm}, and the relation between C_{ww} and C_{ppm} is

$$C_{ppm} = 10^6 \cdot C_{ww}. \qquad (12B1.4)$$

The **volumetric concentration**, C_{vv} [L^3 L^{-3}], is

$$C_{vv} \equiv \frac{\text{volume of sediment}}{\text{volume of (water + sediment)}}. \qquad (12B1.5)$$

The relation between C_{ww} and C_{vv} is

$$C_{ww} = \frac{C_{vv} \cdot G_S}{1 + (G_S - 1) \cdot C_{vv}}, \qquad (12B1.6a)$$

$$C_{vv} = \frac{C_{ww}}{G_S - (G_S - 1) \cdot C_{ww}}, \qquad (12B1.6b)$$

where G_S is the specific weight of sediment (ratio of the weight density of sediment to weight density of water), γ_S/γ. The standard value of G_S for natural sediments is taken as the value for quartz, $G_S = 2.65$. The relation between C_{vw} and C_{ww} is

$$C_{vw} = \frac{\gamma_S \cdot C_{ww}}{G_S - (G_S - 1) \cdot C_{ww}}. \qquad (12B1.7)$$

Table 12B1.1 gives equivalent concentrations assuming $G_S = 2.65$. Note that the discharge that is measured by the standard techniques described in section 2.5.3 and denoted Q throughout this text is actually the volume flow rate of water plus its contained sediment. Comparing the values in the $C_{\text{mg/L}}$ and C_{vv} columns, we see that the sediment makes up less than 5% of the total volume until the volumetric concentration exceeds 132,000 mg/L. Concentrations greater than $C_{vv} = 0.05$ are called **hyperconcentrations**.

Table 12B0.1

C_{vv}	C_{ww}	C_{ppm}	C_{wv}	$C_{mg/L}$
1.00×10^{-4}	2.65×10^{-4}	2.65×10^{2}	2.65×10^{-1}	2.65×10^{2}
2.00×10^{-4}	5.30×10^{-4}	5.30×10^{2}	5.30×10^{-1}	5.30×10^{2}
5.00×10^{-4}	1.32×10^{-3}	1.32×10^{3}	1.33×10^{0}	1.33×10^{2}
1.00×10^{-3}	2.65×10^{-3}	2.65×10^{3}	2.65×10^{0}	2.65×10^{3}
2.00×10^{-3}	5.28×10^{-3}	5.28×10^{3}	5.30×10^{0}	5.30×10^{3}
5.00×10^{-3}	1.31×10^{-2}	1.31×10^{4}	1.33×10^{1}	1.33×10^{4}
1.00×10^{-2}	2.61×10^{-2}	2.61×10^{4}	2.65×10^{1}	2.65×10^{4}
2.00×10^{-2}	5.13×10^{-2}	5.13×10^{4}	5.30×10^{1}	5.30×10^{4}
5.00×10^{-2}	1.22×10^{-1}	1.22×10^{5}	1.32×10^{2}	1.32×10^{5}
1.00×10^{-1}	2.27×10^{-1}	2.27×10^{5}	2.65×10^{2}	2.65×10^{5}

Particulate load is material present as discrete particles large enough to be unaffected by Brownian motion. Nominally, these are particles with diameters greater than about 10^{-4} mm (see figure 12.1b). This material typically consists largely of the mineral quartz (SiO_2), with varying proportions of other minerals depending on the geological setting. Particles of sand size and smaller (diameters <2 mm) are usually of a single mineral species; larger particles may often be rock particles consisting of several mineral types (with quartz typically dominant except in limestone terranes).

Wash load is the portion of particulate load that is not present in the channel bed and banks. Wash load is finer material (clay or fine silt) that is contributed to the stream by overland flow or from glaciers ("rock flour") and that remains in suspension for long periods even at very low flows. Because its presence is almost independent of hydraulic conditions, it is not included in the developments in this chapter.

Bed-material load is the portion of particulate load that is present in the channel bed and banks. This is the portion of sediment load that is subject to entrainment, transport, and deposition, depending on hydraulic conditions. Bed-material load consists of two components, which are the focus of this chapter: (a) **Bed load** is the portion of bed-material load that travels within a few grain diameters above the channel bed. Gravel-size particles move primarily by rolling, whereas sand generally moves as a sliding sheet a few grain diameters thick. (b) **Suspended load** is the portion of bed-material load that is lifted by turbulent eddies to travel within the flow at levels higher than a few grain-diameters above the bed.

In many situations, suspended sediment may include particles that are aggregations of microbes, organic material, and mineral sediments, called **flocs**. These particles undergo continuous changes that complicate the measurement and characterization

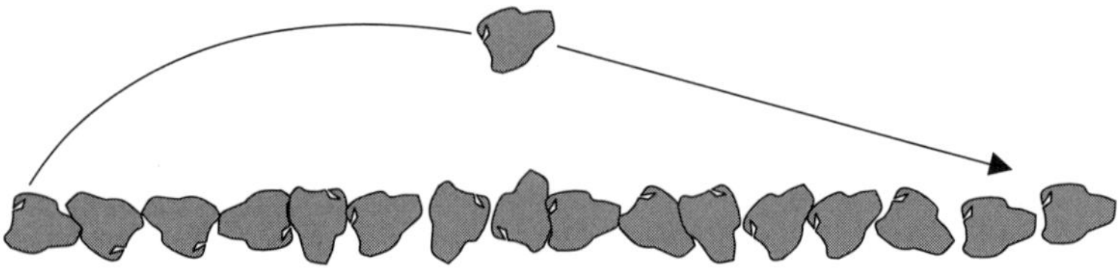

Figure 12.2 Grain saltation.

of suspended sediment, and are not well understood or accounted for in standard descriptions of sediment transport (Droppo 2001).

The bed-load and suspended-load components are generally discussed separately. However, sand-sized and fine-gravel-sized particles may sometimes travel as bed load and sometimes as suspended load as they are affected by the bursts and sweeps of turbulence described in section 3.3.4.1. This process is called **saltation** (figure 12.2).

12.1.2 Measurement

This section describes the basic techniques for measurement of particulate-sediment loads. More detailed discussion of suspended-load and bed-load sediment-sampling instruments and techniques is given by Edwards and Glysson (1999).

12.1.2.1 Bed Load

Obtaining a representative measurement of bed load is extremely difficult because any device placed on the stream bed may disturb the flow and the rate of bed-load movement, and because near-bed velocities and bed-load transport rates vary strongly in space and time.

The most commonly used sampler is the Helley-Smith bed-load sampler (figure 12.3). Its basic design consists of a frame with a 76-mm × 76-mm or 152-mm × 152-mm square entrance nozzle, to which is attached a removable mesh sample bag. The nozzle is designed so that the entrance velocity is equal to the ambient stream velocity. The sampler is made in various weights; the lightest models have a vertical rod and can be held in place by a wading observer, but heavier models for deeper and faster streams require use of a cable and winch. Care must be taken to remove the sampler before the bag is full, and obviously the device cannot capture particles larger than the diameter of the opening or smaller than the sample-bag mesh (typically 0.25 mm). Vericat et al. (2006) evaluated the Helley-Smith sampler and recommended using the larger sized opening to reduce sampling bias. If properly operated, this type of sampling can be highly efficient.

Another technique for measuring bed load is to construct pits or slots in the channel bed that collect the sediment. This type of sampler collects bed load with high efficiency, but the collected material must be continually or frequently removed

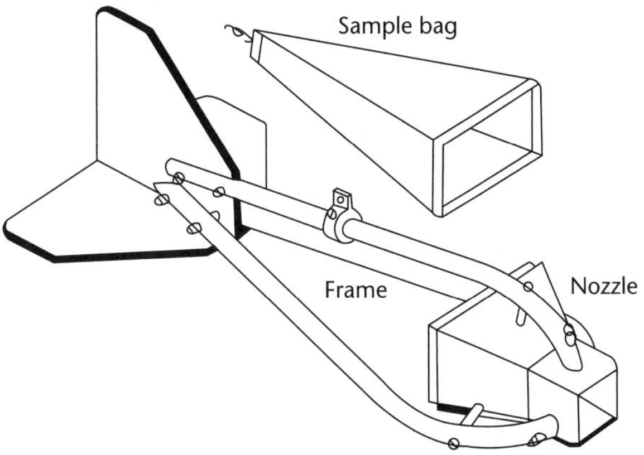

Figure 12.3 Helley-Smith bed-load sampler. From Edwards and Glysson (1999).

either by hand or via elaborate conveyor-belt devices. A promising new nonintrusive technique involves the use of acoustic Doppler current profilers (Gaeuman and Jacobson 2006).

12.1.2.2 Suspended Load

The basic procedure for measuring the concentration of suspended sediment in a flow is to 1) take a representative sample of the water-sediment mixture, 2) measure the volume of the mixture, 3) filter the mixture, 4) dry the solid material collected on the filter, and 5) weigh the dried solids. In practice, the suspended load is determined as the portion of load that is retained by a filter with openings of 0.45 μm (4.5×10^{-4} mm). The concentration is then determined as the weight of the solids divided by the volume of the mixture,[1] and is usually expressed in mg/L.

Since, as we will see below, suspended-sediment concentration generally varies with depth and with distance from the banks, taking a representative sample requires using a depth-integrating sampler (figure 12.4), which is lowered and raised through the flow at a constant rate, and taking such depth-integrated samples at several locations in a cross section. Depth-integrating samplers come in a variety of sizes and weights and may be suspended on a rod and operated by hand or suspended on a cable and raised and lowered by a crane and winch. The nozzle is designed to allow water and its contained sediment to enter without a change in speed or direction. The sample is collected in a removable bottle, and care must be taken to complete the downward and upward transit before the bottle fills.

It is important to be aware that such sampling generally underestimates the amount of material in suspension because 1) the construction of the sampler does not permit it to sample all the way to the bottom (the smallest commonly used sampler, the DH-48 model, leaves an 89-mm unsampled zone above the bed), and 2) the sample cannot collect material larger than the diameter of the intake nozzle (typically 6.35 mm). And, of course, the sample collected includes both wash load and bed-material load;

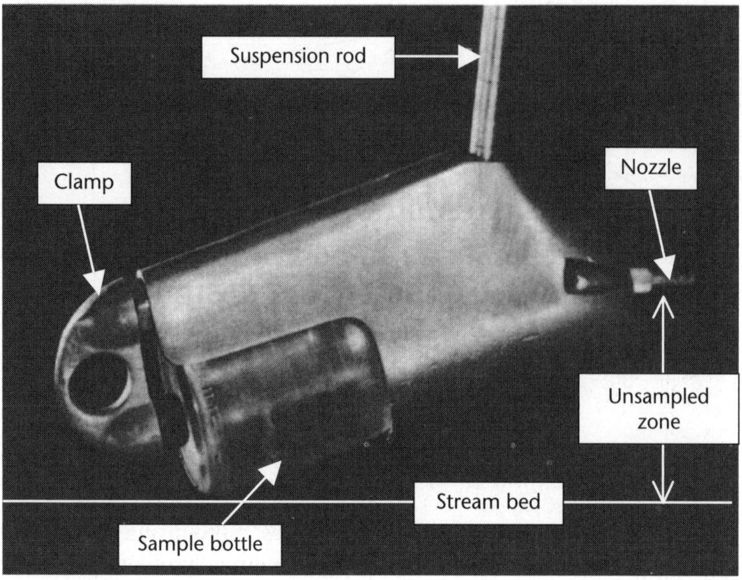

Figure 12.4 DH-48 type rod-suspended depth-integrating sediment sampler. The nozzle is 6.4 mm in diameter. The sample is collected in the glass bottle, which is held in the sampler by a spring-loaded clamp. The solid line indicates the stream bed; the unsampled zone is 89 mm deep. Photo from Guy and Norman (1970).

it may be possible to distinguish the two components by doing a grain-size analysis of the material collected on the filter.

Note that using the Helley-Smith sampler and a suspended-sediment sampler in combination usually leaves an unsampled zone in which the sediment concentration may be high (figure 12.5).

12.2 Sediment Transport and Geomorphological Concepts

Measurements of fluvial sediment load provide the principal source of information about two important geomorphological concepts. First, because rivers are the principal routes by which the products of continental erosion are delivered to the oceans, fluvial sediment loads are the primary sources of estimates of current rates of continental denudation. This information allows comparison of current and past denudation rates; estimation of the role of geology, topography, climate, and land use on erosion rates; and assessment of the relative importance of chemical versus physical erosion processes.

Second, study of the time distribution of fluvial sediment loads provides important insight into the relation between the magnitude of events that accomplish geomorphic work and the frequency with which those events occur. This insight provides

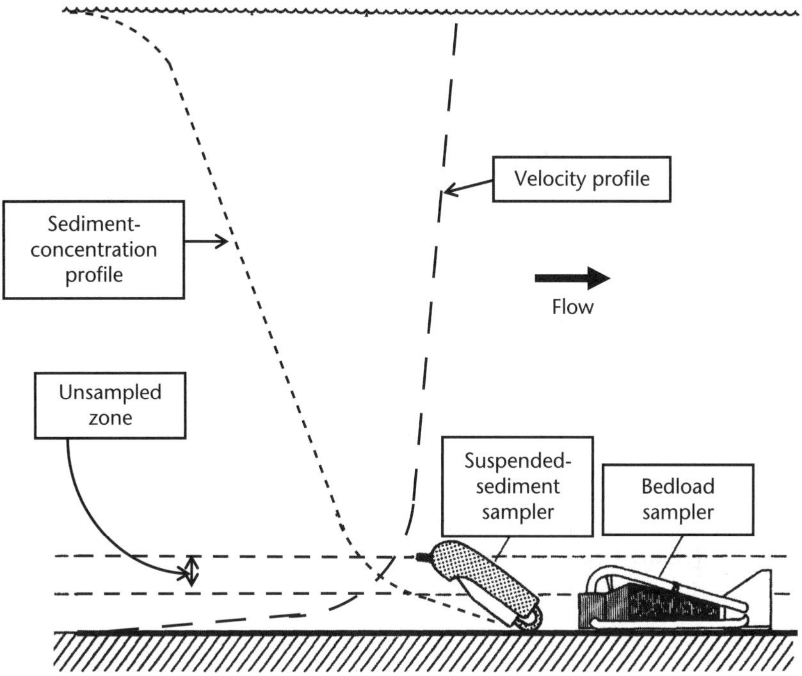

Figure 12.5 Deployment of sediment samplers. The lowest sampling elevation for the DH-48 suspended-sediment sampler is 89 mm above the bottom. The standard opening for the Helley-Shaw bed-load sampler is 76 mm. Using those two samplers, the unsampled zone is 13 mm deep.

perspective about the relative importance of rare catastrophic events versus more common lower intensity events in shaping the earth's land surface.

We begin by exploring the relations between sediment concentrations, loads, and discharge, because these are central to both concepts.

12.2.1 Empirical Concentration–Load–Discharge Relations

For suspended material, the instantaneous load is determined by sampling and measuring concentration as described in section 12.1.2, measuring the concurrent discharge, and using equation 12.1. Since suspended-sediment concentration in a given reach is virtually always a strong function of discharge, equation 12.1 can be written as

$$L_S(Q) = C_S(Q) \cdot Q. \tag{12.2}$$

The relation between suspended-sediment concentration and discharge can usually be well approximated by a power-law relation of the form

$$C_S(Q) = a_S \cdot Q^{bs}, \tag{12.3}$$

where C_S is suspended-sediment concentration, and a_S and b_S are reach-specific empirical values determined by logarithmic regression analysis (section 4.8.3.1); the exponent in such relations is almost always >1. In a comparative study of 59 drainage basins, Syvitski et al. (2000) found that a_s is inversely related to long-term average discharge and b_s is correlated with average air temperature and topographic relief. An example for the Boise River near Twin Springs, Idaho, is shown in figure 12.6a, for which the best-fit equation is

$$C_S = 6.51 \times 10^{-4} \cdot Q^{2.41}, \tag{12.4}$$

where C_S is in mg/L and Q is in m^3/s.

Because load is the product of concentration and discharge (equation 12.1), the suspended-load–discharge relation can also be represented by a power-law relation:

$$L_S = a_S \cdot Q^{b_S} \cdot Q = a_S \cdot Q^{b_S+1} = c_S \cdot Q^{d_S}, \tag{12.5}$$

where $d_S = b_S + 1$.[2] The suspended-load–discharge relation is often called the **suspended-sediment rating curve**; the curve for the Boise River site is shown in figure 12.6b, for which the best-fit relation is

$$L'_S = 5.51 \times 10^{-5} \cdot Q^{3.41}, \tag{12.6a}$$

where Q is in m^3/s and L'_S is the regression estimate of load in tons/day. However, as explained in box 12.2, the best-fit regression relation of equation 12.6a gives a low-biased estimate of the average load associated with a given discharge. Adjusting for this bias, the appropriate relation for estimating average suspended load at this site is

$$L_S = 6.14 \times 10^{-5} \cdot Q^{3.41}. \tag{12.6b}$$

As described in section 12.1.2.1, bed load is usually measured directly by means of a bed-load sampler. The bed-load–discharge relation (**bed-load rating curve**) can also usually be well approximated by an empirical power-law relation,

$$L_B = c_B \cdot Q^{d_B}, \tag{12.7}$$

with d_B typically >1. The best-fit bed-load relation for the Boise River site is shown in figure 12.7; its equation is

$$L'_B = 5.98 \times 10^{-4} \cdot Q^{2.55}, \tag{12.8a}$$

where L'_B is in tons/day. Again adjusting for bias (box 12.2), the appropriate prediction relation is

$$L_B = 7.03 \times 10^{-4} \cdot Q^{2.55}. \tag{12.8b}$$

Now we can combine equations 12.6b and 12.8b to estimate the total particulate load L associated with a given discharge at the Boise River site:

$$L = L_S + L_B = 6.14 \times 10^{-5} \cdot Q^{3.41} + 7.03 \times 10^{-4} \cdot Q^{2.55}. \tag{12.9}$$

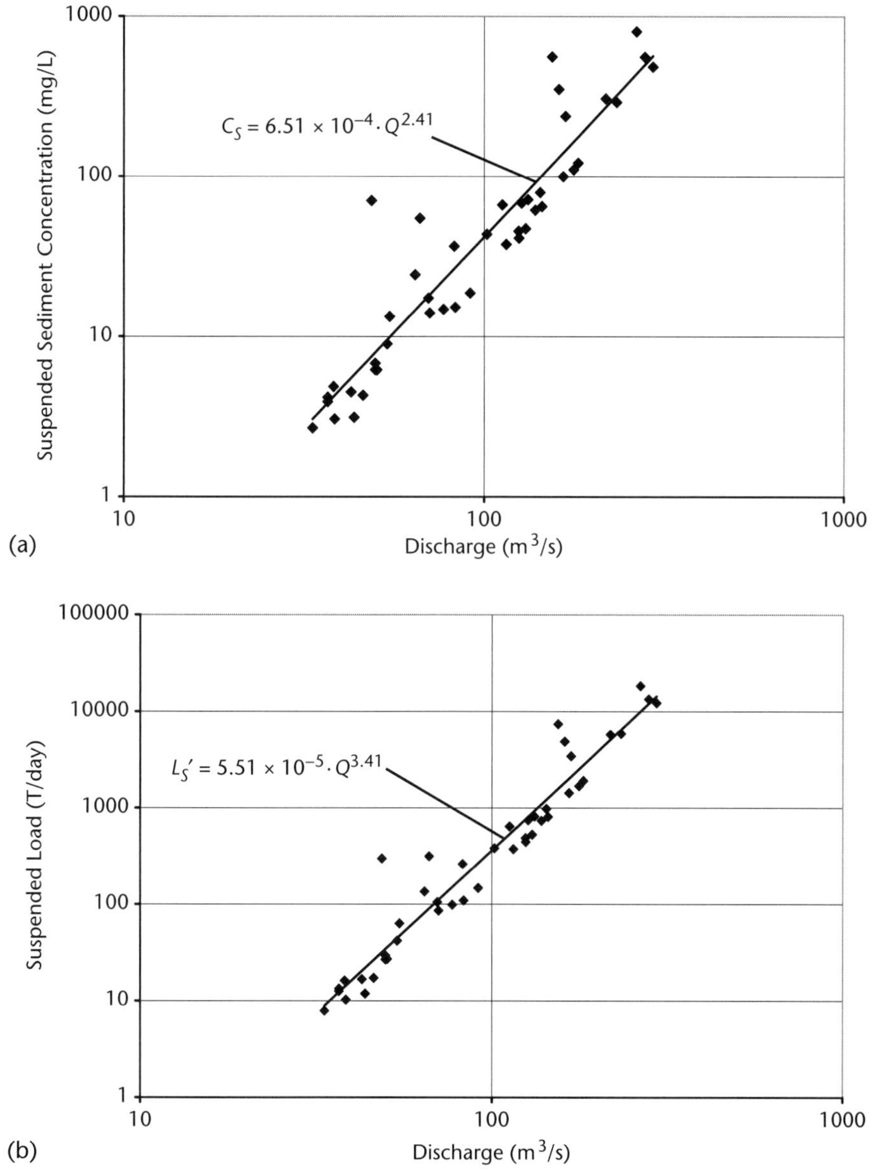

Figure 12.6 Suspended-sediment–discharge relations for the Boise River near Twin Springs, ID. (a) Concentration–discharge relation (equation 12.4). (b) Load–discharge relation (unadjusted suspended-sediment rating curve, equation 12.6a). Data from King et al. (2004).

BOX 12.2 Bias Adjustment for Sediment-Load Estimates

Our objective in using regression equations to relate sediment load to discharge via power-law equations is to estimate the long-term average sediment load at a reach. The power-law equations such as equations 12.3 and 12.7 are developed by regression analyses using the logarithms of both the predictor variable, discharge (Q), and the dependent variable, load (L) (section 4.8.3.1).

As explained by Helsel and Hirsch (1992, pp. 256–260), when a regression procedure is carried out with logarithms, the resulting equation provides an estimate of the mean of the logarithm of the dependent variable (load) associated with a given value of the logarithm of the predictor variable (discharge). However, when retransformed to the original values (e.g., equations 12.6a and 12.8a), this estimate is always less than the mean of the retransformed values. Because we want the best estimate of the long-term average of the load values, we must adjust the low-biased estimates provided by the regression procedure.

Helsel and Hirsch (1992) recommend the following procedure for developing unbiased load estimates. First, transform the N measured load and discharge values to logarithms (base 10 is assumed here), complete a standard regression analysis using these logarithms as outlined in section 4.8.3.1, and retransform the log-regression relation to power-law form as

$$L_i' \equiv c \cdot Q_i{}^d, \qquad (12B2.1)$$

where L_i' is the biased estimate of suspended or bed load associated with the ith discharge value, and $c = c_S$ or c_B, $d = d_S$ or d_B of equations 12.3 and 12.7. Now compute the **bias-adjustment factor** $\mathcal{B}$ as

$$\mathcal{B} = \frac{1}{N} \cdot \sum_{i=1}^{N} 10^{e_i}, \qquad (12B2.2)$$

where

$$e_i \equiv \log_{10}(L_i') - \log_{10}(L_i), \qquad (12B2.3)$$

where L_i is the corresponding measured load value. This bias-adjustment factor $\mathcal{B} > 1$ is then multiplied by each estimate given by the original regression equation to give a new unbiased estimate L_i of the load associated with a given discharge:

$$L_i \equiv \mathcal{B} \cdot c \cdot Q_i{}^d \qquad (12B2.4)$$

For the Boise River data discussed in the text and shown in figures 12.6 and 12.7, the original biased regression equations, bias-adjustment factors, and adjusted load-prediction equations are in table 12B2.1.

Table 12B1.2

Original biased regression equation	Bias-adjustment factor, $\mathcal{B}$	Unbiased load-prediction equation
$L'_S = 5.51 \times 10^{-5} \cdot Q^{3.41}$	1.114	$L_S = 6.14 \times 10^{-5} \cdot Q^{3.41}$
$L'_B = 5.98 \times 10^{-4} \cdot Q^{2.55}$	1.176	$L_B = 7.03 \times 10^{-4} \cdot Q^{2.55}$

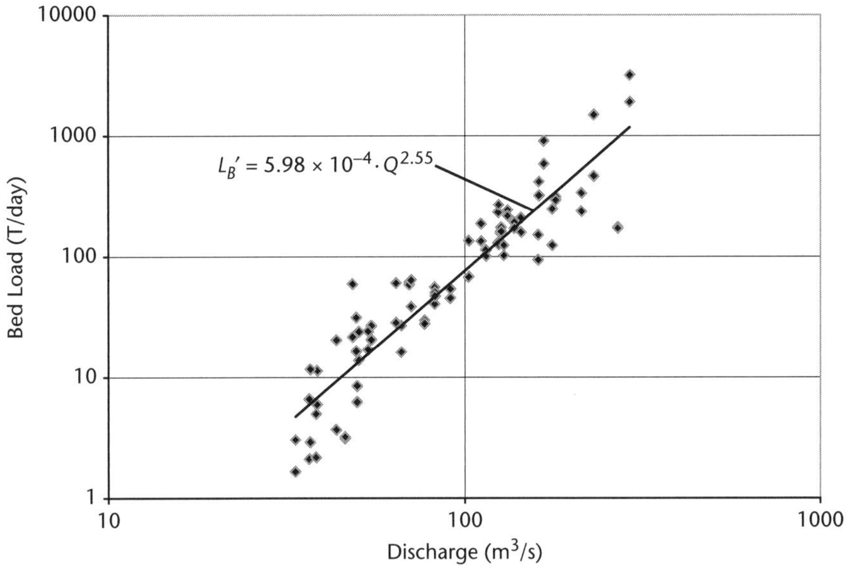

Figure 12.7 Bed-load–discharge relation for the Boise River near Twin Springs, Idaho (unadjusted bed-load rating curve, equation 12.8a). Data from King et al. (2004).

A graph of this relation is shown in figure 12.8.[3]

Sediment-rating curves such as those in figures 12.6–12.8 typically show a great deal of scatter that may be due to a number of causes:

1. Watershed susceptibility to erosion may vary seasonally. For example, rain-on-snow events and heavy rains in late summer when vegetation is well established tend to produce less sediment than do spring rains.
2. In smaller watersheds particularly, the peak sediment discharge tends to precede the peak water discharge because sediment sources are close to the stream

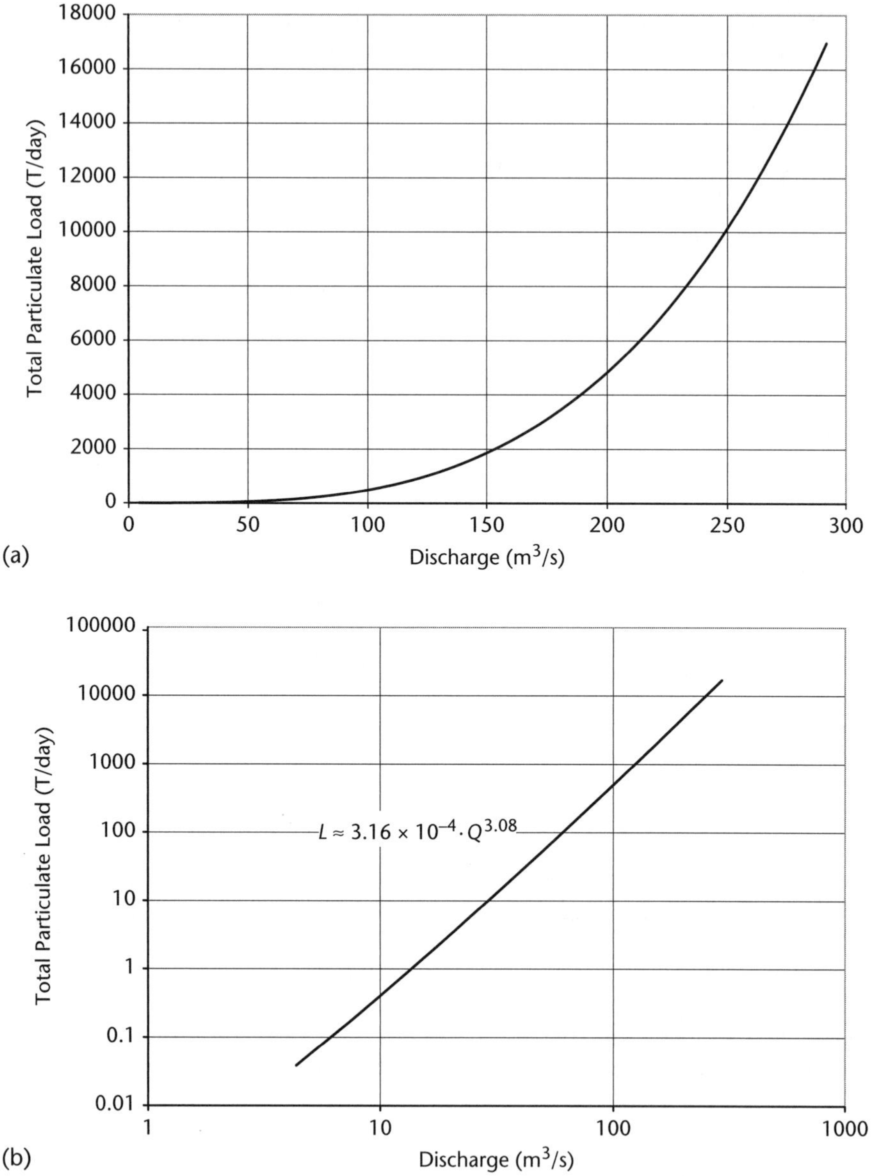

Figure 12.8 Total particulate load as a function of discharge for the Boise River at Twin Springs, Idaho. The curve is given by equation 12.9. (a) Arithmetic plot. (b) Log-log plot. Note that the sum of the power-law relations for suspended load (equation 12.6b) and bed load (equation 12.8b) is very close to a power-law relation: $L = 3.16 \times 10^{-4} \cdot Q^{3.08}$. Data from King et al. (2004).

and the readily available sediment is removed early in a storm. In this case, the sediment-rating curve is **looped**: at a given discharge, the sediment concentration is higher when the hydrograph is rising ($\partial Q/\partial t > 0$) than when it is receding ($\partial Q/\partial t < 0$).

3. The flood wave usually travels faster than the sediment-laden water itself (section 11.5), and the difference becomes more pronounced as travel time increases. In larger watersheds, this may produce a looped rating curve in which the sediment concentration is lower when the hydrograph is rising and higher when it is receding.
4. In larger watersheds, storms confined to more erodible or less erodible tributary watersheds will result in different sediment concentrations at a given discharge at downstream measurement sites.
5. Secular changes in watershed land use (deforestation, reforestation, construction activities) may contribute to the scatter.
6. Secular climate changes that affect watershed vegetation and/or changes in the amount or intensity of precipitation may contribute to the scatter.

Identifying the season and year of measurement for each point plotted may help explain at least some of the scatter in a given sediment rating curve.

The following sections will show how empirical load–discharge relations such as equation 12.9 are used to 1) estimate sediment yields and denudation rates and 2) to reveal relations between the magnitudes of geomorphic work done by events of various frequencies of occurrence.

12.2.2 Sediment Yield and Denudation Rate

As noted above, rivers carry the products of both chemical and physical denudation processes as dissolved and particulate loads, respectively. Both processes contribute significantly to the lowering of the earth's continental surfaces (table 12.1), but here we focus only on the particulate portion of total sediment load. In addition to its reflection of denudation by physical processes, long-term average particulate loads are of great interest in forecasting the sedimentation of reservoirs as well as the development of deltas and other geological processes.

As described in the preceding section, the total particulate load L at a reach is the sum of the component loads and can generally be modeled as the sum of power-law functions of discharge for suspended and bed load as in equation 12.9:

$$L(Q) = c_S \cdot Q^{d_S} + c_B \cdot Q^{d_B} \qquad (12.10)$$

Discharge, of course, varies strongly over time as reflected in the flow-duration curve (section 2.5.6.2). Thus, conceptually, the long-term average particulate sediment load at a particular cross section, $\bar{L}$, is given by

$$\bar{L} = \frac{1}{T} \cdot \int_T [\, c_S \cdot Q(t)^{d_S} + c_B \cdot Q(t)^{d_B}] \cdot \mathrm{d}t, \qquad (12.11)$$

Table 12.1 Sediment load, sediment yield, and denudation rate values for the continents and some of their major rivers.

Continent/river	Area (10^3 km^2)	Discharge (10^3 m^3/s)	Particulate load (10^3 T/day)	Dissolved load (10^3 T/day)	Particulate yield (T/year km^2)	Dissolved yield (T/year km^2)	Total yield (T/year km^2)	Physical denudation rate (mm/10^3 year)	Total denudation rate (mm/10^3 year)	% Particulate
Africa	20,100	108	1,452	551	18	7	24	6.5	9.0	73
Congo	3,700	41.1	132	101	13	10	23	4.8	8.5	56
Niger	2,240	4.9	68	38	11	6	17	4.1	6.4	64
Nile	3,830	1.2	5	33	1	3	4	0.2	1.4	14
Orange	940	0.4	2	4	1	2	2	0.3	0.9	30
Zambezi	1,990	2.4	55	68	10	13	23	3.7	8.4	44
Asia	44,400	387	17,630	4,360	145	36	181	53.7	66.9	80
Ganges[a]	1,630	30.8	4,580	414	1,026	93	1,119	380	414	92
Huang Ho	890	1.1	2,466	60	1,007	25	1,031	373	382	98
Indus	1,140	7.5	274	216	87	69	157	32.4	58.0	56
Irrawaddy	430	13.6	726	249	616	212	828	228	307	74
Lena	2,420	16.0	33	153	5	23	28	1.8	10.4	18
Mekong	770	21.1	438	162	207	76	283	76.6	105	73
Ob	2,570	13.7	36	126	5	18	23	1.9	8.5	22
Yenesei	2,580	17.6	41	164	6	23	29	2.2	10.8	20
Australia[b]	7,800	6.3	8,390	803	393	38	430	145	159	91
Murray[c]	1,030	0.7	82	22	29	8	37	10.8	13.7	79
Europe	10,100	88.8	630	1,164	23	42	65	8.4	24.0	35
Danube	790	6.5	227	145	105	67	173	39.0	63.9	61
Dnieper	510	1.7	6	30	4	22	26	1.5	9.2	16

(Continued)

Table 12.1 *Continued*

Continent/river	Area (10^3 km^2)	Discharge (10^3 m^3/s)	Particulate load (10^3 T/day)	Dissolved load (10^3 T/day)	Particulate yield (T/year km^2)	Dissolved yield (T/year km^2)	Total yield (T/year km^2)	Physical denudation rate (mm/10^3 year)	Total denudation rate (mm/10^3 year)	% Particulate
Rhône	99	1.9	110	153	404	566	970	150	359	42
Volga	1,460	7.7	74	148	18	37	55	6.8	20.5	33
N. America	24,100	187.1	4,005	2,077	61	31	92	22.5	34.1	66
Columbia	720	5.8	38	58	19	29	48	7.2	17.9	40
Mackenzie	1,710	7.9	274	121	58	26	84	21.6	31.1	69
Mississippi	3,200	18.4	575	389	66	44	110	24.3	40.7	60
St. Lawrence	1,270	13.1	14	2,110	4	608	612	1.5	227	1
Yukon	850	6.7	164	93	70	40	110	26.1	40.9	64
S. America	17,900	352.0	4,899	1,652	100	34	134	37.0	49.5	75
Amazon	5,850	200.0	2,466	795	154	50	203	56.9	75.3	76
Magdalena	260	6.8	603	55	846	77	923	313	342	92
Orinoco	1,040	36.0	411	85	152	30	174	56.1	67.7	83
Paraná	2,660	15.0	219	104	30	14	44	11.1	16.4	68

[a] Includes Brahmaputra.

[b] Area and discharge do not include values for New Zealand and other large Pacific Islands, but load, yield, and denudation rates do. Sediment values given for the Murray River are more typical of Australia; the much higher values shown for Australia are strongly influenced by very high erosion rates of New Zealand, New Guinea, and other mountainous islands in the region.

[c] Includes Darling River.

Data from Knighton (1998), Dingman (2002), and Vörösmarty et al. (2000a).

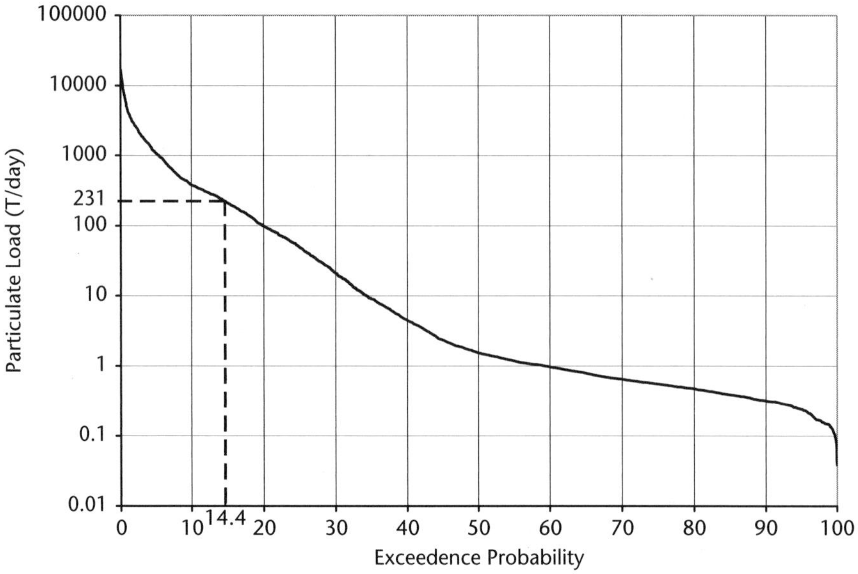

Figure 12.9 The particulate-load duration curve for the Boise River at Twin Springs, Idaho. Note the logarithmic scale for load. The curve was computed by selecting discharges over the range of measured flows, computing the load corresponding to each discharge via equation 12.9, and plotting that load against the exceedence probability associated with the discharge. (The exceedence probability for discharge at this site is shown in the flow-duration curve plotted as figure 2.36.) The long-term average value of particulate load (231 tons/day; exceedence probability 14.4%; shown by the dashed line) is found by numerical integration of this curve.

where the averaging period T is long enough to include the entire range of flows. In practice, $\bar{L}$ is found by constructing and integrating the sediment-load duration curve as described in box 2.5:

$$\bar{L} = \int_0^1 L[Q(EP)] \cdot \mathrm{d}EP, \qquad (12.12)$$

where EP is exceedence probability.

Figure 12.9 shows the **particulate-load duration curve** for the Boise River site, constructed by applying equation 12.9 for discharge values over the range of flows at the site and plotting the computed load against the exceedence probability of each discharge. (The flow-duration curve for the site is shown in figure 2.36.) Approximating equation 12.12 numerically, we find the long-term average particulate load for this site $\bar{L} = 231$ tons/day.

To compare the long-term average loads (and hence physical erosion rates) for different drainage basins, we calculate the **sediment yield**, $\mathcal{Y}$ (weight of sediment per unit drainage area A_D and unit time; [F L^{-2} T^{-1}]):

$$\mathcal{Y} = \frac{\bar{L}}{A_D}. \tag{12.13}$$

Sediment-yield values are typically calculated to compare the effects of geology, topography, climate, and land-use practices on sediment production and are usually expressed in units of tons/year · km^2. The drainage area of the Boise River at Twin Springs, Idaho, is 2,150 km^2, so its particulate sediment yield is (231 tons/day × 365 day/year)/2,150 km^2 = 39.2 tons/year · km^2. This value can be compared with data for the continents and some of the major rivers of the world in table 12.1.

Total sediment yield, in turn, can be used to calculate the **denudation rate**, $\mathcal{D}$ (rate of lowering of the land surface; [L T^{-1}]) as

$$\mathcal{D} = \frac{\mathcal{Y}}{k_{\mathcal{Y}} \cdot \gamma_S}, \tag{12.14a}$$

where γ_S is the density of the eroded material and $k_{\mathcal{Y}}$ is a proportionality constant that reflects adjustments that may be required to account for nonequilibrium conditions of soil formation, storage of sediment on the watershed, and other factors. For particulate load where soil formation rates are in equilibrium with denudation, $k_{\mathcal{Y}} \approx 1$, and it is customary to assume a value of $\gamma_S = 2700$ kg m^{-3} for average continental rocks (Summerfield 1991, p. 382).[4] The customary units for $\mathcal{D}$ are mm/1,000 year; using these units, equation 12.14a becomes

$$\mathcal{D} = 0.370 \cdot \mathcal{Y}, \tag{12.14b}$$

where $\mathcal{Y}$ is in tons/year · km^2 and $k_{\mathcal{Y}} = 1$. Thus, for the Boise River, we find a physical denudation rate $\mathcal{D} = 0.370 \times 39.2$ tons/year · km^2 = 14.5 mm/1,000 year, which can be compared to global values in table 12.1.

12.2.3 Magnitude–Frequency Relations

The debate as to whether landscape evolution (denudation) occurs principally as a result of rare catastrophic erosional events or as the cumulative effect of smaller events that operate more or less continuously is a long-standing issue in earth sciences. This debate was cogently and quantitatively addressed in a seminal paper by Wolman and Miller (1960) that relies strongly on sediment-load as the measure of geomorphic work. Here we follow their approach using the sediment-load data for the Boise River developed in sections 12.2.1 and 12.2.2. First, note from figure 12.8 that sediment loads at the highest discharges are several orders of magnitude larger than for the lower discharges. However, we see from figure 12.9 that the highest loads occur relatively rarely (e.g., loads greater than 100 tons/day occur only 20% of the time).

The contribution of flows in various ranges to the long-term transport of particulate sediment is essentially equal to the product of the load carried by flows in each range times the frequency with which flows in each range occur.

The magnitude-frequency computations can be followed in table 12.2. We first divide the total range of discharges measured at the Boise River site into five equal ranges, designated $q = 1, 2, 3, 4, 5$. The upper limits of these ranges are designated $Q_{\max q}$ and shown in column 2 of the table. The exceedence probabilities associated with these flows (column 3) are determined from the flow-duration curve for the site, plotted in figure 2.36. The average flows $\bar{Q}_q$ in column 4 are the midpoints of the flow ranges, and the time percentages $f_T(\bar{Q}_q)$ of column 5 are the portions of time that each average flow prevails, equal to the differences between the exceedence frequencies of the flows that define each range. The average sediment load $\bar{L}(Q_q)$ associated with each range (column 6) is computed by inserting $\bar{Q}_q$ in equation 12.9. The cumulative load contributed by flows in each range (column 7) is then determined by multiplying $\bar{L}(Q_q)$ by the fraction of time it applies, $f(\bar{Q}_q)$ (we transform the tons/day units by multiplying by 365 to give tons/year); these values are plotted in figure 12.10. Finally, the fraction of the total load that is transported by flows in each flow range is shown in column 8.

Table 12.2 shows that the flows in the midrange, which occur with moderate frequency and carry substantial loads, accomplish the most geomorphic work over time. Flows in the lower ranges occur very frequently (>90% of the time; column 5) but carry relatively small loads and contribute only about one-quarter of the total transport. Flows in the highest ranges carry very high loads but occur so rarely that their net contribution is less than that of midrange flows. The results of this computation are quite typical (although not universal); similar results have been

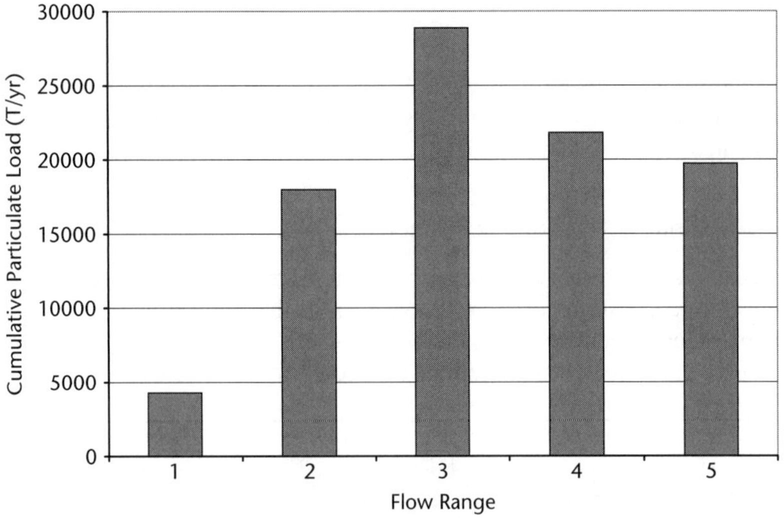

Figure 12.10 Cumulative particulate loads contributed by flows in five ranges for the Boise River near Twin Springs, Idaho. These are the values computed in column 7 of table 12.2.

Table 12.2 Computation of cumulative sediment-transporting work done by flows of various magnitudes for the Boise River at Twin Springs, Idaho.[a]

(1) Discharge range, q	(2) Maximum discharge, $Q_{\mathrm{max}q}$ (m^3/s)	(3) Exceedence probability, EP($Q_{\mathrm{max}q}$) (%)	(4) Average discharge, $\bar{Q}_q$ (m^3/s)	(5) Fraction of time, f_T ($\bar{Q}_q$) (%)/(day/year)	(6) Sediment load, $L(\bar{Q}_q)$ (T/day)	(7) Total transport, $f_T(\bar{Q}_q) \cdot L(\bar{Q}_q) \cdot 365$ (T/year)	(8) Fraction of transport
1	61.8	19.78	33.1	80.22/293	14.6	4,270	4.6
2	119	6.01	90.5	13.77/50	358	18,000	19.4
3	177	1.58	148	4.43/16	1,790	28,900	31.2
4	234	0.45	205	1.13/4	5,280	21,800	23.5
5	292	0	263	0.45/2	12,000	19,700	21.3

[a]The range of average daily discharges recorded at the site is 4.33–292 m^3/s. The discharge data and exceedence probabilities (columns 1–5) are from the flow-duration curve of figure 2.35. $L(\bar{Q}_q)$ (column 6) is computed via equation 12.11.

found for particulate and dissolved loads of streams in various climatic and geological settings (e.g., Torizzo and Pitlick 2004), as well as for erosion by raindrops and ocean waves. Thus, we conclude that, generally, events of moderate size and moderate frequency account for the largest proportion of sediment transport (geomorphic work) over time.

12.3 Forces on Sediment Particles

12.3.1 Relative Motion of a Sphere in a Fluid

Understanding the relative motion of a sphere in a fluid provides basic insight into the forces on particles on the stream bed that cause sediment entrainment and the forces that affect the settling of entrained sediment particles. It is the balance of these forces that determines the size of particle that can be entrained (the **competence** of the flow) and the particulate load that can be carried in suspension (the **capacity** of the flow).

Figure 12.11 shows a sphere moving slowly (we will define "slowly" more precisely shortly) through a fluid—or a fluid moving slowly around a sphere. Because of the no-slip condition, the relative motion causes a velocity gradient in the fluid near the particle (figure 12.11a). This in turn produces a viscous drag force on and parallel to the entire surface of the particle (as in figures 3.15 and 5.3), distributed as shown in figure 12.11c. The relative motion also produces a dynamic pressure force[5] that acts normal to the surface, called the **pressure drag** or **form drag**, distributed as in figure 12.11b. The vector sum of these two forces, integrated over the surface of the sphere, is the **drag force** that the fluid exerts on the particle (and the particle on the fluid).

- If the particle is resting on the bed, the drag force is the force exerted by the flow that tends to move the particle downstream and upward, which is opposed by the weight of the particle as described more fully later.
- If the particle is settling through the fluid at constant velocity, the drag force is balanced by and equal to the submerged weight of the particle.

Thus, understanding how drag force varies with flow conditions is central to understanding sediment movement.

We get the essential insight by conducting a dimensional analysis of the problem, which is carried out in box 12.3. That analysis identifies two dimensionless variables pertinent to the problem:

1. The **drag coefficient**, C_D, defined as

$$C_D \equiv \frac{F_D}{\rho \cdot (U^2/2) \cdot A_S} = \frac{8 \cdot F_D}{\pi \cdot \rho \cdot U^2 \cdot d^2}, \tag{12.15}$$

   where F_D is the drag force on the sphere, ρ is mass density of the fluid, U is the relative velocity of the sphere and fluid, d is the particle diameter, and A_S is the cross-sectional area of the sphere ($= \pi \cdot d^2/4$).

2. The **particle Reynolds number,** defined as

$$Re_p \equiv \frac{\rho \cdot U \cdot d}{\mu}, \tag{12.16}$$

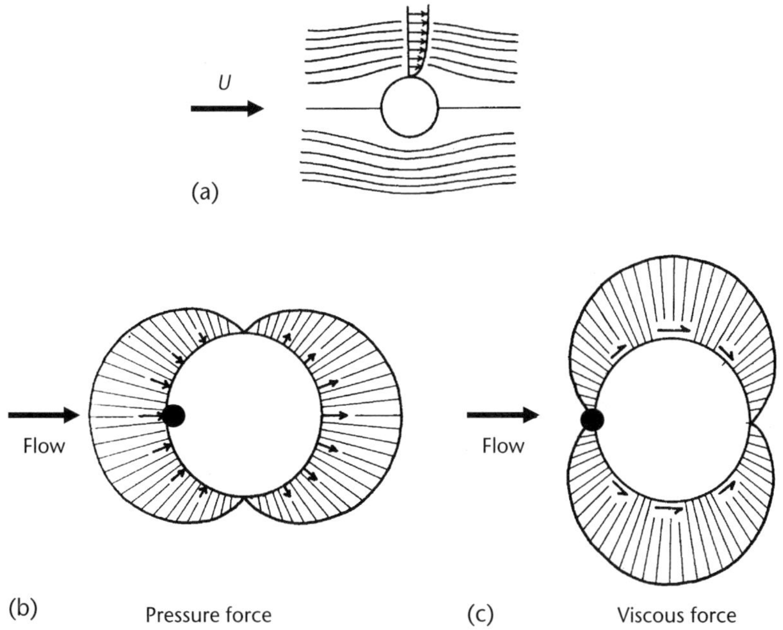

Figure 12.11 Forces on a spherical particle undergoing "slow" (i.e., $Re_p < 1$) relative motion in a fluid. (a) Stream lines and the velocity gradient and boundary layer induced by the no-slip condition shown at one location. U is the "free-stream" relative velocity, that is, the velocity beyond the boundary layer. (b) Distribution of pressure force over the sphere. Pressure is maximum at the stagnation point (black dot) and zero at the "top" and "bottom" of the sphere; a downstream-directed pressure gradient is induced. c) Distribution of viscous force over the sphere. This force is at its maximum at the "top" and "bottom" of the sphere where the induced velocity gradient is strongest, and zero at and opposite the stagnation point, where the velocity is zero. After Middleton and Southard (1984).

where μ is the dynamic viscosity. As we saw in section 3.4.2, the Reynolds number reflects the ratio of turbulent resistance to viscous resistance in a flow.

Now we can conduct experiments to determine the relation between C_D and Re_p. Note that, because the dimensional analysis is universal, we can do the experiments with different fluids with different densities and viscosities (e.g., water, air, molasses) and particles of varying size and densities that can either be suspended in the flowing fluid or allowed to settle through the stationary fluid. In the latter case, as noted above, the drag force is equal to the immersed weight of the spherical particle, F_G:

$$F_D = F_G = \frac{\pi}{6} \cdot (\rho_s - \rho) \cdot g \cdot d^3 = \frac{\pi}{6} \cdot (\gamma_s - \gamma) \cdot d^3, \tag{12.17}$$

where g is gravitational acceleration and ρ_s, γ_s and ρ, γ are the mass and weight densities of the particle and the fluid, respectively.

Figure 12.12 shows the results of such experiments. The curve reflects the nature of the boundary layer produced by the relative motion as described in table 12.3

BOX 12.3 Dimensional Analysis of Relative Motion of a Sphere in a Fluid

The general procedure for dimensional analysis is described in box 4.1. The first step is to identify all the variables involved. In the situation depicted in figure 12.11, one of these is the drag force, F_D. As we saw in section 3.3.3.3, the shear force exerted by the velocity gradient depends on the velocity of the particle relative to the fluid, U, and on viscosity, μ. The fluid density, ρ, is also important because it determines the forces associated with the fluid accelerations. Finally, the only geometric variable is the particle diameter, d. Following the steps in box 4.1, we identify the dimensions of these variables and assign them to one of the following categories:

Geometric: Particle diameter, d [L]

Kinematic/dynamic: Drag force, F_D [M L T^{-2}]; relative velocity of fluid and particle, U [L T^{-1}].

Fluid properties: mass density, ρ [M L^{-3}]; viscosity, μ [M L^{-1} T^{-1}]

Thus, we have five variables and three dimensions, so we can form two dimensionless variables with three common variables. As indicated in box 4.1, we select one common variable from each of the three categories: d, U, and ρ. Then, following through the remaining steps of box 4.1, we identify the dimensionless variables as

$$\Pi_1 = \frac{F_D}{\rho \cdot U^2 \cdot d^2} \tag{12B3.1}$$

and

$$\Pi_2 = \frac{\rho \cdot U \cdot d}{\mu}. \tag{12B3.2}$$

Conventionally, Π_1 is written in a slightly different form, which is called a **drag coefficient**, C_D:

$$C_D \equiv \frac{F_D}{\rho \cdot (U^2/2) \cdot A_S}, \tag{12B3.3}$$

where A_S is the cross-sectional area of the sphere $= \pi \cdot d^2/4$. Note that this is still dimensionless, contains the same variables as Π_1, and differs numerically from it by the factor $8/\pi$. There are two reasons for the modified form of Π_1: 1) The drag coefficient is used to characterize objects of any shape (e.g., automobiles), and it is more general to use the cross-sectional area of the object, measured perpendicularly to the flow direction, than the diameter; and 2) $\rho \cdot U^2/2$ is the dynamic pressure force at the stagnation point (black dot on figure 12.11).

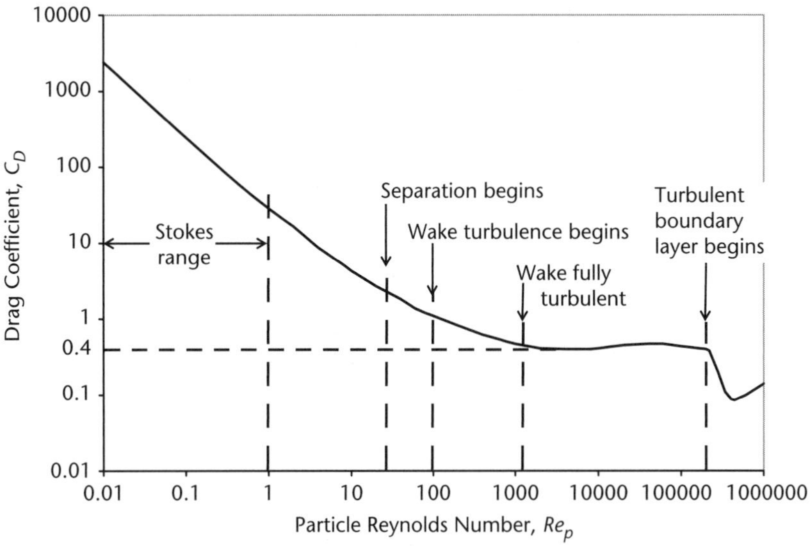

Figure 12.12 Drag coefficient, C_D, as a function of particle Reynolds number, Re_p, for spheres. The curve was determined by experimental results, some of which involved settling of spheres in a still fluid, and others, flow past a sphere at rest. See table 12.3 and figure 12.13 for explanation of phenomena involved in different ranges of Re_p. After Middleton and Southard (1984).

and illustrated in figure 12.13. The following two sections show how these phenomena are involved in determining the forces on particles settling in the fluid and on particles on the bed.[6]

12.3.2 Particles Settling in a Fluid: Fall Velocity

A body falling in a vacuum continuously accelerates at the gravitational acceleration rate, g. As we have just seen, a body falling through a fluid is subject to pressure forces and viscous forces that oppose its motion. These forces increase with increasing velocity, so the body eventually reaches a velocity at which the opposing forces just balance the force due to gravity, after which it descends at a constant **terminal velocity.** In water, this velocity is reached very quickly, and the brief period of acceleration can be ignored for purposes of analysis. Thus, the **fall velocity** of a particle is its terminal settling velocity.

In 1851, the English physicist G.G. Stokes (1819–1903) derived the expression for the total drag force on a sphere at very low particle Reynolds numbers ($Re_p < 1$) by integrating the viscous and pressure force distributions shown in figure 12.11 over the entire sphere surface to give

$$F_D = 3 \cdot \pi \cdot \mu \cdot U \cdot d. \tag{12.18}$$

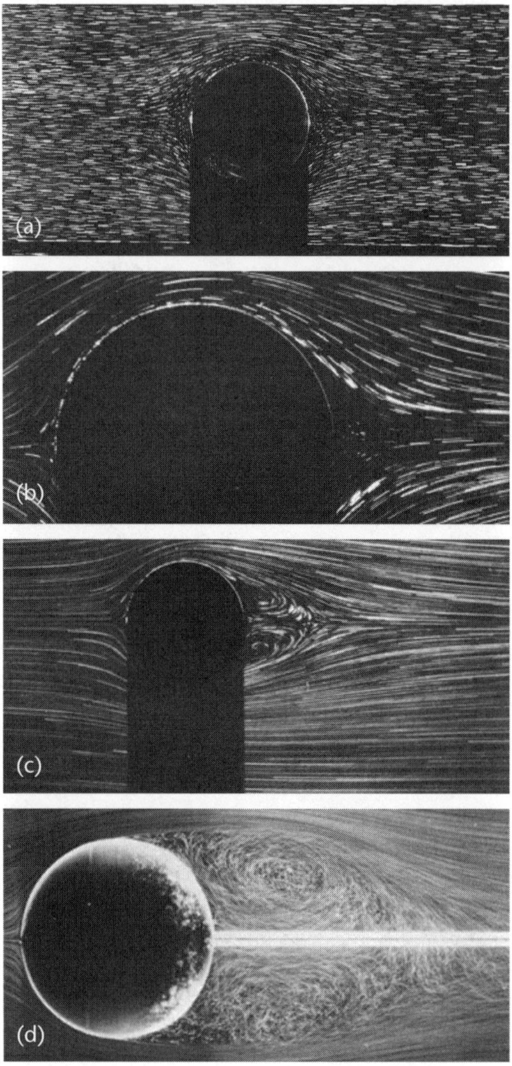

Figure 12.13 Flow around spheres at increasing particle Reynolds number, Re_p (see table 12.3). Flow is from left to right. In photos a–c, flow patterns are visualized by time exposures of tracer particles illuminated from above; the sphere casts a shadow below. Laminar flow exists where tracer lines are quasi parallel. (a) $Re_p = 0.10$: Stokes flow (creeping motion); flow pattern is symmetrical. (b) $Re_p = 9.8$: flow is still attached (no separation), but flow lines are distinctly asymmetrical. (c) $Re_p = 56.5$: Separation has occurred at an angle of about 145°, but flow pattern in wake is regular (turbulence is not present). (d) $Re_p = 15{,}000$: separation occurs at an angle of about 80°; the wake is turbulent. (e) Laminar (upper) and turbulent boundary layers. The laminar boundary layer separates near the crest of the sphere, but when $Re_p \approx 2 \times 10^5$, the boundary layer becomes turbulent and separates farther back, reducing drag. All photos reproduced from Van Dyke (1982); panels a–c reproduced with permission of L'Académie des Sciences Française; panel d reproduced with permission of ONERA, the French Aerospace Laboratory; panel e, original photo by M. R. Head (1980). (*Continued*)

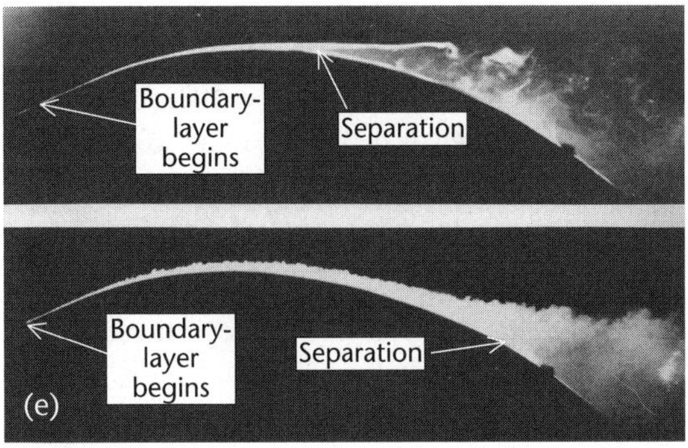

Figure 12.13 *Continued*

Table 12.3 Phenomena responsible for relation between drag coefficient, C_D, and particle Reynolds number, Re_p (see figures 12.12 and 12.13).

Re_p	Phenomena	d (mm)[a]
<1	Stokes flow or creeping flow. Streamlines are symmetrical (figures 12.11, 12.13a). Pressure force is one-third of total drag force; viscous force is two-thirds. In this range, $C_D = 24/Re_p$, so $F_D = (\pi/3) \cdot \mu \cdot U \cdot d$.	<0.05
10	Streamlines become increasingly asymmetrical but remain attached (figure 12.13b). Pressure force becomes increasingly important.	0.25
24	Separation begins: Laminar wake forms and becomes larger as the point of separation moves forward with increasing Re_p (figure 12.13c). Pressure force exceeds viscous force and becomes more important with increasing Re_p.	0.38
100	Turbulence forms in wake and grows as Re_p increases.	0.80
1,000	Entire wake is turbulent, with separation occurring at an angle of 80° (i.e., 10° forward of midpoint) (figure 12.13d). Flow pattern changes little with increasing Re_p, and C_D remains essentially constant at ≈ 0.4, so $F_D = (\pi/20) \cdot \rho \cdot U^2 \cdot d^2$.	4
2×10^5	Boundary layer becomes turbulent and separation point moves suddenly to rear, reducing pressure drag and total drag (figure 12.13e).	120

[a]d is the approximate diameter of a quartz sphere corresponding to the particle Reynolds number.

In the case of a settling object, U is the fall velocity, which we designate v_f. Then, equating 12.17 and 12.18, we find that

$$v_f = \frac{(\rho_s - \rho) \cdot g \cdot d^2}{18 \cdot \mu} = \frac{(\gamma_s - \gamma) \cdot d^2}{18 \cdot \mu}. \tag{12.19}$$

equation 12.19 is known as **Stokes' law**, and situations in which $Re_p < 1$ are said to be in the **Stokes range**. Note that for typical natural sediment particles (quartz spheres), the upper limit of the Stokes range is at a diameter of about 0.1 mm.

Although the boundary layer remains laminar until $Re_p \approx 2 \times 10^5$, the flow pattern above the Stokes range becomes increasingly complicated as Re_p increases, and the drag force cannot be determined analytically. As we see in table 12.3, the pressure drag becomes increasingly important as Re_p increases above the Stokes range, with turbulence appearing in the wake at $Re_p \approx 10^2$. Once the entire wake becomes turbulent at $Re_p \approx 10^3$, the drag coefficient remains constant at $C_D \approx 0.4$ until the boundary layer becomes turbulent at $Re_p \approx 2 \times 10^5$. Using equation 12.15, with $C_D = 0.4$, and equation 12.17, we see that in the range $10^3 < Re_p < 2 \times 10^5$,

$$v_f = \left[\frac{8 \cdot (\rho_s - \rho) \cdot g \cdot d}{3 \cdot \rho}\right]^{1/2}. \tag{12.20}$$

Thus, in the Stokes range, the fall velocity is proportional to d^2 and depends on viscosity (equation 12.19), whereas in the upper range it is proportional to $d^{1/2}$ and does not depend on viscosity.

Ferguson and Church (2004) used equations 12.19 and 12.20 and dimensional analysis to derive an expression for fall velocity as a function of diameter for the entire range $Re_p < 2 \times 10^5$:

$$v_f = \frac{R_S \cdot g \cdot d^2}{C_{D1} \cdot \nu + (0.75 \cdot C_{D2} \cdot R_S \cdot g \cdot d^3)^{1/2}}, \tag{12.21}$$

where $R_S \equiv (\rho_s - \rho)/\rho$, ν is kinematic viscosity ($\equiv \mu/\rho$), and C_{D1} and C_{D2} are drag coefficients. For spheres, $C_{D1} = 18$ (from Stokes' law) and $C_{D2} = 0.4$, the value of C_D in the range $10^3 < Re_p < 2 \times 10^5$ (figure 12.12). After comparison of equation 12.21 with experimental data from several sources, Ferguson and Church (2004) recommended using $C_{D1} = 18$ and $C_{D2} = 1.0$ when sieve diameters are used to characterize d, and $C_{D1} = 20$ and $C_{D2} = 1.1$ when nominal diameters are used to characterize d. (See section 2.3.2.1 for definitions of "sieve diameter" and "nominal diameter.")

Figure 12.14 shows v_f as a function of sieve diameter as given by 12.21 with the typical natural sediment value of $R_S = 1.65$. For $d < 0.1$ mm, fall velocity increases approximately as the square of diameter; for $d > 2$ mm, it increases approximately as the square root of diameter. Viscosity (and hence temperature) affects fall velocity for $d < 1$ mm.

12.3.3 Particles on the Channel Bed

Figure 12.15 shows the forces on a particle resting on a horizontal bed. Each particle on is subject to three forces: 1) the gravitational force, F_G, equal to its submerged weight, which acts vertically downward; 2) the downstream-directed drag force, F_D, due to the viscous drag and pressure drag as described in section 12.3.1; and 3) an upward-directed lift force, F_L, that is due to a) the acceleration of the water flowing over a grain that extends above the general bed level (indicated by the more closely spaced streamlines in figure 12.15; this is the same force that occurs due to air flow over an airplane wing), and b) the upward-directed eddies that tend to form in the lee of the particle (see figure 12.13c,d) and exert an additional upward viscous drag on it.

If the particle diameter is less than about 0.06 mm, it is usually subject to a second force tending to keep it in place: intergranular cohesion due to electrostatic attraction. (Organic or mineral cementation may also act to resist forces tending to cause particle

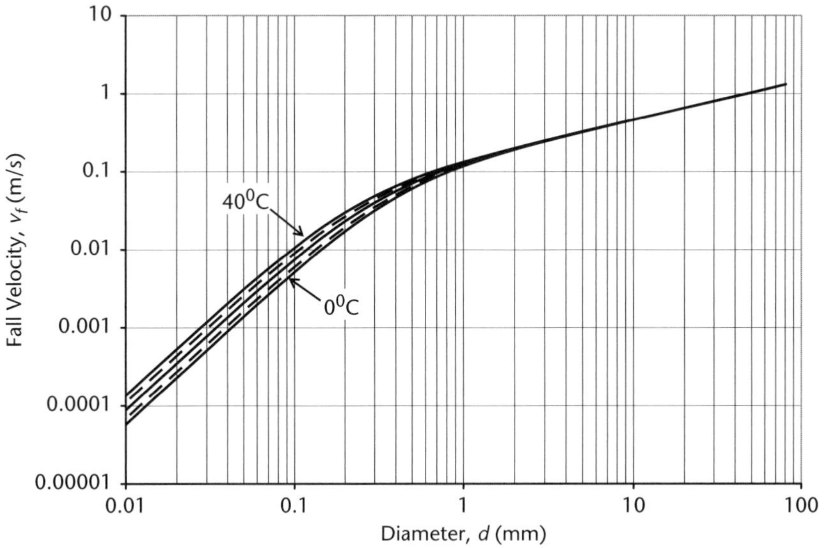

Figure 12.14 Fall velocity as a function of sieve diameter, d, as given by Ferguson and Church (2004) for natural sediment particles (equation 12.21 with $R_S = 1.65$, $C_{D1} = 18$, and $C_{D2} = 1$). For $d < 0.1$ mm, fall velocity increases approximately as the square of diameter; for $d > 2$ mm, it increases approximately as the square root of diameter. Viscosity (and hence temperature) affects fall velocity for $d < 1$ mm; curves are shown for temperatures of 0, 10, 20, 30, and 40°C in this range.

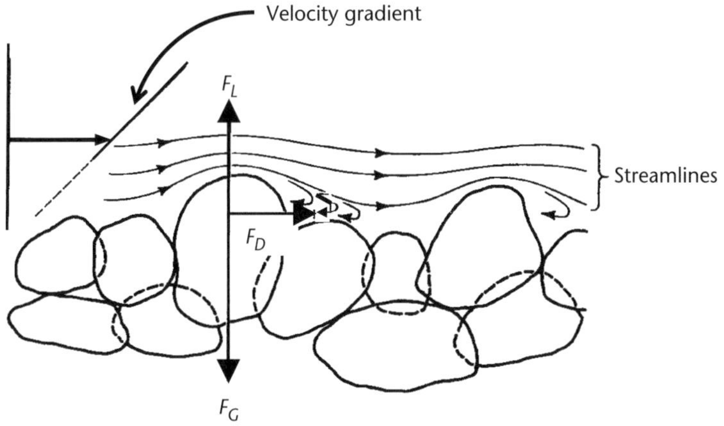

Figure 12.15 Forces on a particle on the stream bed. F_G is the gravitational force, equal to the submerged weight of the particle. F_D is the drag force due to friction and to the pressure difference between the upstream and downstream sides of the particle, as in figure 12.11. F_L is the lift force due to the acceleration over the particle and to upward-directed eddies in the lee of the particle.

movement.) However, we will restrict our analysis to cohesionless sediments and consider only the gravitational force.

Because natural particles are not strictly spherical and vary in size in a reach, we express the gravitational force, F_G, acting on a typical bed particle by generalizing equation 12.17 as

$$F_G = K_G \cdot (\rho_s - \rho) \cdot g \cdot {d_p}^3, \tag{12.22}$$

where K_G depends on particle shape (we see from equation 12.17 that $K_G = \pi/6$ for a spherical particle), and d_p is a characteristic grain diameter. The characteristic diameter is typically chosen to be the diameter that exceeds that of a given percentage, p, of the local bed material, such as d_{84}, d_{75}, or d_{50} (see figure 2.17b).

As we saw in equation 12.15 and figure 12.11, the total downstream-directed drag force exerted by water flowing over a particle due to viscous friction and pressure is given by $F_D = C_D \cdot \rho \cdot (U^2/2) \cdot A_S$. The lift force is also proportional to $\rho \cdot (U^2/2) \cdot A_S$ (Engelund and Hansen 1967), so we can express the total erosive force tending to move a typical particle, F_E, by generalizing equation 12.15 and expressing its time- and space-averaged value as

$$F_E = K_D \cdot \rho \cdot U^2 \cdot {d_p}^2, \tag{12.23a}$$

where K_D is a generalized drag coefficient that reflects variability in grain shape and exposure (and absorbs the constant 1/2 in equation 12.15).

Because we know that velocity varies with distance from the bed, the question arises as to what value to use for U in equation 12.23a. The logical choice is the shear velocity, u_*, which we saw in section 5.3.1.3 can be thought of as a characteristic near-bed velocity in a turbulent flow. Thus, we write 12.23a as

$$F_E = K_D \cdot \rho \cdot u_*^2 \cdot {d_p}^2. \tag{12.23b}$$

Recall that the shear velocity can be determined from macroscopic flow parameters as

$$u_* = (g \cdot Y \cdot S_S)^{1/2}, \tag{12.24}$$

where Y is depth and S_S is water-surface slope (equation 5.24) and can also be expressed in terms of the boundary shear stress, τ_0 ($\equiv \gamma \cdot Y \cdot S_S$):

$$\tau_0 = \rho \cdot u{*}^2. \tag{12.25}$$

From equations 12.23b and 12.25, we can also write the total drag force as being proportional to the product of the boundary shear stress and the area of the particle (which is proportional to ${d_p}^2$):

$$F_E = K_D \cdot \tau_0 \cdot {d_p}^2. \tag{12.26}$$

12.4 When Does Sediment Transport Begin?

12.4.1 Critical Boundary Shear Stress: The Shields Diagram

The seminal work of Shields (1936) used the approach of dimensional analysis followed by experiment to quantitatively address the question of when the forces acting on bed particles are sufficient to cause sediment movement. He identified two dimensionless variables, the first of which reflects the ratio of the total erosive force acting on a particle, F_E, to the gravitational force resisting movement, F_G:

$$\frac{F_E}{F_G} \propto \frac{\rho \cdot u_*^2 \cdot d_p^2}{(\rho_s - \rho) \cdot g \cdot d_p^3} = \frac{\rho \cdot u_*^2}{(\rho_s - \rho) \cdot g \cdot d_p} = \frac{\tau_0}{(\gamma_s - \gamma) \cdot d_p} \equiv \theta, \tag{12.27}$$

where θ is the **dimensionless shear stress**. Note that equation 12.27 can be derived directly from equations 12.22, 12.23b, and 12.25 if the proportionality constants K_G and K_D are absorbed into the value of θ.

Shields's second dimensionless variable was the **boundary Reynolds number**, Re_b:

$$Re_b \equiv \frac{u_* \cdot d_p}{\nu}, \tag{12.28}$$

where ν is kinematic viscosity. Recall that Re_b (with $y_r = d_p$) was introduced in section 5.3.1.6 (equation 5.31) as the parameter that defines smooth ($Re_b < 5$), transitional ($5 < Re_b < 70$), and rough ($Re_b > 70$) turbulent flows. As shown in figure 5.7, in smooth flows the boundary roughness elements—that is, the particles on the bed—are completely within the laminar sublayer, whereas in rough flow they extend through the sublayer.

Shields (1936) undertook studies in a laboratory flume to define the critical value of θ at which particle motion begins, θ^*, as a function of Re_b and summarized his findings in a graph. The critical, or threshold, dimensionless shear stress, θ^*, at which particle motion just begins is now called the **Shields parameter**, and diagrams of θ^* versus Re_b are called **Shields diagrams**.

Since Shields's original work, which is thoroughly reviewed by Buffington (1999), many studies have explored the relation between θ^* and Re_b using a wide range of experimental conditions. The various results show considerable scatter due to the use of different sediment mixtures, different flow configurations, and various approaches for identifying when initial particle motion occurs. Buffington and Montgomery (1997) have reviewed the many incipient-motion studies, and their summary graph is shown in figure 12.16a. The "average" Shields diagram proposed in an earlier review by Yalin and Karahan (1979) fits their central values (figure 12.16b) quite well and can be taken as representative of the relation, with the understanding that it applies for d_{50} and that there is considerable scatter. Note that the curves for laminar and turbulent flows coincide for smooth turbulent flows but differ in the transitional range. The value of θ^* dips to a minimum in the transitional range and then rises to a constant value for rough turbulent flows.[7] Unfortunately, the data are sparse in this range, which is where most natural flows would plot; a value of $\theta^* = 0.06$ is often used, but Buffington and Montgomery reported a range of $0.030 \leq \theta^* \leq 0.073$ for studies in which initial motion was identified visually. We use the value $\theta^* = 0.045$ as suggested by Yalin and Karahan (1979).

Using the procedure described in box 12.4, the Shields diagram can be used to construct a graph (figure 12.17) that expresses stream competence in terms of directly measurable quantities: the critical depth-slope product $(Y \cdot S)^*$. Alternatively, competence can be expressed as the critical boundary shear stress, $\tau_0^* = g \cdot (Y \cdot S)^*$.

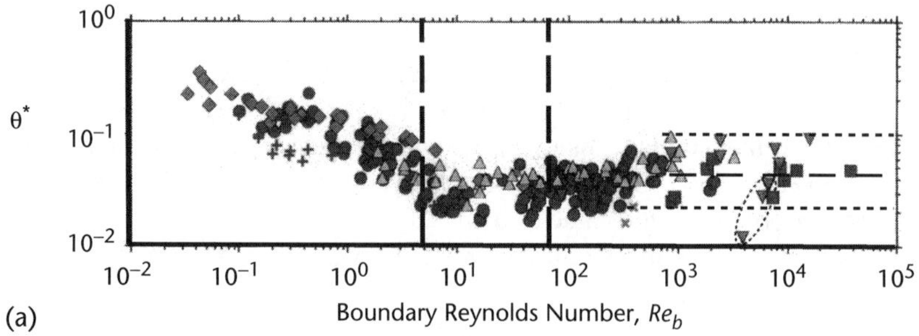

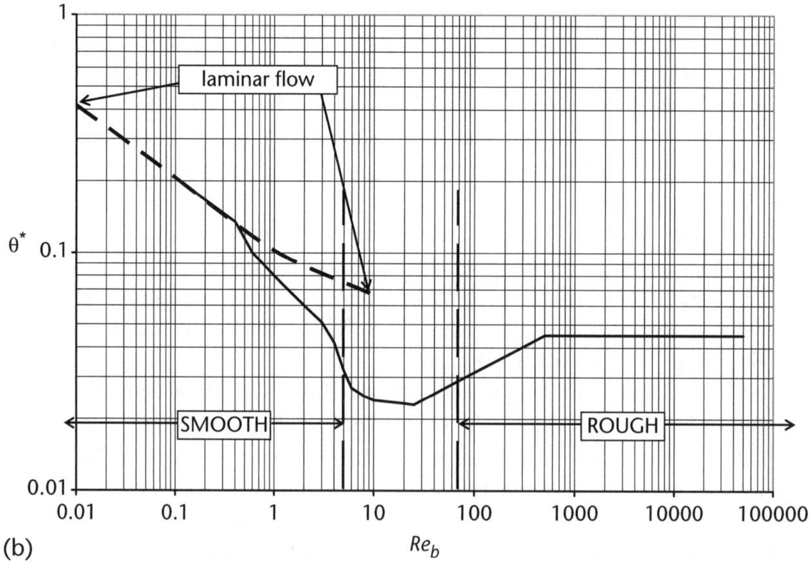

Figure 12.16 Shields diagrams. Vertical dashed lines separate smooth ($Re_b < 5$), transitional ($5 < Re_b < 70$), and rough ($Re_b > 70$) turbulent flows. (a) Values summarized by Buffington and Montgomery (1997). These data are for initial motion of surface particles for relatively well-sorted sediments in flows with relative roughness $d_{50}/Y \leq 0.2$. θ^* and Re_b are defined for $d_p = d_{50}$. The horizontal dotted lines show the range of values reported for fully rough flow: $0.021 \leq \theta^* \leq 0.1$. The horizontal dashed line is $\theta^* = 0.045$, the value recommended by Yalin and Karahan (1979) (see graph b). The diamond-shaped points are for laminar flows. Points in dotted oval are from a field study (see Buffington and Montgomery 1997). (b) A "median" Shields diagram estimated by eye from (a). The dashed curve is for laminar flows. This graph is similar to the summary relation of Yalin and Karahan (1979) and assumes $\theta^* = 0.045$ for $Re_b > 500$.

BOX 12.4 Relationship between Particle Diameter and Critical Depth-Slope Product (Shear Stress)

Here we develop the relationship between median particle diameter, d_{50}, and the critical depth-slope product or boundary shear stress at which transport begins, $(Y \cdot S)^*$ (figure 12.17). [The relation between d_{50} and critical boundary shear stress, $\tau_0{}^*$, may also be readily determined using the fact that $\tau_0{}^* = \gamma \cdot (Y \cdot S)^*$.]

From equation 12.26, the critical dimensionless shear stress (Shields parameter), θ^*, is

$$\theta * \equiv \frac{\tau_0{}^*}{(\gamma_S - \gamma) \cdot d_{50}} = \frac{\gamma \cdot (Y \cdot S)^*}{(\gamma_S - \gamma) \cdot d_{50}} = \frac{(Y \cdot S)^*}{\Gamma_S \cdot d_{50}}, \qquad (12B4.1)$$

where $\Gamma_S \equiv (\gamma_S - \gamma)/\gamma$. Thus,

$$(Y \cdot S)^* = \theta^* \cdot \Gamma_S \cdot d_{50}. \qquad (12B4.2)$$

From equation 12.27,

$$Re_b \equiv \frac{u_* \cdot d_{50}}{\nu} = \frac{(g \cdot Y \cdot S)^{1/2} \cdot d_{50}}{\nu}. \qquad (12B4.3)$$

Substituting equation 12B4.2 into equation 12B4.3,

$$Re_b = \frac{[g \cdot (\theta^* \cdot \Gamma_S \cdot d_{50})]^{1/2} \cdot d_{50}}{\nu} = \frac{\theta^{*1/2} \cdot g^{1/2} \cdot \Gamma_S{}^{1/2} \cdot d_{50}{}^{3/2}}{\nu}. \qquad (12B4.4)$$

Solving equation 12B4.4 for d_{50},

$$d_{50} = \left(\frac{\nu^2 \cdot Re_b{}^2}{g \cdot \Gamma_S \cdot \theta^*} \right)^{1/3}. \qquad (12B4.5)$$

The curve in figure 12.17 was generated by selecting points (Re_b, θ^*) along the curve on the Shields diagram (figure 12.16) and entering them into equation 12B4.5 with $\nu = 1.31 \times 10^{-6}$ m^2/s (its value at 10°C) and $\Gamma_S =$ 1.65 (the value for quartz) to find d_{50}. The corresponding value of $(Y \cdot S)^*$ was then found from equation 12B4.2; the corresponding critical boundary shear stress $\tau_0{}^*$ was found as $\tau_0{}^* = \gamma \cdot (Y \cdot S)^*$, where $\gamma = 999$ kg$_f$/m^3, its value at 10°C.

In the range in which θ^* has a constant value of 0.045 (i.e., $d_{50} > 8$ mm), the critical values of $(Y \cdot S)^*$ in m and $\tau_0{}^*$ in kg$_f$/m^2 can be found directly from equation 12B4.2:

$$(Y \cdot S)^* = 0.045 \cdot 1.65 \cdot (d_{50}/1{,}000) = (7.43 \times 10^{-5}) \cdot d_{50}, \qquad (12B4.6a)$$

or, since $\tau_0{}^* = \gamma \cdot (Y \cdot S)^*$,

$$\tau_0{}^* = 999 \cdot (7.43 \times 10^{-5}) \cdot d_{50} = 0.0742 \cdot d_{50}, \qquad (12B4.6b)$$

where d_{50} is in mm.

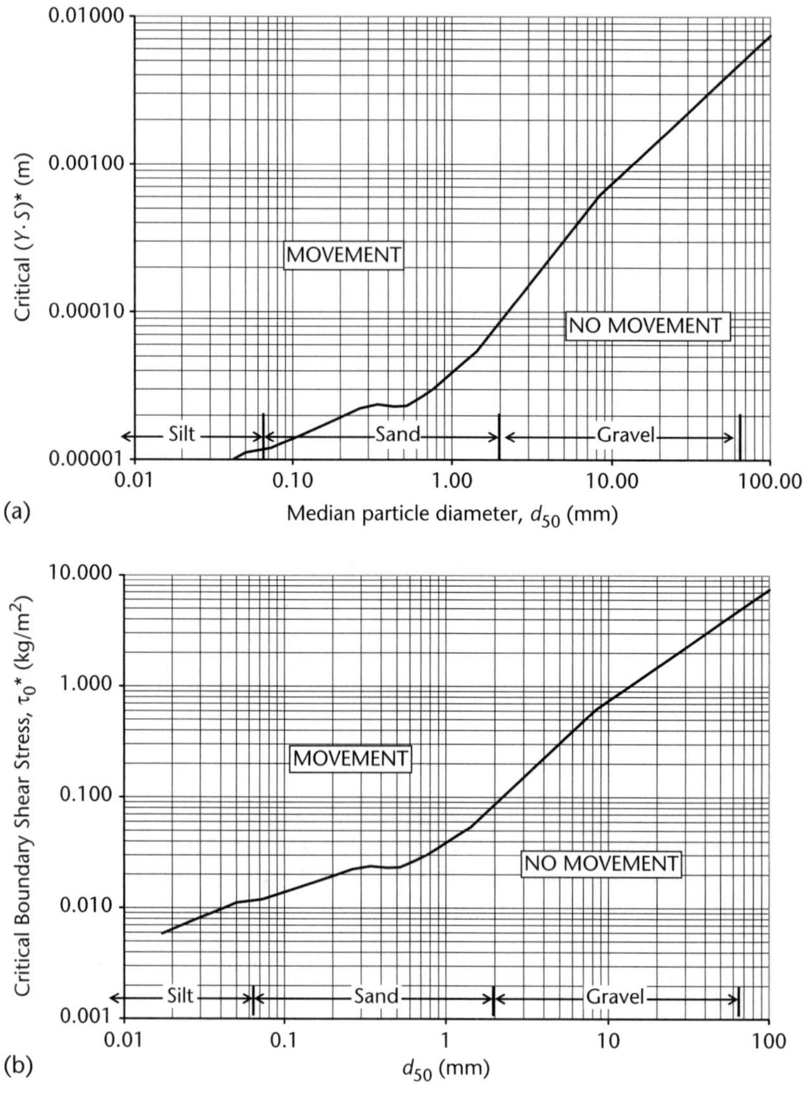

Figure 12.17 (a) Relation between median particle diameter d_{50} and depth-slope product required for initiation of motion, $(Y \cdot S)^*$. (b) Relation between median particle diameter d_{50} and boundary shear stress $\tau_0{}^*$ required for initiation of motion. The curves were generated from the Shields diagram (figure 12.16) as described in box 12.4.

12.4.2 Critical Velocity: The Hjulström Curves

It is also possible to express competence in terms of a **critical erosion velocity**, U^*, which is the cross-section mean velocity at which bed-particle movement begins. The Swedish geomorphologist Filip Hjulström (1935; 1939) devoted particular attention to this relation, and graphs of U^* versus particle size are known as **Hjulström curves**. Figure 12.18a shows the original curve presented in Hjulström (1935), which has been reprinted in many references. Due to "uncertainty of the data," the relation is shown as a wide band for flows of depth greater than 1 m. The entrainment curve (curve A) has a minimum near $d = 0.5$ mm; this is because smaller particles are increasingly affected by cohesive forces (due to electrostatic attraction and organic material) that resist entrainment. Note that there is a curve B separating "transportation" and "deposition"; this is based on the observation that once sediment has been set in motion, it continues to move even when the velocity decreases below the critical velocity. According to Hjulström (1939), deposition occurs when the velocity falls to about $(2/3) \cdot U^*$, and that is the basis for curve B.

We can derive Hjulström-type curves using the relation between $(Y \cdot S)^*$ and d_{50} (figure 12.17) and the expression for mean velocity derived from the Prandtl-von Kármán velocity-profile equation (equation 5.36b):

$$U = 2.5 \cdot u_* \cdot \left[\ln\left(\frac{Y}{y_0}\right) - 1 \right], \qquad (12.29)$$

where

$$y_0 = \frac{\nu}{9 \cdot u_*} \text{ for smooth flows, } Re_b \leq 5; \qquad (12.30a)$$

$$y_0 = \frac{d_{50}}{30} \text{ for transitional and rough flows, } Re_b > 5; \qquad (12.30b)$$

and 12.29 applies to "wide" channels with low relative roughness ($Y >> y_0$).

In order to pursue this approach, we must specify a particular depth, Y, and slope, S, separately, instead of using their product as a single independent variable. Once Y and S are specified, we compute u_* and $(Y \cdot S)^*$ and use figure 12.17 to find the corresponding d_{50}. We then compute Re_b, determine whether to use equation 12.30a or 12.30b to compute y_0, and then use equation 12.29 to compute U^* for the specified depth.

The results are shown in figure 12.18b, with separate curves shown for depths ranging from 0.1 to 10 m. The curves plot somewhat above those plotted by Hjulström (1939), but are consistent with those computed by Sundborg (1956), which were calculated via an approach similar to the one used here but for maximum rather than average velocity. In using the curves of figure 12.18b, one must keep in mind the scatter of experimental results on which they are based (figure 12.16a). The curves are not extended into the cohesive range, because the experiments used only noncohesive sediments.

12.4.3 Erosion of Cohesive Sediments

In particles smaller than about 0.06 mm diameter, significant interparticle cohesion is present due electrostatic forces and perhaps organic material, so that critical values of

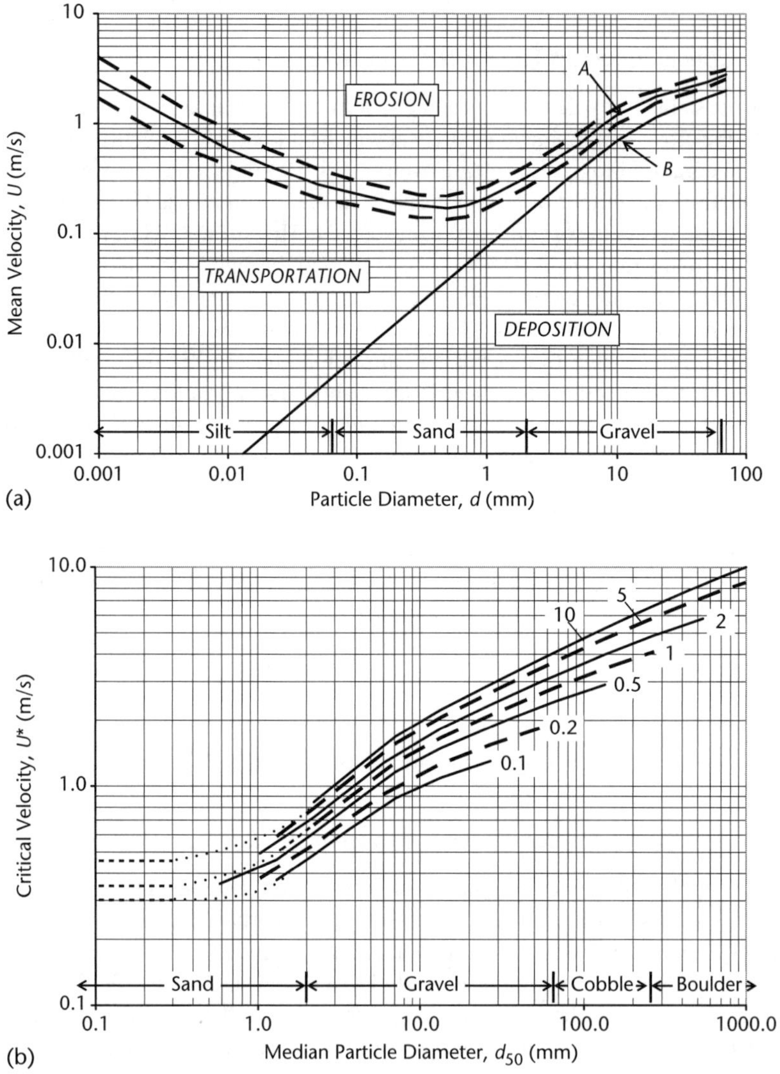

Figure 12.18 (a) Curve A is "approximate curves for erosion of uniform material" for flows of depth greater than 1 m presented by Hjulström (1939) and widely reprinted. Due to "uncertainty of the data," the relation is shown as a wide band (dashed lines). The curve has a minimum critical erosion velocity U^* near $d = 0.5$ mm; this is because smaller particles are increasingly affected by cohesive forces (due to electrostatic attraction and organic material) that resist entrainment. Curve B separates "transportation" and "deposition"; this is based on the observation that once sediment has been set in motion, it continues to move even when the velocity decreases below the critical velocity. According to Hjulström (1939), deposition occurs when the velocity falls to about $(2/3) \cdot U^*$. (b) Relation between critical average velocity U^* and median particle diameter d_{50} for wide channels with low relative roughness computed from figure 12.17 and the Prandtl-von Kármán velocity distribution (equation 12.29). The curve parameter is the average depth, Y, in meters. The dotted curve extensions are based on Sundborg (1956).

shear stress and velocity do not decrease with particle size in this range (figure 12.18). It is not possible to determine universal relations for critical erosion shear stress or velocity of cohesive material, because it is typically eroded not particle by particle, as is noncohesive sediment, but in aggregates of particles. As described by Sundborg (1956, p. 173), "These aggregates vary in size, sometimes attaining centimetre or even decimetre size, in which case weak zones along bedding planes or cracks and surfaces of sliding allow the lumps of clay to break away. It is also likely that corrasion [i.e., abrasion] by coarser particles, sand or gravel … also plays an important part in the erosion of fine sediment."

12.4.4 Bedrock Erosion

Channel reaches formed in bedrock occur where sediment-transport capacity exceeds sediment supply (table 2.4). Such reaches are common in mountainous and tectonically active regions, and bedrock erosion is a significant process governing the form and dynamics of those regions. However, the processes by which streams erode bedrock are not sufficiently well known to allow the development of quantitative relations between flow parameters (e.g., shear stress, velocity) and lithological characteristics. We can, however, provide a summary overview of the state of knowledge of bedrock-erosion processes, based on the comprehensive review of Whipple et al. (1999).

Two processes are known to erode bedrock: 1) plucking, the removal of rock fragments from the bed; and 2) abrasion by suspended- and bed-load particles. A third process, cavitation (described more fully below), may also be effective in some situations, but the evidence for its efficacy is not clear. These processes are briefly described in the following subsections.

12.4.4.1 Plucking

Plucking (figure 12.19) is the dominant bedrock-erosion process where the bedrock has **joints** (quasi-regular cracks due to cooling or pressure release, fractures, or bedding planes) that are relatively closely spaced (less than about 1 m), regardless of bedrock type. The fracture and loosening of joint blocks occurs by 1) chemical and physical weathering within the joints, 2) hydraulic clast wedging by finer sediment particles carried into the cracks, 3) crack propagation induced by the impacts of large sediment particles, and 4) crack propagation induced by pressure fluctuations associated with intense turbulent flows. After reviewing theoretical considerations and limited observational evidence, Whipple et al. (1999) concluded that the erosion rate due to plucking, E_P, should be related to boundary shear stress as

$$E_P \propto (\tau_0 - \tau_0{}^*)^j, \tag{12.31}$$

where τ_0 is boundary shear stress, $\tau_0{}^*$ is a critical value of boundary shear stress, and j is an exponent ≈ 1. However, because of the complex set of processes involved, appropriate values for τ_0 and $\tau_0{}^*$ cannot be specified, and a definitive relation for predicting the rate of erosion by plucking cannot be developed.

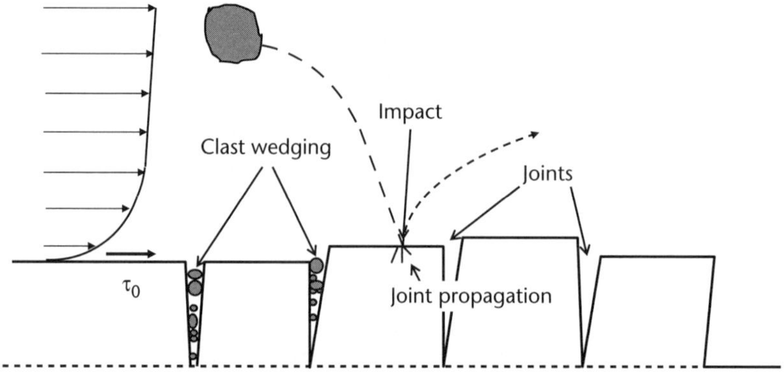

(a)

(b)

Figure 12.19 (a) Processes and forces contributing to erosion by plucking. Impacts by large saltating particles cause crack propagation that loosens joint blocks. Hydraulic wedging by smaller clasts further opens cracks. Surface drag and differential forces across the block tend to lift loosened blocks. Once the downstream neighbor of a block has been removed, rotation and sliding can occur, greatly facilitating block removal. From Whipple et al. (1999); reproduced with permission of Geological Society of America. (b) Extensive plucking has occurred in the highly jointed bedrock on the bed of the Swift Diamond River, New Hampshire. Photo by the author.

12.4.4.2 Abrasion

Abrasion (figure 12.20) is the dominant bedrock-erosion process in massive bedrock, that is, where joints are widely spaced or absent. The flux of kinetic energy impacting the rock surface depends on the kinetic energy of the particles and the role of inertia in determining how particles are "coupled" to the flow:

> The largest particles have large inertia and thus strike any bedrock protuberance on the upstream side, polishing the surface but accomplishing little erosion.
>
> Intermediate-sized particles more closely follow fluid streamlines, striking gently curved obstructions on the upstream side and abrupt obstructions on the downstream side, effecting significant erosion.
>
> The smallest particles closely follow the streamlines and do little abrading.

Field observations show that abrasion is usually greatest on the downstream side of obstructions, where powerful vortices occur, and commonly produces potholes that may coalesce and completely remove even very hard rock. A detailed study by

(a)

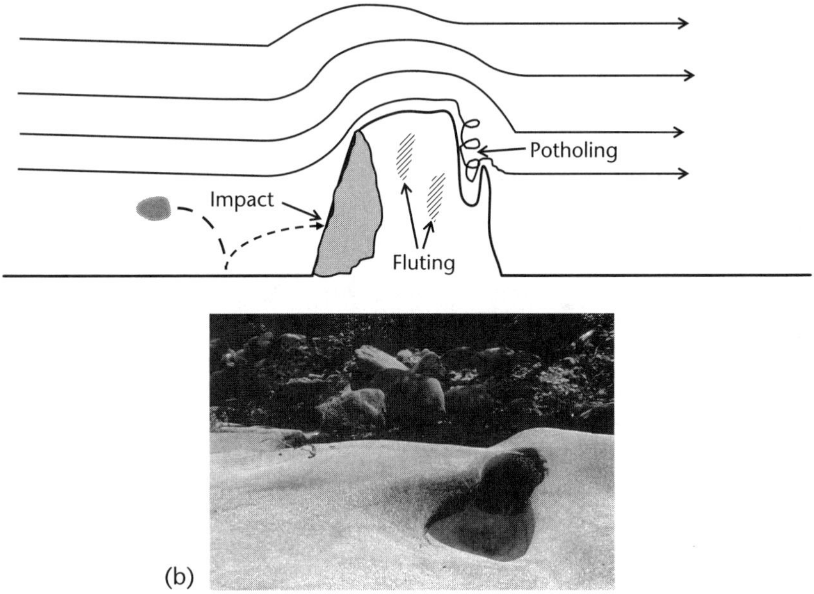

(b)

Figure 12.20 (a) Processes contributing to bedrock erosion by abrasion. Large particles in bed load and suspended load are decoupled from the flow and impact upstream faces of protuberances, polishing the surface (shaded area) but causing little erosion. Intermediate-sized particles produce small-scale flutes and ripples on the flanks and large, often coalescing potholes on the lee sides of obstructions. "The complete obliteration of massive, very hard rocks in these potholed zones testifies to the awesome erosive power of the intense vortices shed in the lee of obstructions" (Whipple et al, 1999, p. 497). Redrawn from Whipple et al. (1999). (b) Abraded massive granite bedrock with a pothole, Lucy Brook, New Hampshire. Photo by the author.

Springer et al. (2006) confirmed that pothole growth is accomplished by suspended sediment in high-speed vortices rather than grinding by the large stones that are often found deposited in them.

Consideration of the dynamics of the abrasion process led Whipple et al. (1999) to conclude that the rate of erosion due to abrasion, E_A, is related to suspended-sediment concentration, C_S, and average velocity or shear stress as

$$E_A \propto C_S \cdot U^3 \propto C_S \cdot \tau_0{}^{3/2}. \tag{12.32a}$$

Because sediment concentration depends approximately on U^2, one may also write the relation as

$$E_A \propto U^5 \propto \tau_0{}^{5/2}. \tag{12.32b}$$

12.4.4.3 Cavitation

Cavitation is the formation of water vapor and air bubbles that occurs when the local fluid pressure drops below the vapor pressure of the dissolved air. When these bubbles are carried into regions of higher pressure, they collapse explosively, generating shock waves that can cause pitting of metal such as turbine blades and rapid destruction of concrete structures. Although direct evidence of cavitation-induced bedrock erosion is lacking, Whipple et al. (1999) conclude that it is likely to be present in many natural streams and may be responsible for some of the fluting and potholing that is usually attributed to abrasion.

The propensity for cavitation is reflected in the **cavitation number**, Ca, given by

$$Ca = \frac{(P_a + \gamma \cdot Y) - P_v}{\rho \cdot (U^2/2)}, \tag{12.33}$$

where P_a is atmospheric pressure; γ and ρ are weight and mass density of water, respectively; Y is flow depth; P_v is the vapor pressure of water; and U is the average velocity (Daily and Harleman 1966). Theoretically, cavitation occurs when $Ca < 1$, but Whipple et al. (1999) note that it is commonly observed at values of Ca as high as 3 in flows with high Reynolds numbers. Thus, they suggest that cavitation is "possible" when $Ca < 4$, and "likely" when $Ca < 2$. The combinations of flow depth and velocity that give these values are shown in figure 12.21; it appears that the conditions are fairly common.

12.5 Sediment Load

Understanding the processes that determine sediment load (sediment discharge) is important for predicting erosion and deposition in natural stream reaches and marine settings, predicting the effects of engineering structures on erosion and deposition, predicting reservoir sedimentation, designing sediment-measurement strategies, and inferring hydraulic and sediment-transport characteristics of ancient environments. This problem has been addressed in many hundreds of empirical and theoretical studies over at least the last 125 years, and many books have been devoted to the subject. Thus, it is not feasible to undertake a review of the various methods and

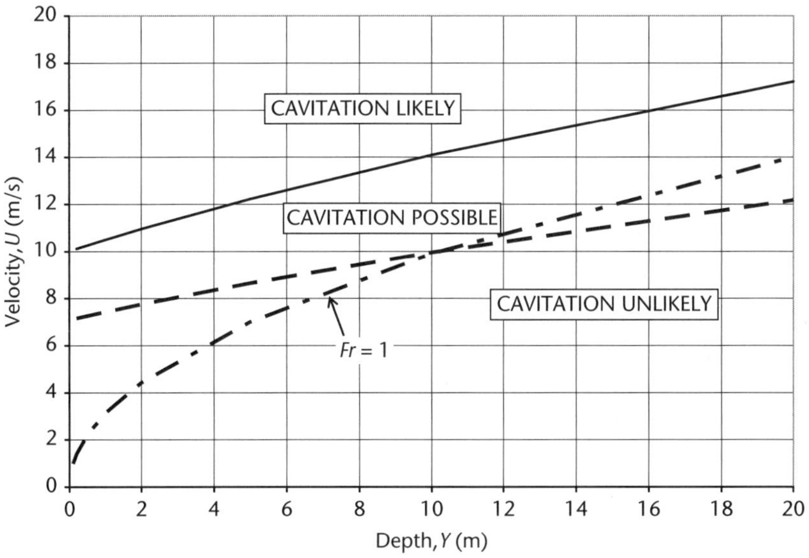

Figure 12.21 Conditions for cavitation as suggested by Whipple et al. (1999). The dashed line denotes cavitation number $Ca = 4$; the solid line denotes $Ca = 2$. These values are calculated assuming a temperature of 10°C. For comparison, the dot-dashed line indicates the conditions at which flow becomes critical; cavitation is "possible" in subcritical flows when $Y > 10$ m.

results here; instead, we explore some well-known approaches based on the hydraulic concepts developed above, with reference citations that can be used to pursue the subject in more detail. We also explore some recent experimental results using new measurement techniques that provide fresh insight into sediment-transport processes. It is likely that such new approaches will greatly increase our understanding of this important process in the near future.

The discussion here treats bed load, suspended load, and total bed-material load in separate sections.

12.5.1 Bed Load

The earliest attempt to predict bed load was developed in 1879 by P.F.D. DuBoys based on an analysis of the balance between the force applied to the surface layer of uniform sediment by the flow and the frictional resistance between the surface layer of particles and the layer just beneath it (for details, see Chang 1988). The resulting equation was

$$l_b = C_{DB} \cdot \tau_0 \cdot (\tau_0 - \tau_0{}^*),\ \tau_0 \geq \tau_0{}^*, \tag{12.34}$$

where l_b is bed load per unit width [F L^{-1} T^{-1}], τ_0 is boundary shear stress, $\tau_0{}^*$ is critical boundary shear stress, and C_{DB} is a coefficient with dimensions [L^3 F^{-1} T^{-1}].

Subsequent experimental work with sands (see Chang 1988) developed empirical relations for C_{DB} and $\tau_0{}^*$ as functions of grain size:

$$C_{DB} = \frac{0.17}{d^{3/4}} \tag{12.35}$$

and

$$\tau_0{}^* = 0.061 + 0.093 \cdot d, \tag{12.36}$$

where d is in mm, C_{DB} is in $m^3/(kg \cdot s)$, and $\tau_0{}^*$ is in kg/m^2. Shields's (1936) work leading to figures 12.16 and 12.17 can also be used to estimate the critical shear stress $\tau_0{}^*$ for bed-load movement. However, the critical shear stress given by equation 12.36 differs considerably from the relation shown in figure 12.17.

The DuBoys approach has been the basis of many subsequent investigations of bed-load transport and has been modified and applied to nonuniform particle-size distributions by Meyer-Peter and Muller (1948), Einstein (1950), and Parker et al. (1982). Other studies have related bed load to different flow variables:

$$l_b = f(Q - Q^*), \tag{12.37}$$

$$l_b = f(U - U^*), \tag{12.38}$$

$$l_b = f(\Pi_A - \Pi_A{}^*), \tag{12.39}$$

where Q is discharge per unit width, U is average velocity, Π_A is stream power per unit bed area ($\equiv U \cdot \tau_0$; see equation 8.27), the asterisk indicates a threshold value for each variable, and f indicates different functional relations for each variable. Several of these approaches are detailed by Chang (1988) and Shen and Julien (1992); the latter writers conclude, "More research is needed to obtain data for the transport of nonuniform sediment size in order to develop a generally acceptable equation" (Shen and Julien 1992, p. 12.29).

12.5.2 Suspended-Sediment Concentration and Load

12.5.2.1 Concentration Profile: Diffusion-Theory Approach

Theoretical Development The most widely accepted approach to predicting suspended-sediment concentration is based on diffusion theory (section 4.6). This approach has been found to capture many aspects of measured concentration profiles, although it is difficult to extend to predictions of sediment concentration and load for entire cross sections.

Diffusion theory was first applied to the problem of predicting the vertical concentration of suspended sediment by the American hydraulic engineer Hunter Rouse (1906–1996) (Rouse 1937). Considering for the moment sediment particles of uniform diameter d, an equilibrium distribution of suspended-sediment concentration at a point in a cross section exists when the downward mass flux $[M\ L^{-2}\ T^{-1}]$ of sediment across a horizontal plane at an arbitrary distance y above the bottom, $\mathcal{F}_{SD}(y)$, equals the upward flux, $\mathcal{F}_{SU}(y)$:

$$\mathcal{F}_{SD}(y) = \mathcal{F}_{SU}(y) \tag{12.40}$$

(figure 12.22).

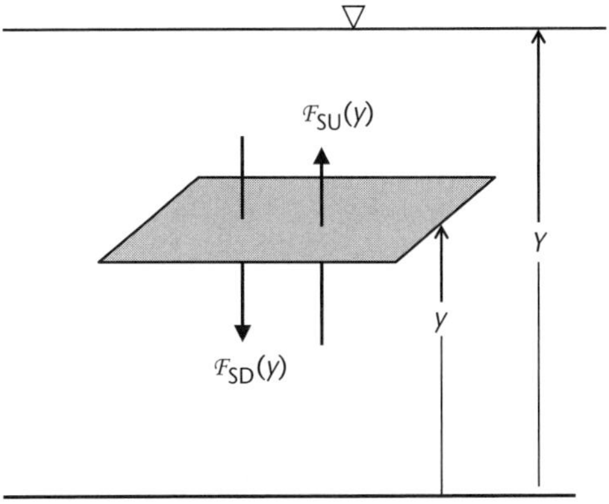

Figure 12.22 Definition diagram for diffusion-theory approach to computing the vertical distribution of suspended-sediment concentration at a vertical (equation 12.45; see text). The shaded area represents a unit area perpendicular to the y-direction.

The downward flux is given by the product of the concentration C_S [M L^{-3}] and the fall velocity, v_f [L T^{-1}], which of course is a function of the particle size (figure 12.14). Because concentration is a function of distance above the bottom, y, we write

$$\mathcal{F}_{SD}(y) = C_S(y) \cdot v_f. \tag{12.41}$$

The upward flux is modeled as a diffusion process (equation 4.46):

$$\mathcal{F}_{SU}(y) = -D_S(y) \cdot \frac{dC_S(y)}{dy}, \tag{12.42}$$

where $D_S(y)$is the vertical diffusivity of suspended sediment in turbulent flows [L^2 T^{-1}] at elevation y. Substituting 12.41 and 12.42 into 12.40 and rearranging yields

$$\frac{dC_S(y)}{C_S(y)} = -\frac{v_f}{D_S(y)} \cdot dy. \tag{12.43}$$

To integrate equation 12.43 and find the form of $C_S(y)$, we must specify the relation between diffusivity and distance above the bottom. Because the upward sediment flux is carried by the vertical component of turbulent eddies, it is reasonable to assume that the turbulent diffusivity of sediment is equal to the turbulent diffusivity of momentum. Beginning with that assumption and using relations developed in section 3.3.4.4, the relation for diffusivity as a function of distance above the bottom is derived in box 12.5:

$$D_S(y) = \kappa \cdot u_* \cdot y \cdot \left(1 - \frac{y}{Y}\right), \tag{12.44}$$

where κ is von Kármán's constant, u_* is shear velocity, and Y is total depth. Figure 12.23 shows measured values of $D_S(y)$ and confirms that equation 12.44 gives at least a reasonable approximation of the vertical distribution of diffusivity.

BOX 12.5 Derivation of Expression for Suspended-Sediment Diffusivity

From the development in section 3.3.4, the vertical flux of momentum due to turbulence, τ_{Tyx}, is given by equation 3.34. That expression can be written as a diffusion relation:

$$\tau_{Tyx} = l^2 \cdot \left|\frac{d\bar{u}_x}{dy}\right| \cdot \frac{d(\rho \cdot \bar{u}_x)}{dy} = -D_M(y) \cdot \frac{d(\rho \cdot \bar{u}_x)}{dy}, \tag{12B5.1}$$

where l is Prandtl's mixing length and $\bar{u}_x$ is the time-averaged downstream velocity, both of which are functions of y; and $D_M(y)$ is the diffusivity of momentum. Note that $D_M(y)$ is identical to the kinematic eddy viscosity, ε, defined in equation 3.36. Thus, the diffusivity of suspended sediment is

$$D_S(y) = \varepsilon = l^2 \cdot \left|\frac{d\bar{u}_x}{dy}\right|. \tag{12B5.2}$$

We saw in equation 3.38 that

$$l = \kappa \cdot y \cdot \left(1 - \frac{y}{Y}\right)^{1/2}, \tag{12B5.3}$$

where κ is von Kármán's constant ($\kappa = 0.4$ generally). The velocity gradient is found from the Prandtl-von Kármán velocity profile (equation 5.23):

$$\left|\frac{d\bar{u}_x}{dy}\right| = \frac{u_*}{\kappa \cdot y}, \tag{12B5.4}$$

where u_* is the friction velocity. Substituting equations 12B5.3 and 12B5.4 in equation 12B5.2, we see that

$$D_S(y) = \kappa \cdot u_* \cdot y \cdot \left(1 - \frac{y}{Y}\right), \tag{12B5.5}$$

which is identical to the vertical distribution of eddy viscosity as given in equation 3.39.

When equation 12.44 is substituted into equation 12.43 and the resulting expression integrated, we find the expression for suspended-sediment concentration as a function of distance above the bottom (i.e., the suspended-sediment-concentration profile):

$$C_S(y) = C_S(y_a) \cdot \left[\left(\frac{Y - y}{y}\right) \cdot \left(\frac{y_a}{Y - y_a}\right)\right]^{v_f/\kappa \cdot u^*}, \tag{12.45}$$

where $C_S(y_a)$ is the concentration at an arbitrary reference level $y = y_a$. Equation 12.45 is often called the **Rouse equation**.

Some cautions about equation 12.45 should be noted:

1. It was derived for steady uniform flow and a single sediment size at a single location in a cross section.

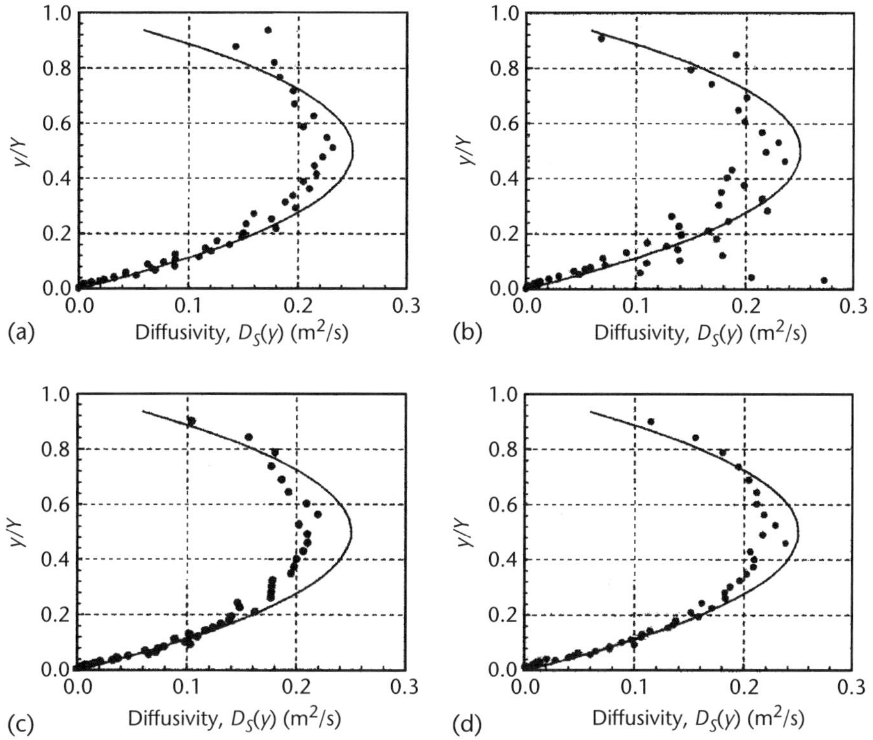

Figure 12.23 The points show the vertical distribution of sediment diffusivity, $D_S(y)$, as a function of relative height, y/Y, as measured in flume experiments by Muste et al. (2005). The curves are the theoretical expression of equation 12.44. (a) Flow of clear water; (b) flow with sand in suspension at a volumetric concentration of 0.00046; (c) flow with sand in suspension at a volumetric concentration of 0.00092; (d) flow with sand in suspension at a volumetric concentration of 0.00162. From Muste et al. (2005).

2. It predicts a zero concentration at the surface and an infinite concentration at the bed, neither of which occurs in nature.[8]
3. As discussed in the following section, laboratory and field measurements show that the value of the exponent that best fits measured profiles generally differs from the value given in equation 12.45, even for steady uniform flow and a single sediment size. This is discussed further in the following section.

Significance of the Exponent (Rouse Number) Before considering the problem of determining y_a and $C_S(y_a)$, we explore the significance of the exponent in equation 12.45, which is called the **Rouse number**, *Ro*:

$$Ro \equiv \frac{v_f}{\kappa \cdot u_*} = \frac{v_f}{\kappa \cdot (\tau_0/\rho)^{1/2}} \tag{12.46}$$

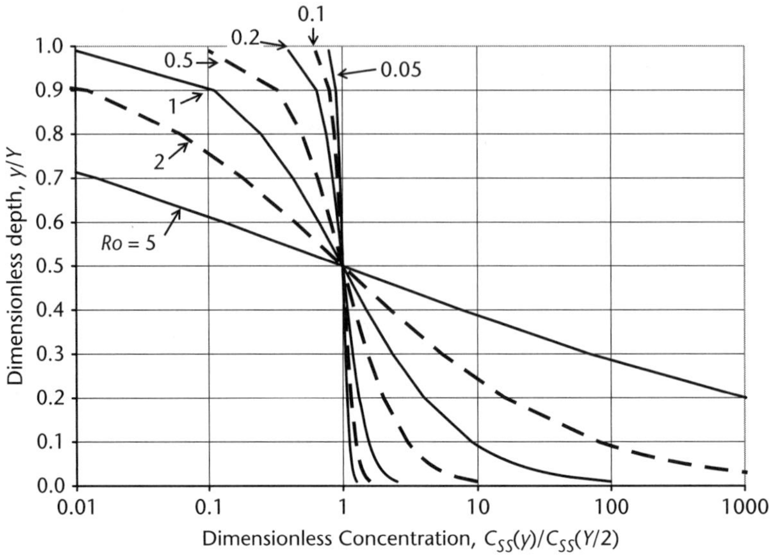

Figure 12.24 Effect of the Rouse number, $Ro \equiv v_f/(\kappa \cdot u_*)$, on suspended-sediment-concentration profile. Curves are computed via equation 12.47 and labeled with the value of *Ro*.

To show the effect of *Ro* on the suspended-sediment-concentration profile, we select $y_a = Y/2$ and use equation 12.45 to express the ratio of concentration at any depth to its value at mid-depth,

$$\frac{C_S(y)}{C_S(Y/2)} = \left(\frac{Y}{y} - 1\right)^{Ro}. \tag{12.47}$$

Figure 12.24 gives plots of equation 12.47 for various values of *Ro*: At small values of *Ro*, particles with a given fall velocity are readily suspended and the concentration profile is nearly uniform; as *Ro* increases, an increasing proportion of the sediment is transported near the bed. Thus, the Rouse number *Ro* reflects the shape of the suspended-sediment-concentration profile for a given particle size *d*; smaller values of *Ro* represent more uniform vertical concentrations.

However, field and laboratory studies show that, although the form of measured concentration profiles is well modeled by equation 12.45, the value of *Ro* that best fits measured profiles is smaller than the value calculated via equation 12.46; that is, actual profiles are more uniform than predicted using the calculated value of *Ro*. To account for this bias, Pizzuto (1984) recommended using an adjusted value, Ro', given approximately by

$$Ro' = 0.740 + 0.362 \cdot \ln(Ro), \qquad Ro > 0.4. \tag{12.48}$$

This relation is shown in figure 12.25. With this adjustment, equation 12.45 provides good predictions over most of the concentration profile, as shown in figure 12.26.

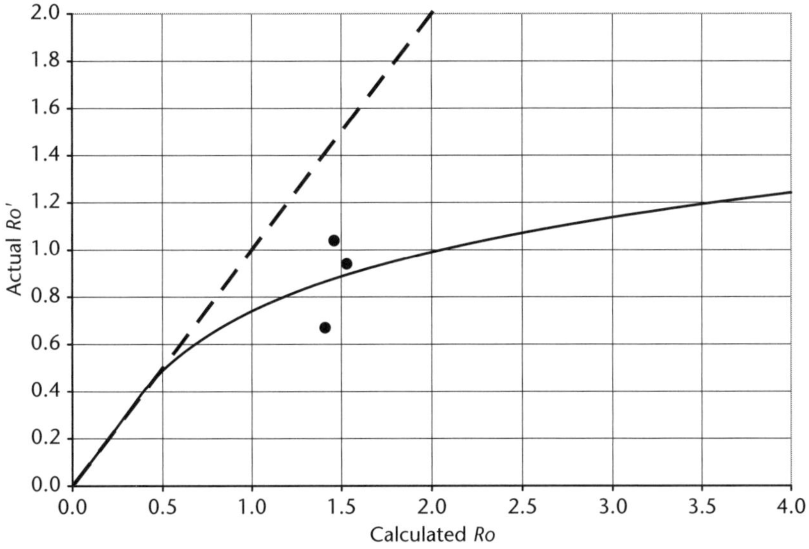

Figure 12.25 Calculated values of *Ro* predict suspended-sediment-concentration profiles that are significantly steeper than measured profiles. Thus, for estimating profiles, *Ro* should be adjusted to *Ro*′ as indicated by equation 12.48 (solid curve). The dashed line is the 1:1 line. The points show *Ro* and *Ro*′ for three profiles measured by Muste et al. (2005) and plotted in figure 12.26.

The Rouse number *Ro* is also significant because it is proportional to the ratio of fall velocity to friction velocity (or to the square root of boundary shear stress) and is thus an expression of the "reluctance" of a particle to be suspended. Because of this, *Ro* can be used as an alternative to critical depth-slope product $(Y \cdot S)^*$ or critical shear stress (τ_0*) to indicate the conditions under which entrainment will occur. The critical value of *Ro** can be determined from figure 12.27, which is a plot of $Y \cdot S$ and *Ro* values for 641 flows in 171 reaches for which the median bed-material size (d_{50}) exceeded 8 mm. Flows for which $(Y \cdot S) > (Y \cdot S)^*$ (figure 12.17a) are indicated by triangles, and all these flows have $Ro < 5.4$. Thus, we conclude that a critical Rouse number $Ro^* = 5.4$ can be used as an alternative to the critical $(Y \cdot S)^*$ values derived from the Shields diagram (figure 12.17a). Note that, for the purposes of determining particle entrainment, the calculated value of *Ro* given by equation 12.46 is used, not the adjusted value R_o' given by equation 12.48.

Reference Level and Reference Concentration The problem of determining the reference level y_a and the reference concentration $C_S(y_a)$ has received much attention (e.g.; Einstein 1950; Graf 1971; Task Force on Preparation of Sediment Manual 1971; Vanoni 1975; Garde and Raga Raju 1978). Pizzuto (1984) compared several of the earlier approaches to field and flume data and found that the best results were obtained

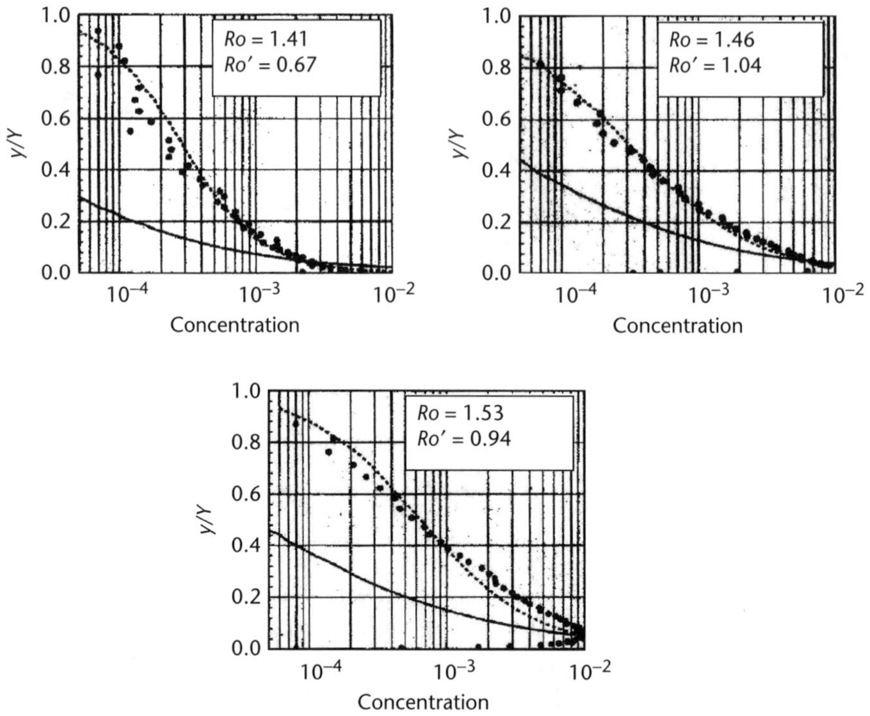

Figure 12.26 Three suspended-sediment-concentration profiles for uniform quartz sand ($d = 0.23$ mm) measured by Muste et al. (2005) in flume experiments. The solid curve is the profile predicted using the computed value of *Ro* in equation 12.45; the dotted curve is the best-fit profile that is given by Ro'. The concentration here is the volumetric concentration (C_{VV} in box 12.1). Note that concentrations reach a maximum somewhat above $y/Y = 0$, which is not predicted by equation 12.45. These values of *Ro* and Ro' are plotted in figure 12.25.

when $y_a = 2 \cdot d_{65}$ and $C_S(y_a) = 22{,}300$ ppm (=22,600 mg/L). With these values, a practical version of equation 12.43 becomes

$$C_S(y) = 22{,}600 \cdot \left[\left(\frac{Y-y}{y}\right) \cdot \left(\frac{2 \cdot d_{65}}{Y - 2 \cdot d_{65}}\right)\right]^{Ro'}, \qquad (12.49a)$$

or, because $2 \cdot d_{65} << Y$,

$$C_S(y) = 22{,}600 \cdot \left[\left(\frac{Y-y}{y}\right) \cdot \left(\frac{2 \cdot d_{65}}{Y}\right)\right]^{Ro'}, \qquad (12.49b)$$

where $C_S(y)$ is in mg/L.

Total Suspended-Sediment Load The depth-averaged suspended-sediment concentration at a given location in the cross section for grain size d, $C_S(d)$, is found by

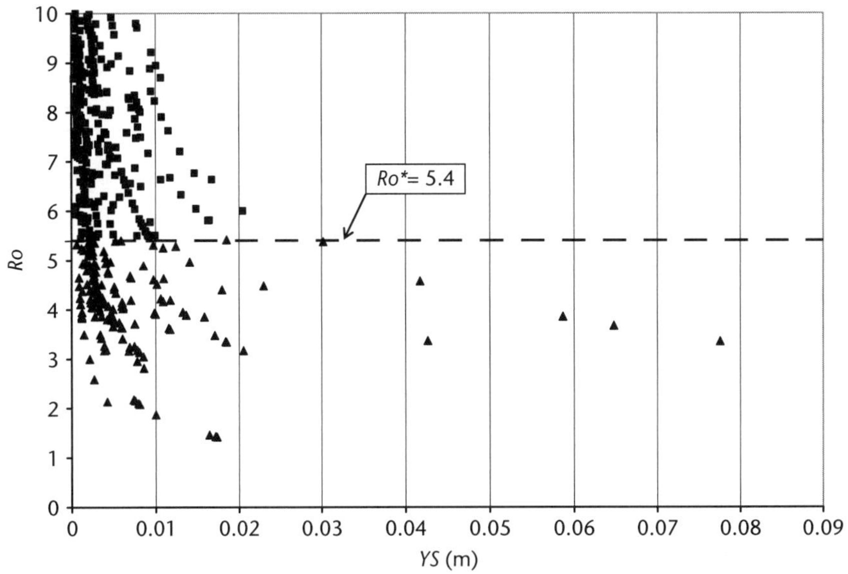

Figure 12.27 The Rouse number, Ro, and depth-slope product $Y \cdot S$ for a database of 641 flows in 171 reaches with $d_{50} > 8$ mm (some of the flows have $Ro > 10$ and do not appear on this graph). Flows for which $Y \cdot S > (Y \cdot S)^*$ as given by figure 12.17a are indicated by triangles. For all these flows, $Ro < 5.4$.

integrating equation 12.49:

$$C_S(d) = 22{,}600 \cdot \left(\frac{2 \cdot d_{65}}{Y}\right)^{Ro'(d)} \left(\frac{1}{Y}\right) \cdot \int_{2 \cdot d_{65}}^{Y} \left(\frac{Y-y}{y}\right)^{Ro'(d)} \cdot dy, \qquad (12.50)$$

where the exponent is written explicitly as a function of diameter, $Ro'(d)$.[9] The sediment load is the product of the discharge and the concentration, and the suspended load of sediment of diameter d per unit width, $l_S(d)$ [F L^{-1} T^{-1}], is given conceptually by

$$l_S(d) = \int_{2 \cdot d_{65}}^{Y} C_S(y) \cdot u(y) \cdot dy, \qquad (12.51)$$

where $C_S(y)$ is given by equation 12.49, and $u(y)$ is the velocity profile (usually taken to be the Prandtl-von Kármán profile of equation 5.34). If equation 12.51 applies for an entire cross section (i.e., if the channel is wide and rectangular), the total suspended load L_S could conceptually be estimated as

$$L_S = W \cdot \sum_{\text{all } d} l_{Sd}, \qquad (12.52)$$

where W is width.

There have been many attempts to develop methods for a priori estimation of suspended load in natural channels based directly on the diffusion approach and equations 12.51 and 12.52. The first and best known of these was by Einstein (1950);

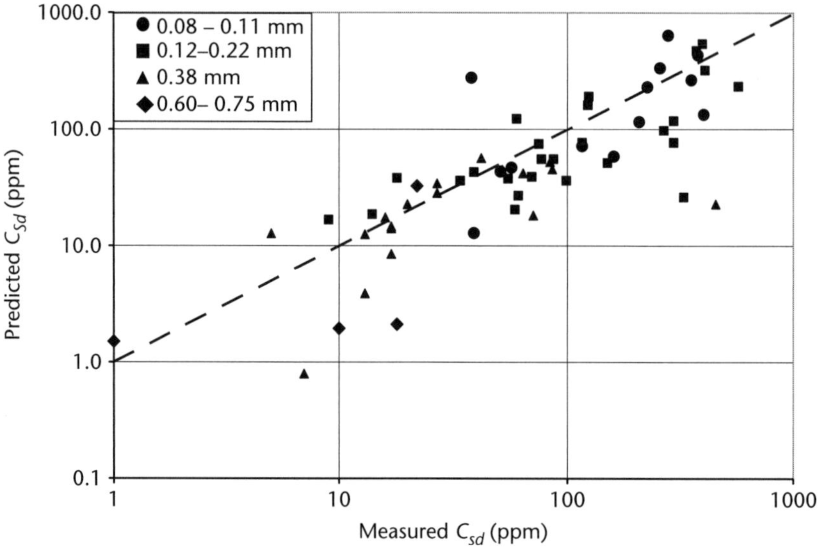

Figure 12.28 Concentration of sand-sized particles predicted by equation 12.53 compared with measured values for 31 flows in natural rivers and canals. Data from Pizzuto (1984).

other variations were given by Engelund and Fredsoe (1976) and Itakura and Kishi (1980). The details of these approaches are too complex to describe here; a clear description of Einstein's method is given in Chang (1988).

Despite extensive research on the problem, predictions of suspended-sediment load by these theoretically based approaches are often considerably in error. For example, Pizzuto (1984) compared the predictions of the Einstein and other diffusion-based methods with actual measurements for sand-sized material and found that all the predictions deviated significantly from measured values. As an alternative approach, he used dimensional analysis to identify dimensionless variables followed by regression analysis to develop an empirical relation that gave more accurate results than any of the theoretical formulas:

$$C_S(d) = 3404 \cdot p(d) \cdot \left(\frac{u_*}{v_f(d)}\right)^2 \cdot \left(\frac{d_{50}}{Y}\right)^{0.60}, \tag{12.53}$$

where $C_S(d)$ is in ppm, and $p(d)$ is the fraction of total bed material of diameter d. figure 12.28 compares values predicted by equation 12.53 with actual values for sand-sized material in natural streams. Although there is still considerable scatter, the equation gives useful predictions.

12.5.2.2 Concentration Profile: Two-Phase Flow?

The "standard" diffusion-theory approach just discussed tacitly assumes that, because of the no-slip condition, the downstream velocities of suspended-sediment particles must equal the downstream water velocity. However, recent experiments by Muste et al. (2005) indicate that this assumption is incorrect.

Muste et al. (2005) carried out their observations of steady, uniform flows in a 0.15-m-wide, 6.0-m-long flume. They used pulsed laser observations of neutrally buoyant particles to measure water velocity, and of quartz-sand particles to measure the velocity and concentration of sediment particles. As shown in figure 12.29, their measurements indicated that sand-particle velocities are up to 5% slower than those for water over most of the flow depth. The most likely explanation for this is that there is a "tendency of the sediment particles to reside in the flow structures [i.e., turbulent eddies] moving with lower velocities" (Muste et al. 2005, p. 8); that is, although the no-slip condition is not violated, the inertia of eddies containing relatively high sediment concentrations causes them to move more slowly. However, sand particles in the region very near the bed travel faster than the water. Muste et al. (2005, p. 8) stated that this "inverse lag" occurs because "sediment particles are not bounded

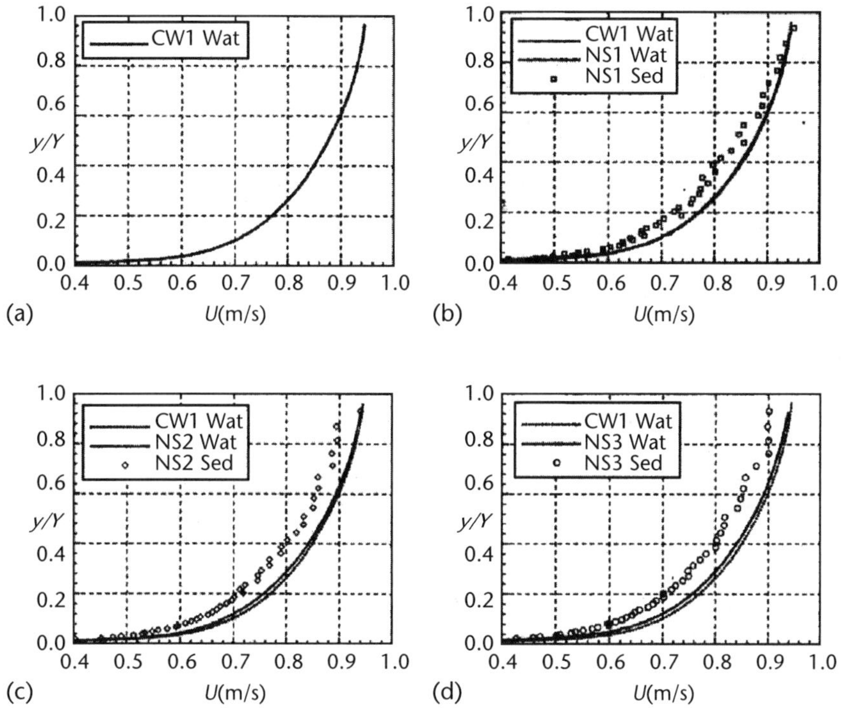

Figure 12.29 Vertical velocity profiles for water and sediment measured in the flume experiments of Muste et al. (2005). (a) "CW1" indicates the measured clear-water profile, plotted in all graphs. (b) "NS1 Wat" and "NS1 Sed" indicate the water- and sediment-velocity profile with a volumetric sediment concentration of 0.00046. (c) "NS2 Wat" and "NS2 Sed" indicate the water- and sediment-velocity profile with a volumetric sediment concentration of 0.00092. (d) "NS3 Wat" and "NS3 Sed" indicate the water- and sediment-velocity profile with a volumetric sediment concentration of 0.00162. In panels c and d, the water-velocity profile slightly lags the clear-water profile. In panels b–d, the sediment-particle velocities lag the water-velocity profiles by up to 5%. From Muste et al. (2005).

by viscosity shear as are fluid particles. Therefore the no-slip condition for water movement at the channel bottom does not apply for the sediment velocity profile."

Thus, Muste et al. (2005) concluded that suspended-sediment transport occurs not as a single-phase mixture moving at the velocity of the water, but as a water phase and a sediment phase moving at slightly different velocities. Thus, the Rouse equation (equation 12.45) qualitatively describes suspended-sediment profiles but departs significantly from measured profiles when the standard value of Ro given by equation 12.46 is used as the exponent. In addition to their finding of two-phase rather than single-phase flow, their experimental results suggest additional discrepancies between the standard theory and actual phenomena:

1. The von Kármán constant decreases from its clear-water value $\kappa = 0.4$ as sediment concentration increases. (This had been suggested by several previous studies, as discussed in section 5.3.1.4.)
2. The vertical velocities of sand particles in turbulent eddies are higher than vertical velocities of water "particles," so that the diffusivities of momentum and sediment may not be equal as assumed in the derivation of equation 12.44 (box 12.5).
3. The assumption of a steadily decreasing concentration with distance above the bed is not generally correct (figure 12.26), leading to difficulties in specifying a reference concentration $C_S(y_a)$.

Overall, Muste et al. (2005, p. 21) concluded that traditional single-phase treatment of suspended-sediment transport is not consistent with actual transport phenomena and that their experimental evidence "proves that use of the traditional formulations, assumptions, and models for suspended sediment transport could be part of the differences, incompleteness, and inconsistency" apparent in the suspended-sediment literature. Further experimental work should lead to improvements in the semiempirical methods used by hydraulic engineers and ultimately to new methods that more completely reflect the physics of two-phase sediment transport.

12.5.3 Total Bed-Material Load

The most widely used approaches for predicting total bed-material load are based on the concept of stream power per unit bed area, $\Pi_A = \tau_0 \cdot U$ (equation 8.27) and, like Pizzuto's (1984) approach (equation 12.53), require separate computations for each component of bed material, d, which are weighted by proportion of the component $p(d)$ and summed to get the total load. These approaches are described and compared by Chang (1988) and are briefly characterized below.

Starting from Einstein's (1950) approach, Colby (1964) developed graphs relating sand-sized bed-material load per unit width to mean velocity and accounting for the effects of depth, particle size, water temperature, and the concentration of wash load.

The method of Engelund and Hansen (1967) was based on Bagnold's (1966) considerations of stream power. Their development led to a dimensionless equation for concentration by weight for each sediment-size class, $C(d)$:

$$C(d) = 0.05 \cdot p(d) \cdot \left(\frac{G_s}{G_s - 1}\right) \cdot \left(\frac{U \cdot S_0}{[(G_s - 1) \cdot g \cdot d]^{1/2}}\right) \cdot \left(\frac{R \cdot S_0}{(G_s - 1) \cdot d}\right), \quad (12.54)$$

where G_s is sediment specific gravity, U is mean velocity, g is gravitational acceleration, R is hydraulic radius ($\approx$ mean depth, Y), and S_0 is slope.

The Ackers and White (1973) approach was also based on Bagnold's (1966) analysis. Their dimensionless relation was of the form

$$C(d) = K_1 \cdot G \cdot \left(\frac{d}{R}\right) \cdot \left(\frac{U}{u_*}\right)^{K_2} \cdot \left(\frac{F_m(u_*, g, d, G_s, U, R, K_3)}{K_4} - 1\right)^{K_5}, \qquad (12.55)$$

where u_* is friction velocity, $F_m(\ldots)$ is a "mobility function," K_1–K_5 are empirical functions of grain size, and the other symbols are as in equation 12.54.

The approach of Yang (1972) (see also Yang 1973, 1984; Yang and Stall 1976; Yang and Molinas 1982) is based on the concept of unit stream power, $\Pi_B \equiv U \cdot S_0$ (equation 8.28), which expresses the time rate of energy dissipation of the flow. The basic relation is of the form

$$C(d) = J_1 \cdot \left(\frac{\Pi_B - \Pi_B{}^*}{v_f}\right)^{J_2}, \qquad (12.56)$$

where $\Pi_B{}^*$ is the critical value of unit stream power, v_f is fall velocity, and J_1 and J_2 are empirically determined values that depend largely on particle diameter and viscosity.

As we have seen, predictions of bed load and suspended load are fraught with uncertainty; thus, it is not surprising that the same is true of attempts to predict total bed-material load. Chang (1988) provided an interesting comparison of the total-load predictions given by three of the above methods for a particular flow: The Engelund and Hansen (1967) method predicted $C = 356$ ppm, $L = 3.47 \times 10^8$ T/day; the Ackers and White (1973) method predicted $C = 866$ ppm, $L = 47.86 \times 10^8$ T/day; and the Yang (1972) method predicted $C = 140$ ppm, $L = 1.27 \times 10^8$ T/day. The highest prediction was more than six times the lowest!

Extensive comparisons of sediment-load predictions with values measured in laboratory flumes and natural rivers were published by Alonso (1980) and Brownlie (1981b). A summary of Brownlie's results is given in figure 12.30, which shows that any given method can give predictions that are many times smaller to many times larger than actual values. A significant part of the discrepancies can be attributed to the difficulties in measurement described in section 12.1, the tremendous spatial and temporal variability of flow conditions and sediment characteristics that make it extremely difficult to characterize flow and sediment conditions and extrapolate samples taken at a few verticals to the entire cross section, and the discrepancies between standard theoretical models of sediment transport and actual transport phenomena described in section 12.5.2.2.

Thus, although the subject is critically important for many practical problems, we must conclude that there is no universally applicable approach to a priori prediction of bed-material load. New experimental techniques such as those used by Muste et al. (2005) will likely improve understanding and predictive ability. Meanwhile, where such predictions are required and basic hydrological and land-use conditions are not changing, it appears that the best approach is to make careful measurements over a wide range of discharges and use empirical (regression) analysis to develop relations of the form of equations 12.5 and 12.7.

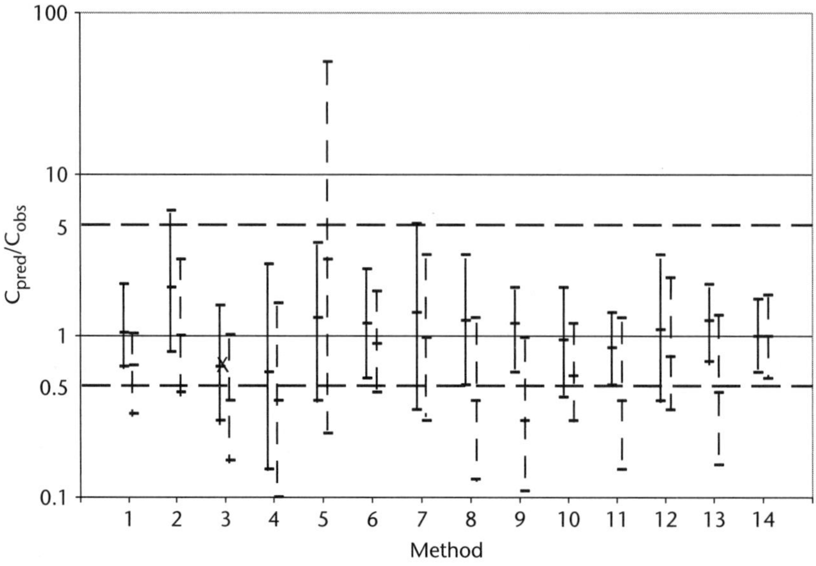

Figure 12.30 Predictive ability of 14 methods for estimating total sediment concentration compared by Brownlie (1981b). The vertical axis is the ratio of predicted to observed concentration. The central dash shows the median value of this ratio for each method; the vertical lines extend from the 16th-percentile value of the ratio to the 84th-percentile value (i.e., 68% of the results for each method fell within the values indicated by the lower and upper ends of the lines). Solid lines show results for flume data; dashed lines, natural-stream data. See Brownlie (1981b) or Chang (1988) for identification and sources of methods. After Chang (1988).

12.5.4 Sediment Transport and Bedforms

As described in section 6.6.4.2, in flows over sand beds there is a typical sequence of bedforms that occurs as discharge increases, proceeding from plane bed to ripples to dunes in the lower flow regime; then from plane bed to antidunes to chutes and pools in the upper flow regime (see table 6.2, figures 6.17–6.20). These forms are intimately related to processes of erosion that begin when the critical threshold for sediment movement is reached, and in turn they strongly influence the velocity and boundary shear stress because of their effects on flow resistance.

An extensive series of flume studies by Simons and Richardson (1966) showed that, for sand-sized particles, the bedform is related to the median fall diameter of the bed material and to stream power per unit bed area, $\Pi_A = \tau_0 \cdot U$ (figure 12.31). For a given median fall diameter, bed-load movement begins when Π_A reaches the critical value represented by the solid curve in figure 12.31. The location of this curve for a given diameter can be computed as the product of critical boundary shear stress $\tau_0{}^*$ from figure 12.17 and critical velocity U^* from figure 12.18. Note that at diameters less than 0.6 mm ripples form initially, and dunes form at higher values of Π_A. With larger sediment, the ripple phased is bypassed and dunes are the initial bedform. According to

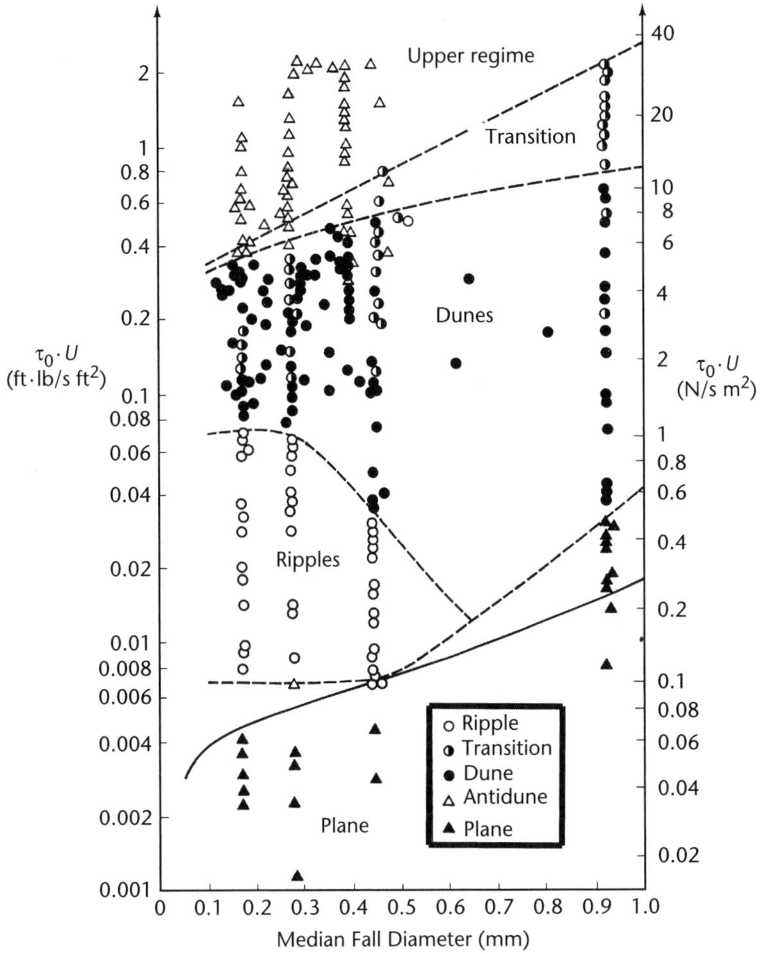

Figure 12.31 Occurrence of bedforms as a function of median bed-material diameter and stream power per unit bed area ($\tau_0 \cdot U$) as determined by flume experiments of Simons and Richardson (1966). The solid curve is the critical value of stream power for initiation of particle motion from figures 12.17 and 12.18. Modified from Simons and Richardson (1966).

Engelund and Hansen (1967), the transition between ripples and dunes as the initial bedform occurs at the transition between hydraulically smooth and hydraulically rough flow. As Π_A increases further, the dunes get "washed out," and a nearly plane bed occurs, marking the boundary between the lower and upper flow regimes.

12.6 The Stable Cross Section

Section 2.4.3.1 introduced the Lane stable channel model. This is a mathematical expression for the shape of the channel cross section (equation 2.19), which was

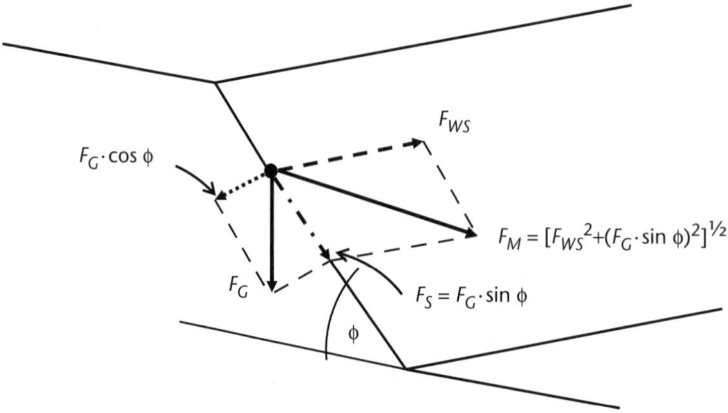

Figure 12.32 Forces on a sediment particle (small circle) on the side of a trapezoidal channel with side-slope ϕ. F_{WS} is the force exerted by the flowing water on a particle on the side-slope; F_G is the submerged weight of the particle; F_S is the downslope force due to gravity; $F_G \cdot \cos \phi$ is the component of particle weight normal to the slope; F_M is the resultant force tending to cause particle movement. After Chang (1988).

derived by hydraulic engineers at the U.S. Bureau of Reclamation (see Lane 1955; Chow 1959; Henderson 1966) assuming that the channel is made of noncohesive material that is just at the threshold of erosion when the flow is bankfull. We can now use the concept of critical boundary shear stress developed from Shields-type experiments to derive this relation. We begin by considering the forces on particles on the bank of a channel with a trapezoidal cross section, and then extend the analysis to a smoothly curved cross section.

12.6.1 Stability of a Trapezoidal Channel

The critical boundary shear stress plotted in figure 12.17b applies to noncohesive particles on the stream bed. If a particle is on a sloping bank, there is an additional gravitational force that tends to move the particle downslope. To develop the force-balance relations, we consider the forces on a particle on the side of a trapezoidal channel with a side-slope ϕ (figure 12.32). The downstream-directed force F_{WS} is due to the boundary shear stress exerted by the flowing water, which is proportional to the product of the channel slope and the local depth. The downslope force F_S is the downslope component of the submerged particle weight, which is

$$F_S = F_G \cdot \sin \phi, \qquad (12.57)$$

where F_G is the submerged weight of the particle. The resultant of these forces is the force tending to cause movement, F_M:

$$F_M = [F_{WS}{}^2 + (F_G \cdot \sin \phi)^2]^{1/2} \qquad (12.58)$$

The concept of angle of repose, Φ, was defined in section 2.3.3 as the maximum slope angle that the bank material can maintain; it reflects the friction among particles and is a function of particle size and shape as shown in figure 2.19. The tangent of the angle of repose is the **coefficient of sliding friction**, and the force on a particle that resists movement on a slope, F_R, is equal to the product of the component of the particle weight that acts normal to the slope, $F_G \cdot \cos \phi$, and that coefficient:

$$F_R = F_G \cdot \cos \phi \cdot \tan \Phi \tag{12.59}$$

The state of incipient motion exists when $F_M = F_R$; equating 12.58 and 12.59 and solving for F_{WS} yields

$$F_{WS} = F_G \cdot [(\cos \phi)^2 \cdot (\tan \Phi)^2 - (\sin \phi)^2]^{1/2}. \tag{12.60a}$$

Using trigonometric identities, equation 12.60a can be written as

$$F_{WS} = F_G \cdot cos\,\phi \cdot \tan \Phi \cdot \left(1 - \frac{(\tan \phi)^2}{(\tan \Phi)^2}\right)^{1/2}. \tag{12.60b}$$

For a particle on the stream bed, $\phi = 0$, so $\tan \phi = 0$, $\cos \phi = 1$, and equation 12.60b gives the force required for incipient motion of a bed particle, F_{WB}, as

$$F_{WB} = F_G \cdot \tan \Phi; \tag{12.61}$$

this force is identical to the critical boundary shear stress $\tau_0{}^*$ shown in figure 12.17b, multiplied by the projected area of the particle, $\pi \cdot d^2/4$. The ratio F_{WS}/F_{WB} is thus equal to the ratio $\tau_{0S}{}^*/\tau_0{}^*$, where $\tau_{0S}{}^*$ is the critical boundary shear stress on a particle on the slope, and we can use equations 12.60b and 12.61 to write

$$\frac{\tau_{0S}{}^*}{\tau_0{}^*} = \frac{F_{WS}}{F_{WB}} = \cos \phi \cdot \left(1 - \frac{(\tan \phi)^2}{(\tan \Phi)^2}\right)^{1/2} = \left(1 - \frac{(\sin \phi)^2}{(\sin \Phi)^2}\right)^{1/2}. \tag{12.62}$$

Figure 12.33 shows values of $\tau_{0S}{}^*/\tau_0{}^*$ as a function of bank angle and angle of repose as given by equation 12.62; the critical shear stress for a sloping bank is less than that for the bed because of the additional gravitational force $F_G \cdot \sin \phi$ that acts on bank particles.

Equation 12.62 can be used to determine the maximum bank angle for stability of a trapezoidal channel, as described by Chow (1959) and Henderson (1966). The procedure requires information about the actual shear stress on the bed and banks, and this information was provided by studies conducted by Olsen and Florey (1952). The pattern of shear-stress distribution depends on the width/depth ratio and the side-slope, but for trapezoidal channels of the shapes ordinarily used the maximum boundary shear stress on the bottom is approximately equal to $\gamma \cdot \Psi \cdot S_0$ and on the sides to $0.75 \cdot (\gamma \cdot \Psi \cdot S_0)$, where Ψ is the maximum depth (Chow 1959; Henderson 1966). A typical shear-stress distribution is shown in figure 12.34.

12.6.2 The Lane Stable Channel

12.6.2.1 Derivation

Referring to figure 12.35, we can now use equation 12.62 to derive an expression for the form of a smoothly curved channel cross section over which the state of incipient

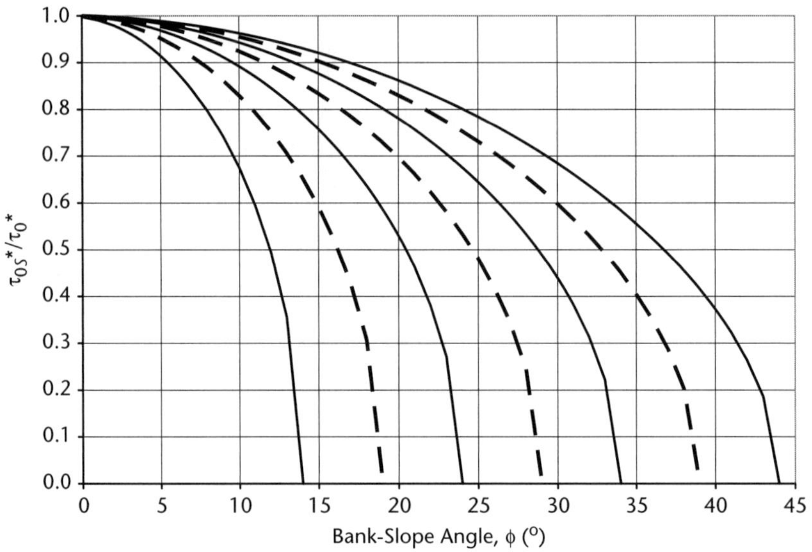

Figure 12.33 Ratio of critical shear stress on a trapezoidal channel bank, τ_{0S}*, to critical shear stress on the channel bed, τ_0* (figure 12.17b), as a function of bank-slope angle, ϕ, for material with various angles of repose, Φ (equation 12.62).

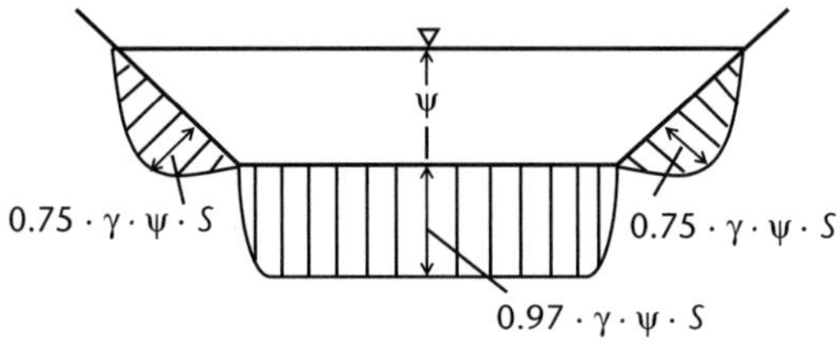

Figure 12.34 Distribution of boundary shear stress in a typical trapezoidal channel as determined in studies by Olsen and Florey (1952).

motion exists when the flow is bankfull. This derivation is carried out in box 12.6, resulting in the expression for the **Lane stable channel section** (Lane 1955):

$$z(w) = \Psi_{BF} \cdot \left[1 - \cos\left(\frac{\tan(\Phi)}{\Psi_{BF}} \cdot w\right)\right], 0 \leq w \leq W_{BF}/2, \tag{12.63}$$

where $z(w)$ is the elevation of the channel bottom at a distance w from the center, Ψ_{BF} is the maximum (central) channel depth, Φ is the angle of repose of the channel material, and W_{BF} is the bankfull channel width.

For given maximum depth Ψ_{BF} or width W_{BF}, the form of the Lane cross section is a function of the angle of repose Φ ; at the channel edge, where $z(W_{BF}/2) = \Psi_{BF}$,

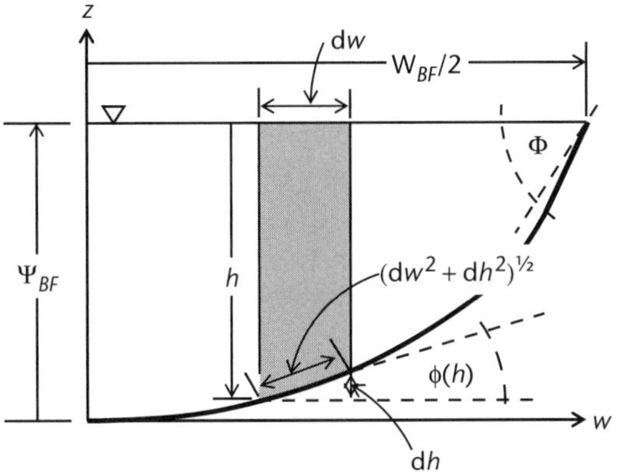

Figure 12.35 Definition diagram for derivation of the Lane stable-channel relation.

BOX 12.6 Derivation of the Lane Stable-Channel Relation

Figure 12.35 shows one-half of an idealized channel cross section at bankfull flow. The shear force per unit downstream distance exerted by the flowing water on the channel bed at the base of an elemental area (shaded) where the distance below the bankfull level is h, $F(h)$, is

$$F(h) = \gamma \cdot h \cdot S_0 \cdot dw, \tag{12B6.1}$$

where γ is the weight density of water and S_0 is the channel slope. $F(h)$ is the downstream component of the weight of the shaded element, per unit downstream distance. This force acts over the perimeter distance $(dw^2 + dh^2)^{1/2}$, and because $dw/(dw^2 + dh^2)^{1/2} = \cos\phi(h)$, the shear stress at this point is

$$\tau_0(h) = \frac{\gamma \cdot h \cdot S_0 \cdot dw}{(dw^2 + dh^2)^{1/2}} = \gamma \cdot h \cdot S_0 \cdot \cos\phi(h). \tag{12B6.2}$$

At the channel center, $h = \Psi_{BF}$, $\phi(\Psi_{BF}) = 0$, and $\cos\phi(\Psi_{BF}) = 1$, so the shear stress is

$$\tau_0(\Psi_{BF}) = \gamma \cdot \Psi_{BF} \cdot S_0. \tag{12B6.3}$$

The ratio of the shear stress at any point in the cross section to that at the center is thus

$$\frac{\tau_0(h)}{\tau_0(\Psi_{BF})} = \frac{\gamma \cdot h \cdot S_0 \cdot \cos\phi(h)}{\gamma \cdot \Psi_{BF} \cdot S_0} = \frac{h \cdot \cos\phi(h)}{\Psi_{BF}}. \tag{12B6.4}$$

This ratio is identical to the ratio derived for a trapezoidal side-slope in equation 12.62, but now ϕ is a function of h. Thus, we can write

$$\frac{h \cdot \cos\phi(h)}{\Psi_{BF}} = \cos\phi(h) \cdot \left(1 - \frac{[\tan\phi(h)]^2}{(\tan\Phi)^2}\right)^{1/2}, \tag{12B6.5}$$

(Continued)

BOX 12.6 *Continued*

which can be solved for $\tan \phi(h)$:

$$tan\,\phi(h) = \tan\Phi \cdot \left[1 - \left(\frac{h}{\Psi_{BF}}\right)^2\right]^{1/2}. \qquad (12B6.6)$$

Note, however, that $\tan \phi(h) \equiv dh/dw$, so equation 12B6.6 can be written as a differential equation:

$$\frac{dh}{dw} = \tan\Phi \cdot \left[1 - \left(\frac{h}{\Psi_{BF}}\right)^2\right]^{1/2}. \qquad (12B6.7)$$

Separating variables,

$$\frac{dh}{\left(1 - \frac{h^2}{\Psi_{BF}{}^2}\right)^{1/2}} = \tan\,\Phi \cdot dw. \qquad (12B6.8)$$

The integral of the left-hand side of equation 12B6.8 can be found from a table of integrals:

$$\int \frac{dh}{\left(1 - \frac{h2}{\Psi_{BF}{}^2}\right)^{1/2}} = \sin^{-1}\left(\frac{h}{\Psi_{BF}}\right). \qquad (12B6.9)$$

The integral of the right-hand side is

$$\int \tan\,\Phi \cdot dw = \tan\,\Phi \cdot w. \qquad (12B6.10)$$

Therefore,

$$\sin^{-1}\left(\frac{h}{\Psi_{BF}}\right) = \tan\,\Phi \cdot w + C, \qquad (12B6.11)$$

where C is a constant of integration. C is evaluated by incorporating the boundary condition $h = \Psi_{BF}$ at $w = 0$, to find

$$C = \frac{\pi}{2}. \qquad (12B6.12)$$

Combining equations 12B6.11 and 12B6.12, we have

$$h = \Psi_{BF} \cdot \sin\left(\frac{\tan\,\Phi}{\Psi_{BF}} \cdot w + \frac{\pi}{2}\right) \qquad (12B6.13a)$$

or, equivalently,

$$h = \Psi_{BF} \cdot \cos\left(\frac{\tan\,\Phi}{\Psi_{BF}} \cdot w\right). \qquad (12B6.13b)$$

Noting that $h = \Psi_{BF} - z(w)$, we have

$$z(w) = \Psi_{BF} \cdot \left[1 - \cos\left(\frac{\tan(\Phi)}{\Psi_{BF}} \cdot w\right)\right], 0 \le w \le W_{BF}/2, \qquad (12B6.14)$$

which is the Lane stable-channel formula.

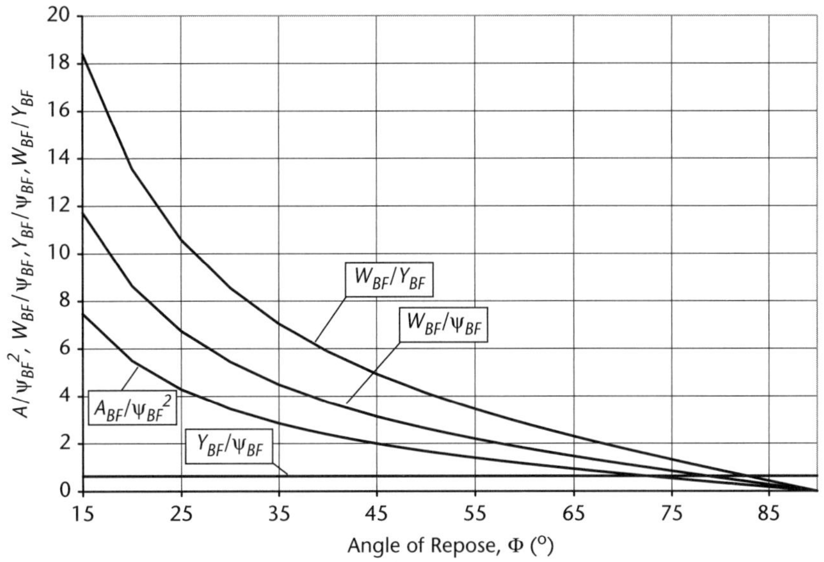

Figure 12.36 Geometry of the Lane stable channel cross section as a function of angle of repose. A_{BF} is area, W_{BF} is width, Y_{BF} is average depth, and Ψ_{BF} is maximum depth. The ratio $Y_{BF}/\Psi_{BF} = 2/\pi = 0.637$, regardless of Φ.

the bank angle $\phi = \Phi$. Because the argument of the cosine function must be $\leq \pi/2$, the relation of equation 12.63 dictates limits on channel geometry:

$$W_{BF} = \frac{\pi \cdot \Psi_{BF}}{\tan \Phi}; \tag{12.64}$$

$$A_{BF} = \frac{2 \cdot \Psi_{BF}{}^2}{\tan \Phi} = \frac{2 \cdot W_{BF}{}^2 \cdot \tan \Phi}{\pi}; \tag{12.65}$$

$$Y_{BF} = \frac{2 \cdot \Psi_{BF}}{\pi} = \frac{2 \cdot W_{BF} \cdot \tan \Phi}{\pi^2}. \tag{12.66}$$

The relations between these limits and Φ are shown in figure 12.36. Note especially that for the range of Φ for natural noncohesive particles the maximum width/depth ratio permitted by equation 12.63 is less than 20, which is smaller than occurs in most natural channels (see section 2.4.2, figure 2.24). As pointed out by Henderson (1966), these limits can be avoided while still satisfying the stability requirements by inserting a rectangular section between two banks having the form dictated by equation 12.63 (figure 12.37); Henderson (1966) called the cross section given by equation 12.63 the "type B" form, and that with an inserted rectangular section the "type A" form. The type A form makes the Lane stable channel model more flexible than first appears and allows it to be used in the design of canals (see Chow 1959).

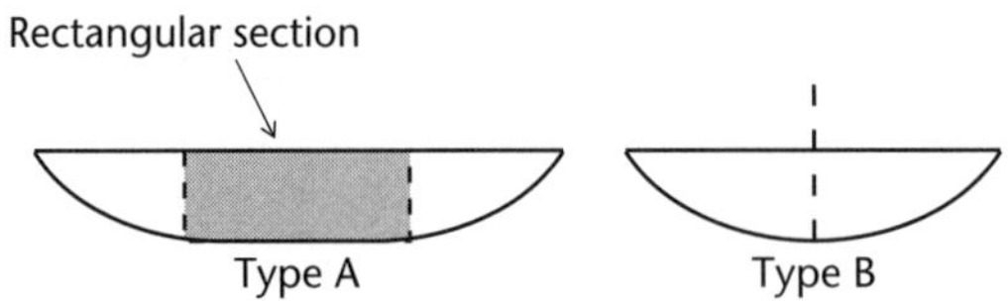

Figure 12.37 The Lane stable channel. Henderson's (1966) "type B" cross section follows equation 12.63 on both sides of the center line, and the dimensions are dictated by the angle of repose as plotted in figure 12.36. In the "type A" section, the sides still follow equation 12.63, but a rectangular section is inserted between them so that values of the form ratios larger than shown in figure 12.36 can be achieved. After Henderson (1966).

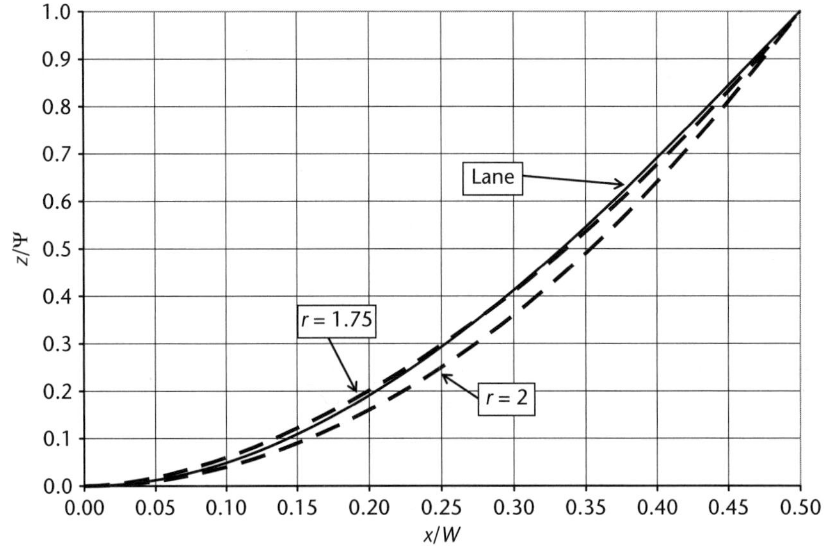

Figure 12.38 Comparison of the Lane stable channel (equation 12.63) (solid curve) with the general power-law cross section (equation 12.67) with $r = 1.75$ and $r = 2$ (parabola).

12.6.2.2 Comparison with Generalized Cross-Section Form

We can compare the form of the Lane stable channel with the generalized cross-section formula given in box 2.4 (equation 2B4.2):

$$z(w) = \Psi_{BF} \cdot \left(\frac{2 \cdot w}{W_{BF}}\right)^r, w \leq W_{BF}/2; \tag{12.67}$$

where $z(w)$ is the elevation above the lowest point at a distance w from the center.[10] Figure 12.38 compares the form given by equation 12.63 with that given by equation 12.67 with $r = 1.75$ and $r = 2$ (a parabola); the Lane curve is very close to that of the general model with $r = 1.75$ (and quite similar to a parabola). Thus the general model with $r = 1.75$ is a mathematically convenient, flexible version of the Lane type A channel that provides a plausible starting point for a physically based model of the form of natural-channel cross sections, as described in the following section.

Appendices

Appendix A. Dimensions, Units, and Numerical Precision

Correct treatment of numerical quantities requires an understanding of the qualitative aspects of numbers—the concepts of dimensions, units, and numerical precision. Engineers and scientists also encounter quantities measured in various unit systems and must become adept at converting measurements and equations made in one system to other systems. In fact, the most common and embarrassing errors you will make in scientific and engineering practice will likely be those involving dimensions, units, and numerical precision.

This appendix summarizes rules for the correct treatment of dimensions and units, relative and absolute measurement precision, and unit and equation conversion.

A.1 Dimensions

Rule 1: The fundamental **dimensional character** of quantities encountered in fluvial hydraulics can be expressed as

$$[\mathrm{M}^a \mathrm{L}^b \mathrm{T}^c \Theta^d] \tag{A.1a}$$

or as

$$[\mathrm{F}^e \mathrm{L}^f \mathrm{T}^g \Theta^h], \tag{A.1b}$$

where [M] indicates the dimension of mass, [F] the dimension of force, [L] the dimension of length, [T] the dimension of time, and [Θ] the dimension of temperature; and the exponents $a, b, \ldots, g$ are rational numbers[1] or zero.

The choice of whether to use force or mass is a matter of convenience. Dimensions expressed in one system are converted to the other system via Newton's second law of motion (section 4.2, equation 4.4):

$$[\mathrm{F}] = [\mathrm{M\,L\,T^{-2}}] \tag{A.2a}$$

$$[\mathrm{M}] = [\mathrm{F\,L^{-1}\,T^{2}}] \tag{A.2b}$$

Rule 2: Dimensionless quantities are obtained as follows:

1. By counting
2. As the ratio of quantities with identical dimensional character (this includes plane angles, which are measured as the ratio of the circular arc length subtended by the angle to circumference of the circle)
3. From pure numbers such as $\pi \equiv 3.14159\ldots$ (actually a ratio) and $e \equiv 2.7182\ldots$
4. By logarithmic, exponential, and trigonometric functions
5. From exponents, except for those that may arise in certain empirical relations[2]

Table A.1 Dimensional classification of quantities encountered in fluvial hydraulics.

Dimensions involved	Classification
[1]	Dimensionless[a]
[L] only	Geometric[a]
[L] and [T] only	Kinematic
[F] or [M]	Dynamic
[Θ]	Thermal[b]

[a]Angle [1] is classified as geometric.
[b]Latent heat $[L^2\ T^{-2}]$ is classified as thermal.

The dimensional quality of dimensionless numbers is expressed as $[1] = [M^0\ L^0\ T^0 \Theta^0] = [F^0\ L^0\ T^0 \Theta^0]$.

Quantities are classified according to their dimensional character, as shown in table A.1. Columns 1–4 of table A.2 give the dimensional character of quantities commonly encountered in fluvial hydraulics.

A.2 Units

Units are the arbitrary standards in which the magnitudes of quantities are expressed. When we give the units of a quantity, we are expressing the ratio of its magnitude to the magnitude of an arbitrary standard with the same fundamental dimension. Table A.3 gives the units of the fundamental dimensions (plus angle) in the three systems of units that are or have been in common use in science and engineering. The **Système International** (SI) is now the international standard for all branches of science; the SI units of quantities commonly encountered in fluvial hydrology are given in column 4 of table A.2. The centimeter-gram-second (cgs) system was an earlier standard, and the "U.S. conventional" system is still widely used in the United States.

A.3 Precision and Significant Figures

Precision is the "fineness" with which a quantity is measured. Precision is determined, at least conceptually, as the repeatability of the results of a given measurement. For example, suppose we make 10 measurements of a stream width with a measurement tape marked in meters and centimeters and obtain the following results (in meters): 10.40, 10.20, 10.32, 10.64, 11.11, 10.94, 10.21, 11.09, 10.85, 11.30. The average of the values is 10.706 m and the range is from 10.20 m to 11.30 m, or 1.10 m. Thus, the precision is approximately 1 m, and we should report the length as 11 m.[3]

Note that precision is distinct from **accuracy**, which is determined as the difference between a measured value and the true value.

The precision of any measured value can be expressed in both absolute and relative terms. **Absolute precision** is expressed in terms like "to the nearest *x*," where *x* is some measurement unit. In the example above, the absolute precision is approximately 1 m.

Table A.2 Dimensions, SI units, and conversion factors (to four significant figures).[a]

Quantity	Dimensions	Classification	SI units	Conversion factors
Acceleration	$[L\,T^{-2}]$	Kinematic	m/s^2	$cm/s^2 \times 10^{-2}$
				$ft/s^2 \times 3.048 \times 10^{-1}$
Angle	[1]	Geometric	rad	degree $\times 1.745 \times 10^{-2}$
Angular acceleration	$[T^{-2}]$	Kinematic	rad/s^2	$degree/s^2 \times 1.745 \times 10^{-2}$
Angular velocity	$[T^{-1}]$	Kinematic	rad/s^1	$degree/s^1 \times 1.745 \times 10^{-2}$
Area	$[L^2]$	Geometric	m^2	acre $\times 4.047 \times 10^3$
				$ft^2 \times 9.290 \times 10^{-2}$
				$cm^2 \times 10^{-4}$
				hectare $\times 10^4$
				$in^2 \times 6.452 \times 10^{-4}$
				$km^2 \times 10^6$
				$mi^2 \times 2.590 \times 10^6$
Density, mass	$[M\,L^{-3}]$	Dynamic property	kg/m^3	1×10^3
	$[F\,L^{-4}\,T^2]$			$g/cm^3 \times 10^3$
				$lb_m/ft^3 \times 1.602 \times 10^1$
				$slug/ft^3 \times 5.154 \times 10^2$
Density, weight	$[M\,L^{-2}\,T^{-2}]$	Dynamic property	N/m^3	9.798×10^3
	$[F\,L^{-3}]$			$g_f/cm^3 \times 9.798 \times 10^3$
				$lb/ft^3 \times 1.571 \times 10^2$
Diffusivity	$[L^2\,T^{-1}]$	Kinematic	m^2/s	$cm^2/s \times 10^{-4}$
				$ft^2/s \times 9.290 \times 10^{-2}$
Discharge	$[L^3\,T^{-1}]$	Kinematic	m^3/s	$cm^3/s \times 10^{-6}$
				$ft^3/s \times 2.832 \times 10^{-2}$
				gal/min $\times 6.309 \times 10^{-5}$
				gal/day $\times 4.381 \times 10^{-8}$
				L/s $\times 10^{-3}$
Energy (work)	$[M\,L^2\,T^{-2}]$	Dynamic	$N \cdot m = J$	Btu $\times 1.055 \times 10^3$
	$[F\,L]$			cal $\times 4.187$
				$ft \cdot lb \times 1.356$
				$kW \cdot hr \times 3.600 \times 10^6$
Energy flux	$[M\,T^{-3}]$	Dynamic	$J/m^2 \cdot s = N/m \cdot s$	$Btu/ft^2 \cdot s \times 1.135 \times 10^4$
	$[F\,L^{-1}\,T^{-1}]$		$= W/m^2$	$cal/cm^2 \cdot s \times 4.187 \times 10^4$
				$lb/ft \cdot s \times 1.460 \times 10^1$
Force (weight)	[F]	Dynamic	N	dyne $\times 10^{-5}$
	$[M\,L\,T^{-2}]$			$kg_f \times 9.807$
				lb $\times 4.448$
Heat capacity	$[L^2\,T^{-2}\Theta^{-1}]$	Thermal property	$J/kg \cdot K$	4.187×10^3
				$cal/g \cdot °C \times 4.187 \times 10^3$
				$Btu/lb_m \cdot °F \times 4.187 \times 10^3$
Latent heat	$[L^2\,T^{-2}]$	Thermal property	J/kg	Freezing $= 3.340 \times 10^5$
				Evaporation $= 2.495 \times 10^6$
				cal/g $\times 4.187 \times 10^3$
				$Btu/lb_m \times 2.326 \times 10^3$
Length	[L]	Geometric	m	cm $\times 10^{-2}$
				ft $\times 3.048 \times 10^{-1}$
				in $\times 2.540 \times 10^{-2}$
				mi $\times 1.609 \times 10^3$

Table A.2 *(Continued)*

Quantity	Dimensions	Classification	SI units	Conversion factors
Mass	[M] [F $L^{-1}T^2$]	Dynamic	kg	g × 10^{-3} lb_m × 4.536 × 10^{-1} slug × 1.459 × 10^1
Momentum	[M L T^{-1}] [F T]	Dynamic	kg · m/s	g · cm/s × 10^{-5} lb_m · ft/s × 1.383 × 10^{-1} slug · ft/s × 5.077 × 10^{-1}
Power	[M L^2 T^{-3}] [F L T^{-1}]	Dynamic	N · m/s = J/s = W	Btu/s × 1.054 × 10^3 cal/s × 4.184 dyne · cm/s × 10^{-7} lb · ft/s × 1.356
Pressure (stress)	[M L^{-1} T^{-2}] [F L^{-2}]	Dynamic	N/m^2 = Pa	atmosphere × 1.013 × 10^5 bar × 10^5 dyne/cm^2 × 10^{-1} ft of water × 2.989 × 10^3 g_f/cm^2 × 9.807 × 10^1 in Hg × 3.386 × 10^3 kg_f/m^2 × 9.807 mb × 10^2 mm Hg × 1.333 × 10^2 lb/ft^2 × 4.788 × 10^1 lb/in^2 × 6.895 × 10^3
Stream power, unit	[L T^{-1}]	Kinematic	m/s	cm/s × 10^{-2} ft/s × 3.048 × 10^{-1}
Stream power, per unit channel length	[M L^2 T^{-3}] [F T^{-1}]	Dynamic	N/s	dyne/s × 10^{-5} lb/s × 4.448
Stream power, per unit bed area	[M L T^{-3}] [F L^{-1} T^{-1}]	Dynamic	N/m · s = J/m^2 · s = W/m^2	dyne/cm · s × 10^{-3} lb/ft · s × 1.459 × 10^1 7.420 × 10^{-2}
Surface tension	[M T^{-2}] [F L^{-1}]	Dynamic property	N/m	dyne/cm × 10^{-3} lb/ft × 1.459 × 10^1
Temperature	[Θ]	Thermal	K	°C + 273.2 °F × (5/9 + 459.7)
Thermal conductivity	[M L $T^{-3}Θ^{-1}$] [F $T^{-1}Θ^{-1}$]	Thermal property	W/m · K	3.474 × 10^{-3} Btu/s · ft ·°F × 3.115 × 10^5 cal/s · cm ·°C × 4.187 × 10^2
Time	[T]	Kinematic	s	day × 8.64 × 10^4 hr × 3.6 × 10^3 min × 6 × 10^1 month (mean) × 2.628 × 10^6 year × 3.154 × 10^7
Velocity	[L T^{-1}]	Kinematic	m/s	cm/s × 10^{-2} ft/s × 3.048 × 10^{-1} km/hr × 2.778 × 10^{-1} mi/hr × 4.470 × 10^{-1}
Viscosity, dynamic	[M L^{-1} T^{-1}] [F T L^{-2}]	Dynamic property	N · s/m^2 = Pa · s	1.307 × 10^{-3} dyne · s/cm^2(poise × 10^{-1}) lb · s/ft^2 × 4.788 × 10^1

(Continued)

Table A.2 *(Continued)*

Quantity	Dimensions	Classification	SI units	Conversion factors
Viscosity, kinematic	$[L^2 T^{-1}]$	Kinematic property	m^2/s	1.307×10^{-6} cm^2/s (stoke $\times 10^{-4}$) $ft^2/s \times 9.290 \times 10^{-2}$
Volume	$[L^3]$	Geometric	m^3	acre · ft $\times 1.233 \times 10^3$ $cm^3 \times 10^{-6}$ $ft^3 \times 2.832 \times 10^{-2}$ gal $\times 3.785 \times 10^{-3}$ L $\times 10^{-3}$

[a]The first four columns give the dimensions and SI units of quantities commonly encountered in fluvial hydraulics. The last column gives conversion factors (four significant figures) from common units to SI units and the values of water properties at 10°C in SI units. g_f, gram force; kg_f, kilogram force; lb_m, pound mass.

Table A.3 Units of the fundamental dimensions (plus angle) in the three unit systems encountered in fluvial hydraulics.

Fundamental dimension	Système International (SI) unit	Centimeter-gram-second (cgs) unit	U.S. conventional unit
Mass	kilogram (kg)	gram (g)	slug
Force	Newton (N)	dyne	pound (lb)
Length	meter (m)	centimeter (cm)	foot (ft)
Time	second (s)	second (s)	second (s)
Temperature	Kelvin (K)	degree Celsius (°C)	degree Fahrenheit (°F)
Angle	radian (rad)	radian (rad)	degree (°)

Relative precision can be expressed as the number of **significant figures** in the numerical expression of a measured quantity; this number is equal to the number of digits beginning with the leftmost nonzero digit and extending to the right to include all digits warranted by the precision of the measurement. In the example above, the relative precision is two significant figures.

> **Rule 3:** All measured quantities have finite precision, which must be appropriately considered in calculations as described below.

A.3.1. Absolute Precision

If we were to measure a distance to the nearest centimeter, we would have to report it as, say, 10.71 m. If we were to report the measurement as 10.706 m, we would be implying that it had been made to the nearest millimeter. If a measurement is given as, say 200 m, the precision is not clear because we do not know if the measurement was made to the nearest meter, 10 m, or 100 m. One way of avoiding this ambiguity is to use scientific notation and express the quantity as 2×10^2 m, 2.0×10^2 m, or 2.00×10^2 m, as appropriate. This is not consistently done, however, so additional information, usually in the form of other analogous measurements, is required to clarify the situation.

In adding or subtracting measured values, we must be concerned with absolute precision, and observe the following rule:

Rule 4: The absolute precision of a sum or difference equals the absolute precision of the least precise number involved in the calculation.

Hydrologists often deal with streamflow data collected by the U.S. Geological Survey (USGS). Based on the variability of repeated measurements of a given flow (see section 2.5.3), the USGS has determined the absolute precision values for discharge measurements shown in table A.4. Below are some examples of how absolute precision should be treated in adding flows.

EXAMPLE A.3.1.1.

Suppose the average flow for two consecutive days is measured as 102 ft^3/s and 3.2 ft^3/s. How should the 2-day total reported?

Adding the reported values gives 105.2 ft^3/s, but because the larger flow was measured only to the nearest 1 ft^3/s, we must report the total as 105 ft^3/s.

EXAMPLE A.3.1.2.

Suppose the flows for two consecutive days were 1020 ft^3/s and 3.2 ft^3/s. What is the total?

Here the sum must be reported as 1020 ft^3/s, because the larger flow was measured to the nearest 10 ft^3/s.

EXAMPLE A.3.1.3.

Given the daily flows 27, 104, 2310, 256, 12, 6.4, and 0.11 ft^3/s, what is the total flow for the 7-day period?

Adding all these values gives 2715.51 ft^3/s, but because the largest value was measured only to the nearest 10 ft^3/s, we must report the sum as 2720 ft^3/s.

A.3.2. Relative Precision

In reporting a measured value, any digits farther to the right than warranted by the measurement precision are **nonsignificant figures**.

Rule 5: Only the significant figures should be included in stating a measured value.

Thus, reporting a measurement as 11, 10.7, and 10.71 m implies two-, three-, and four-significant-figure precision, respectively.

Table A.4 Absolute precision of streamflow (discharge) data reported by the USGS (which still uses the U.S. conventional unit system).

Discharge range (ft^3/s)	Precision (ft^3/s)
<1	0.01
1.0–9.9	0.1
10–999	1
$>1,000$	Three significant figures

In multiplication and division, we must be concerned with relative precision, and observe the following rule:

Rule 6: The number of significant figures of a product or quotient equals the number of significant figures of the least relatively precise number involved in the calculation.

The following examples show how relative precision should be treated in multiplication.

EXAMPLE A.3.2.1.

Discharge, Q, equals the product of width, average depth, and average velocity at a given stream cross section. Suppose the water-surface width of a stream is measured as 20.4 m, the average depth as 1.2 m, and the average velocity as 1.7 m/s. What is the discharge?

Multiplying the measured values:

$$Q = 20.4\ \text{m} \times 1.2\ \text{m} \times 1.7\ \text{m/s} = 41.616\ \text{m}^3/\text{s}.$$

Following of rules 5 and 6, we report the discharge to two significant figures as $Q = 42\ \text{m}^3/\text{s}$.

EXAMPLE A.3.2.2.

To estimate the average depth of a channel cross section, we measure the following depths in m at 10 equally spaced locations: 0.23, 0.65, 0.98, 1.25, 1.03, 1.64, 0.94, 0.76, 0.44, 0.19. The sum of these values is 8.11 m, and the average is $8.11/10 = 0.811$ m. However, because we only measured all depths to the nearest 0.01 m, we must follow rule 6 and report the average depth as 0.81 m.

Rule 7: Unless it is clear that greater precision is warranted, assume no more than three-significant-figure precision in field measurements of fluvial-hydraulic quantities.

As noted in table A.4, there are many cases where only two-significant-figure precision is warranted. The precision of measurements in laboratory flumes may be greater than three significant figures.

A.4 Unit Conversion

Because of the common use of three systems of units and the proliferation of units within each system, hydrologists must become expert at converting from one set of units to another.

Column 5 of table A.2 gives factors for converting common non-SI units to SI units. Conversion factors are used as either numerators or denominators in fractions whose actual physical value is exactly 1, but whose numerical value is some other number. For example, in terms of actual lengths,

$$\frac{1\ \text{ft}}{0.3048\ldots\ \text{m}} = 1.000\ldots; \quad \frac{0.3048\ldots\ \text{m}}{1\ \text{ft}} = 1.000\ldots$$

Rule 6 must be followed in all unit conversions. However, because all conversion factors have infinite precision, it is only the precision of the measured quantities—not the conversion factors—that determines the significant figures of the converted value. Thus, the following rule must be observed in doing unit conversions:

Rule 8: In unit conversions, the number of digits retained in the conversion factors must be greater than the number of significant digits in any of the measured quantities involved.

Except for commonly used temperature units (discussed below), a zero value in one unit system is a zero value in the other systems. Conversion in these cases is simply a matter of multiplying by the appropriate conversion factor, and the decision of whether to put the factor in the numerator or denominator is determined by the direction of the conversion. Below are some examples of unit conversions.

EXAMPLE A.4.1.

Suppose a distance is measured as 9.6 mi. How is that same distance expressed in meters?

Table A.2 indicates that we multiply 9.6 mi times 1609 ... m/mi:

$$9.6\,\text{mi} \times \frac{1609\ldots\,\text{m}}{1.000\ldots\,\text{mi}} = 15,446.4\,\text{m} \rightarrow 15,000\,\text{m}\ldots$$

Note that the conversion factor has four digits, so rule 8 is observed. Following rule 6, we round the converted value to two significant figures.

Clearly, it would be misleading to express the result as 15,446.4 m, because this would imply that we know the distance to a precision of 0.1 m, whereas the original measurement was known only to 0.1 mi or about 161 m. However, in following rule 6 we have in fact lost some absolute precision: stating the distance as 15,000 m implies an absolute precision of 1000 m, which is considerably less precise than the original precision of 161 m. Still, this is the correct procedure—if we had instead stated the converted distance as 15,400 m, we would be exaggerating the true precision of the originally measured value. Generally, we accept the loss in absolute precision that results from applying rule 6. An alternative that more accurately conveys the precision of the original measurement is to state the converted value with an explicit absolute precision—in the given example, as $15,400 \pm 161$ m. This is seldom done, however.

EXAMPLE A.4.2.

Express the measured distance of 855.26 m in kilometers (a), and miles (b).

Observing rules 6 and 8,

$$855.26\,\text{m} \times \frac{1.000\ldots\,\text{km}}{1000.\ldots\,\text{m}} = 0.85526\,\text{km} \quad \text{(a)}$$

$$855.26\,\text{m} \times \frac{1.000.\ldots\,\text{mi}}{1609.34\ldots\,\text{m}} = 0.53144\,\text{mi}. \quad \text{(b)}$$

Note that in equation b there is again a loss of precision, because the original measurement was to the nearest 0.01 m, whereas 0.00001 mi $\approx$ 0.016 m.

EXAMPLE A.4.3.

This example applies rules 6 and 8 in a case where two unit conversions are required.
Convert 19 mi/hr to m/s:

$$19\text{ mi hr}^{-1} \times \left(\frac{1609.\ldots\text{ m}}{1.000\ldots\text{ mi}}\right) \times \left(\frac{1.000\ldots\text{ hr}}{3600\ldots\text{ s}}\right) = 8.4919..\text{ m/s} \rightarrow 8.5\text{ m/s}$$

Rule 9: Conversion of actual temperatures from one system to another involves addition or subtraction because the zero points differ.

The examples below illustrate the procedure. Note that actual Celsius and Fahrenheit temperatures are written here with the degree sign before the letter symbol (read "degree celsius" or "degree fahrenheit"), whereas temperature differences—distances on the temperature scale—for each system are written with the symbol after the letter (read "celsius degree" or "fahrenheit degree"). The zero point for the Kelvin scale is absolute zero, so the degree sign is not used in that system.

Rule 10: Conversion of temperature differences does not involve addition or subtraction because we are dealing only with distances on the temperature scales.

Thus, conversion of temperature differences follows the same procedures illustrated above.

The following examples use rules 6, 8, and 9.

EXAMPLE A.4.4.

To convert −37°F to °C:

$$(-37°\text{F} - 32.000\ldots°\text{F}) \times \frac{1.000\ldots\text{ C}°}{1.800\ldots\text{ F}°} = -38.33\ldots°\text{C} \rightarrow -38°\text{C}$$

EXAMPLE A.4.5.

To convert −37°C to °F:

$$(-37°\text{C}) \times \frac{1.800\ldots\text{ F}°}{1.000\ldots\text{ C}°} + 32.000\ldots°\text{F} = -34.6\ldots°\text{F} \rightarrow -35°\text{F}$$

EXAMPLE A.4.6.

To convert −37°C to K:

$$(-37°\text{C}) \times \frac{1.000\ldots\text{ K}}{1.000\ldots\text{ C}°} + 273.16\ldots\text{K} = 236.16\text{ K} \rightarrow 236\text{ K}$$

EXAMPLE A.4.7.

To convert 295 K to °C:

$$(295\text{ K}) \times \frac{1.000\ldots\text{ C}°}{1.000\ldots\text{ K}} - 273.2\text{ K} = 21.8°\text{C} \rightarrow 22°\text{C}$$

The following examples use rules 6, 8, and 10.

EXAMPLE A.4.8.

Convert a temperature difference of 3.4 F° to C°:

$$3.4\,\mathrm{F}^\circ \times \frac{1.000\ldots\,\mathrm{C}^\circ}{1.800\ldots\,\mathrm{F}^\circ} = 1.888\ldots\mathrm{C}^\circ \rightarrow 1.9\,\mathrm{C}^\circ$$

EXAMPLE A.4.9.

Convert a temperature difference of 3.4 C° to F°:

$$3.4\,\mathrm{C}^\circ \times \frac{1.800\ldots\,\mathrm{F}^\circ}{1.000\ldots\,\mathrm{C}^\circ} = 6.12\ldots\,\mathrm{F}^\circ \rightarrow 6.1\,\mathrm{F}^\circ$$

EXAMPLE A.4.10.

Convert a temperature difference of 3.4 C° to K.

$$3.4\,\mathrm{C}^\circ \times \frac{1.000\ldots\,\mathrm{K}}{1.000\ldots\,\mathrm{C}^\circ} = 3.4\,\mathrm{K}$$

One should also observe the following rules concerning significant figures:

> **Rule 11:** In unit conversions, statistical computations, and other computations involving several steps, do not round off to the appropriate number of significant figures until you get to the final answer.

As noted by Harte (1985, p. 4), "Non-significant figures have a habit of accumulating in the course of a calculation, like mud on a boot, and you must wipe them off at the end. It is still good policy to keep one or two non-significant figures during a calculation, however, so that the rounding off at the end will yield a better estimate."

> **Rule 12:** Computers and calculators do not know anything about significant figures.

The numbers on computer printouts and calculator displays almost always have more digits than is warranted by the precision of measured hydrologic quantities. Thus, you are seldom justified in simply reporting the numbers directly as given by those devices without appropriate rounding off.

A.5 Equations: Dimensional Properties and Conversion

A.5.1. Dimensional Properties of Equations

> **Rule 13:** An equation that completely and correctly describes a physical relation has the same dimensions on both sides of the equal sign. Such equations are **dimensionally homogeneous**.

A corollary of this statement is that only quantities with identical dimensional quality can be added or subtracted.

Although there are no exceptions to rule 13, there are some important qualifications:

> **Rule 13a:** A dimensionally homogeneous equation may not correctly and completely describe a physical relation. Do not assume that every equation you encounter in a book or paper is correct! Typos and other errors are surprisingly common.

Rule 13b: Equations that are not dimensionally homogeneous can be useful approximations of physical relationships.

The magnitudes of hydrologic quantities are commonly determined by the complex interaction of many factors, and it is often virtually impossible to formulate the physically correct equation or to measure all the relevant independent variables. Thus, hydrologists are often forced to develop and rely on relatively simple empirical equations, especially statistical (regression) equations, that may be dimensionally inhomogeneous (see section 4.8.3).

An example of rule 13b is the Manning equation relating the velocity, U [L T^{-1}], of a stream to its average depth, Y [L], and water-surface slope, S_S, expressed as the tangent of the slope angle [1]:

$$U = \frac{Y^{2/3} \cdot S^{1/2}}{n_M} \tag{A.3}$$

In this equation, n_M is a factor reflecting the frictional resistance to flow offered by the channel bed and banks, and it is treated as a dimensionless number; that is, it has the same numerical value in all unit systems. This inhomogeneous empirical relation is commonly taken as the equation of motion for open-channel flows (equation 6.40 and tables 6.3–6.5). (The nature of equation A.3 is discussed more fully in section 6.8 and example A.5.2.1 below.)

Rule 13c: Equations can be dimensionally homogeneous but not unitarily homogeneous. (However, all unitarily homogeneous equations are of course dimensionally homogeneous.)

This situation can arise because each system of units includes "superfluous" units, such as miles (= 5,280 ft), kilometers (= 1,000 m), acres (= 43,560 ft^2), hectares (= 10^4 m^2), liters (= 10^{-3} m^3), and so forth. Thus, the equation

$$Q = 1000 \cdot U \cdot A, \tag{A.4}$$

where Q is streamflow rate in L/s, U is stream velocity in m/s, and A is stream cross-sectional area in m^2, is dimensionally homogeneous but not unitarily homogeneous. Clearly, the multiplier 1,000 in equation A.4 is a unit-conversion factor (L/m^3) required to make the equation correct for the specified units.

As noted above, dimensionally and/or unitarily inhomogeneous empirical equations are frequently encountered. It is extremely important that the practicing scientist cultivate the habit of checking every equation for dimensional and unitary homogeneity because

Rule 14a: If an inhomogeneous equation is given, the units of each variable in it *must* be specified.

This rule is one of the main reasons you should train yourself to examine each equation you encounter for homogeneity, because if you use an inhomogeneous equation with units other than those for which it was given, you will get the wrong answer. Surprisingly, it is not uncommon to encounter in the earth-sciences and engineering literature, including textbooks, inhomogeneous equations for which units are not specified—so caveat calculator!

Rule 14a has an equally important corollary:

Rule 14b: At least one of the coefficients or additive numbers in a unitarily inhomogeneous equation must change when the equation is to be used with different systems of units.

A.5.2. Equation Conversion

In practice, there are two situations in which you may need to convert inhomogeneous equations developed in one set of units for use with another set:

1. In making a series of calculations (as in writing a computer program), you often want to use an inhomogeneous equation with quantities measured in units different from those used in developing the equation.
2. You may want to compare inhomogeneous empirical equations that were developed for differing sets of units.

The guiding principle in equation conversion is as follows:

Rule 15: In equations, the dimensions and units of quantities are subjected to the same mathematical operations as the numerical magnitudes.

Careful execution of the following steps will assure that equation conversion is done correctly.

1. Write out the equation with the *new* units next to each term.
2. Next to each new unit, write the factor for converting the *new* units to the *old* units. (This may seem backward, but it is not.)
3. Perform the algebraic manipulations necessary to consolidate and simplify back to the original form of the equation.

In executing steps 2 and 3, note that exponents are not changed in equation conversion and that the conversion factors are subject to the same exponentiation as the variables they accompany. An example of equation conversion is given in example A.4.2.1.

One should always check to make sure a conversion was done correctly. To do this, follow these steps:

1. Pick an arbitrary set of values in the original units for the variables on the right-hand side of the equation, enter them in the original equation, and calculate the value of the dependent variable in the original units.
2. Convert the values of the independent variables to the new units. (Dimensionless quantities do not change value.)
3. Enter the converted independent variable values from step 2 into the converted equation and calculate the value of the dependent variable in the new units.
4. Convert the value of the dependent variable calculated in step 3 back to the old units and check to see that it is identical to that calculated in step 1.

The following example shows these steps.

Example A.5.2.1.

Conversion of Inhomogeneous Equations

Convert the inhomogeneous equation A.3, which is written for U in m/s and Y in m, for use with U in ft/s and Y in ft.

Following the steps of section A.5.2:

$$1. (U \text{ ft/s}) = \frac{(Y \text{ ft})^{2/3} \cdot S_S{}^{1/2}}{n_M}.$$

$$2. U \text{ ft/s} \cdot \frac{0.3048\ldots \text{ m}}{1.000\ldots \text{ ft}} = \frac{\left(Y \text{ ft} \cdot \frac{0.3048\ldots \text{ m}}{1.000\ldots \text{ ft}}\right)^{2/3} \cdot S_S{}^{1/2}}{n_M}.$$

$$3. 0.3048\ldots \cdot U = \frac{Y^{2/3} \cdot 0.4529\ldots \cdot S^{1/2}}{n_M},$$

$$U = \frac{1.49 \cdot Y^{2/3} \cdot S^{1/2}}{n_M}.$$

Thus, the implicit coefficient 1.000 … in equation A.3 is changed to 1.49 for use with the new units. Note that although this coefficient has infinite precision, it is usually expressed to three significant figures in conformance with rule 6.

Now we must check our conversion:

1. Enter the arbitrary values $Y = 2.40$ m, $S_S = 0.00500$, and $n_M = 0.040$ into the original equation and calculate U in m/s:

$$U = \frac{2.40^{2/3} \times 0.00500^{1/2}}{0.040} = 3.17 \text{ m/s} \quad (A.5)$$

2. Convert: The S_S and n_M values do not change because they are dimensionless. (Although the true dimensions of n_M are $[L^{1/6}]$ (Chow 1959, pp. 98n–99n), n_M values such as given in table 6.5 are taken to be the same in all unit systems, so in practice n_M is treated as though it is dimensionless.)
3. Substitute the converted values into the new equation:

$$U = \frac{1.49 \times 7.87^{2/3} \times 0.00500^{1/2}}{0.040} = 10.42 \text{ ft/s} \quad (A.6)$$

4. Convert this value of U back to the old units and compare with the value in step 1:

$$10.42 \text{ ft/s} \times \frac{0.3048\ldots \text{ m}}{1.000\ldots \text{ ft}} = 3.18 \text{ m/s} \quad (A.7)$$

The difference between this value and the original value is due only to round-off error.

Appendix B. Description of Flow Database Spreadsheet

The EXCEL spreadsheet HydData.xls, accessible at the text website http://www.oup.com/us/fluvialhydraulics, contains data for 931 flows in 171 natural river reaches taken from Barnes (1967), Jarrett (1985), Hicks and Mason (1991), and Coon (1998). The data are collated for ready access to allow students and researchers to explore hydraulic relations in natural channels (table B.1).

Table B.1

Source	No. of reaches	No. of flows
Barnes (1967)	51	62
Jarrett (1985)	21	85
Hicks and Mason (1991)	78	559
Coon (1998)	21	235

Table B.2

Quantity	Symbol	Units[a]
Discharge	Q	m^3/s
Water-surface slope	S_S	m/m
Friction slope	S_f	m/m
Cross-sectional area	A	m^2
Hydraulic radius	R	m
Average depth	Y	m
Water-surface width	W	m
Average velocity	U	m/s
50th percentile bed-material diameter	d_{50}	mm
84th percentile bed-material diameter	d_{84}	mm

[a]Units have been converted to SI for the Barnes, Jarrett, and Coon data.

In the spreadsheet, each flow is identified by

Reach identification number (by source)

Flow identification number (consecutive 1–931)

River and station location

For each flow, the information in table B.2 is given as presented in the original source (not all information is available for all reaches).

Appendix C. Description of Synthetic Channel Spreadsheet

C.1 Overview

The Synthetic Channel EXCEL Spreadsheet, accessible at the book's website http://www.oup.com/us/fluvialhydraulics, simulates the hydraulic behavior of an ideal channel cross section. The user specifies the channel shape, bankfull dimensions, slope, and bed-material size and then can examine characteristics of flows within that channel by specifying a range of central (maximum) flow depths, which are equivalent to water-surface elevations or stages. The basic model is on the worksheet labeled "SynChan" and the model output can be assembled for tabular or graphical presentation on the worksheet labeled "GraphData."

The model can be used to explore the general nature of important hydraulic relations and ways in which these relations change with channel shape, dimensions, slope, and bed-material size, including:

1. At-a-station hydraulic-geometry relations
2. Flow resistance–discharge relation
3. Discharge (or depth) at which erosion begins
4. Stage-discharge (rating-curve) relation
5. Froude-number–discharge relation
6. Reynolds-number–discharge relation
7. Cross-channel distribution of surface velocity
8. Distribution of velocity throughout the flow
9. Effects on hydraulic characteristics of assuming various vertical-velocity profiles
10. Effects of channel shape on hydraulic relations
11. Effects of water temperature on hydraulic relations

The hydraulic relations computed by the synthetic channel model are similar in form to corresponding relations in natural channels, as can be verified by examining data over a range of discharges at a single reach on the HydData.xls spreadsheet (appendix B). However, the model does not simulate the exact quantitative relations of actual channels and should not be used to predict those relations.

The essential aspects of the model are described in the following sections of this appendix; further description is given on the *Fluvial Hydraulics* website.

C.2 Basic Approach

C.2.1. Channel Shape

The channel cross section is symmetrical with its shape determined by the user-specified value of the exponent r in the general cross-section model (equation 2.20) described in Section 2.4.3.2:

$$z(w) = \Psi_{BF} \cdot \left(\frac{2 \cdot w}{W_{BF}}\right)^r, 0 \leq w \leq W_{BF}/2, \tag{C.1}$$

where $z(w)$ is the elevation of the channel bottom at cross-channel distance w from the center, Ψ_{BF} is the user-specified maximum (central) bankfull depth, and W_{BF} is the user-specified bankfull width. For a triangular channel, $r = 1$; for the Lane stable channel, $r = 1.75$; for a parabolic channel, $r = 2$; and the channel shape approaches a rectangle as $r \to \infty$. (A rectangle can be approximated by using a large value for r, say $r = 10,000$.) Values of $r < 1$ ("convex channels") can also be specified.

C.2.2. Velocity

In the model, rectangular elements of one-half of the symmetrical cross section are represented by spreadsheet cells. The width of each element is equal to $W_{BF}/200$, and the height is equal to $\Psi_{BF}/100$.

Each cell that is below the water surface and above the channel bottom displays the local velocity; other cells are blank. In the default version of the model, the local velocities $u_w(y)$ are computed by the Prandtl-von Kármán (P-vK) velocity profile for turbulent flow (equation 5.21),

$$u_w(y) = \left(\frac{1}{\kappa}\right) \cdot (g \cdot Y_w \cdot S_S)^{1/2} \cdot \ln\left(\frac{y}{y_{0w}}\right), \tag{C.2}$$

where y is distance above the channel bed, κ is von Kármán's constant ($\kappa = 0.4$), g is gravitational acceleration ($g = 9.81$ m/s^2), Y_w is the local water depth, S_S is the user-specified water-surface slope. As described in section 5.3.1.6 (equation 5.32), the value of y_0 is determined by the value of the local boundary Reynolds number, Re_{bw},

$$Re_{bw} \equiv \frac{u_{*w} \cdot y_r}{\nu} = \frac{(g \cdot Y_w \cdot S_S)^{1/2} \cdot y_r}{\nu}, \tag{C.3}$$

where u_{*w} is the local friction velocity, y_r is the effective height of bed roughness elements, and ν is kinematic viscosity:

$$\text{if } Re_b \leq 5 \text{ (smooth flow)}, \; y_{0w} = \frac{\nu}{9 \cdot u_{*w}}; \tag{C.4a}$$

$$\text{if } Re_{bw} > 5 \text{ (transitional or rough flow)}, \; y_{0w} = \frac{y_r}{30}, \tag{C.4b}$$

and y_r is considered equal to the user-specified 84th-percentile of the bed-material grain size, d_{84}.

Note that it is a simple matter to replace the Prandtl-von Kármán profile by one of the other profiles discussed in sections 5.3.2–5.3.5.

C.2.3. Water Properties

The values of water properties mass density, ρ; weight density, γ; dynamic viscosity, μ; and kinematic viscosity, ν, are required to compute some of the flow characteristics. These properties are functions of the user-specified water temperature, T, and are computed via equations 3.11 and 3.20.

C.3 Displays

The model computes and displays the following cross-section-averaged or -totaled quantities of interest characterizing each flow (user-specified values are indicated with an asterisk):

Symbol	Quantity
Ψ_{BF}	Bankfull maximum depth*
W_{BF}	Bankfull water-surface width*
S_S	Water-surface slope*

Symbol	Quantity
Ψ	Maximum depth*
d_{84}	84th percentile bed-material diameter*
v_f	Bed material fall velocity
Q	Discharge
A	Cross-sectional area
W	Water-surface width
P_w	Wetted perimeter
Y	Average depth
R	Hydraulic radius
W/Y	Width/depth ratio
U	Average velocity
$u*$	Friction velocity
Ω	Resistance
$\Omega*$	Baseline resistance
$(\Omega - \Omega*)/\Omega$	Relative excess resistance
n_M	Manning's *n*
C	Chézy's *C*
Ro	Rouse number
Fr	Froude number
Re	Reynolds number

These values are displayed so that graphs relating the various quantities can be readily constructed.

Appendix D. Description of Water-Surface Profile Computation Spreadsheet

The EXCEL spreadsheet WSProfile.xls, accessible at the book's website http://www.oup.com/us/fluvialhydraulics, allows computation and plotting of water-surface profiles for a rectangular channel according to the standard step method described in section 9.4.2.2 (figure 9.8). The intent of the spreadsheet is to provide students with a hands-on introduction to the basic aspects of profile computation. Samples of an M1 and an M2 profile are shown, and an instructor can readily develop exercises by modifying channel elevations and characteristics and providing an initial water-surface elevation.

Notes

Chapter 1

1. Evapotranspiration is the sum of water use by plants (about 97% of the total globally) and direct evaporation from open-water surfaces.
2. A generally small proportion of the $P - ET$ residual for a region may be in the form of groundwater discharge. Globally, groundwater discharge is <5% of Q.
3. See the passage from Thomas Mann's "A Man and His Dog" at the front of this book.
4. Satellite images of current locations of riverine flooding around the globe can be viewed at the Dartmouth Flood Observatory Web site (www.dartmouth.edu/~floods/index.html, G. R. Brakenridge principal investigator, Department of Geography, Dartmouth College).

Chapter 2

1. Sellmann and Dingman (1970) found that drainage densities measured on standard U.S. Geological Survey topographic maps at a scale of 1:24,000 were close to true values observed in the field for perennial channels.
2. As noted by Gordon et al. (1992), defining valley length is subjective, and "in practice, straight-line segments which follow the broad-scale changes in channel direction can be used as a measure of valley length" (p. 313).
3. It is important to note that at-a-station hydraulic geometry relations as commonly applied are valid only for within-bank flows (i.e., $Q \leq Q_{BF}$). Garbrecht (1990) expanded the concept by showing that two empirical power functions could be connected to apply to in-bank and overbank flows at a given section. However, the discussion here is limited to in-bank flows.

Chapter 3

1. By convention, the atomic weight is written to the upper left of the element symbol.
2. If the liquid contains no impurities and is not in contact with preexisting ice, it is possible to supercool it to temperatures as low as –41°C.
3. A "free surface" is a surface of liquid water at atmospheric pressure. In diagrams, such a surface is designated by the inverted triangular **hydrat** symbol, ∇.
4. You can capture the essence of this experiment by placing two corks next to each other on the surface of a stream and noting how they separate with time.
5. Note that the y-direction, rather than the z-direction, is oriented vertically. This makes the notation consistent with subsequent developments in which the y-direction is normal to the bottom.
6. Theodore von Kármán (1881–1963) was a Hungarian-born American physicist and aeronautical engineer who made major contributions to the study of turbulence (see chapter 1).
7. That is, equation 3.39 is a "heuristic" equation, as described in section 4.8.4.
8. Since in this section we consider only velocities in the x-direction, we can drop the directional subscripts.

Chapter 4

1. Comte Joseph Louis Lagrange (1736–1813) was a French astronomer and mathematician; Leonhard Euler (1701–1783) was a Swiss mathematician. Lagrange succeeded Euler at the University of Berlin in 1776; each was considered the greatest mathematician of his day. Despite the association of each with one viewpoint for analyzing fluid flows, they both used both viewpoints in their analyses (Rouse and Ince 1963).

2. This is justified because if dX is infinitesimally small, then $(\mathrm{d}X)^2$ is vanishingly small.

3. Strictly speaking, U should be multiplied by a "momentum coefficient" greater than 1 to account for the nonuniform distribution of velocities in real channels (see box 8.1). However, this coefficient is usually close to 1, and we can neglect it for the time being.

4. This is known as the Bernoulli equation after Daniel Bernoulli (section 1.3).

5. To simplify the notation, we henceforth write u in place of $u_x(z)$. Note also that $\mathcal{F}_z(\mathrm{M})$ here is identical to $\mathcal{F}_{\mathrm{M}z}$ in equations 3.22 and 3.24.

6. If we pick one variable, say, X_1, to be the dependent variable, we could write equation 4.57 as $X_1 = f(X_2, X_3, \ldots, X_{N-1})$. The notation used here and in equation 4.58 is equivalent but more general, because we do not specify which X or Π term is considered dependent.

7. There are many excellent statistical texts that describe these methods, including Draper and Smith (1981) and Helsel and Hirsch (1992).

8. Note that the notation in this section deviates from that used earlier in the chapter and in most of the text. Here, Y and X_j denote any dependent and independent variable, respectively, and y_i and x_{ji} refer to individual measured values of Y and X_j.

9. There are regression techniques that produce invertible equations (Helsel and Hirsch 1992), but these are generally not optimum for prediction and are rarely used.

Chapter 5

1. Strictly speaking, the distance y is measured normal to the bottom, but since channel slopes ($\sin\theta$) are almost always less than 0.01 and $\cos\theta$ greater than 0.999, we often refer to these as "vertical" velocity profiles.

2. Recall from section 3.4.2 (equation 3.44) that the general definition of a Reynolds number is the product of a characteristic velocity times a characteristic length divided by kinematic viscosity.

3. If the power-law velocity profile of equation 5.45 applies in a wide rectangular channel, the exponent J in equation 5.50 is identical to the exponent m_{PL} in equation 5.45.

Chapter 6

1. We can safely ignore the remote possibility that Y and W change in a way that precisely balances the downstream change in velocity.

2. We will see shortly that, for turbulent flow, this resistance is proportional to the square of the velocity.

3. In this text, "state" is determined by Reynolds number and "regime" by Froude number. This differs from Chow (1959, p.14), who uses the term "regime" to apply to conditions determined by both the Reynolds number (flow state) and the Froude number, that is, "turbulent-subcritical" and "laminar-supercritical," corresponding to the four fields shown in figure 6.4.

4. A summary of the circumstances that led Chézy to the formula named for him and a translation of his derivation are given by Rouse and Ince (1963).

5. Note that this shear stress operates on a plane normal to the y-direction and is directed in the x-direction, and thus would be designated $\tau_{yx}(y)$ following the convention used in equations 3.19 and 3.29. In this chapter, we can drop the subscripts without confusion.

6. The quantity Y/y_r is also called the relative submergence.

7. Recall that τ_0 is directly proportional to depth (equation 5.7). We explore the hydraulic conditions under which various bedforms occur in more detail in section 12.5.4.

8. Manning himself did not think that an equation of the form of 6.40 could be correct because it is dimensionally inhomogeneous, and later (Manning 1895) put forward an alternative that was homogeneous. However, the physical basis for the alternative was questionable, and it was never adopted (see Dooge 1992).

9. In addition to being based more directly on the concepts underlying the Chézy equation, the method of section 6.7 does not require estimation of some uncertain energy parameters.

Chapter 7

1. Named after Gaspard Gustave de Coriolis (1792–1843), a French hydraulic engineer.

2. Recall that we assume the channel is "wide" and neglect the frictional effects of the sides.

3. We will see in the following section that the Coriolis acceleration is negligible for all but the very largest open-channel flows.

Chapter 8

1. Comparisons with gravitational head are not meaningful because Z_i is measured relative to an arbitrary datum.

2. The development here applies to any cross section, so we drop the subscript, and to simplify the development, we assume $\cos\theta_0 = 1$.

3. Relations of this type can be determined by applying the techniques of regression analysis with logarithmic transforms, as discussed in section 4.8.3.1 (equation 4.67).

Chapter 9

1. In this section, we simplify the notation by dropping the subscript identifying a particular cross section.

2. Average friction slope may also be computed as the geometric mean, $\overline{S_{fi}} = (S_{fi-1} \cdot S_{fi})^{1/2}$, or the harmonic mean, $\overline{S_{fi}} = 2(S_{fi-1} \cdot S_{fi})/(S_{fi-1} + S_{fi})$, but if $X_i - X_{i-1}$ or $|Y_i - Y_{i-1}|$ are small, $\overline{S_{fi}}$ values computed by the various formulas differ little (Chaudhry 1993).

Chapter 10

1. A circular hydraulic jump can be readily observed where water from a faucet strikes a sink surface: The water initially flows out radially at a low depth and high velocity ($Fr > 1$); the velocity and Froude number decrease with distance, and when $Fr = 1$ there is a sudden increase in depth (and decrease in velocity) to form a standing wave. The location of the jump is a function of the discharge from the faucet and the slope and resistance of the sink surface.

2. We do not need to use the partial-differential notation of equations 4.22 and 4.25 because we are considering changes only with respect to X.

3. It is interesting that, although equation 10.10 looks rather nonlinear, Y_D/Y_U plots as very nearly a linear function of Fr_U.

4. If the computations were incorporated in water-surface-profile computations as described in chapter 9, we would be proceeding in the upstream direction in subcritical flow, and the downstream direction in supercritical flow.

5. This description and figure 10.22 assume that the water-surface elevation downstream of the weir is not maintained at a high enough elevation to submerge the nappe.

Chapter 11

1. Equation 11.13 is derived starting with energy considerations (equation 11.4) and is written in terms of head (energy-per-weight) gradients. However, we could arrive at the same relations if we start with the one-dimensional momentum equation (equation 8.32) (as long as the velocity distribution is uniform, and there are no eddy losses), and equation 11.13 or 11.16 is usually called the one-dimensional **momentum equation**. It seems preferable to use the term "dynamic equation" to reflect the fact that the relation can be developed from either energy or force considerations, as done by Chow (1959).

2. However, as shown in box 2.4, the coefficients and exponents in these empirical relations can be rationally related to channel geometry and hydraulics.

3. Seiches are periodic waves in lakes or enclosed bays, such as can be produced by "sloshing" in a bathtub. They may be caused by storms, tsunamis, or other disturbances.

4. Excellent reviews of the theory and practical aspects of oscillatory waves can be found in Bascom (1980) and Brown et al. (1999).

5. The kinematic wave is also called the monoclinal rising wave or the uniformly progressive wave (Chow 1959; Henderson 1966).

6. A heuristic equation is one that, although not derived from basic physics or based on statistical analysis of observations, seems physically plausible and is generally consistent with observations (see section 4.8.4).

Chapter 12

1. Because geological interest is usually only in the suspended mineral solids, it may be necessary to treat the sample with an oxidant such as hydrogen peroxide in order to eliminate organic particles before filtering.

2. As equation 12.6 is written, it appears that $c_S = a_S$. However, in practice, the numerical values of the two coefficients differ because of changes in units.

3. Although not strictly true mathematically, the relation of equation 12.9 can be closely approximated by a simpler power-law relation: $L = 3.16 \times 10^{-4} \cdot Q^{3.08}$, and this relation could be used in place of equation 12.9 (see figure 12.8).

4. A portion of dissolved load typically includes atmospheric gases; this portion must be deducted when calculating chemical denudation rates.

5. The velocity increases as it moves from the stagnation point to the "top" (and bottom) of the particle, as reflected in the smaller distance between streamlines in figure 12.11a. Thus, some of the pressure potential energy is converted to kinetic energy and the pressure decreases. This pressure force is relative to the ambient hydrostatic pressure in the fluid.

6. Note that figure 12.12 applies to spheres. The curves for objects of other shapes differ in detail but have the same general pattern (see Middleton and Southard 1984).

7. It is interesting that there is a general similarity between the $\theta^* - Re_b$ relation and both the $C_D - Re_p$ relation (figure 12.12) and the $\mathbf{\Omega} - Re$ relation (figure 6.8).

8. In fact, the measured profiles in figure 12.26 show a maximum concentration at $y/Y > 0$.

9. Because $[(Y - y)/y]^{Ro'(d)}$ is not analytically integrable, the integration must be done numerically.

10. We have dropped the subscript BF notation used in chapter 2, because all channel dimensions considered here are for the bankfull channel.

Appendix A

1. Rational numbers are the positive and negative integers and ratios of integers.

2. The arguments of logarithmic, exponential, and trigonometric functions can be dimensional; however, the value of the function, though dimensionless, then depends on the units of measurement.

3. Note that many of the quantities of interest in fluvial hydraulics are averages of measured values (e.g., average velocity or depth in a cross section or reach), and for these, statistical considerations are also involved in determining precision.

References

Ackers, P., and W.R. White (1973) Sediment transport, a new approach and analysis. *Proceedings of the American Society of Civil Engineers, Journal of the Hydraulics Division* 99(HY11): 2041–2060.

Ackers, P., W.R. White, J.A. Perkins, and A.J.M. Harrison (1978) *Weirs and Flumes for Flow Measurement*. New York: John Wiley.

Alonso, C.V. (1980) Selecting a formula to estimate sediment transport capacity in non-vegetated channels. In *CREAMS (A Field Scale Model for Chemicals, Runoff, and Erosion from Agricultural Management Systems)*. W.G. Knisel, ed. Conservation Research Report 26, U.S. Department of Agriculture, chap. 5.

Andersen, V.M. (1978) Undular hydraulic jump. *Journal of Hydraulic Engineering* 104: 1185–1188.

Arcement, G.J., and V.R. Schneider (1989) Guide for selecting Manning's roughness coefficients for natural channels and flood plains. Water-Supply Paper 2339, U.S. Geological Survey.

Ashton, G.D., and J.F. Kennedy (1972) Ripples on underside of river ice covers. *Proceedings of the American Society of Civil Engineers, Journal of the Hydraulics Division* 98(HY9): 1603–1624.

Bagnold, R.A. (1960) Some aspects of the shape of river meanders. Professional Paper 282-E, U.S. Geological Survey.

Bagnold, R.A. (1966) An approach to the sediment transport problem from general physics. Professional Paper 422-J, U.S. Geological Survey.

Bailey, J.F., and H.A. Ray (1966) Definition of stage-discharge relation in natural channels by step-backwater analysis. Water-Supply Paper 1869-A, U.S. Geological Survey.

Ball, P. (1999) *Life's Matrix: A Biography of Water*. New York: Farrar, Strauss and Giroux.

Barnes, H.H. (1967) Roughness characteristics of natural channels. Water-Supply Paper 1849, U.S. Geological Survey.

Bascom, W. (1980) *Waves and Beaches*. New York: Doubleday Anchor.

Bates, P.D., K.J. Marks, and M.S. Horritt (2003) Optimal use of high-resolution topographic data in flood inundation models. *Hydrological Processes* 17: 537–557.

Bathurst, J.C. (1985) Flow resistance estimation in mountain rivers. *Journal of Hydraulic Engineering* 111(4): 625–643.

Bathurst, J.C. (1993) Flow resistance through the channel network. In *Channel Network Hydrology*. K. Beven and M.J. Kirkby, eds. New York: John Wiley & Sons, chap. 4.

Bathurst, J.C. (2002) At-a-site variation and minimum flow resistance for mountain rivers. *Journal of Hydrology* 269: 11–26.

Beltaos, S. (2000) Advances in river ice hydrology. *Hydrological Processes* 14: 1613–1625.

Berner, E.K., and R.A. Berner (1987) *The Global Water Cycle*. Englewood Cliffs, NJ: Prentice-Hall.

Birge, E.A. (1910) The apparent sinking of ice in lakes. *Science* 32: 81–82.

Bjerklie, D.M. (2007) Estimating the bankfull velocity and discharge for rivers using remotely sensed river morphology information. *Journal of Hydrology* 341: 144–155.

Bjerklie, D.M., S.L. Dingman, C.J. Vörösmarty, C.H. Bolster, and R.G. Congalton. (2003) Evaluating the potential for measuring river discharge from space. *Journal of Hydrology* 278: 17–38.

Bjerklie, D.M., D. Moller, L.C. Smith, and S.L. Dingman (2005a) Estimating river discharge using remotely-sensed hydraulic information. *Journal of Hydrology* 309: 191–209.

Bjerklie, D.M., S.L. Dingman, and C.H. Bolster (2005b) Comparison of constitutive flow resistance equations based on the Manning and Chézy equations applied to natural rivers [technical note]. *Water Resources Research* 41:W11502, doi:10.1029/2004WR003776.

Bledsoe, B.P., and C.C. Watson (2001) Logistic analysis of channel pattern thresholds: meandering, braided, and incising. *Geomorphology* 38: 281–300.

Boise Adjudication Team (2004) *Boise River Report.* http://www.fs.fed.us./rm/boise/teams/soils/Bat%20River.htm.

Bolster, C.H., and J.E. Saiers (2002) Development and evaluation of a mathematical model for surface-water flow within the Shark River Slough of the Florida Everglades. *Journal of Hydrology* 259: 221–235.

Bridge, J.S. (1993) The interaction between channel geometry, water flow, sediment transport and deposition in braided rivers. In *Braided Rivers.* J.L. Best and C.S. Bristow, eds. Special Publications of the Geological Society of London, vol. 75, pp. 13–71.

Bridge, J.S. (2003) *Rivers and Floodplains*. Oxford: Blackwell Science.

Bridge, J.S., and S.J. Bennett (1992) A model for the entrainment and transport of sediment grains of mixed sizes, shapes, and densities. *Water Resources Research* 28: 337–363.

Brown, E., A. Colling, D. Park, J. Phillips, D. Rothery, and J. Wright (1999) *Waves, Tides, and Shallow-Water Processes*, 2nd ed. Oxford: Butterworth-Heinemann.

Brownlie, W.R. (1981a) Re-examination of Nikuradse roughness data. *Proceedings of the American Society of Civil Engineers, Journal of the Hydraulics Division* 107 (HY1): 115–119.

Brownlie, W.R. (1981b) Prediction of flow depth and sediment discharge in open channels. Report KH-R-43A, W.M. Keck Laboratory of Hydraulics and Water Resources, California Institute of Technology.

Brunner, G.W. (2001a) HEC-RAS River Analysis System User's Manual. Report CPD-68, U.S. Army Corps of Engineers Hydrologic Engineering Center.

Brunner, G.W. (2001b) HEC-RAS River Analysis System Hydraulic Reference Manual. Report CPD-69, U.S. Army Corps of Engineers Hydrologic Engineering Center.

Buchanan, T.J., and Somers, W.P. (1969) Discharge measurement at gaging stations. Techniques of Water-Resources Investigations, bk. 3. U.S. Geological Survey, chap. A8.

Buckingham, E. (1915) Model experiments and the form of empirical equations. *Transactions of the American Society of Mechanical Engineers* 37: 263–292.

Buffington, J.M. (1999) The legend of A.F. Shields. *Journal of Hydraulic Engineering* 125(4): 376–387.

Buffington, J.M., and D.R. Montgomery (1997) A systematic analysis of eight decades of incipient motion studies, with special reference to gravel-bedded rivers. *Water Resources Research* 33(8): 1993–2029.

Bunte, K., and S.R. Abt (2001) Sampling surface and subsurface particle-size distributions in wadable gravel- and cobble-bed streams for analyses in sediment transport, hydraulics, and streambed monitoring. General Technical Report RMRS-GTR-74, U.S. Forest Service Rocky Mountain Research Station.

Carter, R.W., and I.E. Anderson (1963) Accuracy of current meter measurements. *Proceedings of the American Society of Civil Engineers; Journal of the Hydraulics Division* 89: 105–115.

Chang, H.H. (1980) Geometry of gravel streams. *Proceedings of the American Society of Civil Engineers; Journal of the Hydraulics Division* 106(HY9): 1443–1456.

Chang, H.H. (1984) Variation of flow resistance through curved channels. *Journal of Hydraulic Engineering* 110(12): 1772–1782.

Chang, H.H. (1988) *Fluvial Processes in River Engineering*. Malabar, FL: Krieger.

Chanson, H. (2000) Boundary shear stress measurements in in undular flows: Application to standing wave bed forms. *Journal of Hydraulic Research* 36: 3063–3076.

Chaudhry, M.H. (1993) *Open-Channel Flow*. Englewood Cliffs, NJ: Prentice-Hall.

Chen, C.L. (1991) Unified theory on power-laws for flow resistance. *Journal of Hydraulic Engineering* 117: 371–389.

Chiu, C.-L., and S.-M. Hsu (2006) Probabilistic approach to modeling of velocity distributions in fluid flows. *Journal of Hydrology* 316: 28–42.

Chow, V.T. (1959) *Open-Channel Hydraulics*. New York: McGraw-Hill.

Church, M., and K. Rood (1983) *Catalogue of Alluvial River Channel Regime Data*. Vancouver, BC: Department of Geography, University of British Columbia.

Colby, B.R. (1964) Discharge of sands and mean velocity relationships in sand-bed streams. Water-Supply Paper 462-A, U.S. Geological Survey.

Coleman, N.L. (1981) Velocity profiles with suspended sediment. *Journal of Hydraulic Research* 19(3): 211–229.

Comiti, F., and M.A. Lenzi (2006) Dimensions of standing waves at steps in mountain rivers. *Water Resources Research* 42:W03411, doi:10.1029/2004WR003898.

Coon, W.F. (1998) Estimation of roughness coefficients for natural stream channels with vegetated banks. Water-Supply Paper 2441, U.S. Geological Survey.

Corbett, D.M. (1945) Stream-gaging procedure. Water-Supply Paper 888, U.S. Geological Survey.

Coulthard, T.J. (2005) Effects of vegetation on braided stream pattern and dynamics. *Water Resources Research* 41:W04003, doi:10.1029/2004WR003201.

Cowan, W.L. (1956) Estimating hydraulic roughness coefficients. *Agricultural Engineering* 37: 473–475.

Cruff, R.W. (1965) Cross-channel transfer of linear momentum in smooth rectangular channels. Water-Supply Paper 1592-B, U.S. Geological Survey.

Cunge, J.A. (1969) On the subject of a flood propagation computation method (Muskingum method). *Journal of Hydraulic Research* 7(2): 205–230.

Dade, W.B. (2000) Grain size, sediment transport, and alluvial channel pattern. *Geomorphology* 25: 119–126.

Daily, J.W., and D.R.F. Harleman (1966) *Fluid Dynamics*. Reading, MA: Addison-Wesley.

Daly, S.J. (2004) Evolution of river ice [PowerPoint presentation]. U.S. Army Corps of Engineers Engineering Research and Development Center.

Davidian, J. (1984) Computation of water-surface profiles in open channels. Techniques of Water-Resources Investigations, bk. 3. U.S. Geological Survey, chap. A15.

Davis, K.S., and J.A. Day (1961) *Water: The Mirror of Science*. New York: Doubleday-Anchor.

Davis, S.N., and E. Murphy, eds. (1987) Dating ground water and the evaluation of repositories for radioactive waste. NUREG/CR-4912, U.S. Nuclear Regulatory Commission.

Davis, W.M. (1899) The geographical cycle. *Geographical Journal* 14(5): 481–504.

Dean, R.G., and R.A. Dalrymple (1991) *Water Wave Mechanics for Engineers and Scientists*. Singapore: World Scientific Publishing.

Dietrich, W.E. (1982) Settling velocity of natural particles. *Water Resources Research* 18(6): 1615–1626.

Dingman, S.L. (1984) *Fluvial Hydrology*. New York: W.H. Freeman and Company.

Dingman, S.L. (1989) Probability distribution of velocity in natural channels. *Water Resources Research* 25(3): 509–518.

Dingman, S.L. (2002) *Physical Hydrology*. Long Grove, IL: Waveland Press.

Dingman, S.L. (2007a) Analytical derivation of at-a-station hydraulic-geometry relations. *Journal of Hydrology* 334: 17–27.

Dingman, S.L. (2007b) Comment on "Probabilistic approach to modeling of velocity distributions in fluid flows" by C.-L. Chiu and S.-M. Hsu. *Journal of Hydrology* 316: 28–42. *Journal of Hydrology* 335: 419–428.

Dingman, S.L., and D.M. Bjerklie (2005) Hydrological application of remote sensing: Surface fluxes and other derived variables—river discharge. In *Encyclopedia of Hydrologic Sciences*. M.G. Anderson, ed.-in-chief. New York: John Wiley & Sons.

Dingman, S.L., and K.P. Sharma (1997) Statistical development and validation of discharge equations for natural channels. *Journal of Hydrology* 199: 13–35.

Dooge, J.C.I. (1992) The Manning formula in context. In *Channel Flow Resistance: Centennial of Manning's Formula*. Yen, B.C., ed. Highlands Ranch, CO: Water Resources Publications, pp. 136–185.

Dooge, J.C.I., W.G. Strupczewski, and J.J. Napiorkowski (1982) Hydrodynamic derivation of storage parameters of the Muskingum model. *Journal of Hydrology* 54: 371–387.

Dorsey, N.E. (1940) *Properties of Ordinary Water Substance in All Its Phases: Water, Water-Vapor, and All the Ices*. New York: Reinhold.

Draper, N.R., and H. Smith (1981) *Applied Regression Analysis*, 2nd ed. New York: Wiley.

Drever, J.I. (1982) *The Geochemistry of Natural Waters*. Englewood Cliffs, NJ: Prentice-Hall.

Droppo, I.G. (2001) Rethinking what constitutes suspended sediment. *Hydrological Processes* 15: 1551–1564.

Dunne, T., and L.B. Leopold (1978) *Water in Environmental Planning*. San Francisco: W.H. Freeman.

Eagleson, P.S., et al (1991) *Opportunities in the Hydrologic Sciences*. Washington, DC: National Academy Press.

Edwards, T.K., and G.D. Glysson (1999) Field methods for measurement of fluvial sediment. Techniques of Water-Resources Investigations, bk. 3. U.S. Geological Survey, chap. C2. (Also available at http://pubs.usgs.gov/twri/twri3c2/html/pdf_new.html.)

Einstein, H.A. (1950) The bedload function for sediment transportation in open channels. Technical Bulletin No. 1026, U.S. Department of Agriculture, Soil Conservation Service.

Einstein, H.A., and N. Chien (1954) Second approximation to the solution of the suspended load theory. Research Report 3, University of California at Berkeley.

Engelund, F., and J. Fredsøe (1976) A sediment transport model for straight alluvial channels. *Nordic Hydrology* 7: 293–306.

Engelund, F., and E. Hansen (1967) *A Monograph on Sediment Transport in Alluvial Streams*. Copenhagen: Laboratory of Technical University of Denmark.

Ettema, R. (2006) Hunter Rouse—his work in retrospect. *Journal of Hydraulic Engineering* 131: 1248–1258.

Faskin, G.B. (1963) *Guide for Selecting Roughness Coefficient "n" Values for Channels*. Lincoln, NE: U.S. Soil Conservation Service.

Ferguson, R.I. (1986) Hydraulics and hydraulic geometry. *Progress in Physical Geography* 10: 1–31.

Ferguson, R.I., and M. Church (2004) A simple universal equation for grain settling velocity. *Journal of Sedimentary Research* 74(6): 933–937.

Ferguson, R.I., T. Hoey, S. Wathen, and A. Werrity (1996) Field evidence for rapid downstream fining of river gravels through selective transport. *Geology* 24(2): 179–182.

Formica, G. (1955) Esperienze preliminari sulle perdite di carico nei canali, dovute a cambiamente di sezione (Preliminary test on head losses in channels due to cross-sectional changes). *L'energia elettria, Milano* 32(7): 554–568. (Reprinted as Memorie e Studi No. 124, Istituto di Idraulica e Construzioni Idrauliche, Milano.)

Fread, D.L. (1992) Flow routing. In *Handbook of Hydrology*. D.R. Maidment, ed. New York: McGraw-Hill, chap. 10.

French, R.H. (1985) *Open-Channel Hydraulics*. New York: McGraw-Hill.

Fritz, P., and J.C. Fontes, eds. (1980) *Handbook of Environmental Isotope Geochemistry*. New York: Elsevier.

Furbish, D.J. (1997) *Fluid Physics in Geology*. New York: Oxford University Press.

Gaeuman, D., and R.B. Jacobson (2006) Acoustic bed velocity and bed load dynamics in a large sand bed river. *Journal of Geophysical Research* 111:F02005, doi:10.1029/2005JF000411.

Garbrecht, J. (1990) Analytical representation of cross-section hydraulic properties. *Journal of Hydrology* 119: 43–56.

Garde, R.J., and K.G. Raga Raju (1978) *Mechanics of Sediment Transportation and Alluvial Stream Problems*. New Delhi: Wiley Eastern.

Gleick, P.H., ed. (1993) *Water in Crisis*. New York: Oxford University Press.

Golubtsev, V.V. (1969) Hydraulic resistance and formula for computing the average flow velocity of mountain rivers. *Soviet Hydrology* 5: 30–41.

Gordon, N.D., T.A. McMahon, and B.L. Finlayson (1992) *Stream Hydrology: An Introduction for Ecologists*. New York: Wiley.

Graf, W.H. (1971) *Hydraulics of Sediment Transport*. New York: McGraw-Hill

Grant, G.E. (1997) Critical flow constrains flow hydraulics in mobile-bed streams: A new hypothesis. *Water Resources Research* 33: 349–358.
Gray, D.M., and J.M. Wigham (1970) Peak flow—rainfall events. In *Handbook on the Principles of Hydrology*. D.M. Gray, ed. Port Washington, NY: Water Information Center, sec. VIII.
Groisman, P.Y., and D.R. Legates (1994) The accuracy of United States precipitation data. *Bulletin of the American Meteorologic Society* 75: 215–227.
Guy, H.P., and V.W. Norman (1970) Field methods for measurement of fluvial sediment. In Techniques of Water-Resources Investigations, bk. 3. U.S. Geological Survey, chap. C2.
Hakenkamp. C.C., H.M. Valett, and A.J. Boulton (1993) Perspectives on the hyporheic zone: Integrating hydrology and biology—concluding remarks. *Journal of the North American Benthological Society* 12: 94–99.
Harrelson, C.C., C.L. Rawlins, and J.P. Potyondy (1994) Stream channel reference sites: An illustrated guide to field technique. General Technical Report RM-245, U.S. Forest Service, Rocky Mountain Forest and Range Experiment Station.
Harte, J. (1985) *Consider a Spherical Cow.* Los Altos, CA: William Kaufmann, Inc.
Heggen, R.J. (1983) Thermal dependent physical properties of water. *Journal of Hydraulic Engineering* 109: 298–302.
Helsel, D.R., and R.M. Hirsch (1992) *Statistical Methods in Water Resources.* Studies in Environmental Science 49. Amsterdam: Elsevier. Also available as Techniques of Water-Resources Investigations, bk. 4, U.S. Geological Survey, chap. A3. http://pubs.usgs.gov/twri/twri4a3/html/pdf_new.html.
Hem, J.D. (1970) Study and interpretation of the chemical characteristics of natural water. Water-Supply Paper 1473, U.S. Geological Survey.
Henderson, F.M. (1961) Stability of alluvial channels. *Proceedings of the American Society of Civil Engineers, Journal of the Hydraulics Division* 87: 109–138.
Henderson, F.M. (1966) *Open Channel Flow*. New York: Macmillan.
Herschel, C. (1897) On the origin of the Chézy formula. *Journal of the Association of Engineering Societies* 18.
Herschy, R.W. (1999a) Flow measurement. In *Hydrometry: Principles and Practices*, 2nd ed. R.W. Herschy, ed. Chichester: John Wiley & Sons, pp. 9–83.
Herschy, R.W. (1999b) Hydrometric instruments. In *Hydrometry: Principles and Practices*, 2nd ed. R.W. Herschy, ed. Chichester: John Wiley & Sons, pp. 85–142.
Hicks, D.M., and P.D. Mason (1991) *Roughness Characteristics of New Zealand Rivers.* Wellington, New Zealand: New Zealand National Institute of Water and Atmospheric Research (Reprint: Water Resources Publications, 1998).
Hjulström, F. (1935) Studies of the morphological activity of rivers as illustrated by the river Fyris. *Bulletin of the Geological Institute of Uppsala*, vol. 25.
Hjulström, F. (1939) Transportation of detritus by flowing water. In *Recent Marine Sediments.* P.D. Trask, ed. Tulsa, OK: American Association of Petroleum Geologists, pp. 5–31.
Hoey, T., and R.I. Ferguson (1994) Numerical simulation of downstream fining by selective transport in gravel bed rivers: Model development and illustration. *Water Resources Research* 30(7): 2251–2260.
Horton, R.E. (1945) Erosional development of streams and their drainage basins: Hydrophysical approach to quantitative morphology. *Geological Society of America Bulletin* 56: 275–370.
Huang, H.Q., and G.C. Nanson (1998) The influence of bank strength on channel geometry. *Earth Surface Processes and Landforms* 23: 865–876.
Huang, H.Q., H.H. Chang, and G.C. Nanson (2004) Minimum energy as the general form of critical flow and maximum flow efficiency and for explaining variations in river channel pattern. *Water Resources Research* 40:W04502, doi:10.1029/2003WR002539.
Hulsing, H., W. Smith, and E.D. Cobb (1966) Velocity-head coefficients in open channels. Water-Supply Paper 1869-C, U.S. Geological Survey.
Hydrologic Engineering Center (1986) Accuracy of computed water surface profiles. Research Document 26, U.S. Army Corps of Engineers Hydrologic Engineering Center.

Itakura, T., and T. Kishi (1980) Open channel flow with suspended sediments. *Proceedings of the American Society of Civil Engineers, Journal of the Hydraulics Division* 106(HY8): 1345–1352.

Jarrett, R.D. (1984) Hydraulics of high-gradient streams. *Journal of Hydraulic Engineering* 110(11): 1519–1539.

Jarrett, R.D. (1985) Determination of roughness coefficients for streams in Colorado. Water-Resources Investigations Report 85-4004, U.S. Geological Survey.

Jellinek, H.H.G. (1972) The ice interface. In *Water and Aqueous Solutions*. R.A. Horne, ed. New York: Wiley Interscience.

Julien, P.Y. (2002) *River Mechanics*. Cambridge: Cambridge University Press.

Karl, T.R., and W.E. Riebsame (1989) The impact of decadal fluctuations in mean precipitation and temperature on runoff: A sensitivity study over the United States. *Climatic Change* 15: 423–447.

Katul, G., P. Wiberg, J. Albertson, and G. Hornberger (2002) A mixing layer theory for resistance in shallow streams. *Water Resources Research* 38(11): 32-1–32-8.

Kennedy, J.F. (1963) The mechanics of dunes and antidunes in erodible-bed channels. *Journal of Fluid Mechanics* 16: 521–544.

Keulegan, G.H. (1938) Laws of turbulent flow in open channels. *U.S. National Bureau of Standards Journal of Research* 21: 707–741.

Kindsvater, C.E., and R.W. Carter (1959) Discharge characteristics of rectangular thin-plate weirs. *American Society of Civil Engineers Transactions* 124: 772–801.

King, J.G., W.W. Emmett, P.J. Whiting, R.P. Kenworthy, and J.J. Barry (2004) Sediment transport data and related information for selected coarse-bed streams and rivers in Idaho. General Technical Report RMRS-GTR-131, U.S. Forest Service Rocky Mountain Research Station.

Kirby, W.H. (1987) Linear error analysis of slope-area discharge determinations. *Journal of Hydrology* 96: 125–138.

Kirkby, M.J. (1993) Network hydrology and geomorphology. In *Channel Network Hydrology*. K. Beven and M.J. Kirkby, eds. New York: John Wiley & Sons, chap. 1.

Kleitz, M. (1877) Note sur la théorie du movement non permanent des liquids et sur application à la propagation des crues des rivières. (Note on the theory of unsteady flow of liquids and on application to the flood propagation in rivers.) *Annales des Ponts et Chausées* Ser. 5, 16(2): 133–196.

Knighton, D. (1998) *Fluvial Forms and Processes*. London: Arnold.

Koloseus, H.J., and J. Davidian (1966) Free-surface instability correlations. Water-Supply Paper 1592-C, U.S. Geological Survey.

Kouwen, N., and R.-M. Li (1980) Biomechanics of vegetative channel linings. *Proceedings of the American Society of Civil Engineers, Journal of the Hydraulics Division* 106(HY6): 1085–1103.

Lai, C. (1986) Numerical modeling of unsteady open-channel flow. *Advances in Hydroscience* 14: 161–333.

Lane, E.W. (1955) Design of stable channels. *Transactions of the American Society of Civil Engineers* 120: 1234–1279.

Langbein, W.B., and L.B. Leopold (1964) Quasi-equilibrium states in channel morphology. *American Journal of Science* 262: 772–801.

Larkin, J., J. McDermott, D.P. Simon, and H.A. Simon (1980) Expert and novice performance in solving physics problems. *Science* 208: 1335–1342.

Larkin, R.G., and J.M. Sharp, Jr. (1992) On the relationship between river-basin geomorphology, aquifer hydraulics, and ground-water flow direction in alluvial aquifers. *Geological Society of America Bulletin* 104: 1608–1620.

Lau, Y.L. (1983) Suspended sediment effect on flow resistance. *Journal of Hydraulic Engineering* 109(5): 757–763.

Lawrence, D.S.L. (2000) Hydraulic resistance in overland flow during partial and marginal surface inundation: Observations and modeling. *Water Resources Research* 36: 2381–2393.

Leliavsky, S. (1955) *An Introduction to Fluvial Hydraulics*. London: Constable.
Leopold, L.B. (1994) *A View of the River*. Cambridge, MA: Harvard University Press.
Leopold, L.B., and W.B. Bull (1979) Base level, aggradation, and grade. *Proceedings of the American Philosophical Society* 123: 168–202.
Leopold, L.B., and T. Maddock (1953) The hydraulic geometry of stream channels and some physiographic implications. Professional Paper 252, U.S. Geological Survey.
Leopold, L.B., and M.G. Wolman (1957) River channel patterns—braided, meandering, and straight. Professional Paper 282-B, U.S. Geological Survey.
Leopold, L.B., R.A. Bagnold, M.G. Wolman, and L.M. Brush, Jr. (1960) Flow resistance in sinuous or irregular channels. Professional Paper 282-D, U.S. Geological Survey.
Leopold, L.B., M.G. Wolman, and J.P. Miller (1964) *Fluvial Processes in Geomorphology*. San Francisco, CA: W.H. Freeman.
Lighthill, M.J., and G.B. Witham (1955) On kinematic waves: 1. Flood movement in long rivers. *Royal Society of London Proceedings Series A* 229: 281–316.
Limerinos, J.T. (1970) Determination of the Manning coefficient from measured bed roughness in natural channels. Water-Supply Paper 1898-B, U.S. Geological Survey.
Liu, K. et al. (1996) Characterization of a cage form of the water hexamer. *Nature* 381: 501–503.
L'vovich, M.I. (1974) *World Water Resources and Their Future*. R.L. Nace, trans. Washington, DC: American Geophysical Union.
Lyn, D.A. (1991) Resistance in flat-bed sediment-laden flows. *Journal of Hydraulic Engineering* 117(1): 94–114.
Mackin, J.H. (1948) Concept of the graded river. *Geological Society of America Bulletin* 59: 463–512.
Manning, R. (1889) On the flow of water in open channels and pipes. *Transactions of the Institution of Civil Engineers of Ireland* 20: 161–195.
Manning, R. (1895) On the flow of water in open channels and pipes. Supplement to 1889 paper. *Transactions of the Institution of Civil Engineers of Ireland* 24: 179–207.
Marcus, W.A., K. Roberts, L. Harvey, and G. Tackman (1992) An evaluation of methods for estimating Manning's n in small mountain streams. *Mountain Research and Development* 12(3): 227–239.
Matthai, H.F. (1967) Measurement of peak discharge at width contractions by indirect methods. Techniques of Water-Resources Investigations, bk. 3. U.S. Geological Survey, chap. A4.
Matthes, G. (1947) Macroturbulence in natural stream flow. *American Geophysical Union Transactions*, vol. 28.
McCuen, R.H. (2005) *Hydrologic Analysis and Design*, 3rd ed. Upper Saddle River, NJ: Pearson Prentice Hall.
Meier, M.F. (1964) Ice and glaciers. In *Handbook of Applied Hydrology*. V.T. Chow, ed. New York: McGraw-Hill.
Meyer-Peter, E., and R. Müller (1948) Formulas for bed-load transport. *Proceedings of the International Association for Hydraulic Research, Second Congress* (Stockholm): 39–65.
Michel, B. (1971) Winter regime of rivers and lakes. Cold Regions Science and Engineering Monograph III-B1a, U.S. Army Cold Regions Research and Engineering Laboratory.
Middleton, G.V., and J.B. Southard (1984) *Mechanics of Sediment Movement*, 2nd ed. Tulsa, OK: Society of Economic Paleontologists and Mineralogists.
Middleton, G.V., and P.R. Wilcock (1994) *Mechanics in the Earth and Environmental Sciences*. Cambridge: Cambridge University Press.
Millar, R.G. (2000) Influence of bank vegetation on alluvial channel patterns. *Water Resources Research* 36: 1109–1118.
Montgomery, D.R., and J.M. Buffington (1997) Channel-reach morphology in mountain drainage basins. *Geological Society of America Bulletin* 109: 596–611.
Moody, L.F. (1944) Friction factors for pipe flow. *Transactions of the American Society of Mechanical Engineers* 66(8).
Moramarco, T., and V.P. Singh (2000) A practical method for analysis of river waves and for kinematic wave routing in natural channel networks. *Hydrological Processes* 14: 51–62.

Morgali, J.R. (1963) Hydraulic behavior of small drainage basins. Technical Report No. 30, Stanford University Department of Civil Engineering.

Morisawa, M. (1985) *Rivers*. London: Longman.

Morlock, S.E. (1996) Evaluation of acoustic Doppler current profiler measurements of river discharge. Water-Resources Investigations Report 95-4218, U.S. Geological Survey.

Moussa, R., and C. Bocquillon (1996) Criteria for the choice of flood-routing methods in natural channels. *Journal of Hydrology* 186: 1–30.

Muste, M., K. Yu, I. Fujita, and R. Ettema (2005) Two-phase versus mixed-flow perspective on suspended sediment transport in turbulent channel flows. *Water Resources Research* 41:W10402, doi:10.1029/2004WR003595.

Nanson, G.C., and A.D. Knighton (1996) Anabranching rivers: Their cause, character, and classification. *Earth Surface Processes and Landforms* 21: 217–239.

National Atmospheric Deposition Program (2008) http://nadp.sws.uiuc.edu

Noman, N.S., E.J. Nelson, and A.K. Zundel (2001) Review of automated floodplain delineation from digital terrain models. *Journal of Water Resources Planning and Management* 127: 394–402.

Olsen, N.R.B. (2004) *Hydroinformatics, Fluvial Hydraulics, and Limnology*. Department of Hydraulic and Environmental Engineering, Norwegian University of Science and Technology. http://folk.ntnu.no/~nilsol/tvm4155/flures5.pdf.

Olsen, R.B., and Q.L. Florey (1952) Sedimentation studies in open channels: Boundary shear and velocity distribution by membrane analogy, analytical, and finite-difference methods. Laboratory Report Sp-34, U.S. Bureau of Reclamation.

Omer, C.R., E.J. Nelson, and A.K. Zundel (2003) Impact of varied data resolution on hydraulic modeling and floodplain delineation. *Journal of the American Water Resources Association* 39: 467–475.

Parker, G. (1976) On the cause and characteristic scales of meandering and braiding. *Journal of Fluid Mechanics* 76: 457–480.

Parker, G., P.C. Klingeman, and D.G. McLean (1982) Bed load and size distribution in paved gravel-bed streams. *Proceedings of the American Society of Civil Engineers; Journal of the Hydraulics Division* 108(HY4): 544–571.

Pizzuto, J.E. (1984) An evaluation of methods for calculating the concentration of suspended bed material in rivers. *Water Resources Research* 20(10): 1381–1389.

Poff, N.L. et al. (1997) The natural flow regime: A paradigm for river conservation and restoration. *BioScience* 47: 769–784.

Prandtl, L. (1926) Über die ausgebildete Turbulenz. (On fully developed turbulence.) *Proceedings of the Second International Congress of Applied Mechanics, Zurich*, pp. 85–92.

Price, R.K. (1974) Comparison of four numerical routing methods. *Proceedings of the American Society of Civil Engineers; Journal of the Hydraulics Division* 100(HY7): 879–899.

Ragan, R.M. (1966) Laboratory evaluation of a numerical routing technique for channels subject to lateral inflows. *Water Resources Research* 2: 111–122.

Reinauer, R., and W.H. Hager (1995) Non-breaking undular hydraulic jumps. *Journal of Hydraulic Research* 33: 1–16.

Rhodes, D.D. (1977) The b-f-m diagram: Graphical representation and interpretation of at-a-station hydraulic geometry. *American Journal of Science* 277: 73–96.

Richardson, L.F. (1926) Atmospheric diffusion shown on a distance-neighbor graph. *Proceedings of the Royal Society Series A* 110: 709–737.

Riggs, H.C. (1976) A simplified slope-area method for estimating flood discharges in natural channels. *U.S. Geological Survey Journal of Research* 4: 285–291.

Riggs, H.C., and K.D. Harvey (1990) Temporal and spatial variability of streamflow. In *Surface Water Hydrology*. M.G. Wolman and H.C. Riggs, eds. Geology of North America, vol. 0–1. Boulder, CO: Geological Society of America.

Rodriguez-Iturbe, I., and A. Rinaldo (1997) *Fractal River Basins: Chance and Self-Organization*. Cambridge: Cambridge University Press.

Rose, C.W., W.L. Hogarth, H. Ghadiri, J.-Y. Parlange, and A. Okom (2002) Overland flow to and through a segment of uniform resistance. *Journal of Hydrology* 255: 134–150.

Rosgen, D. (1996) *Applied River Morphology*. Lakewood, CO: Wildland Hydrology.

Rouse, H. (1937) Modern conceptions of mechanics of fluid turbulence. *American Society of Civil Engineers Transactions* 102: 463–505.

Rouse, H. (1938) *Fluid Mechanics for Hydraulic Engineers*. Reprint, New York: Dover, 1961.

Rouse, H. (1965) Critical analysis of open-channel resistance. *Proceedings of the American Society of Civil Engineers; Journal of the Hydraulics Division* 91(HY4): 1–25.

Rouse, H., and S. Ince (1963) *History of Hydraulics*. New York: Dover Publications. (Originally published by Iowa Institute of Hydraulic Research, 1957.)

Rowinski, P.M., and J. Kubrak (2002) A mixing-length model for predicting vertical velocity distribution in flows through emergent vegetation. *Hydrological Sciences Journal* 47: 893–904.

Rozovskii, I.L. (1957) *Flow of Water in Bends of Open Channels*. Academy of Sciences of the Ukrainian S.S.R. Translated from the Russian by Israel Program for Scientific Translations 1961; available from U.S. Department of Commerce Office of Technical Services, PST Catalog No. 363, OTS 60-51133.

Russell, J.S. (1844) Report on waves. *Report of the British Association for the Advancement of Science,* 311–390.

Sankarasubramanian, A., R.M. Vogel, and J.F. Limbrunner (2001) Climate elasticity of streamflow in the United States. *Water Resources Research* 37: 1771–1781.

Sarma, K.V.N., and P. Syamala (1991) Supercritical flow in smooth open channels. *Journal of Hydraulic Engineering* 117(1): 54–63.

Savini, J., and Bodhaine, G.L. (1971) Analysis of current-meter data at Columbia River gauging stations, Washington and Oregon. Water-Supply Paper 1869-F, U.S. Geological Survey.

Scales, J.A., and R. Snieder (1999) What is a wave? *Nature* 401: 739–740.

Schlichting, H. (1979) *Boundary Layer Theory*. New York: McGraw-Hill.

Schumm, S.A. (1956) The evolution of drainage systems and slopes in badlands at Perth Amboy, New Jersey. *Geological Society of America Bulletin* 67: 597–646.

Schumm, S.A. (1981) Evolution and response of the fluvial system—sedimentologic implications. *Society of Economic Paleontologists and Mineralogists Special Publication* 31: 19–29.

Schumm, S.A. (1985) Patterns of alluvial rivers. *Annual Review of Earth and Planetary Sciences* 13: 5–27.

Seddon, J.A. (1900) River hydraulics. *American Society of Civil Engineers Transactions* 43: 179–229.

Seife, C. (1996) On ice's surface, a dance of molecules. *Science* 274: 2012.

Sellmann, P.V., and S.L. Dingman (1970) Prediction of stream frequency from maps. *Journal of Terramechanics* 7: 101–115.

Shearman, J.O. (1990) User's manual for WSPRO, a computer model for water surface profile computation. Report FHWA-IP-89-027, U.S. Federal Highway Administration.

Shen, H.W., and P.Y. Julien (1992) Erosion and sediment transport. *Handbook of Hydrology*. D.R. Maidment, ed. New York: McGraw-Hill, chap. 12.

Shields, A. (1936) Anwendung der Ähnlichkeitsmechanik und der Turbulenzforschung auf die Geschiebewegung. (Application of similitude and turbulence experiments to bed-load movement.) *Mitteilungen der Prüssischen Versuchanstalt für Wasserbau und Schiffbau* (Berlin), Heft 26.

Shiklomanov, I.A. (1993) World fresh water resources. In *Water in Crisis*. Gleick, P.H., ed. New York: Oxford University Press, chap. 2.

Simons, D.B., and E.V. Richardson (1966) Resistance to flow in alluvial channels. Professional Paper 422-J, U.S. Geological Survey.

Simons, D.B., E.V. Richardson, and W.L. Haushild (1963) Some effects of fine sediment on flow phenomena. Water-Supply Paper 1498-G, U.S. Geological Survey.

Simpson, M.R., and Oltman, R.N. (1992) Discharge-measurement system using an acoustic Doppler current profiler with applications to large rivers and estuaries. Open-File Report 91-487, U.S. Geological Survey.

Smart, G.M., M.J. Duncan, and J.M. Walsh (2002) Relatively rough flow resistance equations. *Journal of Hydraulic Engineering* 128(6): 568–576.

Smith, C.R. (1997) Turbulence. *Geotimes*, Sept., 15–18.

Song, C.C.S., and C.T. Yang (1980) Minimum stream power: Theory. *Proceedings of the American Society of Civil Engineers; Journal of the Hydraulics Division* 106(HY9): 1477–1487.

Springer, G.S., S. Tooth, and E.E. Wohl (2006) Theoretical modeling of stream potholes based upon observations of the Orange River, Republic of South Africa. *Geomorphology* 82: 160–176.

Stefan, J. (1889) Uber die Theorien des Eisbildung insbesondere uber die Eisbildung in Polarmeere. *Sitzungsbericht Wien Akademie Wissenschaften*, ser. A, 42(pt. 2): 965–983.

Stillinger, F.H. (1980) Water revisited. *Science* 209: 451–455.

Strahler, A.N. (1952) Hypsometric (area-altitude) analysis of erosional topography. *Geological Society of America Bulletin* 63: 1117–1142.

Strelkoff, T. (1970) Numerical solution of St. Venant equations. *Proceedings of the American Society of Civil Engineers; Journal of the Hydraulics Division* 96(HY1): 223–252.

Strickler, A. (1923) Beitrage zur Frage der Geschwundigkeitsformel und der Rauhigkeitszahlen für Strome, Kanale, und geschlossene Leitungen. (Contributions to the question of the velocity formula and roughness factors for streams, canals, and closed conduits.) *Mitteilungen des eidgenossischen Amtes für Wasserwirtschaft* (Bern, Switzerland), No. 16.

Summerfield, M.A. (1991) *Global Geomorphology*. Essex, UK: Longman.

Sundborg, A. (1956) The River Klarälven: A study of fluvial processes. *Geografiska Annaler* 38: 127–316.

Syvitski, J.P., M.D. Morehead, D.B. Bahr, and T. Mulder (2000) Estimating fluvial sediment transport: The rating parameters. *Water Resources Research* 36(9): 2747–2760.

Tal, M., and C. Paola (2007) Dynamic single-thread channels maintained by the interaction of flow and vegetation. *Geology* 35(4): 347–350.

Task Force on Bed Forms in Alluvial Channels (1966) Nomenclature of bed forms in alluvial channels. *Proceedings of the American Society of Civil Engineers, Journal of the Hydraulics Division* 92(HY3): 51–64.

Task Force on Preparation of Sediment Manual (1971) Sediment transportation mechanics: Fundamentals of sediment transportation. *Proceedings of the American Society of Civil Engineers, Journal of the Hydraulics Division* 97(HY12): 1979–2022.

Tooth, S., and G.C. Nanson (2004) Forms and processes of two highly contrasting rivers in arid central Australia, and the implications for channel-pattern discrimination and prediction. *Geological Society of America Bulletin* 116: 802–816.

Torizzo, M., and J. Pitlick (2004) Magnitude-frequency of bed load transport in mountain streams in Colorado. *Journal of Hydrology* 290: 137–151.

Tracy, H.J. (1957) Discharge characteristics of broad-crested weirs. Circular 397, U.S. Geological Survey.

Tracy, H.J., and C.M. Lester (1961) Resistance coefficients and velocity distribution, smooth rectangular channel. Water-Supply Paper 1592-A, U.S. Geological Survey.

Tsang, G. (1982) Resistance of Beuharnois Canal in winter. *Proceedings of the American Society of Civil Engineers, Journal of the Hydraulics Division* 108(HY2): 167–186.

Turk, J.T. (1983) An evaluation of trends in the acidity of precipitation and the related acidification of surface water in North America. Water-Supply Paper 2249, U.S. Geological Survey.

Twidale, C.R. (2004) River patterns and their meaning. *Earth Science Reviews* 67: 159–218.

U.S. Army Corps of Engineers (1969) Water surface profiles. HEC Training Document, U.S. Army Corps of Engineers Hydrologic Engineering Center.

U.S. Army Corps of Engineers (1986) Accuracy of computed water surface profiles.U.S. Army Corps of Engineers Hydrologic Engineering Center, National Technical Information Service ADA176314.

University Corporation for Atmospheric Research (2003) *Flood damages in the United States: A reanalysis of National Weather Service data.* http://www.flooddamagedata.org/cgi/national.cgi

Vallé, B.L., and G.B. Pasternack (2006) Submerged and unsubmerged natural hydraulic jumps in a bedrock step-pool mountain channel. *Geomorphology* 82: 146–159.

van den Berg, J.H. (1995) Prediction of alluvial channel pattern of perennial rivers. *Geomorphology* 12: 259–279.

van der Link, G., et al. (2004) Why the United States is becoming more vulnerable to natural disasters. http://www.agu.org/sci_soc/articles/eisvink.html.

Van Dyke, M. (1982) *An Album of Fluid Motion.* Stanford, CA: Parabolic Press.

van Hylckama, T.E.A. (1979) Water, something peculiar. *Hydrological Sciences Bulletin* 24: 499–507.

Vanoni, V.A., ed. (1975) *Sedimentation Engineering.* Manual of Engineering Practice No. 54, American Society of Civil Engineers

Vericat, D., M. Church, and R.J. Batalla (2006) Bed load bias: Comparison of measurements obtained using two (76 and 152 mm) Helley-Smith samplers in a gravel-bed river. *Water Resources Research* 42:W01402, doi:10.1029/2005WRR004025.

Vogel, R.M., I. Wilson, and C. Daly (1999) Regional regression models of annual streamflow for the United States. *Journal of Irrigation and Drainage Engineering* 125: 148–157.

Vörösmarty, C.J., B.M. Fekete, M. Meybeck, and R.B. Lammers (2000a) Geomorphic attributes of the global system of rivers at 30-minute spatial resolution. *Journal of Hydrology* 237: 17–39.

Vörösmarty, C.J., Green, P., Salisbury, J., and Lammers, R.B. (2000b) Global water resources: Vulnerability from climate change and population growth. *Science* 289: 281–288.

Vörösmarty, C.J., B.M. Fekete, M. Meybeck, and R.B. Lammers (2000c) Global system of rivers: Its role in organizing continental land mass and defining land-to-ocean linkages. *Global Biogeochemical Cycles* 14: 599–621.

Walker, J.F. (1988) General two-point method for determining velocity in open channel. *Proceedings of the American Society of Civil Engineers; Journal of the Hydraulics Division* 114: 801–805.

Walling, D.E., and B.W. Webb (1987) Material transport by the world's rivers: Evolving perspectives. In *Water for the Future: Hydrology in Perspective.* Publication No. 164. Wallingford, U.K.: International Association for Hydrological Sciences, pp. 313–329.

Wang, G.-T., and S. Chen (2003) A semianalytical solution of the Saint-Venant equations for flood routing. *Water Resources Research* 39(4): 1076, doi:10.1029/2002WR001690.

Whipple, K.X., G.S. Hancock, and R.S. Anderson (1999) River incision into bedrock: Mechanics and relative efficacy of plucking, abrasion, and cavitation. *Geological Society of America Bulletin* 112(3): 490–503.

White, K.D. (1999) Hydraulic and physical properties affecting ice jams. CRREL Report 99–11, U.S. Army Cold Regions Research and Engineering Laboratory.

White, K.E. (1978) Dilution methods. In *Hydrometry: Principles and Practices*, 2nd ed. R.W. Herschy, ed. Chichester, U.K.: John Wiley & Sons..

Wiberg, P.L., and J.D. Smith (1991) Velocity distribution and bed roughness in high-gradient streams. *Water Resources Research* 27(5): 825–838.

Wigley, T.M.L., and P.D. Jones (1985) Influence of precipitation changes and direct CO_2 effects on streamflow. *Nature* 314: 149–151.

Williams, G.P. (1966) Freeze-up and break-up of fresh-water lakes. In *Proceedings of the Conference on Ice Pressures against Structures.* Technical Manual 92. Ottawa: National Research Council of Canada, pp. 203–215.

Williams, G.P. (1978) Bankfull discharge of rivers. *Water Resources Research* 14: 1141–1158.

Wilson, C.A.M.E., and M.S. Horritt (2002) Measuring the flow resistance of submerged grass. *Hydrological Processes* 16: 2589–2598.

Winter, T.C. (1981) Uncertainties in estimating the water balance of lakes. *Water Resources Bulletin* 17: 82–115.

Wohl, E., and D. Merritt (2005) Prediction of mountain stream morphology. *Water Resources Research* 41:W08419, doi:1029/2004WR003779.

Wollheim, W.M. (2005) *The Controls of Watershed Nutrient Export.* Ph.D. Thesis (Earth Sciences), University of New Hampshire, Durham, NH.

Wolman, M.G., and J.P. Miller (1960) Magnitude and frequency of geomorphic processes. *Journal of Geology* 68: 54–74.

Yalin, M.S., and E. Karahan (1979) Inception of sediment transport. *Proceedings of the American Society of Civil Engineers, Journal of the Hydraulics Division* 105(HY11): 1433–1443.

Yang, C.T. (1972) Unit stream power and sediment transport. *Proceedings of the American Society of Civil Engineers, Journal of the Hydraulics Division* 98(HY10): 1805–1826.

Yang, C.T. (1973) Incipient motion and sediment transport. *Proceedings of the American Society of Civil Engineers, Journal of the Hydraulics Division* 99(HY10): 1679–1704.

Yang, C.T. (1984) Unit stream power equation for gravel. *Proceedings of the American Society of Civil Engineers, Journal of the Hydraulics Division* 110(HY12): 1783–1798.

Yang, C.T., and A. Molinas (1982) Sediment transport and unit stream power function. *Proceedings of the American Society of Civil Engineers, Journal of the Hydraulics Division* 108(HY6): 776–793.

Yang, C.T., and J.B. Stall (1976) Applicability of unit stream power equation. *Proceedings of the American Society of Civil Engineers, Journal of the Hydraulics Division* 102(HY5): 559–568.

Yarnell, D.L. (1934) Bridge piers as channel obstructions. Technical Bulletin 442, U.S. Department of Agriculture.

Yen, B.C. (2002) Open channel flow resistance. *Journal of Hydraulic Engineering* 128(1): 20–39.

Zingg, T., W.C. Krumbein, and L.L. Sloss (1963) *Stratigraphy and Sedimentation.* San Francisco: W.H. Freeman and Co.

Index

f indicates figure; t indicates table.